Hiroshi Sakai   Mihir Kumar Chakraborty
Aboul Ella Hassanien
Dominik Ślęzak   William

Rough Sets, Fuzzy Sets, Data Mining and Granular Computing

12th International Conference, RSFDGrC 2009
Delhi, India, December 15-18, 2009
Proceedings

 Springer

Series Editors

Randy Goebel, University of Alberta, Edmonton, Canada
Jörg Siekmann, University of Saarland, Saarbrücken, Germany
Wolfgang Wahlster, DFKI and University of Saarland, Saarbrücken, Germany

Volume Editors

Hiroshi Sakai
Kyushu Institute of Technology, Tobata, Kitakyushu, Japan
E-mail: sakai@mns.kyutech.ac.jp

Mihir Kumar Chakraborty
University of Calcutta, Kolkata, India
E-mail: mihirc4@gmail.com

Aboul Ella Hassanien
University of Cairo, Orman, Giza, Egypt
E-mail: aboitcairo@gmail.com

Dominik Ślęzak
University of Warsaw & Infobright Inc., Poland
E-mail: slezak@infobright.com

William Zhu
University of Electronic Science and Technology of China, Chengdu, China
E-mail: williamfengzhu@gmail.com

Library of Congress Control Number: 2009940136

CR Subject Classification (1998): I.2, H.2.8, H.2.4, H.3, F.1, F.4, H.4, G.2

LNCS Sublibrary: SL 7 – Artificial Intelligence

ISSN   0302-9743
ISBN-10   3-642-10645-5 Springer Berlin Heidelberg New York
ISBN-13   978-3-642-10645-3 Springer Berlin Heidelberg New York

springer.com

© Springer-Verlag Berlin Heidelberg 2009
Printed in Germany

Typesetting: Camera-ready by author, data conversion by Scientific Publishing Services, Chennai, India
Printed on acid-free paper   SPIN: 12809714   06/3180   5 4 3 2 1 0

Preface

Welcome to the 12th International Conference on Rough Sets, Fuzzy Sets, Data Mining and Granular Computing (RSFDGrC 2009), held at the Indian Institute of Technology (IIT), Delhi, India, during December 15-18, 2009. RSFDGrC is a series of conferences spanning over the last 15 years. It investigates the meeting points among the four major areas outlined in its title. This year, it was co-organized with the Third International Conference on Pattern Recognition and Machine Intelligence (PReMI 2009), which provided additional means for multi-faceted interaction of both scientists and practitioners. It was also the core component of this year's Rough Set Year in India project. However, it remained a fully international event aimed at building bridges between countries.

The first sectin contains the invited papers and a short report on the above-mentioned project. Let us note that *all* the RSFDGrC 2009 plenary speakers, Ivo Düntsch, Zbigniew Suraj, Zhongzhi Shi, Sergei Kuznetsov, Qiang Shen, and Yukio Ohsawa, contributed with the full-length articles in the proceedings.

The remaining six sections contain 56 regular papers that were selected out of 130 submissions, each peer-reviewed by three PC members. We thank the authors for their high-quality papers submitted to this volume and regret that many deserving papers could not be accepted because of our urge to maintain strict standards. It is worth mentioning that there was quite a good number of papers on the foundations of rough sets and fuzzy sets, many of them authored by Indian researchers. The fuzzy set theory has been popular in India for a longer time. Now, we can see the rising interest in the rough set theory.

The success of the conference would be impossible without the people acknowledged on the following pages. We would like to express our gratitude in particular to Lotfi A. Zadeh, who agreed to serve as Honorary Chair. Furthermore, on behalf of all the rough set researchers, we would like to thank all the PReMI organizers for a very fruitful cooperation. We would also like to acknowledge all the organizations that supported us during our preparations: International Rough Set Society, International Fuzzy Systems Association, Indian Unit for Pattern Recognition and Artificial Intelligence, Indian Statistical Institute in Calcutta, Machine Intelligence Research Labs, Springer, Chinese Rough Set and Soft Computing Society, Special Interest Group on Rough Sets in Japan, Egyptian Rough Sets Working Group, and Infobright. Special thanks go once more to IIT Delhi for providing the basis for both PReMI and RSFDGrC 2009.

October 2009

Hiroshi Sakai
Mihir Kumar Chakraborty
Aboul Ella Hassanien
Dominik Ślęzak
William Zhu

Organization

P. Synak	S. Tsumoto	W.-Z. Wu
A. Szałas	J.V. Valdés	Y. Wu
M. Szczuka	G. Wang	Y. Xiang
N. Takagi	X. Wang	J.T. Yao
D. Talia	J. Watada	W. Ziarko

Table of Contents

Rough Set Algorithms and Applications

Fuzzy Set Foundations and Applications

Data Mining and Knowledge Discovery

Clustering and Current Trends in Computing

Information Retrieval and Text Mining

Affordance Relations*

Ivo Düntsch[1], Günther Gediga[2], and Adam Lenarcic[1]

[1] Dept of Computer Science, Brock University, St. Catharines, ON, Canada
duentsch@brocku.ca, al04uh@brocku.ca
[2] Dept of Psychology, Universität Münster, Fliednerstr. 21, D–48149 Münster
guenther@gediga.de

Abstract. Affordances are a central concept of J.J. Gibson's approach to visual perception. We describe and discuss the concept of affordances with a brief look at its application to robotics, as well as provide an overview of several existing formalizations. It turns out that a representation of affordances can be based on a certain hierarchy of Pawlak's approximation spaces. We also outline how concepts could be used in a theory of affordances, and how affordances might be recognized in simple perceiving situations.

1 Introduction

Over a period of fifty years, J. J. Gibson developed an "Ecological Approach to Visual Perception" [1,2] that was radically different from the prevailing views of the time - and, to some extent, from those of today.

> "To perceive is to be aware of the surfaces of the environment and of oneself in it. The interchange between hidden and unhidden surfaces is essential to this awareness. These are existing surfaces; they are specified at some points of observation. Perceiving gets wider and finer and longer and richer and fuller as the observer explores the environment. The full awareness of surfaces includes their layout, their substances, their events and their affordances."

The term *ecological* in the sense used by Gibson pertains to the natural environment, to the "everyday things" [3] of the acting individual:

- 'We are told that vision depends on the eye which is connected to the brain. I shall suggest that natural vision depends on the eyes in the head on a body supported by the ground, the brain being the central organ of a complete visual system." [2, p. 1]

---

* Equal authorship is implied. Ivo Düntsch gratefully acknowledges support from the Natural Sciences and Engineering Research Council of Canada. Günther Gediga is also adjunct professor in the Department of Computer Science, Brock University.

H. Sakai et al. (Eds.): RSFDGrC 2009, LNAI 5908, pp. 1–11, 2009.

Gibson's two main propositions are as follows [2]:

1. Objects are perceived directly, "not mediated by retinal pictures, neural pictures, or mental pictures". Perception is regarded as the act of extracting information from a changing context. While synthesis is a conceptualization from the parts to the whole, perception proceeds from the whole to parts and features. Each feature then can be viewed as an object of further investigation. Visual perception takes place in a nested environment, and thus, granular computing, in particular the rough sets model, are a natural environment to model visual perception.

2. The observer and the observed are an inseparable pair, related by *affordances*.

> "The affordances of the environment are what it offers the animal, what it provides or furnishes, either for good or ill. The verb to afford is found in the dictionary, but the noun affordance is not. I have made it up. I mean by it something that refers to both the environment and the animal in a way that no existing term does. It implies the complementarity of the animal and the environment." [2]

Numerous experiments have been performed to test Gibson's theory, most notably the early seminal experiments of stair–climbing by Warren [4]. For a more complete account, the reader is invited to consult [5] for a succinct introduction to Gibson's philosophy and Volume 15(2) of the journal *Ecological Psychology* (2002) for an account of recent developments.

2   Affordance Relations

The complementarity of the animal and the environment that it perceives is central to Gibson's world view and he elaborates

– "An affordance is neither an objective property nor a subjective property; or it is both if you like. An affordance cuts across the dichotomy of subjective–objective and helps us to understand its inadequacy. It is equally a fact of an environment and a fact of behavior. It is both physical and psychical, yet neither." [2, p. 129]

As an example, consider the experiments by Warren [4], who associates the affordance "climb–able" with the ratio $\frac{p}{q}$ between the stair riser height (p) and the agent's leg length (q). The affordance "climb-able" then is given, when $\frac{p}{q} \leq$ 0.88. The ratio $\frac{p}{q}$ is regarded as an "ecological invariant": For any agent Z and any set X of stairs, "climb–able" is afforded when $\frac{p}{q} \leq 0.88$; in yet another form, $\frac{p}{q} \leq 0.88$ offers the action "climb". A stair – climbing affordance, then, is a pair

(2.1) $\langle \underbrace{\frac{\text{stair height}}{\text{leg length}} \text{ is favourable}}_{\text{environment}}, \underbrace{\text{can climb}}_{\text{organism}} \rangle.$

The agent perceives the environment according to what it offers him - a path is "walk–able", stairs are "climb–able", a rabbit is "hug–able". Affordances may change over time, and, depending on the state of the agent, the same physical objects are perceived differently – stairs can be "paint–able", and a rabbit "eat – able"; an elephant, viewed from afar may be "wonder–able", yet, when it gallops towards us, it will afford danger and, possibly, harm. Also, a flying goose may afford a feeling of beauty in one observer, while another one sees an object for the dinner table. Perception is thus "economical" and only that information is extracted which is necessitated by the affordance.

It is important to note that in Gibson's world the objects in the environment are not conceptual in the first place ("This is a bucket") but obtained from the concrete physical features of the visual field (after recognizing invariants etc.). The name "bucket" is just a label arising from an affordance.

Even though affordances are a central construct of Gibson's theory, there is surprisingly little agreement on an operational, let alone ontological, definition of the affordance concept. Various proposals have been made to model affordances. Below, we give two examples:

- "Affordances . . . are relations between the abilities of organisms and features of the environment. Affordances, that is, have the structure Affords–ϕ (feature,ability)." (Chemero [6], p. 189)
- "Let W_{pq} (e.g. a person-climbing-stairs system) $= (X_p, Z_q)$ be composed of different things Z (e.g. person) and X (e.g. stairs). Let p be a property of X and q be a property of Z. The relation between p and q, $\frac{p}{q}$, defines a higher order property (i.e., a property of the animal – environment system), h. Then h is said to be an affordance of W_{pq} if and only if
  1. $W_{pq} = (X_p, Z_q)$ possesses h.
  2. Neither Z nor X possesses h." (Stoffregen [7], p. 123).

-

For an overview of operationalizations of affordances the reader is invited to consult [8]. Here, we take the view that

- An affordance is a relation R between states (or intentionalities) of an agent (animal, human, robot) and certain properties of its environment.

This is not as simple as it looks: The environment has infinitely many features, and an affordance selects a set of features according to the agent's "affordance state" or "intentionality", see Figure 1. These states form a nested set of overlapping possibilities. It is important to note that the agent is a part of the environment, and that affordances depend on the agent's properties as well as those offered by the visual field as the stair–climbing environment shows:

"Gibson argued that the proper "objects" of perceiving are the same as those of activity. Standing still, walking, and running are all relations between an animal and its supporting surface." [9, p 239]

Furthermore, affordances need to be modelled in a changing environment.

Fig. 1. Affordances I

3   Affordances and Robots

Just as Gibson suggests that natural vision depends on the eyes, head, body, and ground, so could one suggest that a camera, fixed to a robot, supported on the ground, might support natural vision as well. Sahin et al. [8] provide a review and subsequently build on both the concept of affordance, as well as the formalization, with a particular idea in mind. They are motivated by their "interest in incorporating the affordance concept into autonomous robot control."[8] The affordance concept can be viewed as a binary relation, between the agent and environment, and though several attempts have been made to strengthen this relation, none have been agreed upon. The challenge is trying to decipher what parts of the environment, are related to what parts of the agent, and what each relation should mean to both the agent and environment. Sahin et al. develop the formalization by first generalizing 'environment' to 'entity', and 'agent' to 'behavior', and then suggesting that the effect of each relation between an entity and a behavior, be explicit in the relation. They refine the formalization in their paper from (environment, agent) to (effect, (entity, behavior)). The creation of sets of ⟨effect⟩, and sets of ⟨agent, behavior⟩ tuples is given as the formalization develops to account for the equivalences in entities, behaviors, affordances, effects, and agents.

One of J.J.Gibson's prominent ideas is the concept of optic flow which he studied originally pertaining to flying/landing aeroplanes. He reasoned that during controlled motion, it is not what the parts of the environment are, that an animal is attentive to, but rather where the parts are in relation to the agent, coupled with their relative velocity. This agrees with Duchon, Warren and Kaelbling [10] who implement this activity in robots. Ecological robotics can be described as applying the theory of ecological psychology to the field of robotics. One consequence of adhering to the ecological approach is the importance Gibson lays on the optic array. Vision is the most prominent way in which agents pick up information about their environment. "The flow of optical stimulation provides a 'continuous feedback' of information for controlling the flow of the motor activity."[11] The robots which Duchon et al. implemented used optic flow to navigate a crowded lab and an atrium. Cameras mounted on top supplied constant feedback to the actuators which used the relative velocities of objects in the environment to control movement. The robots navigate around obstacles successfully without modeling the obstacles internally. The authors claim to have recreated a 'how' pathway in an autonomous robot, hoping to support the suggestion that the brain has separate 'how' and 'what' pathways by simulating one of them.

In order to represent visual field data Yin et al [12] use extended quadtrees. The quadtree representation offers a hierarchical decomposition of visual information which can be implemented in a system or a robot. More generally, the quadtree serves well as a representation of a hierarchy of approximation spaces in the sense of Pawlak ([13,14]). As there is a huge number of possibilities to construct the representation, the paper deals with the *focus problem*, which turns out to be the mathematical problem of how to choose an optimal root node for a quadtree to minimize the roughness of the representation. With the solution to that problem, we are able to analyze the information of the data structure at any stage of the hierarchy by an optimal rough set representation.

Certainly, this approximation is no more than a first step towards an affordance based object representation using rough sets, as the guidance of the perceiving act is something like an orientation affordance. But as the focus mechanism is described, it can be used in more complex and more dynamic situations. Obviously, a system using this technology has no a priori layout of objects which classical rough set based robot systems have (e.g. [15]), but offers a way how to construct affordance based objects when perceiving a posterior.

4   Affordances and Concepts

A participatory view of "concepts" naturally leads to an affordance relation. In machine learning, membership in a concept – or, more precisely, in a category – may be "learned" by a sequence of examples and counterexamples; each finite stage of this process may be called an approximation of the category. However, if a concept is to be approximated, then there must be a notion of a "true concept" that can be approximated. The background is the hypothesis that

– *Membership in the category is defined by a common set of attributes the presence of which is necessary and sufficient for membership.*

In other words, machine learning classifiers are based on an extensional understanding of a concept; it may be noted that often no distinction is made between a category and its corresponding concept. A typical situation is an "object – attributes" relationship in which objects are described by feature vectors, and concepts are formed by aggregating the object – attributes pairs into classes; rough sets and formal concept analysis are typical examples of such procedure. Aggregation algorithms are often purely syntactical, and semantical concerns are taken care of in a pre–processing stage such as choice of attributes, dependencies, weightings, prior probabilities etc. Once this is done, the meaning of a concept is strictly truth–functional, i.e. the meaning of a composite expression can be obtained from the meaning of its parts.

While the classical view of understanding a concept based on a set of defining features and an extensional interpretation is appropriate in delineated contexts, it falls short where human cognition is concerned. This has been recognized for some time, and various approaches to graded membership in conceptual categories have been proposed, for example, fuzzy sets [16,17,18], prototype theory [19,20], or the rough set variable precision model [21]. Each of these approaches have their own problems; for example, handling aggregated concepts within fuzzy set theory suffers from the strict conditions imposed by t–norms, which are mathematically expedient, but not necessarily suitable in everyday situations [22,23].

Yet another aspect is the fact that (the meaning of) a concept may change under varying contexts; Rosch [24] argues convincingly that

> "Concepts occur only in actual situations in which they function as participating parts of the situation rather than as either representations or as mechanisms for identifying objects; concepts are open systems by which creatures can learn new things and can invent; and concepts exist in a larger context – they are not the only form in which living creatures know and act".

The contextual aspects adds to the discussion the pragmatic dimension well known from linguistics. It seems sensible to regard a concept as a relation between elements of a set S of *states* or *exemplars* of a concept and elements of a set of *situations* or *contexts*, where the situation affords a certain state of the concept. For example, the concept TREE can have the states "fig tree", "oak", "maple", but also "artificial Christmas tree", or a "connected acyclic simple graph". For another example, a state of the concept UMBRELLA may be "closed" in the context "dry weather", and may change to "open" when it starts to rain. Such change may or may not occur, for example, if I have to walk only a short distance in the rain I may not bother to open the umbrella.

One way to describe the inner object relationship given concept and context information is to assume hidden (and unknown) attributes, which serve as a basis for a state based description of the conditional inner object relationship. In this direction, a formalism based on principles of quantum mechanics which claims to represent concepts in a state – context – property relation (SCOP) was presented by Aerts and Gabora [25,26]. Their theory assumes very strong conditions – for example, the set of contexts needs to be a complete ortholattice – which need further justification. To describe the concept "Pet" the SCOP formalism requires a Hilbert Space with 1400 largely unspecified dimensions[1].

An alternative which avoids adding hidden attributes, uses only the observables and coalitions of objects as a basis of a state description of the conditional inner object relationship given a context. In terms of theories [27,28,29], the Galois connection between the set of contexts and a certain collection of objects sets (we call object states) enables us to use observables without assumptions of hidden states.

A nice way to interpret contexts is to assume that each is a description of an affordance structure, containing affording objects interacting with the perceiver.

---

[1] It turns out that the SCOP formalism is largely tautological; we analyze the SCOP formalism in a separate paper.

Table 1. Exemplars and contexts

Exemplar	e1	e2	e3	e4	e5	e6	1
Rabbit	4	7	15	5	1	0	7
Cat	25	13	22	3	3	1	12
Mouse	3	6	8	11	1	0	5
Bird	2	8	2	4	17	1	8
Parrot	2	16	1	4	63	1	7
Goldfish	1	2	0	2	0	48	10
Hamster	4	7	6	4	1	0	7
Canary	1	7	1	2	7	1	8
Guppy	1	2	0	2	0	46	9
Snake	2	2	1	22	0	1	3
Spider	1	1	3	23	0	0	2
Dog	50	19	24	3	6	0	12
Hedgehog	2	2	8	12	0	0	3
Guinea pig	3	7	9	4	1	0	7

Contexts

e1 The pet is chewing a bone
e2 The pet is being taught
e3 The pet runs through the garden
e4 Did you see the type of pet he has?
   This explains that he is a weird person
e5 The pet is being taught to talk
e6 The pet is a fish
 1 The pet is just a pet

Using the Galois connection among the set of contexts and object states, it is possible to describe concepts and a concept hierarchy by integrating relations which are governed by *basic affordances* (or *simple contexts*) to more complex affordances (or contexts) which govern a certain structure of objects.

As an example for this idea we re-analyse the data presented in [25,26]. The concept to be modelled is PET; 81 respondents were given 14 exemplars a_i (states) of the concept PET and 7 contexts e_i. For each pair $\langle a_i, e_j \rangle$ they were asked to rate the frequency with which exemplar a_i appears in context e_j; the responses and the contexts are shown in Table 1. Each context e_i defines a quasiorder on the set of exemplars by setting

$$a_n \preceq_{e_i} a_m \Longleftrightarrow (a_n, e_i) \leq (a_m, e_i).$$

We exhibit these quasiorders in Table 2; The rows indicate the position of the exemplar in the quasiorder induced by the context named in the first column. We assume that 1 is the most general context ("The pet is just a pet"). The main aim of our approach is now to set up a relationship of concepts with varying contexts. To this end we introduce *contrast concepts* $-e_i$ by reversing the order of the quasi-order of e_i. In the spirit of rough sets, the approximation of 1 is done by finding that relation e_i (or $-e_i$) which is compatible with 1 for most of the elements. In the second step, the compatible elements are removed and the best context with respect to the remaining prototypes will be computed. The iteration will come to an end, when either no context is left or the quasi-order of any remaining contexts are incompatible with 1 on the remaining prototypes.

Table 2. The quasiorders

	1	2	3	4	5	6	7	8	9 & 10
1	Cat Dog	Goldfish	Guppy	Bird Canary	Rabbit Parrot Hamster Guinea pig	Mouse	Snake Hedgehog	Spider	
e1	Dog	Cat	Rabbit Hamster	Mouse Guinea Pig	Bird Parrot Snake Hedgehog	Goldfish Canary Guppy Spider			
e2	Dog	Parrot	Cat	Bird	Rabbit Hamster Canary Guinea pig	Mouse	Goldfish Guppy Snake Hedgehog	Spider	
e3	Dog	Cat	Rabbit	Guinea pig	Mouse Hedgehog	Hamster	Spider	Bird Snake	Parrot(9) Goldfish(10) Guppy(10)
e4	Spider	Snake	Hedgehog	Mouse	Rabbit	Bird Parrot Hamster Guinea pig	Dog Cat	Canary Guppy	
e5	Parrot	Bird	Canary	Dog	Cat	Mouse Hamster Guinea pig	Goldfish Guppy Snake Spider Hedgehog		
e6	Goldfish	Guppy	Cat Bird Parrot Canary Snake	Rabbit Mouse Hamster Spider Dog Hedgehog Guinea pig					

The algorithms leads to following result:

$1 \leftarrow (-e4, e1, e6)$ with an approximation success of 80%, due to case that 76 of the 95 elements of the quasi-order 1 is recovered by $(-e4, e1, e6)$. Furthermore, it is easy to show that

- The contrast concept $-e4$ is indispensable for the approximation of 1.
- The concept $e1$ can be replaced by $e3$.
- The concept $e6$ is only applicable, after $e1$ or $e3$ are applied (conditional approximation). But $e6$ is indispensable.
- The concepts $e2$ and $e5$ are not compatible with 1.
- There are six prototypes for which the corresponding relations (governed by 1) cannot be substituted by any combination relation of the basic concepts. For these prototypes we need other contexts to approximate 1.

We see that this simple approach leads us to quite good and reasonable results. The main assumption was that we may use conditional relationships among objects, which are governed by certain affordances.

5 A Closer Look at the Affordance Relations

In order to describe the affordance relation, many entities are used which have a certain understanding for a perceiver (like a human), but are not self-contained, when we wish to describe perceiving. Consider Figure 2 and the terms used in it.

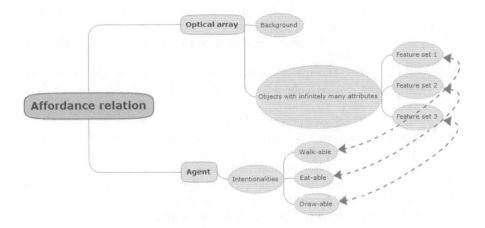

Fig. 2. Affordances II

Optical array. This seems as one would expect – there are some kind of sensors which describe the outer world for the agent. Nevertheless, there is no good reason to restrict ourselves to the term "optical". An agent may have sensors for infrared "light" or for the consistency of surfaces (think of cleaning robots).

Background and objects. It seems clear, the background is the residual of object definitions in the optical array. But there are some problems as well, because affordances establish the background-object relation as well.

Consider the following scenario: You are in a garden with wonderful apple trees and nice apples. Obviously, the apple trees and the apples form objects. The "eat–able" relation establishes the object formation; everything else – e.g. the pathway to a house – is the background for perceiving apples given the "eat–able" affordance. Obviously, one or more apples are objects for this certain act of perceiving, because they are "pick–able" and afterwards "eat–able".

Things change dramatically, when a lion crosses your way. The "is–safe" affordance governs the scene now, and one has to look for a path to the house (which "is–safe"), and where we know that the path is "walk–able" and that it is connected to an object which "is–safe". Now, the object is the path and certainly no apple is the object of perceiving in this moment, and even the wonderful apple tree forms part of the background.

Note that even in very simple perceiving situations the object-background-relations flip-flop dramatically as some paintings by Escher demonstrate.

Intentionalities. Whereas we – as humans – seem to know what "intentionalities" mean, it is somewhat problematic to assume intentionalities for robots. First of all, we note that this term is an abbreviation for one or more very complex systems. One part of the system must refer to the agent as aiming to control its future by applying transformations based on affordances. For example, the "walk–able" affordances may be triggered by certain intentionalities in the following way:

- A pre–outer-part: An object approaches which is connected to the affordance "hurt–able".
- An inner part: The intentionality "have to escape" is triggered by the objects. Now, an "is–safe" object combined with a "walk–able" object is needed.
- An expected-post–outer part: This describes how certain change patterns defined on the "optical array" appear, and what change pattern is connected with an object with an "is–safe" affordance.

The example offers a view of intentionalities which are trigged by an event from outside the agent. This is not true in general – internal "events" like hunger or thirst may set intentionalities as well. The more complex the structure of the agent the more "hidden" intentionalities may exist.

6   Summary

Affordances are a basic concept of Gibson's theory regarding how we view the world. In a broader context, concepts can be viewed as affordances, robotic vision can be modelled based on an affordance concept, and autonomous robots could use the concept to act and react in their environments. A formalization of affordance relations needs to provide crisp and fuzzy structures, mechanisms for spatial and temporal change, as well as contextual modeling. Besides rough sets [13], knowledge structures [28] and formal concept analysis [27], Barwise and Perry's situation theory [30,31] seems to be an appropriate tool for modeling affordances (as suggested by Greeno [32]).

References

1. Gibson, J.J.: The Senses Considered as Perceptual Systems. Greenwood Press, New York (1983); Originally published 1966 by Houghton Mifflin, Westport
2. Gibson, J.J.: The ecological approach to visual perception, 2nd edn. Lawrence Erlbaum, Hillsdale (1986)
3. Norman, D.A.: The Design of Everyday Things. Basic Books (2002)
4. Warren, W.: Perceiving affordances: Visual guidance of stair climbing. Journal of Experimental Psychology 10(5), 371–383 (1984)
5. Mace, W.L., James, J.: Gibson's ecological approach: Perceiving what exists. Ethics and the Environment 10(2), 195–215 (2005)
6. Chemero, A.: An outline of a theory of affordances. Ecological Psychology 15(2), 181–195 (2003)
7. Stoffregen, T.: Affordances as properties of the animal environment system. Ecological Psychology 15(2), 115–134 (2003)
8. Sahin, E., Cakmak, M., Dogar, M., Ugur, E., Ücoluk: To afford or not to afford: A new formalization of affordances toward affordance–based robot control. Adaptive Behavior 15, 447–472 (2007)
9. Turvey, M., Shaw, R., Reed, E., Mace, W.: Ecological laws of perceiving and acting: In reply to Fodor and Pylyshyn 1981). Cognition 9, 237–304 (1981)

10. Duchon, A., Warren, W., Kaelbling, L.: Ecological robotics. Adaptive Behavior 6(3), 473–507 (1998)
11. Gibson, J.J.: The optical expansion–pattern in aerial locomotion. American Journal of Psychology 68, 480–484 (1955)
12. Yin, X., Düntsch, I., Gediga, G.: Choosing the rootnode of a quadtree. In: Lin, T.Y., Hu, X., Xia, J., Hong, T.P., Shi, Z., Han, J., Tsumoto, S., Shen, X. (eds.) Proceedings of the 2009 IEEE International Conference on Granular Computing, pp. 721–726 (2009)
13. Pawlak, Z.: Rough sets. Internat. J. Comput. Inform. Sci. 11, 341–356 (1982)
14. Pawlak, Z.: Rough sets: Theoretical aspects of reasoning about data. System Theory, Knowledge Engineering and Problem Solving, vol. 9. Kluwer, Dordrecht (1991)
15. Dai, S., Ren, W., Gu, F., Huang, H., Chang, S.: Implementation of robot visual tracking system based on rough set theory. In: Fuzzy Systems and Knowledge Discovery, Fourth International Conference, pp. 155–160. IEEE Computer Society, Los Alamitos (2008)
16. Klaua, D.: Über einen Ansatz zur mehrwertigen Mengenlehre. Monatsb. Deutsch. Akad. Wiss. 7, 859–867 (1965)
17. Zadeh, L.A.: Fuzzy sets. Information and Control 8, 338–353 (1965)
18. Smithson, M., Verkuilen, J.: Fuzzy Set Theory - Applications in the Social Sciences. Quantitative Applications in the Social Sciences, vol. 147. Sage Publictions, Thousand Oaks (2006)
19. Rosch, E.: Natural categories. Cognitive Psychology 4, 328–350 (1973)
20. Rosch, E.: Prototypes. In: Hogan, P.C. (ed.) The Cambridge Encyclopedia of the Language Sciences. Cambridge University Press, Cambridge (in press, 2010)
21. Ziarko, W.: Variable precision rough set model. Journal of Computer and System Sciences 46 (1993)
22. Zimmermann, H.J., Zysno, P.: Latent connectives in human decision making. Fuzzy Sets and Systems 4, 37–51 (1980)
23. Gediga, G., Düntsch, I., Adams-Webber, J.: On the direct scaling approach of eliciting aggregated fuzzy information: The psychophysical view. In: Dick, S., Kurgan, L., Musilek, P., Pedrycz, W., Reformat, M. (eds.) Proceedings the 2004 Annual Meeting of the North American Fuzzy Information Processing Society, pp. 948–953 (2004)
24. Rosch, E.: Reclaiming concepts. Journal of Consciousness Studies 6, 61–77 (1999)
25. Aerts, D., Gabora, L.: A theory of concepts and their combinations - I - The structure of the sets of contexts and properties. Kybernetes 34, 167–191 (2005)
26. Aerts, D., Gabora, L.: A theory of concepts and their combinations - II - A Hilbert space and representation. Kybernetes 34, 192–221 (2005)
27. Wille, R.: Restructuring lattice theory: An approach based on hierarchies of concepts. In: Rival, I. (ed.) Ordered sets. NATO Advanced Studies Institute, vol. 83, pp. 445–470. Reidel, Dordrecht (1982)
28. Doignon, J.P., Falmagne, J.C.: Knowledge spaces. Springer, Berlin (1999)
29. Gediga, G., Düntsch, I.: Skill set analysis in knowledge structures. British Journal of Mathematical and Statistical Psychology 55, 361–384 (2002)
30. Barwise, J., Perry, J.: Situations and Attitudes. MIT Press, Cambridge (1983)
31. Devlin, K.: Logic and Information. Cambridge University Press, Cambridge (1991)
32. Greeno, J.G.: Gibson's affordances. Psychological Review 101(2), 336–342 (1994)

Discovering Concurrent Process Models in Data: A Rough Set Approach

Zbigniew Suraj

Chair of Computer Science
University of Rzeszów
Dekerta 2, 35-030 Rzeszów, Poland
zsuraj@univ.rzeszow.pl

Abstract. The aim of the lecture is to provide a survey of state of the art related to a research direction concerning relationships between rough set theory and concurrency in the context of process mining in data. The main goal of this review is the general presentation of the research in this area. Discovering of concurrent systems models from experimental data tables is very interesting and useful not only with the respect to cognitive aspect but also to possible applications. In particular, in Artificial Intelligence domains such as e.g. speech recognition, blind source separation and Independent Component Analysis, and also in other domains (e.g. in biology, molecular biology, finance, meteorology, etc.).

Keywords: Knowledge discovery, data mining, process mining, concurrent systems, rough sets, Petri nets.

1 Introduction

Data Mining and Knowledge Discovery [3],[8],[9],[34],[36] is a very dynamic research and development area that is reaching maturity. Discovering unsuspected relationships between data and hidden (intrinsic) models belong to main tasks of Machine Learning [7]. Data are often generated by concurrent processes, and discovering of concurrent system models may lead to better understanding the nature of modeled systems, i.e., their structures and behaviors [10],[11]-[12],[16],[17],[19],[20]-[22],[23]-[27],[28],[29]-[30],[31].

A concept of concurrent systems can be understood widely. In general case, a concurrent system consists of processes, whose local states can coexist together and they are partly independent. For example, as concurrent systems we can treat systems consisting of social processes, economic processes, financial processes, biological processes, genetic processes, meteorological processes, etc.

Subject matter of this lecture concerns methods of concurrent system modeling on the basis of observations or specifications of their behaviors given in the form of different kinds of data tables. Data tables can include results of observations or measurements of specific states of concurrent processes. In this case, created models of concurrent systems are useful for analyzing properties of modeled systems, discovering the new knowledge about behaviors of processes, etc.

H. Sakai et al. (Eds.): RSFDGrC 2009, LNAI 5908, pp. 12–19, 2009.

Data tables can also include specifications of behaviors of concurrent processes. Then, created models can be a tool for verification of those specifications, e.g. during designing concurrent systems. Methods presented in this lecture can be used, for example, in system designing or analyzing, data analysis, forecasting.

The aim of the lecture is to provide a survey of state of the art related to a research direction concerning relationships between rough set theory and concurrency. The idea of this research direction has been proposed by Z. Pawlak in 1992 [14]. In the last two decades we have witnessed an intensive development of this relatively new scientific discipline by among others A. Skowron, Z. Suraj, J.F. Peters, R. Swinarski, K. Pancerz et al. [10],[11]-[12],[16],[17],[19],[20]-[22],[23]-[27],[28],[29]-[30],[31],[33],[34].

In general, this research direction concerns the following problems: (i) discovering concurrent system models from experimental data represented by information systems, dynamic information systems or specialized matrices, (ii) reconstruction of concurrent models, (iii) prediction of concurrent models change in time, (iv) a use of rough set methods for extracting knowledge from data, (v) a use of rules for describing system behaviors, (vi) modeling and analyzing of concurrent systems by means of Petri nets on the basis of extracted rules.

2 Data Representation and Interpretation

In the research, data tables (information systems in Pawlak's sense [13]) are created on the basis of observations or specifications of process behaviors in the modeled systems. The data table consists of a number of rows (each representing an object). A row in the data table contains the results of sensory measurements represented by the values of vector of attributes (a pattern). Values of attributes can be interpreted as states of local processes in the modeled system of concurrent processes. However, we interpret the rows of data table as global states of the system composed with local states of concurrent processes.

Sometimes during the design phase, it is beneficial to transform the original experimental data table (with original attributes) into the transformed data table containing projected attributes represented in possibly better attribute space. In addition, frequently the attribute selection process follows, when only the most relevant features are taking to form a final feature vector (a pattern). These preprocessing steps are necessary when the resulting concurrent model, constructed directly from the original data table is to complex and dimensionality of model variables is too high. Based on the OccamŠs razor (and Risannen minimum description length paradigm) [3], in order to obtain the best generalizing design system, the model and its variable should be as simple as possible (preserving system functionality). The phase of attribute transformation and relevant attribute selection is yet another difficult data mining step [3],[32],[34]. The input for our approach consists of the data table (if necessary, preprocessed in a way described above).

3 Research Methodology and Knowledge Representation

Proposed methods of discovering concurrent system models from data tables are based on rough set theory and colored Petri net theory. Rough set theory introduced by Z. Pawlak [13] provides advanced and efficient methods of data analysis and knowledge extraction. Petri nets are the graphical and mathematical tool for modeling of different kinds of phenomena, especially those, where actions executed concurrently play a significant role. As a model for concurrency we choose coloured Petri nets proposed by K. Jensen [6]. They allow to obtain coherent and clear models suitable for further computer analysis and verification. Analysis of net models can reveal important information about the structure and dynamic behavior of the modeled system. This information can be also used to evaluate the modeled system and suggest improvements or changes [10],[11]-[12].

Model construction is supported by methods of Boolean reasoning [2]. Boolean reasoning makes a base for solving a lot of decision and optimization problems. Especially, it plays a special role during generation of decision rules [18]. Data describing examined phenomena and processes are represented by means of information systems [14], dynamic information systems [26] or specialized matrices of forbidden states and matrices of forbidden transitions [10],[12]. An information system can include the knowledge about global states of a given concurrent system, understood as vectors of local states of processes making up the concurrent system, whereas a dynamic information system can include additionally the knowledge about transitions between global states of the concurrent system. The idea of representation of concurrent system by information system is due to Z. Pawlak [14].

Nowadays, discovery of process models from data becomes a hot topic under the name process mining (see, e.g. [1],[5],[8],[12],[16],[19],[27],[36]).

Specialized matrices are designed for specifying undesirable states of a given concurrent systems (i.e., those states, which cannot hold together) and undesirable transitions between their states. Decomposition of data tables into smaller subtables connected by suitable rules is also possible. Those subtables make up modules of a concurrent system. Local states of processes represented in a given subtable are linked by means of functional dependencies [21],[27],[12].

4 Maximal Consistent Extensions of Information Systems

Approaches considered in this lecture are based on the assumption that data collected in data tables include only the partial knowledge about the structure and the behavior of modeled concurrent systems. Nevertheless, such partial knowledge is sufficient to construct a suitable mathematical models. The remaining knowledge (or - in the sequel - a part of it) can be discovered on the basis of created models.

The knowledge - about the modeled systems - encoded in a given data table can be represented by means of rules which can be extracted from the data table. We consider deterministic rules and inhibitory rules. In contrast to deterministic (standard) rules which have the relation attribute = value on the right-hand side, inhibitory rules have on the right-hand side the relation attribute ≠ value. The use of inhibitory rules allows us to represent more knowledge encoded in data tables. As a result, concurrent models based on inhibitory rules are often more compact than models based on deterministic rules. Besides ŞexplicitŤ global states, corresponding to objects, the concurrent system generated by the considered data table can also have ŞhiddenŤ global states, i.e., tuples of attribute values not belonging to a given data table but consistent with all the rules. Such ŞhiddenŤ states can also be considered as realizable global states. This was a motivation for introducing in [20] maximal consistent extensions of information systems with both ŞexplicitŤ and ŞhiddenŤ global states. Such extension includes all possible global states consistent with all rules of a given kind extracted from the original data table. More precisely, the maximal consistent extension of an information system relative to the set of given kind of rules is the set of all objects from the Cartesian product of ranges of attributes from the information system, for which each rule from the set of rules is true. They play important role in investigations at the intersection of the rough set theory and the theory of concurrent systems [14],[20],[22],[17],[29]. The theoretical backgrounds for the maximal consistent extensions of information systems as well as the algorithmic problems such as: (i) the membership to the extension, (ii) the construction of the extension, (iii) the construction of rule system describing the extension, are presented in [4]. The obtained results and published in [4] show that the inhibitory rules provide an essentially more power tool for knowledge representation than the deterministic rules. These results will be useful in applications of data tables for analysis and design of concurrent systems specified by data tables.

In this lecture, an approach to consistent extensions of information systems and dynamic information systems is also presented. Especially, we are interested in partially consistent extensions of such systems. Methods for computing such extensions are given. In the proposed approach, global states of a modeled system and also transitions between states (in the case of dynamic information systems) can be consistent only partially with the knowledge included in the original information system or dynamic information system describing a modeled system. The way of computing suitable consistency factors of new global states or new transitions between states with the original knowledge about systems is provided (see [12]).

5   Structures of Concurrent Models

Two structures of concurrent system models are considered, namely synchronous and asynchronous [12]. In the case of modeling based on information systems, a created synchronous model enables us to generate the maximal consistent extension of a given information system. An asynchronous model enable us to find all possible transitions between global states of a given concurrent system, for which only one process changes its local state. A model created on the basis of dynamic information system enables us to generate a maximal consistent extension of that system. In this case, such an extension includes all possible global states consistent with all rules extracted from the original data table and all possible transitions between global states consistent with all transition rules generated from the original transition system.

In this lecture, the problems of reconstruction of models and prediction of their changes in time is also taken up. Those problems occur as a result of appearing the new knowledge about modeled systems and their behaviors. The new knowledge can be expressed by appearing new global states, new transitions between states, new local states of individual processes or new processes in modeled systems. In our approach, the concurrent model can be built on the basis of a decomposed data table describing of a given system. If the description of a given concurrent system changes (i.e., a new information system is available), we have to reconstruct the concurrent model representing the old concurrent system (described by the old data table). The structure of a constructed concurrent model is determined on the basis of components of a data table (an information system). Some methods for the reconstruction of concurrent models according to such idea are presented in [24],[25],[30].

One of the important aspects of data mining is analysis of data changing in time (i.e., temporal data). Many of the systems change their properties as time goes. Then, models constructed for one period of time must be reconstructed for another period of time. In the research, we assume that concurrent systems are described by temporal information systems (data tables include consecutive global states). In such a case, we observe behavior of modeled systems in consecutive time windows that temporal information systems are split into. Observation of changes enables us to determine the so-called prediction rules that can be used to predict future changes of models. For representing prediction rules, both prediction matrices [12] and Pawlak's flow graphs are used [15].

6   Computer Tool

In the todayŠs computer science development, the usefulness of proposed methods and algorithms for real-life data is conditioned by existing suitable computer tools automating computing processes. Therefore, in this lecture the ROSECON system is presented. ROSECON system is a computer tool supporting users in automated discovering net models from data tables as well as predicting their changes in time. The majority of methods and algorithms presented in this lecture is implemented in ROSECON [11]. Results of experiments with using proposed methods and algorithms on real-life data coming from finance are presented [12].

7   Applications

The considered research problems in this lecture belongs to emerged Artificial Intelligence directions, and it is very important not only with the respect to cognitive aspect but also to the possible applications. Discovering of concurrent systems models from experimental data tables is very interesting and useful for a number of application domains. In particular, in Artificial Intelligence domains, e.g. in speech recognition [3], blind source separation and Independent Component Analysis. In biology and molecular biology; for example, in obtaining the answer concerning the following question: How the model of cell evolution depends on change of the gene codes (see e.g. [35], pp. 780-804).

In the light of the our research findings [10],[11]-[12],[17],[19],[20]-[22],[23]-[27],[31] we can conclude that the rough set theory is suitable for solving problems mentioned above.

8   Concluding Remarks

The presented research in the lecture allows us to understand better the structure and behavior of the modeled system. Due to this research, it is possible to represent the dependencies between the processes in information system and their dynamic interactions in graphical way. This approach can be treated as a kind of decomposition of a given data table. Besides, our methodology can be applied for automatic feature extraction. The processes and the connections between processes in the system can be interpreted as new features of the modeled system [23]. Properties of the constructed concurrent systems model (e.g. their invariants) can be understand as higher level laws of experimental data. As a consequence, this approach seems to be useful also for state identification in real-time [20],[22],[27].

In the next paper, we will consider the prediction problem of property changing net models in non-stationary data systems. Such problem arises when experimental data tables change with time and the constructed net needs to be modified by applying some strategies discovered during the process of changes. We also pursuit application of the presented method to blind separation of sources (for example concurrent time series represented by the dynamic discrete data tables (contained sequential data). Practical applications comprise separation of mixed continuous speech data, and model switching detection in time series and in sequential data given by discrete data tables.

Acknowledgment

The research has been partially supported by the grant N N516 368334 from Ministry of Science and Higher Education of the Republic of Poland and by the grant "Decision support - new generation systems" of Innovative Economy Operational Programme 2008-2012 (Priority Axis 1. Research and development of new technologies) managed by Ministry of Regional Development of the Republic of Poland.

References

1. Bridewell, W., Langley, P., Todorovski, L., Dzeroski, S.: Inductive process modeling. Machine Learning 71(1-32) (2008)
2. Brown, E.M.: Boolean Reasoning. Kluwer, Dordrecht (1990)

3. Cios, J.K., Pedrycz, W., Swiniarski, R.W., Kurgan, L.: Data Mining. A Knowledge Discovery Approach. Springer, Heidelberg (2007)
4. Delimata, P., Moshkov, M., Skowron, A., Suraj, Z.: Inhibitory Rules in Data Analysis. A Rough Set Approach. Springer, Heidelberg (2009)
5. Janowski, A., Peters, J., Skowron, A., Stepaniuk, J.: Optimization in Discovery of Compound Granules. Fundamenta Informaticae 85, 249–265 (2008)
6. Jensen, K.: Coloured Petri Nets. Basic Concepts, Analysis Methods and Practical Use, vol. 1. Springer, Heidelberg (1992)
7. Kodratoff, Y., Michalski, R. (eds.): Machine Learning, vol. 3. Morgan Kaufmann Publishers, CA (1990)
8. Kurgan, L.A., Musilek, P.: A survey of Knowledge Discovery and Data Mining process models. The Knowledge Engineering Review 21(1), 1–24 (2006)
9. de Medeiros, A.K.A., Weijters, A.J.M.M., van der Aalst, W.M.P.: Genetic Process Mining: An Experimental Evaluation. Data Mining and Knowledge Discovery 14, 245–304 (2007)
10. Pancerz, K.: An Application of rough sets to the identification of concurrent system models. Ph.D. Thesis, Institute of Computer Science PAS, Warsaw (2006) (in Polish)
11. Pancerz, K., Suraj, Z.: Discovering Concurrent Models from Data Tables with the ROSECON System. Fundamenta Informaticae 60(1-4), 251–268 (2004)
12. Pancerz, K., Suraj, Z.: Rough Sets for Discovering Concurrent System Models from Data Tables. In: Hassanien, A.E., Suraj, Z., Slezak, D., Lingras, P. (eds.) Rough Computing. Theories, Technologies, and Applications, IGI Global, 2008, pp. 239–268 (2008)
13. Pawlak, Z.: Rough Sets - Theoretical Aspects of Reasoning About Data. Kluwer, Dordrecht (1991)
14. Pawlak, Z.: Concurrent Versus Sequential the Rough Sets Perspective. Bulletin of the EATCS 48, 178–190 (1992)
15. Pawlak, Z.: Flow graphs and decision algorithms. LNCS (LNAI), vol. 2639. Springer, Heidelberg (2003)
16. Peters, J.F., Skowron, A., Suraj, Z., Pedrycz, W., Ramanna, S.: Approximate Real-Time Decision Making: Concepts and Rough Fuzzy Petri Net Models. International Journal of Intelligent Systems 14(8), 805–839 (1999)
17. Rzasa, W., Suraj, Z.: A New Method for Determining of Extensions and Restrictions of Information Systems. In: Alpigini, J.J., Peters, J.F., Skowron, A., Zhong, N. (eds.) RSCTC 2002. LNCS (LNAI), vol. 2475, pp. 197–204. Springer, Heidelberg (2002)
18. Skowron, A.: Boolean reasoning for decision rules generation. In: Komorowski, J., Raś, Z.W. (eds.) ISMIS 1993. LNCS, vol. 689, pp. 295–305. Springer, Heidelberg (1993)
19. Skowron, A.: Discovery of Process Models from Data and Domain Knowledge: A Rough-Granular Approach. In: Ghosh, A., De, R.K., Pal, S.K. (eds.) PReMI 2007. LNCS, vol. 4815, pp. 192–197. Springer, Heidelberg (2007)
20. Skowron, A., Suraj, Z.: Rough Sets and Concurrency. Bulletin of the Polish Academy of Sciences 41(3), 237–254 (1993)
21. Skowron, A., Suraj, Z.: Discovery of Concurrent Data Models from Experimental Tables: A Rough Set Approach. In: Fayyad, U.M., Uthurusamy, R. (eds.) Proceedings of Knowledge Discovery and Data Mining, First International Conference, KDD 1995, Montreal, Canada, August 1995, pp. 288–293. The AAAI Press, Menlo Park (1995)

22. Skowron, A., Suraj, Z.: A Parallel Algorithm for Real-Time Decision Making: A Rough Set Approach. Journal of Intelligent Information Systems 7, 5–28 (1996)
23. Suraj, Z.: Discovery of Concurrent Data Models from Experimental Tables: A Rough Set Approach. Fundamenta Informaticae 28(3-4), 353–376 (1996)
24. Suraj, Z.: An Application of Rough Set Methods to Cooperative Information Systems Reengineering, in: Rough Sets, Fuzzy Sets and Machine Discovery. In: Proceedings of 4th International Workshop, RSFD 1996, November 1996, pp. 364–371. Tokyo University, Tokyo (1996)
25. Suraj, Z.: Reconstruction of Cooperative Information Systems under Cost Constraints: A Rough Set Approach. Information Sciences: An International Journal 111, 273–291 (1998)
26. Suraj, Z.: The Synthesis Problem of Concurrent Systems Specified by Dynamic Information Systems. In: Polkowski, L., Skowron, A. (eds.) Rough Sets in Knowledge Discovery, vol. 2, pp. 418–448. Physica-Verlag (1998)
27. Suraj, Z.: Rough Set Methods for the Synthesis and Analysis of Concurrent Processes. In: Polkowski, L., Tsumoto, S., Lin, T.Y. (eds.) Rough Set Methods and Applications, pp. 379–488. Springer, Heidelberg (2000)
28. Suraj, Z., Owsiany, G., Pancerz, K.: On Consistent and Partially Consistent Extensions of Information Systems. In: Ślęzak, D., Wang, G., Szczuka, M.S., Düntsch, I., Yao, Y. (eds.) RSFDGrC 2005. LNCS (LNAI), vol. 3641, pp. 224–233. Springer, Heidelberg (2005)
29. Suraj, Z., Pancerz, K.: A new method for computing partially consistent extensions of information systems: a rough set approach. In: Proceedings of Information Processing and Management of Uncertainty in Knowledge-based Systems 3, International Conference, IPMU 2006, Paris, France, Editions EDK, pp. 2618–2625 (2006)
30. Suraj, Z., Pancerz, K.: Reconstruction of concurrent system models described by decomposed data tables. Fundamenta Informaticae 71(1), 101–119 (2006)
31. Suraj, Z., Swiniarski, R., Pancerz, K.: Rough Sets and Petri Nets Working Together. In: Nguyen, H.S., Szczuka, M. (eds.) Rough Set Techniques in Knowledge Discovery and Data Mining, International Workshop, RSKD 2005, Hanoi, Vietnam, pp. 27–37 (2005)
32. Swiniarski, R., Hunt, F., Chalvet, D., Pearson, D.: Feature Selection Using Rough Sets and Hidden Layer Expansion for Rupture Prediction in a Highly Automated Production System. Systems Science 23(1), 203–212 (1997)
33. Swiniarski, R.: Application of Petri Nets to Modeling and Describing of Microcomputer Sequential Control Algorithms. International Journal Advances in Modeling, and Simulation 14(3), 47–56 (1988)
34. Swiniarski, R., Skowron, A.: Rough Sets Methods in Feature Selection and Recognition. Pattern Recognition Letters 24(6), 833–849 (2003)
35. Tsumoto, S., Slowinski, R., Komorowski, J., Grzymala-Busse, J.W.: RSCTC 2004. LNCS (LNAI), vol. 3066. Springer, Heidelberg (2004)
36. Unnikrishnan, K.P., Ramakrishnan, N., Sastry, P.S., Uthurusamy, R.: 4th KDD Workshop on Temporal Data Mining: Network Reconstruction from Dynamic Data. The Twelfth ACM SIGKDD International Conference on Knowledge Discovery and Data, Philadelphia, USA (2006),
    http://people.cs.vt.edu/~ramakris/kddtdm06/cfp.ktml

Intelligent Science

Zhongzhi Shi

Key Laboratory of Intelligent Information Processing, Institute of Computing Technology
Chinese Academy of Science, Kexueyuan Nanlu #6, Beijing, 100190, China
shizz@ics.ict.ac.cn

Abstract. Intelligence Science is an interdisciplinary subject which dedicates to joint research on basic theory and technology of intelligence by brain science, cognitive science, artificial intelligence and others. Brain science explores the essence of brain, research on the principles and models of natural intelligence at molecular, cellular, and behavior levels. Cognitive science studies human mental activities, such as perception, learning, memory, thinking, consciousness etc. Artificial intelligence attempts simulation, extension, and expansion of human intelligence using artificial methods and technologies. Researchers specialized in above three disciplines work together to explore new concepts, theories, and methodologies. If successful, it will create a brilliant future in 21st century. The paper will outline the framework of intelligence science and present its ten big challenges. Tolerance Granular Space Model (TGSM) will be discussed as one of helpful approaches.

Keywords: Intelligence Science, Brain Science, Cognitive Science, Artificial Intelligence, Tolerance Granular Space Model.

1 Introduction

Intelligence is the ability to think and learn. How to create intelligence from matter? It is a valuable and extractive problem but it is also a tough problem. Since 1956 artificial intelligence is formally founded and has enjoyed tremendous success over the past fifty years. Its achievements and techniques are in the mainstream of computer science and at the core of many systems. For example, the computer beats the world's chess champ, commercial systems are exploiting voice and speech capabilities, there are robots running around the surface of Mars and so on. We have made significant headway in solving fundamental problems in knowledge representing, symbolic reasoning, machine learning, and more.

During the past fifty years, the Turing test and physical symbolic system hypothesis play important roles to push research on artificial intelligence. Alan Turing claimed that it was too difficult to define intelligence. Instead he proposed Turing test in 1950 [1]. But the Turing test does not constitute an appropriate or useful criterion for human-level artificial intelligence. Nilsson suggested we replace the Turing test by the "employment test" [2]. To pass the employment test, AI programs must be able to perform the jobs ordinarily performed by humans. Systems with true human-level intelligence should be able to perform the tasks for which humans get paid. One can

H. Sakai et al. (Eds.): RSFDGrC 2009, LNAI 5908, pp. 20–32, 2009.

hope that the skills and knowledge gained by a system's education and experience and the habile-system approach toward human-level AI can be entered at whatever level.

The 1975 ACM Turing Award was presented jointly to Allen Newell and Herbert A. Simon at the ACM Annual Conference in Minneapolis, October 20. They delivered the 1975 ACM Turing Award Lecture and proposed physical symbolic system hypothesis: "A physical symbol system has the necessary and sufficient means for intelligent action; it consists of a set of entities, called symbols, which are physical patterns that can occur as components of another type of entity called an expression" [3]. Traditional artificial intelligence follows the principle of physical symbolic system hypothesis to get great successes, particular in knowledge engineering.

During the 1980s Japan proposed the fifth generation computer system (FGCS). It suggested expecting knowledge information processing to form the main part of applied artificial intelligence and to become an important field of information processing in the 1990s. The key technologies of FGCS seem to be VLSI architecture, parallel processing such as data flow control, logic programming, knowledge base based on relational database, applied artificial intelligence and pattern processing. Inference machines and relational algebra machines are typical of the core processors which constitute FGCS. After ten years research and development FGCS project did not reach the expected goal and caused many to reflect over the strategy and methodology of artificial intelligence.

In 1991, Kirsh pointed out five foundational issues for AI: (1) Core AI is the study of conceptualization and should begin with knowledge level theories. (2) Cognition can be studied as a disembodied process without solving the symbol grounding problem. (3) Cognition is nicely described in propositional terms. (4) We can study cognition separately from learning. (5) There is a single architecture underlying virtually all aspects of cognition [4]. Minsky argued that intelligence is the product of hundreds, probably thousands of specialized computational mechanisms he terms agents in Society of Mind [5]. There is no homogenous underlying architecture. In the society of mind theory, mental activity is the product of many agents of varying complexity interacting in hundreds of ways. The purpose of the theory is to display the variety of mechanisms that are likely to be useful in a mind-like system, and to advocate the need for diversity. There is no quick way to justify the assumption of architecture homogeneity.

Humans are the best example of human-level intelligence. McCarthy declared the long-term goal of AI is human-level AI [6]. Recent works in multiple disciplines of cognitive science and neuroscience motivate new computational approaches to achieving human-level AI. In the book On Intelligence, Hawkins proposed machine intelligence meets neuroscience [7]. Granger presented a framework for integrating the benefits of parallel neural hardware with more serial and symbolic processing which motivated by recent discoveries in neuroscience [8]. Langley proposed a cognitive architecture ICARUS which uses means-ends analysis to direct learning and stores complex skills in a hierarchical manner [9]. Sycara proposed the multi-agent systems framework which one develops distinct modules for different facets of an intelligent system [10]. Cassimatis and his colleagues investigate Polyscheme which is a cognitive architecture designed to model and achieve human-level intelligence by integrating multiple methods of representation, reasoning and problem solving [11].

Through more than ten years investigation, particular encouraged by bioinformatics which is a paragon combining biological science and information science in the end of 20 century, I think artificial intelligence should change the research paradigm and learn from natural intelligence. The interdisciplinary subject entitled Intelligence Science is promoted. In 2002 the special Web site called Intelligence Science and Artificial Intelligence has been appeared on Internet [12], which is constructed by Intelligence Science Lab of Institute of Computing Technology, Chinese Academy of Sciences. A special bibliography entitled Intelligence Science written by author was published by Tsinghua University Press in 2006 [13]. The book shows a framework of intelligence science and points out research topics in related subject.

In order to resolve the challenge in information science and technology, that is, high performance computers with extremely low intelligence level, scientists research on brain-like computer. IBM has received a $4.9 million grant from DARPA to lead an ambitious, cross-disciplinary research project to create a new computing platform: electronic circuits that operate like a brain. Along with IBM Almaden Research Center and IBM T. J. Watson Research Center, Stanford University, University of Wisconsin-Madison, Cornell University, Columbia University Medical Center, and University of California-Merced are participating in the project. Henry Markram who is Director of the Center for Neuroscience & Technology and co-Director of EPFL's Brain Mind Institute involves unraveling the blueprint of the neocortical column, chemical imaging and gene expression.

2   A Framework of Intelligence Science

Intelligence science is an interdisciplinary subject mainly including brain science, cognitive science, and artificial intelligence. Brain science explores the essence of brain, research on the principle and model of natural intelligence in molecular, cell and behavior level. Cognitive science studies human mental activity, such as perception, learning, memory, thinking, consciousness etc. In order to implement machine intelligence, Artificial intelligence attempts simulation, extension and expansion of human intelligence using artificial methodology and technology [12].

Brain can perceive the outside world through our senses, such as eye, ear, nose, skin, each of which sends patterns corresponding to real-time environment. Sensory input provides abundant information about certain physical properties in the surrounding world. Reception, processing, and transmitting such information are often framed as a neural bottom-up process. The neural correlates of each can be studied in their own right by suitable experimental paradigms, and functional magnetic resonance imaging (fMRI) has proven very valuable in humans.

The brain has trillions of neurons, with complicated branching dendrites, and dozens of different types of ion-selective channels. Brain science, particularly computational neuroscience focuses on making detailed biologically realistic models which can be simulated by computer. It points out that perceptive lobes have special function separately, the occipital lobe processes visual information, the temporal lobe processes auditory information, the parietal lobe processes the information from the somatic sensors. All of three lobes deal with information perceived from the physical world. Each lobe is covered with cortex where the bodies of neurons are located. Cortex consists of primary, intermediate and advanced areas at least. Information is processed in the primary area first, then is passed to intermediate and advanced area.

Cognitive science is interdisciplinary study of mind and intelligence that embraces philosophy, psychology, artificial intelligence, neuroscience, linguistics, and anthropology. Cognitive scientists study the nature of intelligence from a psychological point of view, mostly building computer models that help elucidate what happens in our brains during problem solving, remembering, perceiving, and other psychological processes. Cognitive science is a study how the mind works, both in its conceptual organization and computational and neural infrastructure. The mind contains perception, rational, consciousness and emotion.

Comparing with computer system, the neural network in brain is the same as hardware and the mind looks like software. Most work in cognitive science assumes that the mind has mental representations analogous to computer data structures, and computational procedures similar to computational algorithms. Connectionists have proposed novel ideas to use neurons and their connections as inspirations for data structures, and neuron firing and spreading activation as inspirations for algorithms. Cognitive science then works with a complex 3-way analogy among the mind, the brain, and computers. Mind, brain, and computation can each be used to suggest new ideas about the others. There is no single computational model of mind, since different kinds of computers and programming approaches suggest different ways in which the mind might work.

Artificial Intelligence develops programs to allow machines to perform functions normally requiring human intelligence, that is, attempts simulation, extension and expansion of human intelligence using artificial methods. Russell points out four approaches to artificial intelligence [14]: Acting humanly: the Turing test approach; Thinking humanly: the cognitive modeling approach; Thinking rationally: the "laws of thought" approach; Acting rationally: the rational agent approach.

Traditional work in AI was based on the physical symbol system hypothesis [3]. In terms of the above hypothesis led to many successes both in creating tools that can achieve elements of intelligent behavior, as well as in illuminating the many components that make up human intelligence. Previous research on artificial intelligence mainly simulates the human intelligence functionally and views the brain as black box. Research scientists of intelligence science are changing the situation and exploring innovative strategy and methodology for investigating the principles and key technology of intelligence from cross multiple subjects. The book titled Intelligence Science presents a primary framework in detail [13].

3   Ten Big Issues of Intelligence Science

Intelligence Science is an interdisciplinary subject which dedicates to joint research on basic theory and technology of intelligence by brain science, cognitive science, artificial intelligence and others. Ten big issues of intelligence science will be discussed in this section.

3.1 Basic Process of Neural Activity

The brain is a collection of about 10 billion interconnected neurons. Neurons are electrically excitable cells in the nervous system that process and transmit information. A neuron's dendritic tree is connected to thousands neighbouring neurons [15]. When one of those neurons is activated, positive or negative charge is received by one of the dendrites. The strengths of all the received charges are added together through the processes of spatial and temporal summation. The aggregate input is then passed to the soma (cell body). The soma and the enclosed nucleus don't play a significant role in the processing of incoming and outgoing data. Their primary function is to perform the continuous maintenance required to keep the neuron functional. The output strength is unaffected by the many divisions in the axon; it reaches each terminal button with the same intensity it had at the axon hillock.

Each terminal button is connected to other neurons across a small gap called a synapse. The physical and neurochemical characteristics of each synapse determine the strength and polarity of the new input signal. This is where the brain is the most flexible, and the most vulnerable. In molecular level neuron signal generation, transmission and neurotransmitters are basic problems attracted research scientists to engage investigation in brain science.

3.2 Synaptic Plasticity

One of the greatest challenges in neuroscience is to determine how synaptic plasticity, learning and memory are linked. Two broad classes of models of plasticity are described by Phenomenological models and Biophysical models [16].

Phenomenological models are characterized by treating the process governing synaptic plasticity as a black box that takes as input a set of variables, and produces as output a change in synaptic efficacy. No explicit modeling of the biochemistry and physiology leading to synaptic plasticity is implemented. Two different classes of phenomenological models, rate based and spike based, have been proposed.

Biophysical models, in contrast to phenomenological models, concentrate on modeling the biochemical and physiological processes that lead to the induction and expression of synaptic plasticity. However, since it is not possible to implement precisely every portion of the physiological and biochemical networks leading to synaptic plasticity, even the biophysical models rely on many simplifications and abstractions. Different cortical regions, such as Hippocampus and Visual cortex have somewhat different forms of synaptic plasticity.

3.3 Perceptual Representation and Feature Binding

The perceptual systems are primarily visual, auditory and kinesthetic, that is, pictures, sounds and feelings. There is also olfactory and gustatory, i.e. smell and taste. The perceptual representation is a modeling approach that highlights the constructive, or generative function of perception, or how perceptual processes construct a complete volumetric spatial world, complete with a copy of our own body at the center of that world. The representational strategy used by the brain is an analogical one; that is, objects and surfaces are represented in the brain not by an abstract symbolic code, or in the activation of individual cells or groups of cells representing particular features detected in the visual field. Instead, objects are represented in the brain by constructing full spatial effigies of them that appear to us for whole world like the objects themselves or at least so it seems to us only because we have never seen those objects in their raw form, but only through our perceptual representations of them.

The binding problem is an important one across many disciplines, including psychology, neuroscience, computational modeling, and even philosophy. Feature binding is the process how a large collection of coupled neurons combines external data with internal memories into coherent patterns of meaning. Due to neural synchronization theory, it is achieved via neural synchronization. When external stimuli come into the brain, neurons corresponding to the features of the same object will form a dynamic neural assembly by temporal synchronous neural oscillation, and the dynamic neural assembly, as an internal representation in the brain, codes the object in the external world.

3.4 Coding and Retrieval of Memory

A brain has distributed memory system, that is, each part of brain has several types of memories that work in somewhat different ways, to suit particular purposes. According to the stored time of contents memory can be divided into long term memory, short term memory and working memory. Research topics in memory relate to coding, extracting and retrieval of information. Current working memory attracts more researchers to involve.

Working memory will provides temporal space and enough information for complex tasks, such as understanding speech, learning, reasoning and attention. There are memory and reasoning functions in the working memory. It consists of three components: that is, central nervous performance system, video space primary processing and phonetic circuit [19].

Memory phenomena have also been categorized as explicit or implicit. Explicit memories involve the hippocampus-medial temporal lobe system. The most common current view of the memorial functions of the hippocampal system is the declarative memory. There are a lot of research issues that are waiting for us to resolve. What is the readout system from the hippocampal system to behavioral expression of learning in declarative memory? Where are the long-term declarative memories stored after the hippocampal system? What are the mechanisms of time-limited memory storage in hippocampus and storage of permanent memories in extra- hippocampal structures?

Implicit memory involves cerebellum, amygdala, and other systems [20]. Cerebellum is necessary for classical conditioning of discrete behavioral responses. It is learning to make specific behavioral responses. Amygdalar system is learning fear and associated autonomic responses to deal with the situation.

3.5 Linguistic Cognition

Language is fundamentally a means for social communication. Language is also often held to be the mirror of the mind. Chomsky developed transformational grammar that cognitivism replaced behaviorism in linguistics [21].

Through language we organize our sensory experience and express our thoughts, feelings, and expectations. Language is particular interesting from cognitive informatics point of view because its specific and localized organization can explore the functional architecture of the dominant hemisphere of the brain.

Recent studies of human brain show that the written word is transferred from the retina to the lateral geniculate nucleus, and from there to the primary visual cortex. The information then travels to a higher-order center, where it is conveyed first to the angular gyrus of the parietal-temporal-occipital association cortex, and then to Wernicke's area, where visual information is transformed into phonetic representation of the word. For the spoken word auditory information is processed by primary auditory cortex. Then information input to higher-order auditory cortex, before it is conveyed to a specific region of parietal-temporal- occipital association cortex, the angular gyrus, which is concerned with the association of incoming auditory, visual, and tactile information. From here the information is projected to Wernicke's area and Broca's area. In Broca's area the perception of language is translated into the grammatical structure of a phrase and the memory for word articulation is stored [22]. Fig. 2 illustrates language processing based on Wernicke-Geschwind model in brain.

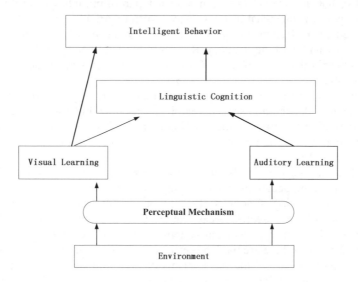

Fig. 1. Language processing in brain

3.6 Learning

Learning is the basic cognitive activity and accumulation procedure of experience and knowledge. Through learning, system performance is improved. Perceptual learning, cognitive learning, and implicit learning are active research topics.

Perceptual learning should be considered as an active process that embeds particular abstraction, reformulation and approximation within the Abstraction framework. The active process refers to the fact that the search for a correct data representation is performed through several steps. A key point is that perceptual learning focuses on low-level abstraction mechanism instead of trying to rely on more complex algorithm. In fact, from the machine learning point off view, perceptual learning can be seen as a particular abstraction that may help to simplify complex problem thanks to a computable representation. Indeed, the baseline of Abstraction, i.e. choosing the relevant data to ease the learning task, is that many problems in machine learning cannot be solve because of the complexity of the representation and is not related to the learning algorithm, which is referred to as the phase transition problem. Within the abstraction framework, we use the term perceptual learning to refer to specific learning task that rely on iterative representation changes and that deals with real-world data which human can perceive.

In contrast with perceptual learning, cognitive learning is a leap in cognition process. It generates knowledge by clustering, classification, conceptualization and so on. In general, there are inductive learning, analogical learning, case-based learning, explanation learning, and evolutional learning connectionist learning.

The core issue of cognitive learning is self-organizing principles. Kohonen has proposed self-organizing maps which is a famous neural network model. Babloyantz applied chaotic dynamics to study brain activity. Haken has proposed a synergetic approach to brain activity, behavior and cognition.

Introspective learning is an inside brain learning, i.e., there is no input from outside environment. We have proposed a model for introspective learning with 7 parts in Figure 3, such as expectant objective, evaluation, explanation, reconstruct strategy, meta cognition, case bases and knowledge base.

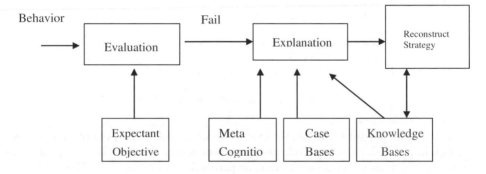

Fig. 2. Introspective learning

The term implicit learning was coined by Reber to refer to the way people could learn structure in a domain without being able to say what they had learnt [23]. Reber proposed artificial grammars to study implicit learning for unconscious knowledge acquisition. It will help to understand the learning mechanism without consciousness. Since middle of 1980's implicit learning become an active research area in psychology.

In the Machine Learning Department within Carnegie Mellon University's School of Computer Science researchers receive $1.1 million from Keck Foundation to pursue new breakthroughs in learning how the brain works. Cognitive neuroscience professor Marcel Just and computer science professor Tom Mitchell have received a three-year grant from the W. M. Keck foundation to pursue new breakthroughs in the science of brain imaging [24].

3.7 Thought

Thought is a reflection of essential attributes and internal laws of objective reality in conscious, indirect and generalization by human brain with consciousness [25]. In recent years, there has been a noteworthy shift of interest in cognitive science. Cognitive process raises man's sense perceptions and impressions to logical knowledge. According to abstraction degree of cognitive process, human thought can be divided into three levels: perception thought, image thought and abstraction thought. A hierarchical model of thought which illustrates the characteristics and correlations of thought levels has been proposed in [26]. Fig 4 shows the hierarchical thought model of brain.

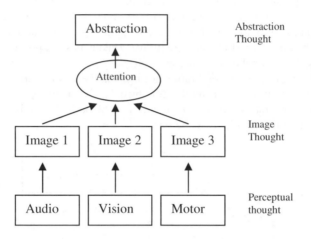

Fig. 3. Hierarchical thought model of brain

Perception thought is the lowest level of thought. Behavior is the objective of research in perception thought. Reflection is a function of stimulus. Perception thought emphasizes stimulus-reflection schema or perception-action schema. The thought of animal and infant usually belong to perception thought because they cannot introspect, and cannot declare empirical consciousness [25]. In perception thought, intelligent behavior takes place without representation and reasoning.

Behavior-based artificial intelligence has produced the models of intelligence which study intelligence from the bottom up, concentrating on physical systems, situated in the world, autonomously carrying out tasks of various sorts. They claim that simple things to do with perception and mobility in a dynamic environment took evolution much longer to perfect. Intelligence in human has been taking place for only a small fraction of our evolutionary lineage. Machine intelligence can take evolution by the dynamics of interaction with the world.

3.8 Emotion

The mental perception of some fact excites the mental affection called the emotion, and that this latter state of mind gives rise to the bodily expression. Emotion is a complex psychophysical process that arises spontaneously, rather than through conscious effort. It may evoke positive or negative psychological responses and physical expressions. Research on emotion at varying levels of abstraction, using different computational methods, addressing different emotional phenomena, and basing their models on different theories of affect.

Since the early 1990s emotional intelligence has been systematically studied [27]. Scientific articles suggested that there existed an unrecognized but important human mental ability to reason about emotions and to use emotions to enhance thought. Emotional intelligence refers to an ability to recognize the meanings of emotion and their relationships, as well as ability to reason and problem solve on the basis of them. Emotional intelligence is involved in the capacity to perceive emotions, assimilate emotion-related feelings, understand information of emotions, and manage them.

3.9 Nature of Consciousness

The most important scientific discovery of the present era will come to answer how exactly do neurobiological processes in the brain cause consciousness? The question "What is the biological basis of consciousness?" is selected as one of the 125 questions formulated for Science's 125th anniversary. Recent scientifically oriented accounts of consciousness emerge from the properties and organization of neurons in the brain. Consciousness is the notion of mind and soul.

Physical basis of consciousness appears as the crucial challenge to scientific, reductionist world view. Francis Crick's book 'The astonishing Hypothesis' is an effort to chart the way forward in the investigation of consciousness [28]. Crick has proposed the basic ideas of researching consciousness: a) It seems probable, however, that at any one moment some active neuronal processes in your head correlate with consciousness, while others do not. What are the differences between them? b) All the different aspect of consciousness, for example pain and visual awareness, employ a basic common mechanism or perhaps a few such mechanisms. If we could understand the mechanisms for one aspect, then we hope we will have gone most of the way to understanding them all.

Chalmers suggests the problem of consciousness can be broken down into several questions. The major one is the neuronal correlate of consciousness (NCC) which focuses on specific processes that correlate with the current content of consciousness [29]. The NCC is the minimal set of neurons, most likely distributed throughout certain cortical and subcortical areas, whose firing directly correlates with the perception of the subject at the time. Discovering the NCC and its properties will mark a major milestone in any scientific theory of consciousness. Several other questions need to be answered about the NCC. What type of activity corresponds to the NCC? What causes the NCC to occur? And, finally, what effect does the NCC have on postsynaptic structures, including motor output.

3.10 Mind Modeling

Mind is a very important issue in intelligence science, and also it is a tuff problem. Mind could be defined as: "That which thinks, reasons, perceives, wills, and feels. The mind now appears in no way separate from the brain. In neuroscience, there is no duality between the mind and body. They are one." in Medical Dictionary [30]. A mind model is intended to be an explanation of how some aspect of cognition is accomplished by a set of primitive computational processes. A model performs a specific cognitive task or class of tasks and produces behavior that constitutes a set of predictions that can be compared to data from human performance. Task domains that have received attention include problem solving, language comprehension, memory tasks, and human-device interaction.

Researchers try to construct mind model to illustrate how brains do. Anderson and colleagues have demonstrated that a production rule analysis of cognitive skill, along with the learning mechanisms posited in the ACT model, provide detailed and explanatory accounts of a range of regularities in cognitive skill acquisition in complex domains such as learning to program Lisp [31]. ACT also provides accounts of many phenomena surrounding the recognition and recall of verbal material, and regularities in problem solving strategies.

In the early 1980's, SOAR was developed to be a system that could support multiple problem solving methods for many different problems [32]. In the mid 1980's, Newell and many of his students began working on SOAR as a candidate of unified theories of cognition. SOAR is a learning architecture that has been applied to domains ranging from rapid, immediate tasks such as typing and video game interaction to long stretches of problem solving behavior. SOAR has also served as the foundation for a detailed theory of sentence processing, which models both the rapid on-line effects of semantics and context, as well as subtle effects of syntactic structure on processing difficulty across several typologically distinct languages.

The Society of Mind offers a revolutionary theory of human thought [5]. Minsky proposes that the mind consists of several kinds of non-thinking entities, called agents. Agents alone repeat their tasks with great acumen, but they execute their work with no understanding of it. Thought occurs when societies of agents interact and relate, much as a jet engine's components work to generate thrust. Human personality is not controlled by a centralized "conductor" in the brain, but rather emerges from seemingly unintelligent and unconnected mental processes, or "agents." With Minsky's theory as a metaphor, participants will reach a new sensitivity to the many different parts of the mind that are engaged when we enjoy and respond to music.

3.11   Tolerance Granular Space Model

Granular computing is an emerging paradigm of information processing. Information granules are collections of entities that usually originate at the numeric level and are arranged together due to their similarity, functional adjacency, indistinguishability, coherency, and so on, which arise in the process of data abstraction and derivation of knowledge from information.

At the present granular computing mainly includes fuzzy set-based computing with words, rough set and quotient space theory. Computing with Words involves computing and reasoning with fuzzy information granules. Rough set focuses on study on how to represent unknown concept (subset) by constructing upper approximation and lower approximation with equivalence classes. But topological structures of spaces consisting of these equivalence classes are hardly taken into account. In contrast, quotient space theory describes the space structure, which focuses on transformation and dependence between different granular worlds.

A new tolerance granular space model is presented [33]. The basic idea of the model is based on the human ability, that is, people can abstract or synthesize the knowledge and data relating with special tasks to different degrees or sizes granules, and accomplish the tasks with the help of the granules and relations among them. The model of tolerance granular spaces has been applied to solve classification, decision-making, image texture recognizing and so on.

4 Perspective on Intelligence Science

The intelligence revolution with the goal to replace human brain work by machine intelligence is the next revolution in human society. The incremental efforts in neuro-science and cognitive science provide us exciting solid foundation to explore brain model and intelligent behavior. We should research on neocortical column, population coding, mind model, consciousness etc. for the human-level intelligence and brain-like computer. We believe that intelligence science will make great progress and new breakthroughs in the coming 50 years. Let us work together to contribute our intellect and capability to promote the development of intelligence science and become a bright spot of human civilization in 21 century.

Acknowledgments. This paper is supported by National Basic Research Programme (No. 2007CB311004), National Natural Science Foundation of China (No. 60775035, 60933004, 60903141), National High-Tech Programme (No. 2007AA01Z132), and National Science and Technology Support Plan (No. 2006BAC08B06).

References

1. Turing, A.M.: Computing Machinery and Intelligence. Mind 59, 433–460 (1950)
2. Nilsson, N.J.: Human-Level Artificial Intelligence? Be Serious! AI Magazine 26(4), 68–75 (2005)
3. Newell, A., Simon, H.A.: Computer science as empirical inquiry: Symbols and search. Communications of the Association for Computing Machinery 19(3), 113–126 (1976) (1975 ACM Turing Award Lecture)
4. Kirsh, D.: Foundations of AI: the big issues. Artificial Intelligence 47, 3–30 (1991)
5. Minsky, M.: The Society of Mind. Simon & Schuster, New York (1985)
6. McCarthy, J.: The future of AI-A manifesto. AI Magazine 26(4), 39 (2005)
7. Hawkins, J., Blakeslee, S.: On Intelligence, Times Books. Henry Holt & Company (2004)
8. Granger, R.: Engines of the Brain. AI Magazine 27(2), 15–31 (2006)
9. Langley, P.: Cognitive Architectures and General Intelligent Systems. AI Magazine 27(2), 33–44 (2006)
10. Sycara, K.: Multi-Agent Systems. AI Magazine 19(2), 79–93 (1998)
11. Cassimatis, N.: A Cognitive Substrate for Achieving Human-Level Intelligence. AI Magazine 27(2), 45–56 (2006)
12. Intelligence Science, http://www.intsci.ac.cn/
13. Shi, Z.: Intelligence Science. Tsinghua University Press, Beijing (2006)

14. Russell, S., Norvig, P.: Artificial Intelligence: A Modern Approach. Prentice Hall, Engle-wood Cliffs (2003)
15. Bear, M.F., Connors, B., Paradiso, M., Bear, M.F., Connors, B.W., Michael, A.: Neurosci-ence: Exploring the Brain, 2nd edn. Lippincott Williams & Wilkins (2002)
16. Shouval, H.Z.: Models of synaptic plasticity. Scholarpedia 2(7), 1605 (2007)
17. Eckhorn, R., Reitboeck, H.J., Arndt, M., Dicke, P.W.: Feature linking via synchronization among distributed assemblies: Simulation of results from cat cortex. Neural Comput. 2, 293–307 (1990)
18. Shi, Z., Liu, X.: A Computational Model for Feature Binding. Science in China, Series F: Information Sciences (to appear, 2009)
19. Dehn, M.J.: Working Memory and Academic Learning: Assessment and Intervention. Wiley, Chichester (2008)
20. Tarsia, M., Sanavio, E.: Implicit and explicit memory biases in mixed anxiety-depression. Journal of affective disorders 77(3), 213–225 (2003)
21. Chomsky, N.: Syntactic Structures. Mouton, The Hague (1957)
22. Mayeux, R., Kandel, E.R.: Disorders of language: the aphasias. In: Kandel, E.R., Schwarz, J.H., Jessell, T.M. (eds.) Principles of Neural Science, 3rd edn., pp. 840–851. Elsevier, Amsterdam (1991)
23. Reber, A.S.: Implicit learning and artificial grammar. Journal of Verbal Learning and Ver-bal Behavior 6, 855–863 (1967)
24. Machine Learning Department (2008), http://www.cald.cs.cmu.edu/
25. Shi, Z.: Cognitive Science. Press of USTC (2008)
26. Shi, Z.: Advanced Artificial Intelligence. Science Press of China (2006)
27. Norman, D.A.: Emotion and design: Attractive things work better. Interactions Maga-zine 9(4), 36–42 (2002)
28. Crick, F.: The Astonishing Hypothesis, Scribner (1995)
29. Chalmers, D.: The Conscious Mind: In Search of a Fundamental Theory. Oxford Univer-sity Press, Oxford (1995)
30. Medical Dictionary, http://www.medterms.com/script/main/hp.asp
31. Anderson, J.R.: The Adaptive Character of Thought. Erlbaum, Hillsdale (1993)
32. Newell, A.: Unified Theories of Cognition. Harvard, Cambridge (1990)
33. Shi, Z., Zheng, Z.: Tolerance Granular Space Model and Its Applications. In: Miao, D., et al. (eds.) Granular Computing: Past, Present and Future, pp. 42–82. Science Press, Beijing (2007)

Pattern Structures for Analyzing Complex Data

Sergei O. Kuznetsov

Department of Applied Mathematics and Information Science,
State University Higher School of Economics,
Myasnitskaya 20, 101000 Moscow, Russia
skuznetsov@hse.ru

Abstract. For data given by binary object-attribute datatables Formal Concept Analysis (FCA) provides with a means for both convenient computing hierarchies of object classes and dependencies between sets of attributes used for describing objects. In case of data more complex than binary to apply FCA techniques, one needs *scaling* (binarizing) data. Pattern structures propose a direct way of processing complex data such as strings, graphs, numerical intervals and other. As compared to scaling (binarization), this way is more efficient from the computational point of view and proposes much better vizualization of results. General definition of pattern structures and learning by means of them is given. Two particular cases, namely that of graph pattern structures and interval pattern structures are considered. Applications of these pattern structures in bioinformatics are discussed.

1   Introduction

Many problems of constructing domain taxonomies and ontologies, as well as finding dependencies in data, can be solved with the use of the models based on closure operators and respective lattices of closed sets within Formal Concept Analysis (FCA) [21,9]. The main definitions of FCA start from a binary relation, coming from applications as a binary object-attribute table. These tables (called *contexts* in FCA) give rise to lattices whose diagrams give nice visualizations of classes of objects of a domain. At the same time, the edges of these diagrams give essential knowledge about objects, by giving the probabilities of cooccurrence of attributes describing objects [17,18,19], this type of knowledge being known under the name of *association rules* in data mining.

However in many real-world applications researchers deal with complex and heteregeneous data different from binary datatables in involving numbers, strings, graphs, intervals, logical formulas, etc. for making descriptions of objects from an application domain. To apply FCA tools to data of these types, one needs binarizing initial data or, in FCA terms, applying *conceptual scaling*. Many types of scaling exist (see [9]), but do not always suggest the most efficient implementation right away, and there are situations where one would choose original or other data representation forms rather than scaled data [7]. Although scaling allows one to apply FCA tools, it may drastically increase the complexity of representation and worsen the visualization of results.

H. Sakai et al. (Eds.): RSFDGrC 2009, LNAI 5908, pp. 33–44, 2009.

Instead of scaling one may work directly with initial data descriptions defining so-called *similarity* operators, which induce semilattice on data descriptions. In recent decades several attempts were done in defining such semilattices on sets of graphs [12,16,13], numerical intervals [12,10], logical formulas [2,3], etc. In [7] a general approach called pattern structures was proposed, which allows one to extend standard FCA approaches to arbitrary partially ordered data descriptions. In this paper we consider pattern structures for several data types and applications, showing their advantages and application potential.

The rest of the paper is organized as follows: In Section 2 we recall basic definitions of FCA, as well as related machine learning and rule mining models. In Section 3 we present pattern structures and respective generalization of machine learning and rule mining models. In Sections 4 and 5 we consider particular pattern structures on sets of graphs and vectors of intervals and discuss their applications in bioinformatics. In Section 6 we discuss computational issues of pattern structures.

2   Concept Lattices and Concept-Based Learning

2.1   Main Definitions

First we introduce standard FCA definitions from [9]. Let G and M be arbitrary sets and $I \subseteq G \times M$ be an arbitrary binary relation between G and M. The triple (G, M, I) is called a *(formal) context*. Each $g \in G$ is interpreted as an object, each $m \in M$ is interpreted as an attribute. The fact $(g, m) \in I$ is interpreted as "g has attribute m". The two following *derivation operators* $(\cdot)'$

$$A' = \{m \in M \mid \forall g \in A : gIm\} \qquad \text{for } A \subseteq G,$$
$$B' = \{g \in G \mid \forall m \in B : gIm\} \qquad \text{for } B \subseteq M$$

define a *Galois connection* between the powersets of G and M. For $A \subseteq G$, $B \subseteq M$, a pair (A, B) such that $A' = B$ and $B' = A$, is called a *(formal) concept*. Concepts are partially ordered by $(A_1, B_1) \leq (A_2, B_2) \Leftrightarrow A_1 \subseteq A_2 (\Leftrightarrow B_2 \subseteq B_1)$. With respect to this partial order, the set of all formal concepts forms a complete lattice called the *concept lattice* of the formal context (G, M, I). For a concept (A, B) the set A is called the *extent* and the set B the *intent* of the concept.

The notion of dependency in data is captured in FCA by means of implications and partial implications (association rules). For $A, B \subseteq M$ the *implication* $A \rightarrow B$ holds if $A' \subseteq B'$ and the *association rule* (called *partial implication* in [17]) $A \longrightarrow_{c,s} B$ with *confidence* c and *support* s holds if $s \geq \frac{|A' \cap B'|}{|G|}$ and $c \geq \frac{|A' \cap B'|}{|A'|}$.

The language of FCA, as we showed in [6], is well suited for describing a model of learning JSM-hypotheses from [4,5]. In addition to the structural attributes of M, consider a *target attribute* $\omega \notin M$. This partitions the set G of all objects into three subsets: The set G_+ of those objects that are known to have the property ω (these are the *positive examples*), the set G_- of those objects of which it is known that they do not have ω (the *negative examples*) and the set G_τ of *undetermined examples*, i.e., of those objects, of which it is unknown if they have property ω or not. This gives three subcontexts of $\mathbb{K} = (G, M, I)$, the first two staying for the training sample:

$$\mathbb{K}_+ := (G_+, M, I_+), \quad \mathbb{K}_- := (G_-, M, I_-), \quad \text{and } \mathbb{K}_\tau := (G_\tau, M, I_\tau),$$

where for $\varepsilon \in \{+, -, \tau\}$ we have $I_\epsilon := I \cap (G_\varepsilon \times M)$ and the corresponding derivation operators are denoted by $(\cdot)^+$, $(\cdot)^-$, $(\cdot)^\tau$, respectively.

Intents, as defined above, are attribute sets shared by some of the observed objects. In order to form hypotheses about structural causes of the target attribute ω, we are interested in sets of structural attributes that are common to some positive, but to no negative examples. Thus, a *positive hypothesis* h for ω (called "counter-example forbidding hypotheses" in the JSM-method [4,5]) is an intent of $\mathbb{K}_+$ such that $h^+ \neq \emptyset$ and $h \not\subseteq g^- := \{m \in M \mid (g, m) \in I_-\}$ for any negative example $g \in G_-$. *Negative hypotheses* are defined similarly. Various classification schemes using hypotheses are possible, as an example consider the following simple scheme from [5]: If the intent

$$g^\tau := \{m \in M \mid (g, m) \in I_\tau\}$$

of an object $g \in G_\tau$ contains a positive, but no negative hypothesis, then g^τ is *classified positively*. Negative classifications are defined similarly. If g^τ contains hypotheses of both kinds, or if g^τ contains no hypothesis at all, then the classification is contradictory or undetermined, respectively. In this case one can apply standard probabilistic techniques known in machine learning and data mining (majority vote, Bayesian approach, etc.). Notwithstanding its simplicity, the model of learning and classification with concept-based hypotheses proved to be efficient in numerous studies in bioinformatics [1,8,15].

A well-known application of concept lattices in data mining use the fact that the edges of the lattice diagram make a basis of association rules for the context [17,18,19]. In fact, each edge of a concept lattice diagram, connecting a higher concept (A', A) and a lower concept (B, B'), corresponds to a set of association rules of the form $(Y) \longrightarrow_{c,s} B$ (where Y is minimal in the set $\{X \subseteq A \mid X'' = A\}$) and all other association rules may be obtained from rules of these type by some inference [11].

2.2 Many-Valued Contexts and Their Interordinal Scaling

Consider an object-attribute table whose entries are not binary. It can be given by a quadruple $\mathbb{K}_1 = (G, S, W, I_1)$, where G, S, W are sets and I_1 is a ternary relation $I_1 \subseteq G \times S \times W$. In FCA terms $\mathbb{K}_1 = (G, S, W, I_1)$ is called a *many-valued context*.

Consider an example of analyzing *gene expression data* (GED) given by tables of values. The names of rows correspond to genes. The names of the columns of the table correspond to *situations* where genes are tested. A table entry is called an *expression value*. A row in the table is called *expression profile* associated to a gene. In terms of many-valued contexts, the set of genes makes the set of objects G, the set of situations makes the set of many-valued attributes S, the set of expression values makes the set $W \subset \mathbb{R}$ and $I_1 \subseteq G \times S \times W$. Then $\mathbb{K}_1 = (G, S, W, I_1)$ is a many-valued context representing a GED. The fact $(g, s, w) \in I_1$ or simply $g(s) = w$ means that gene g has an expression value w for situation s. The objective is to extract formal concepts (A, B) from $\mathbb{K}_1$, where $A \subseteq G$ is a subset of genes sharing "similar values" of W, i.e. lying in a same interval. To this end, we use an appropriate binarization (scaling) technique to build a formal context $\mathbb{K}_2 = (G, S_2, I_2)$, called *derived context* of $\mathbb{K}_1$.

A scale is a formal context (cross-table) taking original attributes of $\mathbb{K}_1$ with the derived ones of $\mathbb{K}_2$. As attributes do not take necessarily same values, each of them is scaled separately. Let $W_s \subseteq W$ be the set of all values of the attribute s. The following *interordinal scale* (see pp. 42 in [9]) can be used to represent all possible intervals of attribute values:

$$\mathbb{I}_{W_s} = (W_s, W_s, \leq) | (W_s, W_s, \geq).$$

The operation of *apposition of two contexts* with identical sets of objects, denoted by $|$, returns the context with the same set of objects W_s and the set of attributes corresponding to the disjoint union of attribute sets of the original contexts. In our case this operation is applied to two contexts $(W_s, W_s, \leq)$ and $(W_s, W_s, \geq))$, the table below gives an example for $W_s = \{4, 5, 6\}$.

	$s_1 \leq 4$	$s_1 \leq 5$	$s_1 \leq 6$	$s_1 \geq 4$	$s_1 \geq 5$	$s_1 \geq 6$
4	×	×	×	×		
5		×	×	×	×	
6			×	×	×	×

The intents given by interordinal scaling are value intervals.

3   Pattern Structures

3.1   Main Definitions and Results

Let G be a set (interpreted as a set of objects), let $(D, \sqcap)$ be a meet-semi-lattice (of potential object descriptions) and let $\delta : G \longrightarrow D$ be a mapping. Then $(G, \underline{D}, \delta)$, where $\underline{D} = (D, \sqcap)$, is called a *pattern structure*, provided that the set

$$\delta(G) := \{\delta(g) \mid g \in G\}$$

generates a complete subsemilattice $(D_\delta, \sqcap)$ of $(D, \sqcap)$, i.e., every subset X of $\delta(G)$ has an infimum $\sqcap X$ in $(D, \sqcap)$ and D_δ is the set of these infima.

Elements of D are called *patterns* and are naturally ordered by subsumption relation $\sqsubseteq$: given $c, d \in D$ one has $c \sqsubseteq d \iff c \sqcap d = c$. A pattern structure $(G, \underline{D}, \delta)$ gives rise to the following derivation operators $(\cdot)^\diamond$:

$$A^\diamond = \prod_{g \in A} \delta(g) \qquad \text{for } A \subseteq G,$$

$$d^\diamond = \{g \in G \mid d \sqsubseteq \delta(g)\} \qquad \text{for } d \in (D, \sqcap).$$

These operators form a Galois connection between the powerset of G and $(D, \sqsubseteq)$. $\sqcap$ is also called a similarity operator. The pairs (A, d) satisfying

$$A \subseteq G, \quad d \in D, \quad A^\diamond = d, \quad \text{and} \quad A = d^\diamond$$

are called the *pattern concepts* of $(G, \underline{D}, \delta)$, with extent A and *pattern intent d*. For $a, b \in D$ the *pattern implication* $a \rightarrow b$ holds if $a^\diamond \subseteq b^\diamond$, and the *pattern association rule* $a \longrightarrow_{c,s} b$ with *confidence c* and *support s* holds if $s \geq \frac{|a^\diamond \sqcap b^\diamond|}{|G|}$ and $c \geq \frac{|a^\diamond \sqcap b^\diamond|}{|a^\diamond|}$. Like in case of association rules, pattern association rules may be inferred from a base that corresponds to the set of edges of the diagram of the pattern concept lattice.

Operator $(\cdot)^{\diamond\diamond}$ is an algebraical closure operator [9] on patterns, since it is

idempotent: $d^{\diamond\diamond\diamond\diamond} = d^{\diamond\diamond}$,
extensive: $d \sqsubseteq d^{\diamond\diamond}$,
monotone: $d^{\diamond\diamond} \sqsubseteq c^{\diamond\diamond}$ for $d \sqsubseteq c$.

In [6] we showed that if $(D, \sqcap)$ is a complete meet-semi-lattice (where infimums are defined for arbitrary subsets of elements), in particular a finite semi-lattice, there is a subset $M \subseteq D$ with the following interesting property: The concepts of the formal context (G, M, I) where I is given as $gIm \colon \Leftrightarrow m \sqsubseteq \delta(g)$, called a *representation context* for $(G, \underline{D}, \delta)$, are in one-to-one correspondence with the pattern concepts of $(G, \underline{D}, \delta)$. The corresponding concepts have the same first components (called *extents*). These extents form a complete lattice, which is isomorphic to the concept lattice of (G, M, I). This result is proved by a standard application of the basic theorem of FCA (which allows one to represent every lattice as a concept lattice) [21,9] and shows the way of binarizing complex data representation given by a pattern structure. The cost of this binarization may be a large amount attributes of the representation context and hence, the space needed for storing this context.

3.2   Learning with Pattern Structures

The concept learning model described in the previous section for standard object-attribute representation (i.e., formal contexts) is naturally extended to pattern structures. Suppose we have a set of positive examples E_+ and a set of negative examples E_- w.r.t. a target attribute.

A pattern $h \in D$ is a *positive hypothesis* iff

$$h^\diamond \cap E_- = \emptyset \text{ and } \exists A \subseteq E_+ : A^\diamond = h.$$

Again, a positive hypothesis is a similarity (or least general generalization of descriptions) of positive examples, which is not contained in (does not cover) any negative example. A *negative hypothesis* is defined analogously, by interchanging $+$ and $-$.

The meet-preserving property of projections implies that a hypothesis H_p in data under projection ψ corresponds to a hypothesis H in the initial representation for which the image under projection is equal to H_p, i.e., $\psi(H) = H_p$.

Hypotheses are used for classification of undetermined examples along the lines of [5]. The corresponding definitions are similar to those from Section 2, one just needs to replace $\subseteq$ with $\sqsubseteq$.

3.3 Projections and Learning in Projections

For some pattern structures (e.g., for the pattern structures on sets of graphs with labeled vertices) even computing subsumption of patterns may be NP-hard. Hence, for practical situations one needs approximation tools, which would replace the patterns with simpler ones, even if that results in some loss of information. To this end we use a mapping $\psi\colon D \to D$ that replaces each pattern $d \in D$ by $\psi(d)$ such that the pattern structure $(G, \underline{D}, \delta)$ is replaced by $(G, \underline{D}, \psi \circ \delta)$. To distinguish two pattern structures, which we consider simultaneously, we use the symbol $\,^\circ$ only for $(G, \underline{D}, \delta)$, not for $(G, \underline{D}, \psi \circ \delta)$. Under some natural algebraic requirements (that hold for all natural projections in particular pattern structures we studied in applications) the meet operation $\sqcap$ is preserved:

$$\psi(X \sqcap Y) = \psi(X) \sqcap \psi(Y).$$

This property of projection allows one to relate hypotheses in the original representation with those approximated by a projection.

This helped us to describe [6] how the lattice of pattern concepts changes when we replace $(G, \underline{D}, \delta)$ by its approximation $(G, \underline{D}, \psi \circ \delta)$. First, we note that $\psi(d) \sqsubseteq \delta(g) \Leftrightarrow \psi(d) \sqsubseteq \psi \circ \delta(g)$. Moreover, for pattern structures $(G, \underline{D}, \delta_1)$ and $(G, \underline{D}, \delta_2)$ one has $\delta_2 = \psi \circ \delta_1$ for some projection ψ of $\underline{D}$ iff there is a representation context (G, M, I) of $(G, \underline{D}, \delta_1)$ and some $N \subseteq M$ such that $(G, N, I \cap (G \times N))$ is a representation context of $(G, \underline{D}, \delta_2)$. Thus, the basic theorem of FCA helps us not only to "binarize" the initial data representation, but to relate binarizations of different projections.

Pattern structures are naturally ordered by projections: $(G, \underline{D}, \delta_1) \geq (G, \underline{D}, \delta_2)$ if there is a projection ψ such that $\delta_2 = \psi \circ \delta_1$. In this case, representation $(G, \underline{D}, \delta_2)$ can be said to be rougher than $(G, \underline{D}, \delta_1)$ and the latter to be finer than the former. In comparable pattern structures implications are related as follows: If $\psi(a) \to \psi(b)$ and $\psi(b) = b$ then $a \to b$ for arbitrary $a, b \in D$. In particular, if $\psi(a)$ is a positive (negative) hypothesis in projected representation, then a is positive (negative) hypothesis in the original representation.

4 Pattern Structures on Closed Sets of Labeled Graphs

In [12,13] we proposed a semi-lattice on sets of graphs with labeled vertices and edges. This lattice is based on a natural domination relation between pairs of graphs with labeled vertices and edges. Consider an ordered set P of connected graphs[1] with vertex and edge labels from the set $\mathcal{L}$ partially ordered by $\preceq$. Each

---

[1] Omitting the condition of connectedness, one obtains a similar, but computationally much harder model.

labeled graph Γ from P is a quadruple of the form $((V, l), (E, b))$, where V is a set of vertices, E is a set of edges, $l: V \to \mathcal{L}$ is a function assigning labels to vertices, and $b: E \to \mathcal{L}$ is a function assigning labels to edges. In $(P, \leq)$ we do not distinguish isomorphic graphs.

For two graphs $\Gamma_1 := ((V_1, l_1), (E_1, b_1))$ and $\Gamma_2 := ((V_2, l_2), (E_2, b_2))$ from P we say that Γ_1 *dominates* Γ_2 or $\Gamma_2 \leq \Gamma_1$ (or Γ_2 is a *subgraph* of Γ_1) if there exists an injection $\varphi: V_2 \to V_1$ such that it

- respects edges: $(v, w) \in E_2 \Rightarrow (\varphi(v), \varphi(w)) \in E_1$,
- fits under labels: $l_2(v) \preceq l_1(\varphi(v))$, if $(v, w) \in E_2$ then $b_2(v, w) \preceq b_1(\varphi(v), \varphi(w))$.

Obviously, $(P, \leq)$ is a partially ordered set. Now a *similarity operation* $\sqcap$ on graph sets can be defined as follows: For two graphs X and Y from P

$$\{X\} \sqcap \{Y\} := \{Z \mid Z \leq X, Y, \forall Z_* \leq X, Y\ Z_* \not\geq Z\},$$

i.e., $\{X\} \sqcap \{Y\}$ is the set of all maximal common subgraphs of graphs X and Y. Similarity of non-singleton sets of graphs $\{X_1, \ldots, X_k\}$ and $\{Y_1, \ldots, Y_m\}$ is defined as

$$\{X_1, \ldots, X_k\} \sqcap \{Y_1, \ldots, Y_m\} := \text{MAX}_{\leq}(\cup_{i,j}(\{X_i\} \sqcap \{Y_j\})),$$

where $\text{MAX}_{\leq}(X)$ returns maximal (w.r.t. $\leq$) elements of X.

The similarity operation $\sqcap$ on graph sets is commutative: $X \sqcap Y = Y \sqcap X$ and associative: $(X \sqcap Y) \sqcap Z = X \sqcap (Y \sqcap Z)$. A set X of labeled graphs from P for which $\sqcap$ is idempotent, i.e., $X \sqcap X = X$ holds, is called a *graph pattern*. For patterns we have $\text{MAX}_{\leq}(X) = X$. For example, for each graph $g \in P$ the set $\{g\}$ is a pattern. On the contrary, for $\Gamma_1, \Gamma_2 \in P$ such that $\Gamma_1 \leq \Gamma_2$ the set $\{\Gamma_1, \Gamma_2\}$ is not a pattern. Denote by D the set of all patterns, then $(D, \sqcap)$ is a semi-lattice with infimum (meet) operator $\sqcap$. The natural subsumption order on patterns is given by $c \sqsubseteq d \Leftrightarrow c \sqcap d = c$.

Let E be a set of object names, and let $\delta: E \to D$ be a mapping, taking each object name to $\{g\}$ for some labeled graph $g \in P$ (thus, g is "graph description" of object e). The triple $(E, (D, \sqcap), \delta)$ is a particular case of a pattern structure.

A set of graphs X is called *closed* if $X^{\diamond\diamond} = X$. This definition is related to the notion of a closed graph in data mining and graph mining, which is important for computing association rules between graphs. Closed graphs are defined in [20] in terms of "counting inference" as follows.

Given a graph dataset E, support of a graph g or *support*(g) is a set (or number) of graphs in E that have subgraphs isomorphic to g. A graph g is called *closed* if no supergraph f of g (i.e., a graph such that g is isomorphic to its subgraph) has the same support.

In terms of pattern structures, E is a set of objects, each object $e \in E$ having a graph description $\delta(e)$, support$(g) = \{e \in E \mid \delta(g) \leq e\}$. Note that the definitions distinguish between a closed graph g and the closed set $\{g\}$ consisting of one graph g. Closed sets of graphs form a *meet semi-lattice* w.r.t. $\sqcap$. Closed graphs do not have this property, since in general, there are closed graphs with no infimums. However, closed graphs and closed sets of graphs are intimately related, as shown in the following

Proposition 1. *Let a dataset described by a pattern structure* $(E, (D, \sqcap), \delta)$ *be given. Then the following two properties hold:*
 1. For a closed graph g *there is a closed set of graphs* G *such that* $g \in G$.
 2. For a closed set of graphs G *and an arbitrary* $g \in G$, *graph* g *is closed.*

Proof

1. Consider the closed set of graphs $G = \{g\}^{\diamond\diamond}$. Since G consists of all maximal common subgraphs of graphs that have g as a subgraph, G contains as an element either g or a supergraph f of g. In the first case, property 1 holds. In the second case, we have that each graph in G that has g as a subgraph also has f as a subgraph, so f has the same support as g, which contradicts with the fact that g is closed. Thus, $G = \{g\}^{\diamond\diamond}$ is a closed set of graphs satisfying property 1.

2. Consider a closed set of graphs G and $g \in G$. If g is not a closed graph, then there is a supergraph f of it with the same support as g has and hence, with the same support as G has. Since G is the set of all maximal common subgraphs of the graphs describing examples from the set $G^{\diamond}$ (i.e, its support), $f \in G$ should hold. This contradicts the fact that $g \in G$, since a closed set of graphs cannot contain as elements a graph and a supergraph of it (otherwise, its closure does not coincide with itself). ☐

Therefore, one can use algorithms for computing closed sets of graphs, e.g., the algorithm described in [13], to compute closed graphs. With this algorithm one can also compute all *frequent* closed sets of graphs, i.e., closed sets of graphs with support above a fixed threshold (by introducing a slightly different backtrack condition).

The learning model based on graph pattern structures along the lines of the previous section was successfully used in series of applications in bioinformatics, namely in problems where chemical substructures causing particular biological activities (like toxicity) were investigated [8,15]. In many cases the proposed graph representation resulted in better predictive accuracy as compared to that obtained with standard attribute-type languages used for the analysis of biological activity of chemicals.

5   Pattern Structures on Intervals

5.1   Main Definitions

To define a semilattice operation $\sqcap$ for intervals that would be analogous to the set-theoretic intersection or meet operator on sets of graphs, one should realize that "similarity" between two real numbers (between two intervals) may be expressed in the fact that they lie within some (larger) interval, this interval being the smallest interval containing both two.

Thus, for two intervals $[a_1, b_1]$ and $[a_2, b_2]$, with $a_1, b_1, a_2, b_2 \in \mathbb{R}$, we define their meet as

$$[a_1, b_1] \sqcap [a_2, b_2] = [min(a_1, a_2), max(b_1, b_2)].$$

This operator is obviously idempotent, commutative and associative, thus defining a pattern structure on intervals. The counterintuitive observation that the meet operator takes two intervals to a larger one (in contrast to set intersection and meet on graph sets which take sets to smaller ones) fails after realizing that a larger interval, like in case of smaller sets and smaller sets of graphs, correspond to a larger set of objects, whose descriptions fall in the interval.

The natural order relation (subsumption) on intervals is given as follows:

$$[a_1, b_1] \sqsubseteq [a_2, b_2]$$
$$\iff [a_1, b_1] \sqcap [a_2, b_2] = [a_1, b_1]$$
$$\iff [min(a_1, a_2), max(b_1, b_2)] = [a_1, b_1]$$
$$\iff a_1 \le a_2 \quad and \quad b_1 \ge b_2.$$

Again, contrary to usual intuition, smaller intervals subsume larger intervals that contain the former. A next step would be considering vectors of intervals. An *interval p-vector* is a p-dimensional vector of intervals. The meet $\sqcap$ for interval vectors is defined by component-wise interval meets. Interval p-vector patterns are p-dimensional rectangular parallelepipeds in Euclidean space. Another step further would be made by allowing any type of patterns for each component. The general meet operator on a vector like that is defined by component-wise meet operators.

5.2   Interval Patterns and Interordinal Scaling

For a many-valued context (G, M, W, I) with $W \subset \mathbb{R}$ consider the respective pattern structure $(G, (D, \sqcap), \delta)$ on interval vectors, the interordinal scaling $\mathbb{I}_{W_s} = (W_s, W_s, \le) \mid (W_s, W_s, \ge)$ from the previous Section, and the context K_I resulting from applying interordinal scaling $\mathbb{I}_{W_s}$ to (G, M, W, I). Consider usual derivation operators $(\cdot)'$ in context K_I. Then the following proposition establishes an isomorphism between the concept lattice of K_I and the pattern concept lattice of $(G, (D, \sqcap), \delta)$.

Proposition 2. *Let $A \subseteq G$, then the following statements 1 and 2 are equivalent:*

1. A is an extent of the pattern structure $(G, (D, \sqcap), \delta)$ and $A^\diamond = \langle [\underline{m}_i, \overline{m}_i] \rangle_{i \in [1,p]}$

2. A is a concept extent of the context K_I so that for all $i \in [1, p]$ $\underline{m}_i$ is the largest number n such that the attribute $s_i \ge n$ is in A' and $\overline{m}_i$ is the smallest number n such that the attribute $s_i \le n$ is in A'.

Proof. 1 $\to$ 2 Let $A \subseteq G$ be a pattern extent. Given $\delta_i(g)$ the mapping that returns the ith interval of the vector describing object g. Since $A^\diamond =$

$\langle[\underline{m}_i, \overline{m}_i]\rangle_{i\in[1,p]}$, for every object $g \in A$ one has $\underline{m}_i \leq \delta_i(g) \leq \overline{m}_i$ and there are objects $g_1, g_2 \in A$ such that $\delta_i(g_1) = \underline{m}_i$, $\delta_i(g_1) = \overline{m}_i$. Hence, in context K_I one has

$$A' = \cup_{i\in[1,p]}\{s_i \geq n_{\min}, \ldots, s_i \geq n_1, s_i \leq n_2, \ldots, s_i \leq n_{\max}\}$$

where

$$n_{\min} \prec \ldots \prec n_1 \leq n_2 \prec \ldots \prec n_{\max}$$

and $n_1 = \underline{m}_i$, $n_2 = \overline{m}_i$. Hence, $\underline{m}_i$ is the largest number n such that the attribute $s_i \geq n$ is in A' and $\overline{m}_i$ is the smallest number n such that the attribute $s_i \leq n$ is in A'. Suppose that A is not an extent of K_I. Hence, $A \subset A''$ and there is $g \in A'' \setminus A$ and $g' \supseteq A'$. This means that for all i $\underline{m}_i \leq \delta_i(g) \leq \overline{m}_i$. Therefore, $g \in A^{\diamond\diamond}$ and $A \neq A^{\diamond\diamond}$, a contradiction. The proof $2 \to 1$ is similar. □

The larger is a pattern concept, the more there are elements in its extent, and the more there are intervals in its intent. However, the main goal in applications like analysis of gene expression data is extracting homogeneous groups of objects (e.g., genes), i.e. groups of objects having similar expression values. Therefore, descriptions of homogeneous groups should be composed of intervals with "small" sizes where $size([a, b]) = b - a$. Consider parameter max_{size} that specifies the maximal admissible size of any interval composing an interval vector. In our gene expression data analysis [10] we restricted to pattern concepts with pattern intents $d = \langle[a_i, b_i]\rangle_{i\in[1,p]} \in (D, \sqcap)$ satisfying the constraint: $\exists i \in [1, p]$ $(b_i - a_i) \leq max_{size}$, for any $a, b \in \mathbb{R}$, or a stricter constraint like $\forall i \in [1, p]$ $(b_i - a_i) \leq max_{size}$, where max_{size} is a parameter. Since both constraints are monotone (if an intent does not satisfy it, than a subsumed intent does not satisfy it too), the subsets of patterns satisfying any of these constraints make an order filter (w.r.t. subsumption on intervals $\sqsubseteq$) of the lattice of pattern intents and can be computed by an ordinary FCA algorithm with a modified backtracking condition.

Interval pattern structures were successfully applied to gene expression data analysis [10], where classes of situations with similar gene expressions were generated.

5.3   Computing in Pattern Structures

Many algorithms for generating formal concepts from a formal context are known, see e.g. a performance comparative [14]. Experimental results of [14] highlight several best algorithms for dense and large contexts, which is the case of interordinal derived formal contexts. Worst-case upper bound time complexity of these algorithms computing the set of all concepts of the context (G, M, I) is $O(|G|^2 \cdot |M| \cdot |L|)$, where L is the set of generated concepts [14].

Several algorithms for computing concept lattices, like NextClosure and CbO, may be adapted to computing pattern lattices in bottom-up way (starting from intersecting individual object descriptions and proceeding by intersecting more and more object descriptions). The worst-case time complexity of computing all pattern concepts of a pattern structure $(G, \underline{D}, \delta)$ in the bottom-up way is

$O((\alpha + \beta|G|)|G||L|)$, where α is time needed to perform $\sqcap$ operation and β is time needed to test $\sqsubseteq$ relation. In case of graphs, even β may be exponential wrt. the number of graph vertices, that is why approximations (like those given by projections) are often needed. In experiments with many chemical rows in [15] we used projections to graphs with about 10 vertices to be able to process datasets with hundreds of chemical substances.

The worst-case time complexity of computing the set of interval pattern structures is $O(|G|^2 \cdot |M| \cdot |L|)$. If a many-valued context (G, M, W, I) is given, the worst-case complexity of computing the set of all concepts of its interordinally scaling is $O(|G|^2 \cdot |W| \cdot |L|)$, which may be fairly large if the cardinality of the set of attribute values $|W|$ is much larger than that of the set of attributes $|M|$. The worst case $|W| = |G| \times |S|$ is attained when attribute values are different for each object-attribute pair. In [10] several algorithms for computing with interval patterns were compared. The experimental comparison shows that when the number of attribute values w.r.t. $|G| \times |S|$ is very low, computing concepts in representation contexts is more efficient. For large datasets with many different attribute values, it is more efficient to compute in pattern structures.

6   Conclusion

Pattern structures propose a universal means of analyzing hierarchies of classes and dependencies in case of data given by complex ordered descriptions. As compared to binarization techniques, computing with pattern structures often gives more efficiency and better vizualization. Pattern projections allows one to reduce representation dimension to attain even better computer efficiency. Future research on pattern structure will be concerned with new complex data types, interesting projections and new applications. The use of pattern structures for mining association rules in complex data will also be studied.

Acknowledgements

The author was partially supported by the project of the Russian Foundation for Basic Research, grant no. 08-07-92497-NTsNIL_a.

References

1. Blinova, V.G., Dobrynin, D.A., Finn, V.K., Kuznetsov, S.O., Pankratova, E.S.: Toxicology analysis by means of the JSM-method. Bioinformatics 19, 1201–1207 (2003)
2. Chaudron, L., Maille, N.: Generalized Formal Concept Analysis. In: Ganter, B., Mineau, G.W. (eds.) ICCS 2000. LNCS, vol. 1867, pp. 357–370. Springer, Heidelberg (2000)
3. Férré, S., Ridoux, O.: A Logical Generalization of Formal Concept Analysis. In: Ganter, B., Mineau, G.W. (eds.) ICCS 2000. LNCS, vol. 1867, Springer, Heidelberg (2000)

4. Finn, V.K.: On Machine-Oriented Formalization of Plausible Reasoning in the Style of F. Backon–J. S. Mill. Semiotika Informatika 20, 35–101 (1983) (in Russian)
5. Finn, V.K.: Plausible Reasoning in Systems of JSM Type. Itogi Nauki i Tekhniki, Seriya Informatika 15, 54–101 (1991) (in Russian)
6. Ganter, B., Kuznetsov, S.: Formalizing Hypotheses with Concepts. In: Ganter, B., Mineau, G.W. (eds.) ICCS 2000. LNCS, vol. 1867, pp. 342–356. Springer, Heidelberg (2000)
7. Ganter, B., Kuznetsov, S.: Pattern Structures and Their Projections. In: Delugach, H.S., Stumme, G. (eds.) ICCS 2001. LNCS (LNAI), vol. 2120, pp. 129–142. Springer, Heidelberg (2001)
8. Ganter, B., Grigoriev, P.A., Kuznetsov, S.O., Samokhin, M.V.: Concept-based Data Mining with Scaled Labeled Graphs. In: Wolff, K.E., Pfeiffer, H.D., Delugach, H.S. (eds.) ICCS 2004. LNCS (LNAI), vol. 3127, pp. 94–108. Springer, Heidelberg (2004)
9. Ganter, B., Wille, R.: Formal Concept Analysis: Mathematical Foundations. Springer, Heidelberg (1999)
10. Kaytoue, M., Duplessis, S., Kuznetsov, S.O., Napoli, A.: Two FCA-Based Methods for Mining Gene Expression Data. In: Ferre, S., Rudoplh, S. (eds.) ICFCA 2009. LNCS (LNAI), pp. 251–266. Springer, Heidelberg (2009)
11. Kryszkiewicz, M.: Concise Representations of Association Rules. In: Hand, D.J., Adams, N.M., Bolton, R.J. (eds.) Pattern Detection and Discovery. LNCS (LNAI), vol. 2447, pp. 92–109. Springer, Heidelberg (2002)
12. Kuznetsov, S.O.: JSM-method as a machine learning method. Itogi Nauki i Tekhniki, ser. Informatika 15, 17–50 (1991)
13. Kuznetsov, S.O.: Learning of Simple Conceptual Graphs from Positive and Negative Examples. In: Żytkow, J.M., Rauch, J. (eds.) PKDD 1999. LNCS (LNAI), vol. 1704, pp. 384–391. Springer, Heidelberg (1999)
14. Kuznetsov, S.O., Obiedkov, S.A.: Comparing performance of algorithms for generating concept lattices. J. Exp. Theor. Artif. Intell. 14(2-3), 189–216 (2002)
15. Kuznetsov, S.O., Samokhin, M.V.: Learning Closed Sets of Labeled Graphs for Chemical Applications. In: Kramer, S., Pfahringer, B. (eds.) ILP 2005. LNCS (LNAI), vol. 3625, pp. 190–208. Springer, Heidelberg (2005)
16. Liquiere, M., Sallantin, J.: Structural Machine Learning with Galois Lattice and Graphs. In: Proc. Int. Conf. Machine Learning ICML 1998 (1998)
17. Luxenburger, M.: Implications partielle dans un contexte. Math. Sci. Hum. (1991)
18. Pasquier, N., Bastide, Y., Taouil, R., Lakhal, L.: Efficient Minining of Association Rules Based on Using Closed Itemset Lattices. J. Inf. Systems 24, 25–46 (1999)
19. Lakhal, L., Stumme, G.: Efficient mining of association rules based on formal concept analysis. In: Ganter, B., Stumme, G., Wille, R. (eds.) Formal Concept Analysis. LNCS (LNAI), vol. 3626, pp. 180–195. Springer, Heidelberg (2005)
20. Yan, X., Han, J.: CloseGraph: Mining closed frequent graph patterns. In: Getoor, L., Senator, T., Domingos, P., Faloutsos, C. (eds.) Proc. 9th ACM SIGKDD Int. Conf. on Knowledge Discovery and Data Mining (KDD 2003), pp. 286–295. ACM Press, New York (2003)
21. Wille, R.: Restructuring Lattice Theory: an Approach Based on Hierarchies of Concepts. In: Rival, I. (ed.) Ordered Sets, pp. 445–470. Reidel, Dordrecht (1982)

Fuzzy Sets and Rough Sets
for Scenario Modelling and Analysis*

Qiang Shen

Dept. of Computer Science, Aberystwyth University, Wales, UK
qqs@aber.ac.uk

Abstract. Both fuzzy set theory and rough set theory play an important role in data-driven, systems modelling and analysis. They have been successfully applied to building various intelligent decision support systems (amongst many others). This paper presents an integrated utilisation of some recent advances in these theories for detection and prevention of serious crime (e.g. terrorism). It is shown that the use of these advanced theories offers an effective means for the generation and assessment of plausible scenarios which can each provide an explanation for the given intelligence data. The resulting systems have the potential to facilitate rapid response in devising and deploying preventive measures. The paper also suggests a number of important further challenges in consolidating and refining such systems.

1 Introduction

Solving complex real-world problems often requires timely and intelligent decision-making, through analysis of a large volume of information. For example, in the wake of terrorist atrocities such as September 11, 2001, and July 7, 2005, intelligence experts have commented that the failure in the detection of terrorist activity is not necessarily due to lack of data, but to difficulty in relating and interpreting the available intelligence on time. Thus, an important and emerging area of research is the development of decision support systems that will help to establish so-called situational awareness: a deeper understanding of how the available data is related and whether or not it represents a threat.

Most criminal and terrorist organisations are embedded within legitimate society and remain secrete. However, organised crime and terrorist activity does leave a trail of information, such as captured communications and forensic evidence, which can be collected by police and security services. Whilst experienced intelligence analysts can suggest plausible scenarios, the amount of intelligence data possibly relevant may well be overwhelming for human examination. Hypothetical (re-)construction of the activities that may have generated the intelligence data obtained, therefore, presents an important and challenging research topic for crime prevention and detection.

* This work was supported by UK EPSRC grants GR/S63267/01-02, GR/S98603/01 and EP/D057086/1. The author is grateful to all members of the project teams for their contributions, but will take full responsibility for the views expressed here.

H. Sakai et al. (Eds.): RSFDGrC 2009, LNAI 5908, pp. 45–58, 2009.

This paper presents a knowledge-based framework for the development of such systems, to assist (but not to replace) intelligence analysts in identifying plausible scenarios of criminal or terrorist activity, and in assessing the reliability, risk and urgency of generated hypotheses. In particular, it introduces an integrated use of some recent advances in fuzzy set [34] and rough set [23,24] theories to build intelligent systems for the monitoring and interpretation of intelligence data. Here, integration of fuzzy and rough techniques does not necessarily imply a direct combination of both, but utilising them within a common framework. It differs from the conventional hybridisation approaches [20,21,26], although part of the work does involve the employment of the combined fuzzy-rough set theory [3,9].

The rest of the paper is organised as follows. Section 2 outlines the underlying approach adopted and describes the essential components of such a system. Section 3 shows particular instantiations of the techniques used to implement the key components of this framework. Essential ideas are illustrated with some simple examples. Section 4 summarises the paper and points out important further research. Due to space limit, this paper concentrates on the introduction of the underlying conceptual approaches adopted, with specific technical and application details omitted (which can be found in the references).

2   Plausible Scenario-Based Approach

In order to devise a robust monitoring system that is capable of identifying many variations on a given type of terrorist activity, this work employs a model-based approach to scenario generation [28]. The knowledge base of such a system consists of generic and reusable component parts of plausible scenarios, called model or scenario fragments (interchangeably). Such fragments include: types of (human and material) resources required for certain classes of organised criminal/terrorist activity, ways in which such resources can be acquired and organised, and forms of evidence that may be obtained or generated (e.g. from intelligence databases) when given certain scenarios.

Note that conventional knowledge-based systems (for instance, rule or case-based) have useful applications in the crime detection area. However, their scope is restricted to either the situations foreseen or those resulting from previously encountered cases. Yet, organised terrorist activity tends to be unique, whilst employing a relatively restricted set of methods (e.g. suicide bombing or bomb threats in public places). A model-based reasoner designed to (re-)construct likely scenarios from available evidence, as combinations of instantiated scenario fragments, seems to be ideally suited to cope with the variety of scenarios that may be encountered. Indeed, the main strength of model-based reasoning is its adaptability to scenarios that are previously unseen [13].

Figure 1 shows the general architecture of the approach taken in this research. Based on intelligence data gathered, the scenario generation mechanism instantiates and retrieves any relevant model fragments from the library of generic scenario fragments, and combines such fragments to form plausible explanations for the data. A description of how such a system is built is given below.

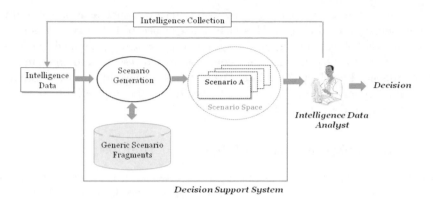

Fig. 1. Architecture of Intelligent Systems for Intelligence Data Analysis

2.1 Flexible Composition Modelling

The central idea is to establish an inference mechanism that can instantiate and then dynamically compose generic model fragments into scenario descriptions, which are plausible and may explain the available data (or evidence). A compositional modelling approach [12] is devised for this purpose. The main potential of using this approach over conventional techniques is its ability to automatically construct many variations of a given type of scenario from a relatively small knowledge base, by combining reusable model fragments on the fly. This ensures the robustness required for the resulting system to tackle the problems at hand.

The compositional modelling approach developed in this research differs from those in the literature in two distinct ways:

1. *Ability* to speculate about plausible relations between different cases. Often, intelligence data will refer to individuals and objects whose identity is only partially specified. For example, when a person is observed on a CCTV camera, some identifying information can be collected, but this may be insufficient for an exact identification. When a person with similar features has been identified elsewhere, it is important that any relation between both sightings is explored. Ideas originally developed in the area of link-based similarity analysis [2,14] are adapted herein for: (a) identifying similar individuals and objects in a space of plausible scenarios, and (b) supporting the generation of hypothetically combined scenarios to explore the implications of plausible matches.

2. *Coverage* to generate scenarios from a wide range of data sources, including factual data, collected intelligence, and hypothesised but unsubstantiated information. This requires matching specific data (e.g. the names of discovered chemicals) with broader (and possibly subjective) knowledge and other vague information contents. Such knowledge and information may be abstractly specified in the knowledge base, e.g. "a chemical being highly explosive". Similarly, matching attributes of partially identified objects and

individuals may involve comparing vague features, such as a person's apparent height, race and age. This suggests the use of a formal mathematical theory that is capable of capturing and representing ill-defined and imprecise linguistic terms, which are common in expressing and inferring from intelligence knowledge and data. Fuzzy systems methods are therefore introduced to compositional modelling to decide on the applicability of scenario fragments and their compositions.

2.2   Plausible Scenario-Based Intelligence Monitoring

Monitoring intelligence data for evidence of potential serious criminal activity, especially terrorist activity, is a non-trivial task. It is not known in advance what aspects of such activity will be observed, and how they will be interconnected. There are nevertheless, many different ways in which a particular type of activity may be arranged. Hence, conventional approaches to monitoring, which aim to identify pre-specified patterns of data, are difficult to adapt to this domain.

Although general and potentially suitable, the model-based approach adopted here may lead to systems that generate a large number of plausible scenarios for a given problem. It is therefore necessary for such a system to incorporate a means to sort the plausible scenarios, so that the generated information remains manageable within a certain time frame. For this purpose, scenario descriptions are presented to human analysts with measurements of their reliability, risk, and urgency. Each of these features may be assessed by a numeric metric. However, intelligence data and hypotheses are normally too vague to produce precise estimates that are also accurate. Therefore, a novel fuzzy mechanism is devised to provide an appropriate method of assessing and presenting these factors. The framework also covers additional tools such as a facility to propose additional information sources (by exploring additional, real or hypothesised, evidence that may be generated in a given scenario).

Figure 2 shows a specification of the general framework given in Fig. 1. Technical modules include:

- Fuzzy Feature Selection carries out semantics-preserving dimensionality reduction (over nominal and real-valued data).
- Fuzzy Learning provides a knowledge modelling mechanism to generalise data with uncertain and vague information into mode fragments.
- Fuzzy Iterative Inference offers a combination of abductive and deductive inferences, capable of reasoning with uncertain assumptions.
- Flexible CSP (constraint satisfaction problem-solver) deals with uncertain and imprecise constraint satisfaction, subject to preference and priority.
- Fuzzy Interpolative Reasoning enables approximate inference over sparse knowledge base, using linear interpolation.
- Flexible ATMS is an extended truth-maintenance system that keeps track of uncertain assumption-based deduction.
- Flexible Coreference Resolution implements a link-based identity resolution approach, working with real, order-of-magnitude, and nominal values.

- Fuzzy Aggregation performs information aggregation by combining uncertain attributes as well as their values.
- Fuzzy Evidence Evaluation performs evidence assessment, including discovery of misleading information, and generates evidence-gathering proposal.
- Fuzzy Risk Assessment computes potential loss-oriented risk evaluation through fuzzy random process modelling.

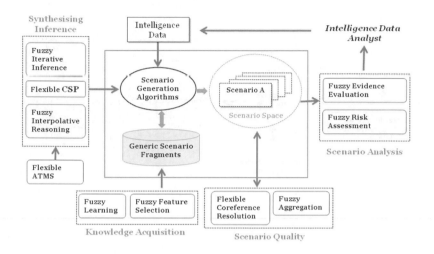

Fig. 2. Instantiated Architecture

Systems built following the approach of Fig. 2 can help to improve the likelihood of discovering potential threats posed by criminal or terrorist organisations. The reasoning of such a system is logical and readily interpretable by human analysts. Thus, it can be very helpful in supporting human analysts when working under time constraints. For instance, this may aid in avoiding premature commitment to certain seemingly more likely but unreal scenarios, minimising the risk of producing incorrect interpretations of intelligence data. This is of particular interest to support staff investigating cases with unfamiliar evidence. The resulting approach may also be adapted to build systems that facilitate training of new intelligence analysts. This is possible because the underlying inference mechanism and the knowledge base built for intelligence data monitoring can be used to artificially synthesise various scenarios (of whatever likelihood), and to systematically examine the implications of acquiring different types of evidence.

3   Illustrative Component Approaches

As a knowledge-based approach to building decision support systems, any implementation of the framework proposed above will require a knowledge base to begin with. The first part of this section will then introduce a number of recent advances in developing data-driven learning techniques that are suitable to derive such required knowledge from potentially very complex data. The second part will describe one of the key techniques that support scenario composition, especially for situations where limited domain knowledge is available. The third and final part of the section will demonstrate how risks of generated scenarios may be estimated. Figure 3 outlines a simplified version of the framework which may be implemented using the techniques described herein.

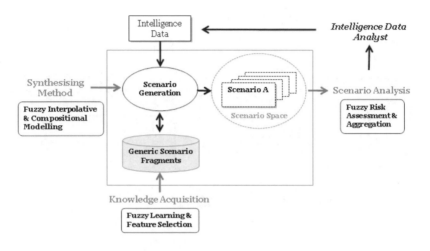

Fig. 3. Focussed Illustration

All of these approaches have been developed using fuzzy and rough methods. These techniques will be introduced at conceptual level with illustrative examples. Mathematical and computational details are omitted, but can be found in the relevant references.

3.1   Fuzzy Learning and Feature Selection

In general, an initial knowledge base of generic scenario fragments is built partly by generalising historical intelligence data through computer-based induction, and partly through manual analysis of past terrorist or criminal activity. This work focusses on the automated induction of model fragments.

Fuzzy Descriptive Learning. Many real-world problems require the development and application of algorithms that automatically generate human interpretable knowledge from historical data. Such a task is clearly not just for learning model fragments.

Most of the methods for fuzzy rule induction from data have followed the so-called precise approach. Interpretability is often sacrificed, in exchange for a perceived increase in precision. In many cases, the definitions of the fuzzy sets that are intended to capture certain vague concepts are allowed to be modified such that they fit the data better. This modification comes at the cost of ruining the original meaning of the fuzzy sets and the loss of transparency of the resulting model. In other cases the algorithms themselves generate the fuzzy sets, and present them to the user. The user must then interpret these sets and the rules which employ them (e.g. a rule like: If *volume* is Tri(32.41, 38.12, 49.18), then *chance* is Tri(0.22, 0.45, 0.78), which may be learned from the data presented in Fig. 4). In some extreme cases, each rule may have its own fuzzy set definition for every condition, thereby generating many different sets in a modest rule base. The greatest disadvantage of the precise approach is that the resulting sets and rules are difficult to match with human interpretation of the relevant concepts.

As an alternative, there exist proposals that follow the descriptive (or linguistic) approach. In such work no changes are made to human defined fuzzy sets. The rules must use the (fuzzy) words provided by the user without modifying them in any way. One of the main difficulties with this approach is that the possible rules available are predetermined, equivalently speaking. This is because the fuzzy sets can not be modified, and only a small number of them are typically available. Although there can be many of these rules they are not very flexible and in many cases they may not necessarily fit the data well (e.g. a rule like: If *volume* is Moderate, then *chance* is High, which may be learned from the data and predefined fuzzy sets given in Fig. 5). In order to address this problem, or at least partially, linguistic hedges (aka. fuzzy quantifiers) are employed.

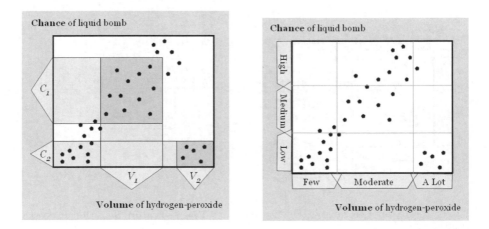

Fig. 4. Precise Modelling   **Fig. 5.** Descriptive Modelling

The concept of linguistic hedges has been proposed quite early on in fuzzy systems research [33]. Application of such a hedge to a fuzzy set produces a new fuzzy set, in a fixed and interpretable manner. The interpretation of the resultant set emanates from the original fuzzy set and a specific transformation that the hedge implies. In so doing, the original fuzzy sets are not changed, but the hedged fuzzy sets provide modifiable means of modelling a given problem and therefore, more freedom in representing information in the domain.

This research adopts the seminal work of [18] which champions this approach. Figure 6 illustrates the ideas: Descriptive fuzzy system models are produced with a two-step mechanism. The first is to use a precise method to create accurate rules and the second to convert the resulting precise rules to descriptive ones. The conversion is, in general, one-to-many. It is implemented by using a heuristic algorithm that derives potentially useful translations and then, by employing evolutionary computation to perform a fine tuning of these translations. Both steps are computationally efficient. The resultant descriptive model is ready to be directly applied for inference; no precise rules are needed in runtime.

Note that Fig. 6 shows the learning of a "model" in a general sense. Such a model may be a set of conventional production fuzzy if-then rules, or one or more generic model fragments which involve not only standard conditions but also assumptions or hypotheses that must be made in order to draw conclusions.

Fuzzy-Rough Feature Selection. Feature selection [9,15] addresses the problem of selecting those characteristic descriptors of a domain that are most informative. Figure 7 shows the basic procedures involved in such a process. Unlike other dimensionality-reduction methods, feature selectors preserve the original meaning of the features after reduction.

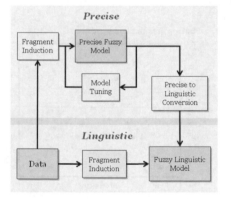

Fig. 6. Two-Step Learning of Descriptive Models

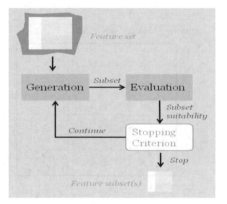

Fig. 7. Feature Selection Process

There are often many features involved in intelligence data, and combinatorially large numbers of feature combinations, to select from. It might be expected that the inclusion of an increasing number of features would increase the likelihood of including enough information to distinguish between classes. Unfortunately, this is not necessarily true if the size of the training dataset does not also increase rapidly with each additional feature included. A high-dimensional dataset increases the chances that a learning algorithm will find spurious patterns that are not valid in general. Besides, more features may introduce more measurement noise and, hence, reduce model accuracy [7].

Recently, there have been significant advances in developing methodologies that are capable of minimising feature subsets in an imprecise and uncertain environment. In particular, a resounding amount of research currently being done utilises fuzzy and rough sets (e.g. [11,16,17,27,30,32]). Amongst them is the fuzzy-rough feature selection algorithm [8,10] that works effectively with discrete or real-valued noisy data (or a mixture of both), without the need for user-supplied information. This approach is suitable for the nature of intelligence data and hence, is adopted in the present work. A particular implementation is done via hill-climbing search, as shown in Fig. 8. It employs the fuzzy-rough dependency function, which is derived from the notion of fuzzy lower approximation, to choose those attributes that add to the current candidate feature subset in a best-first fashion. The algorithm terminates when the addition of any remaining attribute does not result in an increase in the dependency. Note that as the fuzzy-rough dependency measure is nonmonotonic, it is possible that the hill-climbing search terminates having reached only a local optimum.

3.2   Fuzzy Interpolative Reasoning

In conventional approaches to compositional modelling, the completeness of a scenario space depends upon two factors: (a) the knowledge base must cover all essential scenario fragments relevant to the data, and (b) the inference mechanism must be able to synthesise and store all combinations of instances of such fragments that constitute a consistent scenario. However, in practice, it is difficult, if not impossible, to obtain a complete library of model fragments. Figure 9 shows an example, where a sparse model library consisting of two simplified model fragments (i.e. two simple if-then rules) is given:

$Rule_i$: If *frequency* is None then *attack* is Unlikely
$Rule_j$: If *frequency* is Often then *attack* is Likely

In this case, with an observation that states "*frequency* is Few", no answer can be found to the question "Will there be an attack"? A popular tool to deal with this type of problem is fuzzy interpretative reasoning [1,31]. In this work, the transformation-based approach as proposed in [5,6] is employed to support model composition, when given an initial sparse knowledge base.

The need for a fuzzy approach to interpolation is obvious: The precision degree of the available intelligence data is often variable. The potential sources of such variability include vaguely defined concepts (e.g. materials that constitute a "high explosive", certain organisations that are deemed "extremist"), quantities (e.g. a "substantial" amount of explosives, "many" people) and specifications of importance and certainty (e.g. in order to deploy a radiological dispersal device, the perpetrator "must" have access to radioactive material and "should" have an ideological or financial incentive). Finding a match between the given data and the (already sparse) knowledge base cannot in general be achieved precisely.

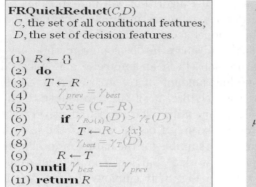
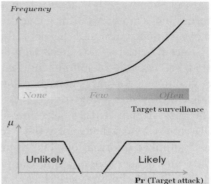

Fig. 8. Fuzzy-Rough Feature Selection   **Fig. 9.** Spare Knowledge Base

Figure 10 illustrates the basic ideas of fuzzy interpolative reasoning. It works through a two-step process: (a) computationally constructing a new inference rule (or model fragment in the present context) via manipulating two given adjacent rules (or related fragments), and (b) using scale and move transformations to convert the intermediate inference results into the final derived conclusions.

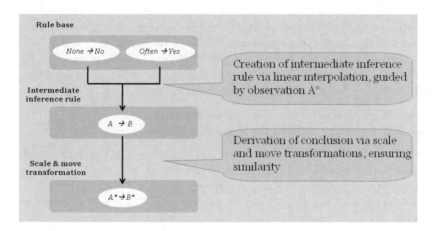

Fig. 10. Transformation-Based Fuzzy Interpolation

3.3 Fuzzy Risk Assessment

In developing intelligent systems for intelligence data monitoring, a trade-off needs to be considered. On the one hand, it is important not to miss out any potentially significant scenarios that may explain the observed evidence. On the other hand, too many unsorted and especially, spurious scenarios may confuse human analysts. Thus, it is desirable to be able to filter the created scenario space with respect to certain objective measures of the quality of the generated scenario descriptions. Fortunately, as indicated previously, preferences over different hypothetical scenarios can be determined on the basis of the reliability, risk and urgency of each scenario.

The *reliability* of a generated scenario may be affected by several distinct factors: the given intelligence data (e.g. the reliability of an informant), the inferences made to abduce plausible scenarios (e.g. the probability that a given money transfer is part of an illegitimate transaction), and the default assumptions adopted (e.g. the likelihood that a person seen on CCTV footage is identified positively). The *urgency* of a scenario is inversely proportional to the expected time to completion of a particular terrorist/criminal activity. Therefore, an assessment of urgency requires a (partial) scenario to be described using the scenario's possible consequences and information on additional actions required to achieve completion. The *risk* posed by a particular scenario is determined by its potential consequences (e.g. damage to people and property). Whilst these are very different aspects that may be used to differentiate and prioritise composed scenarios, the underlying approaches to assess them are very similar. In this paper, only the scenario risk aspect is discussed.

Risk assessment helps to efficiently devise and deploy counter measures, including further evidence gathering of any threat posed by the scenario concerned. However, estimating the risk of a plausible event requires consideration of variables exhibiting both randomness and fuzziness, due to the inherent nature of intelligence data (and knowledge also). Having identified this, in the present work, risk is estimated as the mean chance of a fuzzy random event [4,29] over a pre-defined confidence level, for each individual type of loss. In particular, plausible occurrence of an event is considered random, while the potential loss due to such an event is expressed as a fuzzy random variable (as it is typically judged linguistically). In implementation, loss caused by an event is modelled by a function mapping from a boolean sample space of {Success, Failure} onto a set of nonnegative fuzzy values. Here, success or failure is judged from the criminal's viewpoint, in terms of whether they have carried out a certain activity or not.

Risks estimated over different types of loss (e.g. range of geometric destruction and number of casualties) can be aggregated. Also, assessments obtained using different criteria (e.g. resource and urgency) may be integrated to form an overall situation risk. Such measures may be utilised as flexible constraints [19] imposed over an automated planning process, say for police resource deployment. This can help to minimise the cost of successful surveillance, for example. To generalise this approach further, order-of-magnitude representation [22,25] may be introduced to describe various cost estimations. Figure 11 shows such an application.

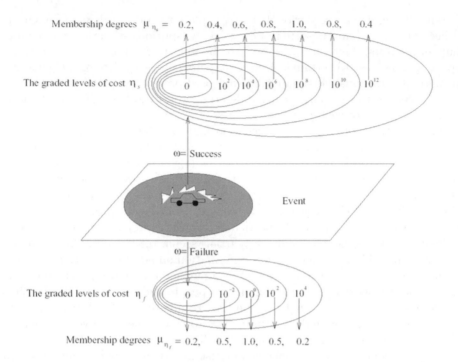

Fig. 11. Risk Assessment

4 Conclusion

This paper has introduced a novel framework upon which to develop intelligent decision support systems, with a focussed application to intelligence data monitoring and interpretation. It has outlined methods that can aid intelligence analysts in considering as widely as possible a range of emerging scenarios which are logically inferred and justified, and which may each reflect organised criminal/terrorist activity. This work has indicated that some of the recent advances in fuzzy and rough techniques are very successful for data-driven systems modelling and analysis in general, and for performing the following tasks in particular:

- Fragment induction
- Feature selection
- Interpolative reasoning
- Model composition
- Constraint satisfaction

- Truth maintenance
- Co-reference resolution
- Information aggregation
- Evidence evaluation
- Risk assessment

However, important research remains. The following lists a number of further issues that are worthy of investigation and/or development:

- Learning hierarchical model fragments
- Hierarchical and ensemble feature selection

- Unification of scenario generation algorithms
- Dynamic co-reference resolution and information fusion
- Evidence-driven risk-guided scenario generation
- Reconstruction of reasoning process
- Discovery of rare cases
- Meta-feature learning and selection for scenario synthesis

Further studies will help to consolidate and broaden the scope of applications of fuzzy set and rough set theories. In particular, the proposed framework and associated techniques can be adapted to perform different tasks in intelligence data modelling and analysis, such as: investigator training, policy formulation, and multi-modal profiling. Additionally, this work may be applied to accomplishing tasks in other domains, such as academic performance evaluation and financial situation forecasting. Finally, it is worth noting that most of the component techniques within the current framework utilise fuzzy set theory as the mathematical foundation. It would be very interesting to investigate if alternative approaches may be developed using rough sets or their extensions in an analogous manner. Also, the employment of directly combined and/or hybrid fuzzy-rough systems may offer even more advantages in copying with complex decision support problems. The research on fuzzy-rough feature selection as adopted within this framework has demonstrated, from one aspect, such potential.

References

1. Baranyi, P., Koczy, L., Gedeon, T.: A generalized concept for in fuzzy rule interpolation. IEEE Transactions on Fuzzy Systems 12(6), 820–837 (2004)
2. Calado, P., Cristo, M., Goncalves, M., de Moura, E., Ribeiro-Neto, E., Ziviani, N.: Link based similarity measures for the classification of web documents. Journal of American Society for Information Science and Technology 57(2), 208–221 (2006)
3. Dubois, D., Prade, H.: Rough fuzzy sets and fuzzy rough sets. International Journal of General Systems 17, 191–209 (1990)
4. Halliwell, J., Shen, Q.: Linguistic probabilities: theory and application. Soft Computing 13(2), 169–183 (2009)
5. Huang, Z., Shen, Q.: Fuzzy interpolative and extrapolative reasoning: a practical approach. IEEE Transactions on Fuzzy Systems 16(1), 13–28 (2008)
6. Huang, Z., Shen, Q.: Fuzzy interpolative reasoning via scale and move transformation. IEEE Transactions on Fuzzy Systems 14(2), 340–359 (2006)
7. Jensen, R., Shen, Q.: Are more features better? IEEE Transactions on Fuzzy Systems (to appear)
8. Jensen, R., Shen, Q.: New approaches to fuzzy-rough feature selection. IEEE Transactions on Fuzzy Systems 17(4), 824–838 (2009)
9. Jensen, R., Shen, Q.: Computational Intelligence and Feature Selection: Rough and Fuzzy Approaches. IEEE and Wiley, Hoboken, New Jersey (2008)
10. Jensen, R., Shen, Q.: Fuzzy-rough sets assisted attribute selection. IEEE Transactions on Fuzzy Systems 15(1), 73–89 (2007)
11. Jensen, R., Shen, Q.: Semantics-preserving dimensionality reduction: Rough and fuzzy-rough approaches. IEEE Transactions on Knowledge and Data Engineering 16(12), 1457–1471 (2004)

12. Keppens, J., Shen, Q.: On compositional modelling. Knowledge Engineering Review 16(2), 157–200 (2001)
13. Lee, M.: On models, modelling and the distinctive nature of model-based reasoning. AI Communications 12, 127–137 (1999)
14. Liben-Nowell, D., Kleinberg, J.: The link-prediction problem for social networks. Journal of American Society for Information Science and Technology 58(7), 1019–1031 (2007)
15. Liu, H., Motoda, H.: Feature Selection for Knowledge Discovery and Data Mining. Springer, Heidelberg (1998)
16. Mac Parthalain, N., Shen, Q.: Exploring the boundary region of tolerance rough sets for feature selection. Pattern Recognition 42(5), 655–667 (2009)
17. Mac Parthalain, N., Shen, Q., Jensen, R.: A distance measure approach to exploring the rough set boundary region for attribute reduction. IEEE Transactions on Knowledge and Data Engineering (to appear)
18. Marín-Blázquez, J., Shen, Q.: From approximative to descriptive fuzzy classifiers. IEEE Transactions on Fuzzy Systems 10(4), 484–497 (2002)
19. Miguel, I., Shen, Q.: Fuzzy rrDFCSP and planning. Artificial Intelligence 148(1-2), 11–52 (2003)
20. Pal, S., Polkowski, L., Skowron, A.: Rough-Neural Computing: Techniques for Computing with Words. Springer, Heidelberg (2004)
21. Pal, S., Skowron, A.: Rough Fuzzy Hybridization: A New Trend in Decision-Making. Springer, Heidelberg (1999)
22. Parsons, S.: Qualitative probability and order of magnitude reasoning. International Journal of Uncertainty, Fuzziness and Knowledge-Based Systems 11(3), 373–390 (2003)
23. Pawlak, Z.: Rough Sets: Theoretical Aspects of Reasoning About Data. Kluwer Academic Publishing, Dordrecht (1991)
24. Pawlak, Z., Skowron, A.: Rudiments of rough sets. Information Sciences 177(1), 3–27 (2007)
25. Raiman, O.: Order-of-magnitude reasoning. Artificial Intelligence 51, 11–38 (1991)
26. Shen, Q., Chouchoulas, A.: A rough-fuzzy approach for generating classification rules. Pattern Recognition 35(11), 2425–2438 (2002)
27. Shen, Q., Jensen, R.: Selecting informative features with fuzzy-rough sets and its application for complex systems monitoring. Pattern Recognition 37(7), 1351–1363 (2004)
28. Shen, Q., Keppens, J., Aitken, C., Schafer, B., Lee, M.: A scenario driven decision support system for serious crime investigation. Law, Probability and Risk 5(2), 87–117 (2006)
29. Shen, Q., Zhao, R., Tang, W.: Modelling random fuzzy renewal reward processes. IEEE Transactions on Fuzzy Systems 16(5), 1379–1385 (2008)
30. Slezak, D.: Rough sets and functional dependencies in data: Foundations of association reducts. Transactions on Computational Science 5, 182–205 (2009)
31. Tikk, D., Baranyi, P.: Comprehensive analysis of a new fuzzy rule interpolation method. IEEE Transactions on Fuzzy Systems 8(3), 281–296 (2000)
32. Tsang, E., Chen, D., Yeung, D., Wang, X., Lee, J.: Attributes reduction using fuzzy rough sets. IEEE Transactions on Fuzzy Systems 16(5), 1130–1141 (2008)
33. Zadeh, L.: The concept of a linguistic variable and its application to approximate reasoning I. Information Sciences 8, 199–249 (1975)
34. Zadeh, L.: Fuzzy sets. Information and Control 8(3), 338–353 (1965)

Innovation Game as a Tool of Chance Discovery

Yukio Ohsawa

Department of Systems Innovation, The University of Tokyo
7-3-1 Hongo, Bunkyo-ku Tokyo, 113-8656 Japan
ohsawa@sys.t.u tokyo.ac.jp

Abstract. We are finding rising demands for chance discovery, i.e., methods for focusing on new events significant for human's decision making. Innovation Game is a tool for aiding innovative thoughts and communication, coming after our 10-year experiences in chance discovery where tools of data visualization have been applied in cases of decision making by business teams. This game enables us to run and accelerate the process of innovation, as well as to train human's talent of analogical and combinatorial thinking. In this paper, it is shown that the effects of Innovation Game are enhanced, especially when suitable communications and timely usage of a tool for visualizing the map of knowledge are executed.

1 Introduction: Chance Discovery as Value Sensing

Since year 2000, we have been developing tools and methods of Chance Discovery, under the definition of "chance" as an event significant for human's decision. We edited books [14], etc. and special issues of journals. We stand on the principle that a decision is to choose one from multiple scenarios of actions and events in the future. Thus, a chance defined above can be regarded as an event at the cross of scenarios, which forces human(s) to choose one of the scenarios. Events bridging multiple clusters of strongly co-related frequent events, as shown by tools such as C4.5, Correspondence Analysis [7], KeyGraph [11], etc., have been regarded as the candidates of "chances" which may have been rare in the past but may become meaningful in the future.

Another aspect for explaining the role of visualization in chance discovery is what has been called value sensing. Value sensing, to feel associated with the something in one's environment, has been regarded as a dimension of human's sensitivity in the literature of developmental psychology [3]. We can interpret this as the cognition of analogy between the target event versus a piece or a combination of pieces of one's knowledge (tacit or explicit). In the real world, a huge number of analogical relationships may exist, from which we should choose one link between the confronted novel event and some part of the knowledge [2].

In this paper, we show Innovation Game, a tool for aiding innovation. This game came from our experiences in applying chance discovery to cases of business decision

H. Sakai et al. (Eds.): RSFDGrC 2009, LNAI 5908, pp. 59–66, 2009.

making. According to the data of communications during the play-times of this game, we show that the thoughts in Innovation Game come to be innovative when players executed suitable communications and timely usage of KeyGraph.

2   Using a Scenario Map for Chance Discovery

In projects of chance discovery we conducted so far with companies, the marketer teams acquired novel awareness of valuable products they had produced but had not taken into consideration so far because of the weak contribution to their sales performance. For acquiring this awareness, KeyGraph assisted business people by showing a diagram as a map of the market having (1) clusters of items frequently bought as a set, i.e., at the same time together, and (2) items bridging the clusters in (1), which may embrace a latent market coming up in the near future.

For example, let us show an example where a diagram obtained by KeyGraph assisted textile marketers seeking new hit products [13]. The marketers started from data collected in exhibition events, where pieces of textile samples had been arranged on shelves for customers representing apparel companies just to see (not to buy yet). In comparison with data on past sales, the exhibition data were expected to include customers' preferences of products not yet displayed in stores. After the exhibition, the marketers of the textile company visualized the data of customers' preferences using decision trees [15], correspondence analysis [7], etc. After all, they reached KeyGraph and obtained the diagram as in Fig. 1(a), where the black nodes linked by black lines show the clusters corresponding to (1) above, and the red nodes and the red lines show the items corresponding to (2) above and their co-occurrence with items in clusters respectively. The marketers, in order to understand this graph, attached real product samples as in Fig. 1(b), in order to sense the smoothness, colors, etc with eyes and fingers.

Then three, of the 10 marketers, who were experts of women's blouse interpreted the cluster at the top of Fig. 1 (b), i.e., of "*dense textile for neat clothes, e.g., clothes for blouse*" and 3 others interpreted the cluster in the right as of business suits. 2 others interpreted the popular item, not linked to any clusters of (1) via black lines, in the left, corresponding to materials of casual clothes. These clusters corresponded to established (already popular) submarkets of the company.

Next, a marketer of 10-years experience paid attention to an item between the item in the left and the large cluster in the right of the graph. This *between* node appeared as a red node, i.e., the *niche* lying between popular clusters, on which the marketers came up with a strategic scenario to design a new semi-casual cloth in which ladies can go both to workplaces and to light dinner after working. As a result, the material of the red node marked a hit – the 13th highest sales among their 800 products. We have other cases of graph-based chance discoveries [8].

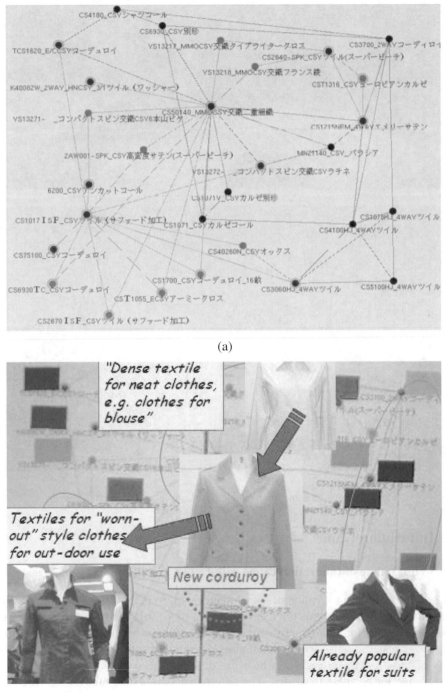

(a)

(b)

Fig. 1. Marketing as value sensing from visualized data

62   Y. Ohsawa

Fig. 2. Innovation game on the game board, made by data crystallization

3 Innovation Game

3.1 The Outline of Innovation Game with Data Crystallization

Innovation, meaning a creation of social trend by inventing a new technology, comes from the combination of existing ideas according to [5] [6]. The Innovation Game is a tool we invented for aiding innovative communications, where combinatorial creativity, i.e., creating a new idea by combining ideas, is activated. The game starts from tens of basic cards, on each of which the title and the summary of some existing knowledge for business is written. The core players are called innovators, who start with the capital of $10. The innovator's main operation is to (1) buy a preferable number of basic cards for $1 per card, (2) combine the cards of one's own or with cards bought/borrowed from other players, and (3) present with an idea created by the combination. Other innovators may propose the presenter to start collaboration, or borrow/buy the new idea, with negotiating the dealing price. At the halting time (2 hours after starting), the richest player, i.e., the player having the largest amount of money comes to be the winner.

Investors and consumers stand around innovators who also start from having 10$. Each investor buys stocks of innovators who seem to be excellent, according to the investor's own sense. The investor having obtained stocks of the highest amount of total price at the halting time comes to be the winning investor. And, consumers may buy ideas for prices determined by negotiation with innovators. The consumer who obtained the idea-set of the highest total price becomes the winning consumer.

Several methods for creative thinking, as the one shown in Section 2, can be positioned in the application of visualization of ideas. For example, Mind maps [1] have been introduced for creating ideas with considering the relevance to the central key-word corresponding to the target problem. The graph obtained from basic cards is used as the game-board of Innovation Game, as in Fig.2. This intuitively visualizes the map of ideas' market, showing the positions of both existing knowledge and latent ideas which does not appear on any basic card but may be created by combinations, by applying Data Crystallization [9][12] to the text of basic cards. Data Crystallization is an extension of KeyGraph, enabling to show latent items of which the frequency in the given data is zero. E.g., a node such as DE58 in Fig.2 means a new idea may emerge at the positions by combining ideas in its neighborhood clusters. The innovators put basic cards on corresponding black nodes when combining the cards for creating an idea, and to write the created idea on a post-it and put it on the corresponding position. If the basic cards combined have been linked to a red ("DE-X") node via lines, then the position is the "DE-X" node.

3.2 Findings from Players' Communications

The players of games we conducted (we organized more than 50 games so far) mention they felt their skills of communication and thought for creating socially useful knowledge in business has been elevated during and after the game. After each game, the quality of created ideas are evaluated by all investors and more objective reviewers, on criteria such as "originality" "cost" "utility" and "reality." We found a significant relevance between the quality of the players' communication and the quality of ideas. According to our data on the utterances by players, we found the originality and the utility of ideas tend significantly to increase (1) after the increase in the empathetic utterances of investors/consumers and (2) before the appearance of a sequence of negative utterances followed by a positive utterance. Here, an empathetic utterance means a comment referring to the context the presented idea may be utilized for future businesses. These results imply an innovative communication comes from (1) the context sharing induced by the visualized graph, and (2) the interest in revising presented ideas, of all participants.

4 Niche of Idea Activations as Source of Innovations: Another Finding from the Innovation Game

We hypothesized and evaluated the effect of *idea niche* on the innovations. An idea niche is a part of the market where outstanding ideas do not exist but is surrounded by

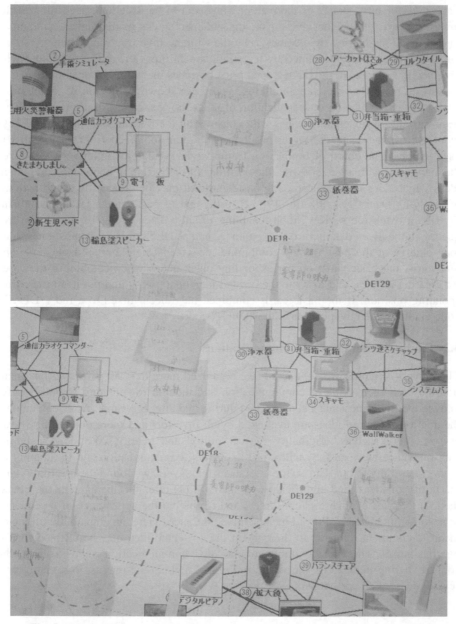

Fig. 3. Free (left)/connected (right) niche ideas: with/without connections via links

existing ideas. We expected humans can activate such a part to have the surrounding ideas to meet and become combined. As Fig.3 shows, we classified idea niches into two types. In the upper picture, a created idea is put on a free space, i.e., not at any node in or linked to clusters. And, in the lower picture, a new idea is presented at a red node, linking between clusters. Let us call the former type of niche, i.e., a node connecting nodes in cluster(s), via links in the graph, a *connected niche*. And, let us call the space on the graph surrounded by but not connected to clusters, a *free niche*. We also classified a niche in another dimension: Whether it is between activated clusters (clusters, all including ideas used already in the game), between partially activated clusters (i.e. some, not all, of which include ideas already used), or between newly activated clusters (i.e. none of which include ideas already used). Thus, the presented ideas in games can be classified into the six classes. Here, all the ideas in each class, the percentage of the 5 highest-score ideas in each game was counted for each type of niches respectively. As a result, we found the following three tendencies:

(1) The ideas at niches "between activated clusters" include the highest-score ideas.
(2) The free niche tends to include especially highly scored ideas, but the deviation of the score is large i.e., the reliability of the ideas presented at free spaces is low although the average quality is high. On the other hand, the connected niche tends to include relatively highly scored ideas, and the deviation is small.
(3) The connected niche tends to create more excellent ideas than a free space, when it is between partially activated clusters.

In summary, we can recommend players in the future to be patient until ideas have been created combining ideas in clusters on the graph, without expecting high scores, in the early stage. And, then, the players will be enabled to create good ideas by focusing on the niches of activated ideas on the graph (on tendency (1)). Here, if the player likes a hit (especially high score), the free space between clusters including activated ideas will be recommended at the risk of large deviation of the quality of the idea. On the other hand, if the player likes reliability (hedging the risk of low score), positioning ideas on nodes or lines on the graph will be better (on tendency (2)). However, in real games in companies, where players join for real innovation, it is not easy to have their patience to wait until clusters become occupied by activated ideas. In such a case, the player should apply tendency (4). That is, if one prefers to combine ideas in clusters without activated ideas and in clusters with activated ideas, it is recommended to create an idea on a node connecting these clusters.

5   Conclusions

Human's insight is a fruit of the interactions of mental process and the social environment [4] [10]. We developed the Innovation Game based on author's experience of applying KeyGraph to chance discovery in business teams, where members interacted in the real team, in the real company, and in the real market.

In this paper, we analyzed the presented ideas and the communications during each game. The potential social contribution of each idea was evaluated on measures as originality, reality, utility, etc. As a result, we are obtaining guidelines for players e.g., to aim at suitable niche in the market of ideas reflecting the situation. These findings partially correspond to known hypotheses about the mechanism of innovation, but the evidences showing how the activation of basic knowledge leads to the creation of ideas in real communication is novel as far as we know.

In the next step, we plan to model our recent experiences to put the created ideas into final decisions and real actions of the company.

References

[1] Buzan, T.: The Mind Map Book. Penguin Books, London (1996)
[2] Dietric, E., Markman, A.B., Stilwell, H., Winkley, M.: The Prepared Mind: The Role of Representational Change in Chance Discovery. In: Ohsawa, Y., McBurney, P. (eds.) Chance Discovery, pp. 208–230. Springer, Heidelberg (2003)
[3] Donaldson, M.: Human Minds: An Exploration. Penguin Books, London (1992)
[4] Csikszentmihalyi, M., Sawyer, K.: Creative insight: The social dimension of a solitary moment. In: Sterngerg, R.J., Davidson, J.E. (eds.) The Nature of Insight, pp. 329–364. MIT Press, Cambridge (1995)
[5] Johansson, F.: The Medici Effect: Breakthrough Insights at the Intersection of Ideas. Harvard Business School Press, Boston (2004)
[6] Goldberg, D.E.: The Design of Innovation: Lessons from and for Competent Genetic Algorithms. Kluwer Academic Publishers, Dordrecht (2002)
[7] Greenacre, M.J.: Correspondence Analysis in Practice (Interdisciplinary Statistics). Chapman & Hall, London (2007)
[8] Horie, K., Ohsawa, Y.: Product Designed on Scenario Maps Using Pictorial KeyGraph. WSEAS Transaction on Information Science and Application 3(7), 1324–1331 (2006)
[9] Maeno, Y., Ohsawa, Y.: Human-Computer Interactive Annealing for Discovering Invisible Dark Events. IEEE Tran. on Industrial Electronics 54(2), 1184–1192 (2007)
[10] Mead, G.H.: Mind, Self, and Society. Univ. of Chicago Press, Chicago (1934)
[11] Ohsawa, Y., Benson, N.E., Yachida, M.: KeyGraph: Automatic Indexing by Co-occurrence Graph based on Building Construction Metaphor. In: Proc. Advanced Digital Library Conference, pp. 12–18. IEEE Press, Los Alamitos (1998)
[12] Ohsawa, Y.: Data Crystallization: Chance Discovery Extended for Dealing with Unobservable Events. New Mathematics and Natural Computation 1(3), 373–392 (2005)
[13] Ohsawa, Y., Usui, M.: Creative Marketing as Application of Chance Discovery. In: Ohsawa, Y., Tsumoto, S. (eds.) Chance Discoveries in Real World Decision Making, pp. 253–272. Springer, Heidelberg (2006)
[14] Ohsawa, Y., McBurney, P.: Chance Discovery. Springer, Heidelberg (2003)
[15] Quinlan, J.R.: C4.5: Programs For Machine Learning. Morgan Kaufmann, Los Altos (1993)

Rough Set Year in India 2009

Manish Joshi[1], Rabi N. Bhaumik[2], Pawan Lingras[3]
Nitin Patil[4], Ambuja Salgaonkar[5], and Dominik Ślęzak[6,7]

[1] Department of Computer Science, North Maharashtra University
Jalgaon, Maharashtra, 425 001, India
joshmanish@gmail.com
[2] Department of Mathematics, Tripura University
Suryamaninagar, Tripura, 799 130, India
bhaumik_r_n@yahoo.co.in
[3] Department of Mathematics and Computing Science, Saint Mary's University
Halifax, Nova Scotia, B3H 3C3, Canada
pawan@cs.smu.ca
[4] Department of Computer Science, University of Pune
Pune, 411 007, India
nitinp@cs.unipune.ernet.in
[5] Department of Computer Science, University of Mumbai
Mumbai, 400 098, India
ambujas@gmail.com
[6] Institute of Mathematics, University of Warsaw
Banacha 2, 02-097 Warsaw, Poland
[7] Infobright Inc., Poland
Krzywickiego 34 pok. 219, 02-078 Warsaw, Poland
slezak@infobright.com

Since its inception in early 80's, the rough set theory has attracted a lot of interest from global research community. It turns out as useful in building classification and prediction models. It complements a number of other soft computing paradigms. It may be combined with fuzzy logic and probabilistic data analysis. It has led towards enhancements of neural networks, genetic algorithms, clustering, support vector machines, regression models, et cetera. Its application domains include pattern recognition, feature selection, information retrieval, bioinformatics, computer vision, multimedia, medicine, retail data mining, web mining, control, traffic engineering, data warehousing, and many others.

A number of researchers have been working on the rough set theory also in India. It is especially important to mention Mihir K. Chakraborty and Sankar K. Pal, who contributed to its foundations and applications, respectively. Nowadays, rough sets are present at the centers of higher learning such as: Calcutta University and ISI (Kolkata), IIT (Delhi, Kanpur, Kharagpur), IISc (Bangalore), Anna University (Chennai), Jawaharalal Nehru University (Delhi), et cetera. In particular, inclusion of rough sets as one of the topics in Schools on Logic held at IIT Kanpur (2008) and IIT Kharagpur (2009) has provided an evidence of its acceptance as an unquestionably important research area in India.

H. Sakai et al. (Eds.): RSFDGrC 2009, LNAI 5908, pp. 67–68, 2009.

The rough set conferences and workshops have been held in Canada, China, Japan, Poland, Sweden, USA, and most recently in Australia and Italy. Also, a number of Indian conferences have considered rough sets as an important topic. However, until this year, there has been no exclusive rough set conference in India. In the end of 2008, preparations for two such events were announced: the *Twelfth International Conference on Rough Sets, Fuzzy Sets, Data Mining and Granular Computing* reported in this volume (see Preface), as well as the *International Conference on Rough Sets, Fuzzy Sets and Soft Computing*, organized at Tripura University, November 5-7, 2009. The aim of that second event was particularly to expose young researchers to the latest trends in fuzzy and rough systems through deliberations by well-known scientists. The committee has chosen over 40 papers to be included into the conference materials published by Serial Publications, New Delhi. It is also important to acknowledge that over 10 invited speakers from both India and abroad attended the conference.

In order to help in linking various rough set research groups in India and enable interaction with international rough set community, we proposed to extend the two above-mentioned events towards a broader initiative named *Rough Set Year in India 2009*.[1] A variety of additional events distributed through the year, in different parts of India were organized successfully. A good example here is the *International Workshop on Rough Sets, Fuzzy Sets and Soft Computing: Theory and Applications*, organized by University of Pune, July 7-8, 2009. This two day workshop was attended by over 60 delegates from 19 different institutions and 7 different states of India. Presentations by Vijay Raghavan (USA), Sushmita Mitra (India), Sonajharia Minz (India), Pawan Lingras (Canada), Dominik Ślęzak (Poland), Yiyu Yao (Canada), and Georg Peters (Germany) described fundamental rough set concepts, as well as their usage in academic research and real world market applications. It is worth noting that the last three of mentioned presentations were delivered online, which shows how new communication technologies can help in building the worldwide scientific network.

Another example of important event specially dedicated to the rough set theory and applications is the *International Symposium on Soft Computing*, organized by Department of Computer Science at University of Mumbai, December 1-2, 2009. Rough set sessions were also present as components of other conferences. For instance, Sonajharia Minz held a special session on rough sets and granular computing at the *National Conference on Computational Mathematics and Soft Computing*, Women's Christian College, Chennai, July 24-25, 2009. Rough set research was also broadly represented at the *4th Indian International Conference on Artificial Intelligence*, Tumkur (near Bangalore), December 16-18, 2009. It is also worth mentioning about the workshop organized by Mohua Banerjee in Delhi, just before the conference that is reported in this volume.

All these multiple activities helped researchers to attend an event based on their temporal and spatial convenience. We are confident that our initiative successfully achieved its objective in developing an interest regarding rough sets and its applications among a larger group of academicians in India.

---

[1] http://cs.smu.ca/~pawan/rsIndia09/index.html

An Algebraic Semantics for the Logic of Multiple-Source Approximation Systems

Md. Aquil Khan and Mohua Banerjee

Department of Mathematics and Statistics,
Indian Institute of Technology,
Kanpur 208 016, India
{mdaquil,mohua}@iitk.ac.in

Abstract. An algebraic semantics for the logic LMSAS, proposed to study the behavior of rough sets in multiple-source scenario, is presented. Soundness and completeness theorems are proved.

1 Introduction

In last three decades, rough set theory [7] has been generalized and extended in many directions enabling it to capture different situations. In [9], a multi-agent scenario is considered where each agent has her own knowledge base represented by equivalence relations, and thus perceives the same domain differently depending on what information she has about the domain. This multi-agent dimension was also considered by Pawlak in [6] although not mentioned explicitly. In [3,4], rough set theory is again explored in this context, although the more general term 'source' is used there instead of 'agent'. A *multiple-source approximation system* is considered to study the behavior of rough set in such a situation.

Definition 1. *[3] A multiple-source approximation system (MSAS) is a tuple* $\mathfrak{F} := (U, \{R_i\}_{i \in N})$, *where* U *is a non-empty set,* N *an initial segment of the set* $\mathbb{N}$ *of positive integers, and each* $R_i, i \in N$, *is an equivalence relation on the domain* U. $|N|$ *is referred to as the* cardinality *of* $\mathfrak{F}$ *and is denoted by* $|\mathfrak{F}|$.

So MSASs are collections of Pawlak approximation spaces over the same domain – the idea being that these approximation spaces are obtained from different sources. The standard concepts such as approximations and definability of sets, membership functions related with the Pawlak approximation spaces are generalized to define these notions on MSASs. The following notions of lower/upper approximations are introduced.

Definition 2 ([3]). *Let* $\mathfrak{F} := (U, \{R_i\}_{i \in N})$ *be a MSAS, and* $X \subseteq U$. *Then* strong lower approximation $\underline{X}_s$, weak lower approximation $\underline{X}_w$, strong upper approximation $\overline{X}_s$, *and* weak upper approximation $\overline{X}_w$ *of* X, *respectively, are defined as follows.*

$$\underline{X}_s := \bigcap \underline{X}_{R_i}; \quad \underline{X}_w := \bigcup \underline{X}_{R_i}; \quad \overline{X}_s := \bigcap \overline{X}_{R_i}; \quad \overline{X}_w := \bigcup \overline{X}_{R_i},$$

H. Sakai et al. (Eds.): RSFDGrC 2009, LNAI 5908, pp. 69–76, 2009.

where $\underline{X}_R$ and $\overline{X}_R$ respectively denotes the lower and upper approximation in the Pawlak approximation space (U, R).

So in a special case when $\mathfrak{F}$ consists of a single relation, weak/strong lower and upper approximations are just the standard lower and upper approximations respectively.

It is not difficult to show that $\underline{X}_s \subseteq \underline{X}_w \subseteq X \subseteq \overline{X}_s \subseteq \overline{X}_w$. Thus given a MSAS $\mathfrak{F} := (U, \{R_i\}_{i \in N})$ and a set $X \subseteq U$, on the basis of possibility of objects to be an element of X, the universe is divided into five disjoint sets namely $\underline{X}_s, \underline{X}_w \setminus \underline{X}_s, \overline{X}_s \setminus \underline{X}_w, \overline{X}_w \setminus \overline{X}_s$ and $(\overline{X}_w)^c$. The elements of these regions are respectively called the *certain positive, possible positive, certain boundary, possible negative* and *certain negative* element of X. Here, we would like to mention that the strong/weak lower and upper approximations are different from the lower and upper approximations of a set X in the approximation space (U, R) of *strong distributed knowledge* R [9], i.e. where $R := \bigcap_{i \in N} R_i$. In fact, we have the inclusion $\underline{X}_w \subseteq \underline{X}_R \subseteq \overline{X}_R \subseteq \overline{X}_s$.

The above notions of approximations along with other concepts related with MSASs are studied in [3,4] in some detail.

The existing logical systems employed to study the Pawlak approximation spaces including the epistemic logic $S5_n$ [2] and one given in [6,9] are not strong enough to express the generalized notions of approximations and definability of sets introduced in [3]. Thus a quantified propositional modal logic LMSAS is introduced in [3], using which we can study the behavior of rough sets in MSASs. In this article, we shall present an algebraic semantics for LMSAS. The soundness and completeness theorem obtained in the process also establishes a strong connection between the MSASs and the algebraic counterpart of LMSAS. In order to obtain the completeness theorem, we have used the technique of completions of algebras (cf. [5]). $Q - filters$ [10] are used instead of ultra-filters, because the embedding given by Jónsson-Tarski Theorem may not preserve infinite joins and meets — which is what we require. Since the embedding is done in some *complex algebra*[1], we also obtain completeness with respect to a class of complex algebras.

The remainder of this paper is organized as follows. In Sect. 2, we present the logic LMSAS. In Sect. 3, we come to the main issue of the article, i.e. the algebraization of LMSAS. Detailed proofs of the results of this section are skipped because of a lack of space. Finally, we conclude the article in Sect. 4.

2   LMSAS

In this section we briefly describe the logic LMSAS.

Syntax

The alphabet of the language of LMSAS contains (i) a non-empty countable set Var of variables, (ii) a (possibly empty) countable set Con of constants, (iii) a non-empty countable set PV of propositional variables and (iv) the propositional constants $\top, \bot$.

The set T of *terms* of the language is given by $Var \cup Con$. Using the standard Boolean logical connectives $\neg$ (negation) and $\wedge$ (conjunction), a unary modal connective $[t]$ (necessity) for each term $t \in T$, and the universal quantifier $\forall$, well-formed formulae (wffs) of LMSAS are defined recursively as:

$$\top \mid \bot \mid p \mid \neg\alpha \mid \alpha \wedge \beta \mid [t]\alpha \mid \forall x\alpha,$$

where $p \in PV$, $t \in T$, $x \in Var$, and α, β are wffs. The set of all wffs and closed wffs of LMSAS will be denoted by $\mathcal{F}$ and $\overline{\mathcal{F}}$ respectively.

Semantics

Definition 3. $\mathfrak{M} := (\mathfrak{F}, V, I)$ *is an* interpretation, *where* $\mathfrak{F} := (U, \{R_i\}_{i \in N})$ *is a MSAS (cf. Definition 1),* $V : PV \rightarrow \mathcal{P}(U)$ *and* $I : Con \rightarrow N$.
    An assignment *for an interpretation* $\mathfrak{M}$ *is a map* $v : T \rightarrow N$ *such that* $v(c) = I(c)$, *for each* $c \in Con$.

Let $\mathfrak{M}$ be an interpretation. As in classical first-order logic, two assignments v, v' for $\mathfrak{M}$ are said to be *x-equivalent* for a variable x, provided $v(y) = v'(y)$, for every variable y, (possibly) other than x.

Definition 4. *The satisfiability of a wff* α *in an interpretation* $\mathfrak{M} := (\mathfrak{F}, V, I)$, *under an assignment* v *is defined inductively. We give the modal and quantifier cases.*

    $\mathfrak{M}, v, w \models [t]\alpha$, *if and only if for all* w' *in* U *with* $w R_{v(t)} w'$, $\mathfrak{M}, v, w' \models \alpha$.
    $\mathfrak{M}, v, w \models \forall x\alpha$, *if and only if for every assignment* v' *x-equivalent to* v, $\mathfrak{M}, v', w \models \alpha$.

α *is* valid, *denoted* $\models \alpha$, *if and only if* $\mathfrak{M}, v, w \models \alpha$, *for every interpretation* $\mathfrak{M} := (\mathfrak{F}, V, I)$, *assignment* v *for* $\mathfrak{M}$ *and object* w *in the domain of* $\mathfrak{F}$.

Given an interpretation $\mathfrak{M} := (\mathfrak{F}, V, I)$ and assignment v, one can extend the map V to the set of all wffs such that $V(\alpha) := \{w \in U : \mathfrak{M}, v, w \models \alpha\}$. Let us recall Definition 2. It is not difficult to prove

Proposition 1

1. *(a)* $V(\langle t \rangle \alpha) = \overline{V(\alpha)}_{R_{\tilde{v}(t)}}$; *(b)* $V([t]\alpha) = \underline{V(\alpha)}_{R_{\tilde{v}(t)}}$.
   For α *which does not have a free occurrence of* x,
2. $V(\forall x[x]\alpha) = \underline{V(\alpha)}_s$; $V(\exists x[x]\alpha) = \underline{V(\alpha)}_w$.
3. $V(\forall x\langle x \rangle \alpha) = \overline{V(\alpha)}_s$; $V(\exists x\langle x \rangle \alpha) = \overline{V(\alpha)}_w$.

We would like to mention here that the epistemic logic $S5_n$ and the logics considered in [6,9] will not suffice for our purpose. The semantics for these logics considers a finite and fixed number of agents, thus giving a finite and fixed number of modalities in the language. But in the case of LMSAS, the number of sources is not fixed. So it is not possible here to refer to all/some sources using only the connectives $\wedge$, $\vee$, and quantifiers $\forall$, $\exists$ are used to achieve the task.
    The following sound and complete deductive system for LMSAS was proposed in [3]. t stands for a term in T.

Axiom schema:
(1) All axioms of classical propositional logic;
(2) $\forall x \alpha \rightarrow \alpha(t/x)$, where α admits the term t for the variable x;
(3) $\forall x(\alpha \rightarrow \beta) \rightarrow (\alpha \rightarrow \forall x \beta)$, where the variable x is not free in α;
(4) $\forall x[t]\alpha \rightarrow [t]\forall x\alpha$, where the term t and variable x are different;
(5) $[t](\alpha \rightarrow \beta) \rightarrow ([t]\alpha \rightarrow [t]\beta)$;
(6) $\alpha \rightarrow \langle t \rangle \alpha$; (7) $\alpha \rightarrow [t]\langle t \rangle \alpha$; (8) $\langle t \rangle \langle t \rangle \alpha \rightarrow \langle t \rangle \alpha$.

Rules of inference:

$$\forall. \quad \frac{\alpha}{\forall x \alpha} \qquad MP. \quad \frac{\alpha \qquad \alpha \rightarrow \beta}{\beta} \qquad N. \quad \frac{\alpha}{[t]\alpha}$$

3 Algebraic Semantics for LMSAS

In this section, we present an algebraic semantics for LMSAS. We begin with the following definition [1].

Definition 5. *A tuple* $\mathfrak{A} := (A, \cap, \sim, 1, f_k)_{k \in \Delta}$ *is said to be a* Boolean algebra with operators *(BAO) if* $(A, \cap, \sim, 1)$ *is a Boolean algebra and each* $f_k : A \rightarrow A$ *satisfies (i)* $f_k(1) = 1$ *and (ii)* $f_k(a \cap b) = f_k(a) \cap f_k(b)$. *Moreover,* $\mathfrak{A}$ *is said to be* complete *if it satisfies the following additional properties for all* $X \subseteq A$:

(CB1) $\bigcap X$ *and* $\bigcup X$ *exist and* **(CB2)** $f_k \bigcap X = \bigcap f_k X$, $k \in \Delta$.
$\bigcap X$ *and* $\bigcup X$, *respectively, denote the g.l.b and l.u.b of the set* X.

In this paper, we are interested only in those complete BAOs (CBAOs) where $\Delta = \mathbb{N}$ and each f_k satisfies the following three additional conditions:

(B1) $f_k a \leq f_k f_k a$; **(B2)** $f_k a \leq a$ *and* **(B3)** $a \leq f_k g_k a$, *where* $g_k := \sim f_k \sim$.

Let us denote this class of complete BAOs by $\mathfrak{C}$. We shall obtain completeness of LMSAS with respect to the class $\mathfrak{C}$.

Definition 6. *Let* $\mathfrak{A} := (A, \cap, \sim, 1, f_k)_{k \in \mathbb{N}}$ *be a BAO satisfying (B1)-(B3). An* assignment *in* $\mathfrak{A}$, *is a function* $\theta : PV \rightarrow A$. θ *can be extended uniquely, in the standard way, to a* meaning function $\tilde{\theta} : \overline{\mathcal{F}} \rightarrow A$ *where in particular,* $\tilde{\theta}([c_i]\alpha) := f_i(\tilde{\theta}(\alpha))$, $i \in \mathbb{N}$ *and* $\tilde{\theta}(\forall x \alpha) := \bigcap\{\tilde{\theta}(\alpha(c_j/x)) : j \in \mathbb{N}\}$, *provided the g.l.b. exists. We define* $\tilde{\theta}(\alpha) := \tilde{\theta}(cl(\alpha))$, $\alpha \in \mathcal{F}$ *and* $cl(\alpha)$ *denotes the closure of* α.

Note that in order to define the natural translation corresponding to all possible assignments from closed LMSAS wffs to elements of a BAO, we only require the existence of joins and meets of the sets of the form $\{\tilde{\theta}(\alpha(c_j/x)) : j \in \mathbb{N}\}$, where α is a LMSAS wff with only one free variable x and θ is an assignment. This motivates us to define a realization for LMSAS in the line of realization of first order formalized languages [8].

Definition 7. *A BAO* $\mathfrak{A} := (A, \cap, \sim, 1, f_k)_{k \in \mathbb{N}}$ *satisfying (B1)-(B3) is said to be a* realization *for LMSAS, if for every assignment* $\theta : PV \rightarrow A$ *the following is satisfied:*

(R1) $\tilde{\theta}(\alpha)$ *exists for all* $\alpha \in \overline{\mathcal{F}}$;
(R2) $f_k \bigcap_j \tilde{\theta}(\alpha(c_j/x)) = \bigcap_j f_k \tilde{\theta}(\alpha(c_j/x))$, *where* α *has only one free variable* x.

Condition (R2) corresponds to the Axiom 4, and is essential to get the soundness theorem. Note that every complete BAO satisfying (B1)-(B3) is a realization for LMSAS. But not all realizations for LMSAS are complete BAOs. For instance, if a BAO $\mathfrak{A}$ satisfying (B1)-(B3), has only one distinct function symbol, then each set $\{\tilde{\theta}(\alpha(c_j/x)) : j \in \mathbb{N}\}$ will be singleton and thus $\mathfrak{A}$ becomes a realization which may not necessarily be a CBAO.

Definition 8. *Let us consider a structure of the form* $\mathfrak{F} := (U, \{R_i\}_{i \in \Delta})$, *where* Δ *is an index set and for each* $i \in \Delta$, $R_i \subseteq U \times U$. *The* complex algebra *of* $\mathfrak{F}$ *(notation* $\mathfrak{F}^+$*) is the expansion of the power set algebra* $\mathcal{P}(U)$ *with operators* $m_{R_i} : 2^U \to 2^U$, $i \in \Delta$, *where*

$$m_{R_i}(X) := \{x \in U : \text{ For all } y \text{ such that } xR_iy, y \in X\}.$$

In the case of MSAS, one can verify (B1)-(B3) to obtain

Proposition 2. *Every complex algebra of a MSAS is a complete BAO satisfying* *(B1)-(B3).*

Let us denote the class of all realizations of LMSAS and complex algebras of MSASs by $\mathfrak{R}$ and $\mathfrak{Cm}$ respectively. So we have $\mathfrak{Cm} \subseteq \mathfrak{C} \subseteq \mathfrak{R}$.

Definition 9. *Let* $\mathfrak{A} := (A, \cap, \sim, 1, f_k)_{k \in \mathbb{N}}$ *be a realization for MSAS. Then we write* $\mathfrak{A} \Vdash \alpha \approx \beta$ *if and only if for every assignment* $\theta : PV \to A$, $\tilde{\theta}(\alpha) = \tilde{\theta}(\beta)$. *We simply write* $\mathfrak{R} \Vdash \alpha$ *if* $\mathfrak{A} \Vdash \alpha \approx \top$ *for all* $\mathfrak{A} \in \mathfrak{R}$. *Similarly we write* $\mathfrak{C} \Vdash \alpha$ *and* $\mathfrak{Cm} \Vdash \alpha$ *according as* $\mathfrak{A} \Vdash \alpha \approx \top$ *for all* $\mathfrak{A} \in \mathfrak{C}$ *or* $\mathfrak{A} \in \mathfrak{Cm}$ *respectively.*

Proposition 3 (Soundness Theorem). *If* $\vdash \alpha$ *then* $\mathfrak{R} \Vdash \alpha$ *and hence* $\mathfrak{C} \Vdash \alpha$ *and* $\mathfrak{Cm} \Vdash \alpha$.

Proposition 4 (Completeness Theorem). *For* $\alpha \in \mathcal{F}$, *if* $\mathfrak{C} \Vdash \alpha$, *then* $\vdash \alpha$.

We begin our journey to prove the above completeness theorem with the Lindenbaum algebra $\mathfrak{Ln}$ for LMSAS. In fact, giving exactly the same argument as in the modal logic case, one can easily show that $\mathfrak{Ln} := (\overline{\mathcal{F}}|_{\equiv}, \cap, \sim, 1, f_k)_{k \in \mathbb{N}}$, where $1 = [\top]$, is a BAO. Moreover, Axioms (6)-(8) give us the properties (B1)-(B3). $\mathfrak{Ln}$ is, in fact, a realization for MSAS. But in order to prove this, we need a few more definitions and results.

Let $p_1, p_2, \ldots$ be an enumeration of the propositional variables and $\theta' : PV \to \overline{\mathcal{F}}|_{\equiv}$ be an assignment. Let $\alpha_1, \alpha_2, \ldots$ be countably many distinct wffs such that $\theta'(p_i) := [\alpha_i]$. For a given wff α, α^* denotes the wff obtained from α by uniform replacement of propositional variables p_i's by α_i's. By induction on the complexity of α, we obtain

Proposition 5. *The wff* $(\alpha(c_j/x))^*$ *is same as the wff* $\alpha^*(c_j/x)$, $j \in \mathbb{N}$.

Proposition 6. *Consider* $\overline{\mathcal{F}}|_{\equiv}$. *Then for any* $\alpha \in \mathcal{F}$ *which has only x as free variable,* $\bigcap_j [\alpha(c_j/x)]$ *exists and is given by* $[\forall x \alpha]$.

We use Propositions 5 and 6 to get

Proposition 7. $\tilde{\theta}'(\alpha) = [\alpha^*]$, *for all* $\alpha \in \overline{\mathcal{F}}$.

This result ensures (R1). Moreover, due to the presence of Axiom 4, we obtain the following, giving (R2).

Proposition 8. *Let* $\theta' : PV \to \overline{\mathcal{F}}|_{\equiv}$ *be an assignment. Then*

$$f_k \bigcap_j \tilde{\theta}'(\alpha(c_j/x)) = \bigcap_j f_k \tilde{\theta}'(\alpha(c_j/x)).$$

From Propositions 7 and 8, we obtain

Proposition 9. $\mathfrak{Ln} := (\overline{\mathcal{F}}|_{\equiv}, \cap, \sim, 1, f_k)_{k \in \mathbb{N}}$ *is a realization for LMSAS.*

Due to Proposition 9, we obtain the completeness theorem with respect to the class of all realizations. But, as mentioned earlier, we want the completeness with respect to the class $\mathfrak{C}$. It can be shown, as in the propositional logic case, that the Lindenbaum algebra $\mathfrak{Ln}$ defined above is *not* a CBAO and so we need to do some more work in order to get the completeness theorem with respect to the class $\mathfrak{C}$. Note that we would achieve our goal if we could embed any LMSAS realization $\mathfrak{A} := (A, \cap, \sim, 1, f_k)_{k \in \mathbb{N}}$ into some complex algebra. At this point one may think of the BAO consisting of all subsets of the set of all ultra-filters of the BAO $\mathfrak{A}$, as described in the Jónsson-Tarski Theorem. But the embedding given in this theorem may not preserve infinite joins and meets. This problem could be overcome if we consider the BAO consisting of all subsets of the set of all Q-filters [10] (defined below) instead of ultra-filters. Here, Q is a countably infinite collection of subsets of A satisfying certain conditions and the embedding obtained in this case preserves all the infinite joins and meets *in* Q. Since this embedding may not preserve *all* existing joins and meets, the question again arises whether even this embedding will be able to give us the desired result? The answer is yes. In the rest of this section, we shall present the result of [10] discussed above and use it to prove the completeness theorem with respect to the class $\mathfrak{C}$.

Definition 10. *Let* $\mathfrak{A} := (A, \cap, \sim, 1)$ *be a Boolean algebra. Let* $Q := \{Q_n \subseteq A : n \in \mathbb{N}\}$, *where each* Q_n *is non-empty. A prime filter F of $\mathfrak{A}$ is called a Q-filter, if it satisfies the following for each* $n \in \mathbb{N}$.

1. *If* $Q_n \subseteq F$ *and* $\bigcap Q_n$ *exists then* $\bigcap Q_n \in F$.
2. *If* $\bigcup Q_n$ *exists and belongs to F then* $Q_n \cap F \neq \emptyset$.

The set of all Q-filters of $\mathfrak{A}$ is denoted by $\mathcal{F}_Q(\mathfrak{A})$.

Let $\mathfrak{A} := (A, \cap, \sim, 1, f_i)_{i \in \Delta}$ be a BAO and $Q := \{Q_n \subseteq A : n \in \mathbb{N}\}$, where each Q_n is non-empty. Let $\mathfrak{A}_Q$ be the structure $(\mathcal{F}_Q(\mathfrak{A}), \{\mathfrak{R}_i\}_{i \in \Delta})$, where $\mathfrak{R}_i \subseteq \mathfrak{A} \times \mathfrak{A}$ such that $(F, G) \in \mathfrak{R}_i$ if and only if $f_i a \in F$ implies $a \in G$. It is not difficult to obtain:

Proposition 10. *If* $\mathfrak{A} := (A, \cap, \sim, 1, f_k)_{k \in \mathbb{N}}$ *be a BAO satisfying (B1)-(B3), then* $\mathfrak{A}_Q$ *is a MSAS.*

Now, we are in the position of defining the important result which we will use to obtain the completeness theorem.

Theorem 1 ([10]). *Let* $\mathfrak{A} := (A, \cap, \sim, 1, f_i)_{i \in \Delta}$ *be a BAO and Q be a countable subset of 2^A. Let $\{X_n\}_{n \in \mathbb{N}}$ and $\{Y_n\}_{n \in \mathbb{N}}$ be an enumeration of the sets $Q_* := \{Q_m \in Q : \bigcap Q_m \in A\}$ and $Q^* := \{Q_m \in Q : \bigcup Q_m \in A\}$. Moreover, suppose that Q satisfies the following conditions for each $i \in \Delta$:*

(QF1) *for any n, $\bigcap f_i X_n$ exists and satisfies that $\bigcap f_i X_n = f_i \bigcap X_m$,*
(QF2) *for any $z \in A$ and n, there exists m such that $\{f_i(z \to x) : x \in X_n\} = X_m$, where $z \to x := \sim z \cup x$,*
(QF3) *for any $z \in A$ and n, there exists m such that $\{f_i(y \to z) : y \in Y_n\} = Y_m$.*

Then the function $r : A \to 2^{\mathcal{F}_Q(\mathfrak{A})}$ defined by $r(a) := \{F \in \mathcal{F}_Q(\mathfrak{A}) : a \in F\}$ is a BAO embedding of $\mathfrak{A}$ into the complex algebra $(\mathfrak{A}_Q)^+$ which also preserves all of $\bigcap X_n$ and $\bigcup Y_n$.

Let us consider the Lindenbaum algebra $\mathfrak{Ln}$ and the canonical assignment θ^c which maps propositional variables to its class, i.e. $\theta^c(p) = [p]$. For each wff α with a single free variable x, let us define the set $Q_\alpha := \{\tilde{\theta}(\alpha(c_j/x)) : j \in \mathbb{N}\}$ and let $Q := \{Q_\alpha : \alpha \text{ has the single free variable } x\}$. Note that Q is countable. Take an enumeration $\{X_n\}_{n \in \mathbb{N}}$ and $\{Y_n\}_{n \in \mathbb{N}}$ of the set Q_* and Q^*. Then it is not difficult to obtain:

Proposition 11. *Q satisfies the condition (QF1)-(QF3).*

Therefore, from Theorem 1, we obtain,

Proposition 12. *There exists a BAO embedding r of $\mathfrak{Ln}$ into $(\mathfrak{Ln}_Q)^+$ such that $r(\bigcap_j \tilde{\theta}^c(\alpha(c_j/x))) = \bigcap_j r(\tilde{\theta}^c(\alpha(c_j/x)))$.*

We note that by Proposition 10, $\mathfrak{Ln}_Q$ is a MSAS and hence by Proposition 2, $(\mathfrak{Ln}_Q)^+$ is a complete BAO satisfying (B1)-(B3). By induction on the complexity of α, we obtain

Proposition 13. *Consider the assignment γ in the BAO $(\mathfrak{Ln}_Q)^+ \in \mathfrak{Cm}$ defined as $\gamma(p) := r([p])$, $p \in PV$. Then $\tilde{\gamma}(\alpha) = r([\alpha])$ for all $\alpha \in \overline{\mathcal{F}}$.*

Proposition 14. *(i) For $\alpha \in \overline{\mathcal{F}}$, $\mathfrak{Cm} \Vdash \alpha$ implies $\vdash \alpha$.*
(ii) For $\alpha \in \mathcal{F}$, $\mathfrak{Cm} \Vdash \alpha$ implies $\vdash \alpha$.

Proof
(i) If possible, let $\nvdash \alpha$. Then $[\alpha] \neq 1$. Now, consider the algebra $(\mathfrak{Ln}_Q)^+ \in \mathfrak{Cm}$ and the assignment γ defined in Proposition 13. Since $[\alpha] \neq 1$, we have $r([\alpha]) \neq 1$. Therefore, $\tilde{\gamma}(\alpha) \neq 1$, a contradiction.
(ii) If possible, let $\nvdash \alpha$. Then $\nvdash cl(\alpha)$ and hence by (i), we obtain a $\mathfrak{A} \in \mathfrak{Cm}$ and an assignment θ in $\mathfrak{Cm}$ such that $\tilde{\theta}(cl(\alpha)) \neq 1$ and thus we obtain $\tilde{\theta}(\alpha) \neq 1$.

So Proposition 4 follows from Proposition 14 and the fact that $\mathfrak{Cm} \subseteq \mathfrak{C}$.

We end this section with the remark that the soundness and completeness theorems establish a strong connection between the MSASs and the class $\mathfrak{C}$ of algebras. It follows that the operators f_i, f_s and f_w are the counterparts of the lower, strong lower and weak lower approximations respectively, where $f_s(a) := \bigcap_{i \in \mathbb{N}} f_i(a)$ and $f_w(a) := \bigcup_{i \in \mathbb{N}} f_i(a)$. Thus one could study the properties of MSASs involving the different notions of lower and upper approximations in the algebras of the class $\mathfrak{C}$ using these operators, and conversely. For instance, the properties $\underline{X^c}_s = (\overline{X}_w)^c$ and $(\overline{X}_w)_w = \overline{(\overline{X}_s)}_w$ of MSAS correspond to the properties $f_s(\neg a) = \neg g_w a$ and $f_w g_w a = g_w g_s a$ of the algebras of $\mathfrak{C}$ respectively, where $g_s(a) := \bigcap_{i \in \mathbb{N}} g_i(a)$ and $g_w(a) := \bigcup_{i \in \mathbb{N}} g_i(a)$.

4   Conclusions

Algebraic semantics of the logic LMSAS is presented and completeness theorem is proved with respect to the class of complete BAOs, complex algebras and LMSAS realizations. It appears from the study that the LMSAS realization is the natural counterpart of LMSAS. However, an independent algebraic characterization of the LMSAS realization is yet to be obtained.

References

1. Blackburn, P., de Rijke, M., Venema, Y.: Modal Logic. Cambridge University Press, Cambridge (2001)
2. Fagin, R., Halpern, J.Y., Moses, Y., Vardi, M.Y.: Reasoning About Knowledge. The MIT Press, Cambridge (1995)
3. Khan, M.A., Banerjee, M.: Formal reasoning with rough sets in multiple-source approximation systems. International Journal of Approximate Reasoning 49(2), 466–477 (2008)
4. Khan, M.A., Banerjee, M.: Multiple-source approximation systems: membership functions and indiscernibility. In: Wang, G., Li, T., Grzymala-Busse, J.W., Miao, D., Skowron, A., Yao, Y. (eds.) RSKT 2008. LNCS (LNAI), vol. 5009, pp. 80–87. Springer, Heidelberg (2008)
5. Ono, H.: Completions of algebras and completeness of modal and substructural logics. In: Balbiani, P., Suzuki, N., Wolter, F., Zakharyaschev, Z. (eds.) Advances in Modal Logic, vol. 4, pp. 335–353. King's College Publications (2003)
6. Orłowska, E., Pawlak, Z.: Expressive power of knowledge representation systems. International Journal of Man-Machine Studies 20(5), 485–500 (1984)
7. Pawlak, Z.: Rough Sets. Theoretical Aspects of Reasoning about Data. Kluwer Academic Publishers, Dordrecht (1991)
8. Rasiowa, H.: An Algebraic Approach to Non-classical Logics. North-Holland Publishing Company, Amsterdam
9. Rauszer, C.M.: Rough logic for multiagent systems. In: Masuch, M., Polos, L. (eds.) Logic at Work 1992. LNCS, vol. 808, pp. 161–181. Springer, Heidelberg (1994)
10. Tanaka, Y., Ono, H.: Rasiowa-Sikorski lemma and Kripke completeness of predicate and infinitary modal logics. In: Zakharyaschev, Z., Segerberg, K., de Rijke, M., Wansing, H. (eds.) Advances in Modal Logic, vol. 2, pp. 401–419. CSLI Publications, Stanford (2001)

Towards an Algebraic Approach for Cover Based Rough Semantics and Combinations of Approximation Spaces

A. Mani

Calcutta Logic Circle
9/1B, Jatin Bagchi Road
Kolkata(Calcutta)-700029, India
a.mani@member.ams.org
http://www.logicamani.co.cc

Abstract. We introduce the concept of a synchronal approximation space (SA) and a AUAI-multiple approximation space and show that they are essentially equivalent to an AUAI rough system. Through this we have established connections between general cover based systems, dynamic spaces and generalized approximation spaces (APS) for easier algebraic semantics. AUAI-rough set theory (RST) is also extended to accommodate local determination of universes. The results obtained are also significant in the representation theory of general granular RST, for the problems of multi source RST and Ramsey-type combinatorics.

1 Introduction

A generalised cover based theory of AUAI rough sets was initiated in [1]. It is relatively more general than most other cover based rough set theories. In the theory, any given generalised cover cannot be associated with a general approximation space or an information system in a unique way without additional assumptions. An axiomatic framework for the concept of granules in general RST is considered in [2] by the present author. Relative this framework, the elements of the cover used do not by themselves constitute the most appropriate granules for the theory. The isolation of usable concepts of granulation in the theory is also complicated by different possible definitions of *rough equalities* and concepts of *definite objects*. Granulation can also be reflected in connections of the theory with other types of RST.

In the next section, we develop a finer characterization of granules in AUAI systems. In the third section, the notion of SA is introduced and shown to be essentially equivalent to AUAI-approximation systems, but with an improved explicit notion of granularity and better semantic properties. We prove representation theorems on the connections between AUAI-RST and a new form of multi source (or dynamic) APS and provide a long example in the fourth section.

In many possible application contexts common universes may not exist and it makes sense to modify the theory for a finite set of universes. This modification

H. Sakai et al. (Eds.): RSFDGrC 2009, LNAI 5908, pp. 77–84, 2009.

is also sensible when *subsets determine their own universes* by way of other semantic considerations. These may also be related to problems of combining general APS. The representation theorems mentioned above extend to the new context with limited modifications. We outline the essentials for this in the fifth section. All of the definitions and theorems (except those in the introduction) are new and due to the present author.

Some of the essential notions are stated below.

Let S be a set and $\mathcal{K} = \{K_i\}_1^n : n < \infty$ be a collection of subsets of it. We will abbreviate subsets of natural numbers of the form $\{1, 2, \ldots, n\}$ by $\mathbf{N}(n)$. If $X \subseteq S$, then consider the sets (with $K_0 = \emptyset$, $K_{n+1} = S$ for convenience):

(i) $X^{l1} = \bigcup\{K_i : K_i \subseteq X, \ i \in \{0, 1, \ldots, n\}\}$
(ii) $X^{l2} = \bigcup\{\cap_{i \in I}(S \setminus K_i) : \cap_{i \in I}(S \setminus K_i) \subseteq X, \ I \subseteq \mathbf{N}(n+1)\}$
(iii) $X^{u1} = \bigcap\{\cup_{i \in I}K_i : X \subseteq \cup_{i \in I} K_i, \ I \subseteq \mathbf{N}(n+1)\}$
(iv) $X^{u2} = \bigcap\{S \setminus K_i : X \subseteq S \setminus K_i, \ i \in \{0, \ldots, n\}\}$

The pair (X^{l1}, X^{u1}) is called an *AU-rough set* by union, while (X^{l2}, X^{u2}) an *AI-rough set* by intersection (in the present author's notation [3]). In the notation of [1] these are $(\mathcal{F}_*^{\cup}(X), \mathcal{F}_{\cup}^*(X))$ and $(\mathcal{F}_*^{\cap}(X), \mathcal{F}_{\cap}^*(X))$ respectively. We will also refer to the pair $\langle S, \mathcal{K} \rangle$ as an AUAI-*approximation system*. By a *partition* of a set S, we mean a pairwise disjoint collection of subsets that covers S.

2   Granules and Equalities in AUAI Rough Set Theory

Possible constructive definitions of granules in a mereology based axiomatic framework are introduced in [2] by the present author. In this section we simply take a *Granule* to mean an element of $\wp(S)$ that is definite in one of the senses defined below and is minimal with respect to being so. The associated *granulation* should also be able to represent any approximation as a set theoretic combination of constituent granules. Concepts of rough equalities are naturally relatable to types of discernibility. [4] suggests another direction.

Definition 1. *In a AUAI system $\langle S, \mathcal{K} \rangle$, the following equalities are definable ($A, B \in \wp(S)$):*

Equality	Defining If and Only If Condition	Type
$A =_z B$	$A^z = B^z$; $z \in \{l1, l2, u1, u2\}$	*Pre-Basic*
$A =_1 B$	$A =_{l1} B$ and $A =_{u1} B$	*Basic*
$A =_2 B$	$A =_{l2} B$ and $A =_{u2} B$	*Basic*
$A =_o B$	$A =_1 B$ and $A =_2 B$	*Derived*
$A =_l B$	$A =_{l1} B$ and $A =_{l2} B$	*Derived*
$A =_u B$	$A =_{u1} B$ and $A =_{u2} B$	*Derived*
$A =_{l-} B$	$A =_{l1} B$ or $A =_{l2} B$	*SubBasic*
$A =_{u-} B$	$A =_{u1} B$ or $A =_{u2} B$	*SubBasic*
$A =_- B$	$A =_{l-} B$ and $A =_{u-} B$	*SubBasic*
$A \asymp B$	$A =_{l-} B$ or $A =_{u-} B$	*SubBasic*

The 'Types' used are relative a natural perspective.

Definition 2. *We define different usable concepts of* definite *objects below:*

Concept	Defining If and Only If Condition	Type
A *is* 1-Definite	*iff* $A^{l1} = A = A^{u1}$	*Balanced*
A *is* 2-Definite	*iff* $A^{l2} = A = A^{u2}$	*Balanced*
A *is* 12-Definite	*iff* $A^{l1} = A = A^{u1} = A^{l2} = A^{u2}$	*Balanced*
A *is* 0-Definite	*iff* $A^{l1} = A^{l2}$ *and* $A^{u1} = A^{u2}$	*Strong*
A *is* x-Definite	*iff* $A^{x} = A$; $x \in \{l1, l2, u1, u2\}$	*One-Sided*

Below we reduce the number of possible concepts of definiteness to six.

Proposition 1. *In the context of the above all of the following hold for any subset A of S:*

1. *If A is $l1$-definite then A is $u1$-definite, but the converse implication may not hold in general.*
2. *If A is $u2$-definite then A is $l2$-definite, but the converse implication may not hold in general.*
3. *A is 12-definite if and only if A is $l1$-definite and $u2$-definite.*

Proof

1. If A is $l1$-definite, then $A = A^{l1} = \bigcup \{K_i : K_i \subseteq A, i \in \{0, 1, ..., n\}\}$ and so A^{l1} is one of the sets being intersected over in $\bigcap \{\cup_{i \in I} K_i : A, \subseteq \cup_{i \in I} K_i, I \subseteq \mathbf{N}(n+1)\}$. Obviously the whole intersection must coincide with A^1. So $A^{u1} = A$.
2. If A is $u2$-definite, then $A = A^{u2} = \bigcap \{S \setminus K_i : A, \subseteq S \setminus K_i, i \in \{0, ..., n\}\}$. But this is then the largest possible set included in the union $\bigcup \{\cap K_i^c : \cap K_i^c \subseteq A\}$. So $A^{l2} = A$.
3. This follows from the two propositions proved above. If A is 12-definite, then it is obviously $l1$- and $u2$-definite. □

The elements of $\mathcal{K}$ used in AUAI-rough set theory can be seen as quasi-inductive granules in a more general sense. This is reinforced by the following proposition:

Proposition 2. *If $K_i \in \mathcal{K}$, then*

1. *K_i is 1-definite, but is not necessarily 2-definite*
2. *K_i need not be a minimal element with the property (in the usual order)*
3. *K_i^c is 2-definite, but is not necessarily 1-definite.*

3   Synchronal Approximation Spaces

We introduce SAs and show them to be essentially equivalent to AUAI-approximation systems, but with improved explicit notion of granularity and semantic features. Basically, these are APS with operators that map equivalence classes into other classes and are otherwise like the identity map.

Definition 3. *By a synchronal approximation space SA, we mean a tuple of the form $\langle \underline{S}, R, \eta_1, \eta_2, \ldots \eta_n \rangle$ satisfying all of*

(i) $\langle \underline{S}, R \rangle$ is an APS with partition $\mathcal{R}$ with η_i being maps : $\wp(S) \longmapsto \wp(S)$

(ii) $(\forall A \in \mathcal{R}) \eta_j(A) \in \mathcal{R}$ or $\eta_j(A) = \emptyset$; $(\forall A \in \wp(S) \setminus \mathcal{R}) \eta_j(A) = A$

(iii) $(\forall A, B \in \mathcal{R})(\eta_i(A), \eta_i(B) \in \mathcal{R} \longrightarrow \exists k, t \, \eta_i^k(A) = \eta_i^t(B))$

(iv) $(\forall \eta_i, \eta_j)(\exists A, B \in \mathcal{R}) \eta_i(B) = \eta_j(A) = \emptyset, \eta_i(A), \eta_j(B) \in \mathcal{R}$

(v) $(\forall \eta_j)(\exists t \in N)(\forall A \in \wp(S)) \eta_j^{t+1}(A) = \eta_j^t(A)$

Definition 4. *By a η_j-connected component of a SA, we mean a subset $C \subseteq \mathcal{R}$ that is maximal with respect to satisfying*

$$(\forall A, B \in C)(\exists k \in N) \, \eta_j^k(A) = B \text{ or } \eta_j^k(B) = A$$

In other words it is a subset satisfying the condition and no proper superset satisfies the same condition.

Definition 5. *On a SA, $\langle \underline{S}, R, \eta_1, \eta_2, \ldots \eta_n \rangle$, apart from the usual lower and upper aproximations of a subset X (denoted by X^l and X^u) of S, we can define the following approximations:*

(i) $X^{l1+} = \bigcup \{B; \, B \in \mathcal{R}, \cup_j \cup_k \eta_j^k(B) \subseteq X\}$

(ii) $X^{u1+} = \bigcap \{\cup_{j \in J} \cup_k \eta_j^k(B) ; \, X \subseteq \cup_{j \in J} \cup_k \eta_j^k(B), \, J \subseteq \mathbf{N}(n+1), \, B \in \mathcal{R}\}$

(iii) $X^{l2+} = \bigcup \{(\cup_{j \in J} \cup_k \eta_j^k(B))^c ; \, (\cup_{j \in J} \cup_k \eta_j^k(B))^c \subseteq X, \, J \subseteq \mathbf{N}(n+1), \, B \in \mathcal{R}\}$

(iv) $X^{u2+} = \bigcap \{(\cup_k \eta_j^k(B))^c ; \, X \subseteq (\cup_k \eta_j^k(B))^c, \, B \in \mathcal{R}\}$

(v) $X^{u0+} = \bigcup_j \{\cup_k \eta_j^k(B) ; \, \eta_j^k(B) \cap X \neq \emptyset, \, B \in \mathcal{R}\}$

Theorem 1. *Any AUAI approximation system $\langle S, \mathcal{K} \rangle$ determines an partition S along with a SA $\langle \underline{S}, R, \eta_1, \eta_2, \ldots \eta_n \rangle$ that is essentially equivalent to the former in that $l1+, l2+, u1+, u2+$-approximations of a subset in the latter are the same as $l1, l2, u1, u2$-approximations in the former respectively. Further, the SA uniquely determines the AUAI approximation system.*

Proof. For the forward transformation:

1. Simply decompose each K_i into $\{K_{ij}\}$ with each subset being disjoint from any other of the form K_{hv} (for any distinct index). Let sets of the form K_{ii} for $i = 1, \ldots, n$ be the ones obtained from K_i by subtracting all other K_js from it.

2. Define the η_is as per Definition 3 so that $\cup_{B \in \mathcal{R}} \cup_k \eta_i^k(B) = K_i$. More concretely, let $\eta_1^j(K_{11})$ take all values in K_{ij} with last class being mapped to $\emptyset$ and $\eta_1(B) = \emptyset$ for other $B \in \mathcal{R}$ and so on.

Given a SA, the union of connected components of each of the η_is that exclude the empty set are precisely the elements of the collection $\mathcal{K}$ of the AUAI-approximation system. This can be checked by substitution in the definition of $l1+, l2+, u1+, u2+$ approximations. The uniqueness part can be verified by a contradiction argument. □

As equivalence classes have better granular properties than the elements of $\mathcal{K}$ (relative approximations l, u), SAs have better granulation than AUAI-rough systems. The lack of uniqueness of definition of η_is in the proof is considered in more detail in a separate paper. Note that we do not need the exact index set over which k takes values in the above, but the combinatorial part is definitely of much interest.

Proposition 3. *In a* SA $\langle \underline{S}, R, \eta_1, \eta_2, \ldots \eta_n \rangle$, *the elements of* $\mathcal{R}$ *are admissible granules for the approximation operators* $l, u, l1+, u1+, l2+, u2+, u0+$. *These possess the following properties:*

(i) $(\forall B \in \mathcal{R}) B^l = B = B^u$

(ii) $(\forall B \in \mathcal{R})(\forall A \in \wp(S))(A \subset B \longrightarrow A \notin \mathcal{R})$

(iii) *All of the above mentioned approximations are representable as set-theoretic combinations of elements of* $\mathcal{R}$.

(iv) $(\forall B \in \mathcal{R})(\exists B_1, \ldots, B_r \in \mathcal{R}) B \cup \bigcup_1^r B_i$ is $1-$ definite $\qquad\square$

4 Multiple Approximation Spaces

Different APS can be derived from an AUAI-approximation system. In this section, we investigate the question of reducibility and equivalence of such systems with special multi-source APS or dynamic spaces (see [5], [6] for example).

Definition 6. *By the* AIAU-Mutation Algorithm *we will mean the following procedure:*

1. *INPUT:* $\mathcal{K}$ *(interpreted as a sequence of sets), for simplicity of notation we will assume that no element is included in another.*
2. *The total orders on the index set* $\{1, 2, \ldots, n\}$ *correspond to bijections on the same set (The set of bijections will be denoted by* $B(n)$*).*
3. *Fix* $\sigma \in B(n)$. *Set* $P_1 = K_{\sigma 1}$
4. *Set* $P_2 = K_{\sigma 2} \setminus P_1$
5. *... $P_s = K_s \setminus \cup_{r<s} P_r$ for* $s = 2 \ldots, n+1$
6. *OUTPUT:* $\mathcal{P}_\sigma = \{P_i\}_1^{n+1}$ *for each* σ. *We need to ignore empty sets in the collection for our partitions.*

Proposition 4. *The collections formed by the* AUAI-*mutation algorithm are partitions of the underlying set* S. *The equivalence corresponding to the partition* $\mathcal{P}_\sigma$ *will be denoted by* R_σ.

Definition 7. *In the above context, by a* concrete AIAU multiple approximation space *CAMS, we will mean a tuple of the form* $\langle \underline{S}, \{R_\sigma\}_{\sigma \in B(n)} \rangle$. *The partitions determined by each* R_σ *will be denoted by* $\mathcal{P}_\sigma$.

Theorem 2. *A CAMS* $\langle \underline{S}, \{R_\sigma\}_{\sigma \in B(n)} \rangle$, *satisfies all of the following:*

(i) $(\forall \sigma \in B(n)) R_\sigma \odot R_\sigma = R_\sigma, R_\sigma^{-1} = R_\sigma, \Delta_S \subseteq R_\sigma$

(ii) $(\forall \sigma, \sigma' \in B(n))(R_\sigma \odot R_{\sigma'})^{-1} = R_\sigma \odot R_{\sigma'}, \Delta_S \subseteq R_\sigma \odot R_{\sigma'}$

(iii) $(\forall \sigma, \tau \in B(n))(\exists A, B \in \mathcal{P}_\tau)(\exists C, E \in \mathcal{P}_\sigma) A \subset C, E \subset B$

(iv) $(\forall k \in \mathbf{N}(n))(\forall \sigma \in B(n))(\exists^{\geq (n-k)-1} \mathcal{P}_\tau)(\exists A_1, \ldots, A_k \in \mathcal{P}_\sigma)(\exists B_1, \ldots, B_k \in \mathcal{P}_\tau) A_1 = B_1, \ldots, A_k = B_k$

Proof

(i) Each R_σ is an equivalence relation
(ii) The composition of two distinct equivalences on the same set is symmetric and reflexive
(iii) Suppose for some i, j, $K_i \in \mathcal{P}_\sigma$ and $K_j \in \mathcal{P}_\tau$. These are guaranteed to exist by the AUAI-mutation algorithm. If $K_i \nsubseteq K_j$ and $K_j \nsubseteq K_i$, then we will be able to find some $E \subset K_j$ in $\mathcal{P}_\sigma$ and a $A \subset K_i$ in $\mathcal{P}_\tau$. If $K_i = K_j$, then we need to consider the classes at some later stage of the mutation process. This proves the statement.
(iv) From the collection $\mathcal{K}$ of $'n'$ number of subsets of S, we can generate at most n number of equivalence relations (and the same number of partitions) by the AUAI-mutation algorithm. Given a specific partition $\mathcal{P}_\sigma$, the number of partitions with exactly k common elements is $(n-k)-1$.

Theorem 3 (Representation Theorem). *Every AUAI approximation system $\langle S, \mathcal{K} \rangle$ determines a unique CAMS $\langle \underline{S}, \{R_\sigma\}_{\sigma \in B(n)} \rangle$, which in turn determines the same (up to a definable isomorphism) AUAI approximation system by a reverse algorithm.*

Proof. The $\Rightarrow$ part of the proof has already been done. For generality, we will assume that $\bigcup \mathcal{K} \neq S$. Given a CAMS $\langle \underline{S}, \{R_\sigma\}_{\sigma \in B(n)} \rangle$:

1. Form the partitions $\mathcal{P}_\sigma$ corresponding to R_σ and group them into hierarchial collections $\{\mathcal{H}_{ij}\}_{j \in \mathbf{N}(n)}$ on the basis of number of common elements by the following rules:
2. For fixed j, any two collections in $\{\mathcal{H}_{ij}\}$ have one common element, while any elements (partitions) of any two collections in $\mathcal{H}_{ij}$ have $i + 1$ common elements within themselves (for $i = 1, 2, \ldots n - 1$)
3. $(y \in x \in \mathcal{H}_{ij} \longrightarrow \exists !^{(n-i)!} z \, x \subset z \in \mathcal{H}_{i-1j})$
4. The elements of $\mathcal{K}$ are the single common elements in $\mathcal{H}_{ij}$ (for each j).   □

This theorem completely describes concrete AUAI multiple APS and can be used as an equivalent representation for AUAI approximation systems. It is also a very intricate new Ramsey-type theorem ! (see the extended version of this paper for details).

Extended Example: Let $S = \{a, b, c, e, f, g\}$, $K_1 = \{a, b, c\}$, $K_2 = \{b, c\}$, $K_3 = \{c, e, f\}$, $K_4 = \{f, g\}$, $K_5 = \{b, e\}$ and let $\mathcal{K} = \{K_i\}_1^5$, then,

- $K_1^{l1} = \{a, b, c\}$, $K_1^{u1} = \{a, b, c\}$, $K_1^{l12} = \{a, b, c\}$, $K_1^{u2} = \{a, b, c, e\}$,. So K_1 is 1-definite, but not 2-definite.
- If $A = \{a, f\}$, then $A^{l1} = \emptyset$, $A^{u1} = \{a, b, c, f\}$, $A^{l2} = \{a\}$, $A^{u2} = \{a, f, g\}$
- If $B = \{a, f, c\}$, then $B^{l1} = \emptyset$, $B^{u1} = \{a, b, c, f\}$, $B^{l2} = \{a, c\}$, $B^{u2} = \{a, c, f, g\}$
- So $A =_1 B$, but $A \neq_2 B$
- AUAI Mutation Algorithm: Applying the algorithm in the order $(4, 1, 3, 2, 5)$ to $\mathcal{K}$, we get $P_1 = \{f, g\}$, $P_2 = \{a, b, c\}$, $P_3 = \{e\}$, $P_4 = \emptyset$, $P_5 = \emptyset$. Naturally we need to ignore the empty partitions.

- If we apply the algorithm in the order $(4, 2, 5, 1, 3)$ to $\mathcal{K}$, then we get $P_1 = \{f, g\}$, $P_2 = \{b, c\}$, $P_3 = \{e\}$, $P_4 = \{a\}$, $P_5 = \emptyset$. Obviously, we can get a large number of distinct partitions by this method.

More examples can be constructed from the ones in [1] and [3].

5   Generalisation of **AUAI**-Approximation Systems

In many applications, it can happen that each granule or union of granules determines its own universe. This may be because the system under consideration is actually the result of combining APS over different universes. The same applies when the relevant universes are locally determined with respect to granules (see [7] by the present author).

Definition 8. *A partial map* $\varphi : \wp(S) \longmapsto \wp(S)$ *will be said to be a* universe determining map *if and only if it is a monotone increasing partial map defined on the set* $\mathbb{B}$ *of granules and unions of granules. s.t.* $dom(\varphi) = \mathbb{B}$; $(x \subset y \longrightarrow \varphi(x) \subseteq \varphi(y))$; $x \subseteq \varphi(x)$ *and* $(\varphi(x) = y \longrightarrow \varphi(\sim x) = y)$

Definition 9. *A tuple of the form* $\langle S, \mathcal{K}, \mathbb{B}, \varphi \rangle$ *will be said to be a Quasi-AUAI-approximation system (or a QAIAU system) if* $\mathbb{B}$ *is the set of granules and unions of granules,* φ *is an universe determining map and the approximations* $l1, l2, u1, u2$ *are defined by conditions similar in form to that of AUAI systems, but complementation is interpreted relative* φ*-determined universes (that is for any subset* A, $A^* = \varphi(A) \setminus A$*).*

From a classicalist perspective, we can define the usual set operations, special complementation, the unary approximation operators and the 0-place operations $\perp, 1, T$ on the power set $\wp(S)$. The resulting structure, $\mathbb{S} = \left\langle \wp(S), \cup, \cap, {}^*, l1, l2, u1, u2, \perp, 1, T, \eta_1, \ldots \eta_{\phi(n)}, \xi_1, \ldots, \xi_{\phi(n)} \right\rangle$ with $\perp = \emptyset$, $T = \wp(S)$, $1 = \bigcup \mathcal{K}$, will be termed a *concrete* QAUAI-*algebra*. The operations η_i, ξ_j have been introduced to ease the expression of the last two conditions of Thm 1.1 of [3]. For even n, $\phi(n) = \frac{n!}{(n/2)!^2}$ and for odd n, $\phi(n) = \frac{n!}{(n+1/2)!(n-1/2)!}$.

A similar structure satisfying all the conditions of Thm 1.1 (of [3]) is definable for AUAI systems. The main differences are in the properties of $l2, u2$, the equalities involving mixed approximations and *difficulty of abstract representation*. It is also possible to represent QAUAI systems as a collection of APS over distinct universes under constraints. Importantly the connections with the different extended APS extend to QAUAI systems in a modified way.

Theorem 4. *The following properties hold in any QAUAI-algebra:*

(i) $x^{**} = x$; $x^* \cap y^* \subseteq (x \cup y)^*$; $x^{*l2} \subseteq x^{u1*}$; $x^{*u1} \subseteq x^{l2*}$
(ii) In general, $(x \cap y)^{l2} \neq (x^{l2} \cap y^{l2})$; $x^{*l1} \neq x^{u2*}$

6 Further Directions

We have shown that SAs and AUAI multiple APS are essentially equivalent to AUAI-approximation systems, but with improved granularity. Moreover they are more amenable from the algebraic point of view through direct methods and decomposition theorems. For forming the logics of *roughly equivalent objects*, we can use typed approaches or deal with the *AI* and *AU* approximations separately (at the semantic level). The connections proved permit us to consider the semantic domains of roughly equivalent objects of the AI, AU type respectively and the classical semantic domain in a dialectical way. This suggests a natural dialectical approach to the semantics and logic (see [8] and [2]).

In the extended version of this paper we associate typed logics with the variant of multiple APS obtained in the above. The objects of the rough semantic domain are described in the same. As the granularity is fairly intricate, we have separate logics for AI, AU and AUAI approximations. We also introduce a relation algebra like approach to describe the semantics from a classicalist and rough perspective of things. The associated structures are partial algebras. For the rough perspective, we use relativisation in the sense of [9] to provide a distinct interpretation. The main problems solved therein are those of selection of operations and domain of definition, axiomatisability and/or direct logic formulation. We will also consider the fine structure of the connections with most other cover based approaches in a separate paper.

Acknowledgement. I would like to thank the anonymous referees for useful suggestions towards improving the readability of the paper.

References

1. Inuiguchi, M.: Generalisation of rough sets and rule extraction. In: Peters, J.F., Skowron, A. (eds.) Transactions on Rough Sets I. LNCS, vol. 3100, pp. 96–119. Springer, Heidelberg (2004)
2. Mani, A.: A program in dialectical rough set theory (preprint, 2009), http://arxiv.org/abs/0909.4876
3. Mani, A.: Esoteric rough set theory-algebraic semantics of a generalized VPRS and VPRFS. In: Peters, J.F., Skowron, A. (eds.) Transactions on Rough Sets VIII. LNCS, vol. 5084, pp. 182–231. Springer, Heidelberg (2008)
4. Pawlak, Z., Novotny, M.: Characterization of rough top and bottom equalities. Bull. Pol. Acad. Sci. (Math) 33, 91–97 (1985)
5. Pagliani, P.: Pretopologies and dynamic spaces. Fundamenta Informaticae 59, 221–239 (2004)
6. Pawlak, Z.: Some issues in rough sets. In: Peters, J.F., Skowron, A., Grzymała-Busse, J.W., Kostek, B.z., Świniarski, R.W., Szczuka, M.S. (eds.) Transactions on Rough Sets I. LNCS, vol. 3100, pp. 1–58. Springer, Heidelberg (2004)
7. Mani, A.: Meaning, choice and similarity based rough set theory. In: Internat. Conf. Logic and Applications, Chennai, pp. 1–12 (2009), http://arxiv.org/abs/0905.1352
8. Mani, A.: Integrated dialectical logics for relativised general rough set theory. In: Internat. Conf. on Rough Sets, Fuzzy Sets and Soft Computing, Agartala, India, p. 6 (2009)
9. Marx, M.: Relativised relation algebras. ILLC Preprint Series (1999)

New Approach in Defining Rough Approximations

E.K.R. Nagarajan and D. Umadevi

Centre for Research and Post Graduate Studies in Mathematics,
Ayya Nadar Janaki Ammal College (Autonomous), Sivakasi-626124,
Tamil Nadu, India
nagarajanekr@yahoo.com, umavasanthy@yahoo.com

Abstract. In this paper, we discuss some structures on the ordered set of rough approximations in a more general setting of complete atomic Boolean lattices. Further, we define an induced map from the map defined from the atoms of complete atomic Boolean lattice ($\mathcal{A}(B)$) to that lattice B. We also study the connections between the rough approximations $x^{\check{}}$, $x^{\wedge}$ defined with respect to the induced map and the rough approximations $x^{\blacktriangledown}$, $x^{\blacktriangle}$ defined with respect to the considered map under certain conditions on the map.

Keywords: Complete atomic Boolean lattices, Complete Ortholattices, linearly ordered set, rough approximations.

1 Introduction

Rough set theory was introduced by Pawlak [10] to deal with uncertainty, where the objects were observed only through the available knowledge represented by the indiscernibility relation. The Rough Set theory approach is based on indiscernibility relation and approximation. Pawlak's rough sets model is based on equivalence relation. According to Slowinski and Vanderpooten [11], "The equivalence relation seems to be a stringent condition that may limit the application domain of the standard rough set model". So the equivalence relation has been relaxed to arbitrary binary relation [13] and Yao [14, 15, 16] introduced the notion of generalized rough approximations. This is one of the ways to generalize the rough set model.

Another way of generalizing the rough set model was done by defining the approximation operators in various algebraic structures such as Boolean algebras [7, 8], Complete distributive lattices [5], Completely distributive Complete lattices [4], lattices, posets [17] etc. The properties of the rough approximations in a more general setting of complete atomic Boolean lattice were studied in [7, 8]. In this paper, we have proved according to the notations in [7, 8] that the ordered sets $(B^{\blacktriangledown}, \leq)$ and $(B^{\blacktriangle}, \leq)$ are complete ortholattices if the map φ is extensive and symmetric. Also, we give a necessary and sufficient condition for the ordered sets $(B^{\blacktriangledown}, \leq)$ and $(B^{\blacktriangle}, \leq)$ to be linearly ordered. Further, we define a new map induced from the considered map φ and study the connection between the rough approximations defined by the map $\langle \varphi \rangle$ and the map φ.

H. Sakai et al. (Eds.): RSFDGrC 2009, LNAI 5908, pp. 85–92, 2009.

2 Preliminaries

An ordered set $(P, \leq)$ is linearly ordered if for every $x, y \in P$, $x \leq y$ or $y \leq x$. A map f: $P \rightarrow P$ is said to be extensive, if $x \leq f(x)$, for all $x \in P$. The map f is order – preserving if $x \leq y$ implies $f(x) \leq f(y)$. The map f: $P \rightarrow P$ is said to be idempotent if $f(f(x))=f(x)$. A closure operator on an ordered set P is an idempotent, extensive and order-preserving self-map. A self-map f is called a topological closure operator (also called a Kura-towski closure operator) on complete lattice L if it is idempotent, extensive, and satis-fies $f(0) = 0$ and $f(a \vee b) = f(a) \vee f(b)$ for all a, b $\in$ L. Further, if f is a closure operator and a complete join-morphism on a complete lattice L, then f is called an Alexandrov closure operator on L. Let $(L, \vee, \wedge, 0, 1)$ be a bounded lattice. Then L is said to be an ortholattice if there exists a unary operation $'$: L→L satisfying the conditions $x \vee x' = 1$, $x \wedge x' = 0$, $x \leq y \Rightarrow y' \leq x'$ and $x'' = x$. For definitions and results in lattice theory not given here the readers are asked to refer [6, 9, 12].

Let us recall some definitions and results given in [7].

If B is a complete atomic Boolean lattice, then $\mathcal{A}(B)$ denote the set of all atoms of B.

Definition 1[7]. Let $(B, \leq)$ be a complete atomic Boolean lattice. A map $\varphi : \mathcal{A}(B) \rightarrow B$ is said to be

    i)   extensive, if $x \leq \varphi(x)$, for all $x \in \mathcal{A}(B)$.

    ii) symmetric, if $x \leq \varphi(y) \Rightarrow y \leq \varphi(x)$, for all $x, y \in \mathcal{A}(B)$.

    iii) closed, if $x \leq \varphi(y) \Rightarrow \varphi(x) \leq \varphi(y)$, for all $x, y \in \mathcal{A}(B)$.

Definition 2[7]. Let $(B, \leq)$ be a complete atomic Boolean lattice and let $\varphi : \mathcal{A}(B) \rightarrow B$ be any map. For any element $x \in B$, let

$$x^{\triangledown} = \vee \{a \in \mathcal{A}(B) \, / \, \varphi(a) \leq x\} \text{ and}$$

$$x^{\triangle} = \vee \{a \in \mathcal{A}(B) \, / \, \varphi(a) \wedge x \neq 0\}. \tag{1}$$

Result 1[7]. *Let $(B, \leq)$ be a complete atomic Boolean lattice and let $\varphi : \mathcal{A}(B) \rightarrow B$ be any map. Then for all $a \in \mathcal{A}(B)$ and $x \in B$,*

    i) $a \leq x^{\triangle} \Leftrightarrow \varphi(a) \wedge x \neq 0$

    ii) $a \leq x^{\triangledown} \Leftrightarrow \varphi(a) \leq x$

The following results are shown in [7]. For any $S \subseteq B$, let $S^{\triangle} = \{x^{\triangle} / x \in S\}$. The or-dered sets $(B^{\triangledown}, \leq)$ and $(B^{\triangle}, \leq)$ are always complete lattices. $(B^{\triangledown}, \leq)$ and $(B^{\triangle}, \leq)$ are distributive sub lattices of $(B, \leq)$ if φ is extensive and closed. If the map φ is exten-sive, symmetric and closed, then the ordered sets $(B^{\triangledown}, \leq)$ and $(B^{\triangle}, \leq)$ are mutually equal complete atomic Boolean lattices. Further the example 3.16 of [7] shows that $(B^{\triangledown}, \leq)$ and $(B^{\triangle}, \leq)$ are not necessarily distributive if φ is extensive and symmetric.

3   Some Structures on the Ordered Set of Rough Approximations

The join ($\bar{\vee}$) and meet ($\bar{\wedge}$) operations in the complete lattice $(B^\blacktriangle, \leq)$ are as follows: Let $S^\blacktriangle \subseteq B^\blacktriangle$. Then $\bar{\vee} S^\blacktriangle = (\vee_{x \in S} x)^\blacktriangle$ and $\bar{\wedge} S^\blacktriangle = \vee \{ \varphi(a) / \varphi(a) \leq \wedge_{x \in S} x^\blacktriangle \}$, where $\vee$ and $\wedge$ are the join and meet operations in $(B, \leq)$ respectively.

Theorem 1. *If φ is extensive and symmetric, then $(B^\blacktriangle, \leq)$ and $(B^\blacktriangledown, \leq)$ are complete ortholattices.*

Proof. For each $x^\blacktriangle \in B^\blacktriangle$, $(x^\blacktriangle)^\perp = x^{\blacktriangle'\blacktriangle}$, where ' is the complementation operator in $(B, \leq)$ is the orthocomplement of $x^\blacktriangle$. Hence the ordered set $(B^\blacktriangle, \leq)$ is a complete ortholattice. Since φ is symmetric, $(B^\blacktriangle, \leq) \cong (B^\blacktriangledown, \leq)$. Hence $(B^\blacktriangledown, \leq)$ is also a complete ortholattice.   □

Let us denote $\varphi(\mathcal{A}(B)) = \{ \varphi(a) / a \in \mathcal{A}(B) \}$.

Theorem 2. *Let $(B, \leq)$ be a complete atomic Boolean lattice. Then $(B^\blacktriangle, \leq)$ is linearly ordered if and only if $(\varphi(\mathcal{A}(B)), \leq)$ is linearly ordered.*

Proof. Assume that $(\varphi(\mathcal{A}(B)), \leq)$ is linearly ordered. Let $x, y \in B^\blacktriangle$. Then $x = u^\blacktriangle$, $y = v^\blacktriangle$, for some $u, v \in B$. Suppose x and y are not comparable in $B^\blacktriangle$. Then there exists $a, b \in \mathcal{A}(B)$ such that $a \leq x = u^\blacktriangle$ and $a \not\leq y = v^\blacktriangle$, $b \leq y = v^\blacktriangle$ and $b \not\leq x = u^\blacktriangle$. Then, we have $\varphi(a) \wedge u \neq 0$ and $\varphi(a) \wedge v = 0$, $\varphi(b) \wedge v \neq 0$ and $\varphi(b) \wedge u = 0$. Since $(\varphi(\mathcal{A}(B)), \leq)$ is linearly ordered, we have $\varphi(a) \leq \varphi(b)$ or $\varphi(b) \leq \varphi(a)$. If $\varphi(a) \leq \varphi(b)$, then $\varphi(a) \wedge u \neq 0$ implies $\varphi(b) \wedge u \neq 0$, which is a contradiction to the hypothesis. Similar contradiction also occurs when $\varphi(b) \leq \varphi(a)$. Hence $(B^\blacktriangle, \leq)$ is linearly ordered. Conversely, assume that $(B^\blacktriangle, \leq)$ is linearly ordered. Suppose $(\varphi(\mathcal{A}(B)), \leq)$ is not linearly ordered. Then there exists $a, b \in \mathcal{A}(B)$ such that neither $\varphi(a) \leq \varphi(b)$ nor $\varphi(b) \leq \varphi(a)$. Then there exists $c, d \in \mathcal{A}(B)$ such that $c \leq \varphi(a)$ and $c \not\leq \varphi(b)$, $d \leq \varphi(b)$ and $d \not\leq \varphi(a)$. Then $c \leq \varphi(a)$ and $c \not\leq \varphi(b)$ implies $\varphi(a) \wedge c \neq 0$ and $\varphi(b) \wedge c = 0$. Then, we have $a \leq c^\blacktriangle$ and $b \not\leq c^\blacktriangle$. Similarly, $d \leq \varphi(b)$ and $d \not\leq \varphi(a)$ implies $b \leq d^\blacktriangle$ and $a \not\leq d^\blacktriangle$. Thus there exists $a, b \in \mathcal{A}(B)$ such that $a \leq c^\blacktriangle$ and $a \not\leq d^\blacktriangle$, $b \leq d^\blacktriangle$ and $b \not\leq c^\blacktriangle$. This implies there exists $c^\blacktriangle, d^\blacktriangle \in B^\blacktriangle$ such that $c^\blacktriangle \not\leq d^\blacktriangle$ and $d^\blacktriangle \not\leq c^\blacktriangle$, which is a contradiction to the hypothesis. Hence $(\varphi(\mathcal{A}(B)), \leq)$ is linearly ordered.   □

Since $(B^\blacktriangle, \leq)$ is dually order isomorphic to $(B^\blacktriangledown, \leq)$, we have the following corollary.

Corollary 1. *Let $(B, \leq)$ be a complete atomic Boolean lattice. Then $(B^\blacktriangledown, \leq)$ is linearly ordered if and only if $(\varphi(\mathcal{A}(B)), \leq)$ is linearly ordered.*

Remark 1. Let us consider $\varphi: U \to \wp(U)$ defined by $\varphi(x) = R(x)$ for all $x \in U$, where $R(x) = \{y \in U / xRy\}$. If R is transitive and connected, then for every $x, y \in U$, $R(x) \subseteq R(y)$ or $R(y) \subseteq R(x)$. Therefore, by the above theorem $(\wp(U)^\blacktriangle, \subseteq)$ and $(\wp(U)^\blacktriangledown, \subseteq)$ are linearly ordered sets.

4 New Approaches in Defining the Rough Approximations

We can define a new map $\langle \varphi \rangle$ induced from the map φ as follows:

$$\langle \varphi \rangle (a) = \bigwedge_{a \leq \varphi(b)} \varphi(b), \text{ for all } a \in \mathcal{A}(B). \tag{2}$$

The idea behind the setting is the map $\langle \varphi \rangle : U \to \wp(U)$ by $\langle \varphi \rangle (x) = \langle x \rangle R$ may be considered to be of the form $\langle \varphi \rangle : \mathcal{A}(B) \to B$, where $(B, \leq)$ equals $(\wp(U), \subseteq)$ resembles the definition of $\langle x \rangle R$ given in [1, 2, 3].

Now, we can define the lower and upper approximation operators on $(B, \leq)$ with respect to the induced map $\langle \varphi \rangle$.

Definition 3. Let $(B, \leq)$ be a complete atomic Boolean lattice and $\varphi : \mathcal{A}(B) \to B$ be any map. For any element $x \in B$, we define

$$x^{\vee} = \vee \{a \in \mathcal{A}(B) / \langle \varphi \rangle (a) \leq x\} \text{ and}$$
$$x^{\wedge} = \vee \{a \in \mathcal{A}(B) / \langle \varphi \rangle (a) \wedge x \neq 0\}. \tag{3}$$

The elements $x^{\vee}$ and $x^{\wedge}$ are the lower and upper approximations of x with respect to $\langle \varphi \rangle$ respectively, where the elements $x^{\bullet}$ and $x^{\ast}$ are the lower and upper approximations of x with respect to φ.

According to lemma 3.3, 3.4 and proposition 3.5 in [7], the following results are true for any map φ. Since $\langle \varphi \rangle$ is also a map, the same results hold for the lower and upper approximations of $x \in B$ with respect to $\langle \varphi \rangle$. So, we omit the proof for the following lemma and proposition.

Lemma 1. *Let B be a complete atomic Boolean lattice and $\varphi : \mathcal{A}(B) \to B$ be any map. Then for any element x, $y \in B$ and $a \in \mathcal{A}(B)$, the following hold*

  i) $a \leq x^{\wedge} \Leftrightarrow \langle \varphi \rangle (a) \wedge x \neq 0$
  ii) $a \leq x^{\vee} \Leftrightarrow \langle \varphi \rangle (a) \leq x$
  iii) $0^{\wedge} = 0$ and $1^{\vee} = 1$
  iv) $x \leq y \Rightarrow x^{\vee} \leq y^{\vee}$ and $x^{\wedge} \leq y^{\wedge}$

For any $S \subseteq B$, denote by $S^{\wedge} = \{x^{\wedge} / x \in S\}$ and $S^{\vee} = \{x^{\vee} / x \in S\}$.

Proposition 1. *Let B be a complete atomic Boolean lattice and $\varphi : \mathcal{A}(B) \to B$ be any map. Then the following hold*

  i)   *The maps $^{\vee} : B \to B$ and $^{\wedge} : B \to B$ are mutually dual.*
  ii) *For all $S \subseteq B$, $\vee S^{\wedge} = (\vee S)^{\wedge}$ and $\wedge S^{\vee} = (\wedge S)^{\vee}$.*
  iii) *$(B^{\wedge}, \leq)$ is a complete lattice with null element 0 and all element $1^{\wedge}$.*
  iv) *$(B^{\vee}, \leq)$ is a complete lattice with null element $0^{\vee}$ and all element 1.*

Proposition 2. *Let B be a complete atomic Boolean lattice and* $\varphi : \mathcal{A}(B) \to B$ *be any map. Then the following hold for every* $x \in B$.

i) $x^{\wedge \blacktriangle} = x^{\blacktriangle}$
ii) $x^{\vee \blacktriangledown} = x^{\blacktriangledown}$

Proof. i) Let $a \in \mathcal{A}(B)$ be such that $a \leq x^{\wedge \blacktriangle}$. Then, we have $\varphi(a) \wedge x^{\wedge} \neq 0$. Then there exists $b \in \mathcal{A}(B)$ such that $b \leq \varphi(a)$ and $b \leq x^{\wedge}$ which implies $\langle \varphi \rangle (b) \wedge x \neq 0$. Since $b \leq \varphi(a)$, we have $\langle \varphi \rangle (b) \leq \varphi(a)$. Thus, $\varphi(a) \wedge x \neq 0$. Then $\varphi(a) \wedge x \neq 0$ implies $a \leq x^{\blacktriangle}$. This implies $\vee \{a \in \mathcal{A}(B)/a \leq x^{\wedge \blacktriangle}\} \leq \vee \{a \in \mathcal{A}(B)/a \leq x^{\blacktriangle}\}$ which implies $x^{\wedge \blacktriangle} \leq x^{\blacktriangle}$.

Let $a \in \mathcal{A}(B)$ be such that $a \leq x^{\blacktriangle}$. Then, we have $\varphi(a) \wedge x \neq 0$. Then there exists $b \in \mathcal{A}(B)$ such that $b \leq \varphi(a)$ and $b \leq x$. Since there exists $b \in \mathcal{A}(B)$ such that $b \leq \varphi(a)$, we have $b \leq \langle \varphi \rangle (b)$. Also, we have $b \leq x$ implies $\langle \varphi \rangle (b) \wedge x \neq 0$. Then, we have $b \leq x^{\wedge}$. So, $b \leq \varphi(a)$ and $b \leq x^{\wedge}$ implies $\varphi(a) \wedge x^{\wedge} \neq 0$. Then, we have $a \leq x^{\wedge \blacktriangle}$. Thus, $\vee \{a \in \mathcal{A}(B) / a \leq x^{\blacktriangle}\} \leq \vee \{a \in \mathcal{A}(B) / a \leq x^{\wedge \blacktriangle}\}$ which implies $x^{\blacktriangle} \leq x^{\wedge \blacktriangle}$. Hence $x^{\wedge \blacktriangle} = x^{\blacktriangle}$, for all $x \in \mathcal{A}(B)$.

ii) By (i), we have $x^{\wedge \blacktriangle} = x^{\blacktriangle}$, for all $x \in \mathcal{A}(B)$. Thus for x', we have $x'^{\wedge \blacktriangle} = x'^{\blacktriangle}$. By duality of the maps $^{\blacktriangle}$, $^{\blacktriangledown}$ and $^{\wedge}$, $^{\vee}$, we have $x'^{\vee \blacktriangle} = x'^{\blacktriangle} \Rightarrow x^{\vee \blacktriangledown \prime} = x^{\blacktriangledown \prime} \Rightarrow x^{\vee \blacktriangledown} = x^{\blacktriangledown}$. $\qquad \square$

The following lemma shows that the map $\langle \varphi \rangle$ is always closed for any map φ.

Lemma 2. *Let B be a complete atomic Boolean lattice. For any map* $\varphi : \mathcal{A}(B) \to B$, *the induced map* $\langle \varphi \rangle$ *is always closed.*

Proof. Let $a, b \in \mathcal{A}(B)$ be such that $a \leq \langle \varphi \rangle (b)$. Then by definition of $\langle \varphi \rangle$, we have

$$a \leq \varphi(c), \text{ for all } c \in \mathcal{A}(B) \text{ such that } b \leq \varphi(c) \qquad (2.1)$$

Let $x \in \mathcal{A}(B)$ be such that $x \leq \langle \varphi \rangle (a)$. Then by definition of $\langle \varphi \rangle$, we have

$$x \leq \varphi(d), \text{ for all } d \in \mathcal{A}(B) \text{ such that } a \leq \varphi(d) \qquad (2.2)$$

Let $c \in \mathcal{A}(B)$ be such that $b \leq \varphi(c)$. This implies $a \leq \varphi(c)$ (by (2.1)) which implies $x \leq \varphi(c)$ (by (2.2)). Therefore $x \leq \varphi(c)$, for all $c \in \mathcal{A}(B)$ such that $b \leq \varphi(c)$. Then $x \leq \bigwedge_{b \leq \varphi(c)} \varphi(c)$ implies $x \leq \langle \varphi \rangle (b)$. Thus $\{x \in \mathcal{A}(B)/x \leq \langle \varphi \rangle (a)\} \subseteq \{x \in \mathcal{A}(B)/x \leq \langle \varphi \rangle (b)\}$ implies $\vee \{x \in \mathcal{A}(B)/x \leq \langle \varphi \rangle (a)\} \leq \vee \{x \in \mathcal{A}(B)/x \leq \langle \varphi \rangle (b)\}$ which implies $\langle \varphi \rangle (a) \leq \langle \varphi \rangle (b)$. Thus $a \leq \langle \varphi \rangle (b)$ implies $\langle \varphi \rangle (a) \leq \langle \varphi \rangle (b)$, for all $a, b \in \mathcal{A}(B)$. Hence the map $\langle \varphi \rangle$ is closed. $\qquad \square$

Symmetry of φ does not imply symmetry of $\langle \varphi \rangle$. But, obviously the extensiveness of φ implies the extensiveness of $\langle \varphi \rangle$. Example 1 shows that, there exist maps φ which are not extensive, but $\langle \varphi \rangle$ are extensive.

Example 1. Let B = {0, *a*, *b*, 1}, the order ≤ be defined as in fig. 1 and let the map $\varphi : \mathcal{A}(B) \to B$ be defined by $\varphi(a) = b$, $\varphi(b) = a$.

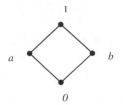

Fig. 1. The ordered set (B, $\overset{\leq}{}$)

Now, $\langle \varphi \rangle (a) = \varphi(b) = a$, $\langle \varphi \rangle (b) = \varphi(a) = b$. We have, $x \leq \langle \varphi \rangle (x)$, for all $x \in \mathcal{A}(B)$. Hence the map $\langle \varphi \rangle$ is extensive, though φ is not extensive. Hence for the extensiveness of the map $\langle \varphi \rangle$, a weaker condition, than the extensive condition on φ is sufficient.

Definition 4. Let (B, ≤) be a complete atomic Boolean lattice. A map $\varphi : \mathcal{A}(B) \to B$ is said to be a cover if $\vee \{ \varphi(a) / a \in \mathcal{A}(B) \} = 1$.

Lemma 3. *Let B be a complete atomic Boolean lattice. Then the following are equivalent:*

  i) φ *is a cover;*
  ii)   for every $a \in \mathcal{A}(B)$, there exists $b \in \mathcal{A}(B)$ such that $a \leq \varphi(b)$;
  iii) $\langle \varphi \rangle$ is extensive.

Proposition 3. *Let B be a complete atomic Boolean lattice. Then the following are equivalent:*

  i) φ *is a cover;*
  ii) $x^{\vee} \leq x$, for all $x \in B$;
  iii) $x \leq x^{\wedge}$, for all $x \in B$.

Proof. The proof follows from the above lemma and proposition 4.2 in [8].   □

Proposition 4. *Let B be a complete atomic Boolean lattice. Then the following are equivalent:*

  i) φ *is a cover;*
  ii) $x^{\vee\vee} = x^{\vee}$, for all $x \in B$;
  iii) $x^{\wedge\wedge} = x^{\wedge}$, for all $x \in B$.

Proof. The proof follows by using lemma 2 and lemma 3 in proposition 4.4 of [8].   □

Corollary 2. *Let B be a complete atomic Boolean lattice. Then the following are equivalent.*

i) φ *is a cover;*

ii) $\check{}: B \to B$ *is a Alexandrov interior operator;*

iii) $\hat{}: B \to B$ *is a Alexandrov closure operator.*

Theorem 3. *Let B be a complete atomic Boolean lattice. Then* $x^{\check{}} \leq x^{\vee} \leq x \leq x^{\wedge} \leq x^{\hat{}}$ *holds for all* $x \in B$ *if and only if* φ *is extensive.*

Proof. Let $a \in \mathcal{A}(B)$ be such that $a \leq x^{\check{}}$. Then we have $\varphi(a) \leq x$. Since $a \leq \varphi(a)$, $\bigwedge\limits_{a \leq \varphi(c)} \varphi(c) \leq \varphi(a) \leq x$. Then $\langle \varphi \rangle(a) \leq x$ implies $a \leq x^{\vee}$. Thus, we have $x^{\check{}} \leq x^{\vee}$, for all $x \in B$. Now for x', we have $x'^{\check{}} \leq x'^{\vee}$. By proposition 3.5(i) of [6] and proposition 1(i), we have $x'^{\check{}} = x^{\hat{}'}$ and $x'^{\vee} = x^{\hat{}'}$, Then $x^{\hat{}'} = x'^{\check{}} < x'^{\vee} = x^{\hat{}'}$ implies $x^{\hat{}} \leq x^{\hat{}}$. Combining proposition 3 with this we have $x^{\check{}} \leq x^{\vee} \leq x \leq x^{\wedge} \leq x^{\hat{}}$, for all $x \in B$. Other part is obvious. □

Example 2. Let us consider a complete atomic Boolean lattice B, the order $\leq$ be defined as in the fig. 2. The map $\varphi: \mathcal{A}(B) \to B$ be defined by $\varphi(a)=c'$, $\varphi(b)=b'$ and $\varphi(c)=b'$. Then obviously φ is a cover but not extensive, for $b \nleq \varphi(b)$.

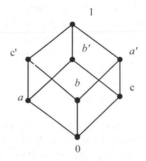

Fig. 2. The ordered set (B, $\leq$)

Here we have that $(b')^{\check{}} \nleq (b')^{\vee}$ and $b^{\hat{}} \nleq b^{\hat{}}$. Thus the chain of inequality in the above theorem does not hold if φ is merely a cover.

Lemma 4. *Let B be a complete atomic Boolean lattice. Then the following holds:*

i)   *If* φ *is extensive and closed, then* $\langle \varphi \rangle = \varphi$.

ii) *If* φ *is symmetric and closed, then* $\langle \varphi \rangle = \varphi$.

Proposition 5. *Let B be a complete atomic Boolean lattice. If* φ *is extensive and closed or* φ *is symmetric and closed, then the rough approximations of the elements of B with respect to* $\langle \varphi \rangle$ *and* φ *are equal.*

Proof. The proof follows from the above lemma. □

The following proposition gives the necessary and sufficient condition for a cover map φ to be equal to its induced map $\langle \varphi \rangle$.

Proposition 6. *Let B be a complete atomic Boolean lattice. Then the map $\varphi: \mathcal{A}(B) \rightarrow B$ is a cover and $\langle \varphi \rangle = \varphi$ if and only if φ is extensive and closed.*

Proof. If φ is extensive and closed, then by lemma 4 $\langle \varphi \rangle = \varphi$. Since φ is extensive, φ is a cover. Conversely, suppose the map $\varphi: \mathcal{A}(B) \rightarrow B$ is a cover and $\langle \varphi \rangle = \varphi$. Then from lemma 3, we have φ is extensive. Also, we have $\langle \varphi \rangle$ is always closed. This implies φ is also closed. Hence φ is extensive and closed.   □

References

[1] Allam, A.A., Bakeir, M.Y., Abo-Tabl, E.A.: New approach for Basic Rough set Concepts. In: Ślęzak, D., Wang, G., Szczuka, M.S., Düntsch, I., Yao, Y. (eds.) RSFDGrC 2005. LNCS (LNAI), vol. 3641, pp. 64–73. Springer, Heidelberg (2005)

[2] Allam, A.A., Bakeir, M.Y., Abo-Tabl, E.A.: Some Methods for Generating Topologies by Relations. Bull. Malays. Math. Sci. Soc. 31(2,1), 35–45 (2008)

[3] Bakhir, M.Y.: On Pre-Granulation (2008) (preprint)

[4] Degang, C.: Rough approximations on a complete completely distributive lattice with applications to generalized rough sets. Information Sciences 176, 1829–1848 (2006)

[5] Gehrke, M., Walker, E.: On the Structure of Rough Sets. Bulletin of the Polish academy of sciences Mathematics 40, 235–245 (1992)

[6] Gratzer, G.: General Lattice Theory. Academic Press, New York (1978)

[7] Jarvinen, J.: On the Structure of Rough Approximations. Fundamenta Informaticae 53, 135–153 (2002)

[8] Jarvinen, J., Kondo, M., Kortelainen, J.: Modal-Like Operators in Boolean Lattices, Galois Connections and Fixed Points. Fundamenta Informaticae 76, 129–145 (2007)

[9] Kalmbach, G.: Orthomodular Lattices. Academic Press Inc., London (1983)

[10] Pawlak, Z.: Rough Sets. International Journal of Computer and Information Sciences 5, 341–356 (1982)

[11] Slowinski, R., Vanderpooten, D.: A generalized definition of rough approximations based on similarity. IEEE Transactions on Knowledge and Data Engineering 12(2), 331–336 (2000)

[12] Szasz, G.: Introduction to Lattice Theory. Academic Press Inc., London (1963)

[13] Zhu, W.: Generalized rough sets based on relations. Information Sciences 177(22), 4997–5011 (2007)

[14] Yao, Y.Y.: Two views of the theory of rough sets in finite universes. International Journal of Approximation Reasoning 15, 291–317 (1996)

[15] Yao, Y.Y.: Relational Interpretations of Neighborhood Operators and Rough Set Approximation Operators. Information Sciences 111(1-4), 239–259 (1998)

[16] Yao, Y.Y.: Constructive and algebraic methods of the theory of rough sets. Information Sciences 109(1-4), 21–47 (1998)

[17] Yao, Y.Y.: On generalizing Pawlak approximation operators. In: Polkowski, L., Skowron, A. (eds.) RSCTC 1998. LNCS (LNAI), vol. 1424, pp. 298–307. Springer, Heidelberg (1998)

Rough Set Approximations Based on Granular Labels

Wei-Zhi Wu

School of Mathematics, Physics and Information Science,
Zhejiang Ocean University, Zhoushan, Zhejiang, 316004, P.R. China
wuwz@zjou.edu.cn

Abstract. In this paper, rough set approximations based on labelled blocks are explored. The concept of labelled blocks determined by a function is first introduced. Lower and upper label-block approximations of sets are then defined. Properties of label-block approximation operators are also examined. Finally, relationship between properties of label-block approximation operators and some essential properties of the corresponding function is characterized.

Keywords: Granular computing, Granules, Labelled blocks, Rough sets.

1 Introduction

Granular computing (GrC) is a basic issue in knowledge representation and data mining. The purpose of GrC is to seek for an approximation scheme which can effectively solve a complex problem, albeit not in the most precise way. Ever since the introduction of the concept of "GrC" [8,22,23], we have witnessed a rapid development of and a fast growing interest in the topic (see e.g. [1,2,5,9,10,12,13,15,16,17,18,19,20,21]).

A granule is a primitive notion in GrC which is a clump of objects (points) drawn together by the criteria of indistinguishability, similarity or functionality [23]. A granule may be interpreted as one of the numerous small particles forming a larger unit. Alternatively, a granule may be considered as a localized view or a specific aspect of a large unit satisfying a given specification. The set of granules provides a representation of the unit with respect to a particular level of granularity. The process of constructing information granules is called information granulation. It granulates a universe of discourse into a family of disjoint or overlapping granules. Thus one of main directions in the study of GrC is to deal with the construction, interpretation, and representation of granules.

Many models and methods of GrC concentrating on concrete models in special contexts have been proposed and studied over the years. Rough set theory is one of the most advanced areas popularizing GrC [5,6,7,9,14,18,19,20,21]. Rough set theory was originally proposed by Pawlak [11] as a formal tool for modelling and processing incomplete information. The basic structure of rough set theory is an approximation space consisting of a universe of discourse and a binary relation

H. Sakai et al. (Eds.): RSFDGrC 2009, LNAI 5908, pp. 93–100, 2009.

imposed on it. Based on the approximation space, the notions of lower and upper approximation operators can be constructed. This model is very useful in the analysis of data in complete information systems/tables [11,24]. The equivalence relation in the Pawlak's rough set model groups together entities which are in some sense indiscernible or similar called equivalence classes. Thus equivalence classes are the basic building blocks for the representation and approximation of any subset of the universe. Based on this observation and by employing the notion of labelled partition, Bittner and Smith [3] proposed the concept of a granular partition. A granular partition can be seen as an extension of the concept of equivalence relation. In [4], Bittner and Stell showed how the technique of making rough descriptions of a subset with respect to an equivalence relation can be extended to descriptions with respect to a granular partition.

A labelled partition of a universe of discourse is a surjective function from the universe to a labelled set. We observe that each attribute in a complete information system [11,24] can be taken as a labelled partition. In this paper, we propose a concept of block-labelled rough set. We introduce the notion of labelled blocks, define rough set approximations based on labelled blocks, and examine their properties.

2   Functions and Labelled Blocks

Let U be a nonempty set, the class of all subsets of U will be denoted by $\mathcal{P}(U)$. For $X \in \mathcal{P}(U)$, we denote by $\sim X$ the complement of X in U.

Let U denote a nonempty set of objects called the universe of discourse and $R \subseteq U \times U$ an equivalence binary relation on U. The equivalence relation R partitions the universe U into disjoint subsets. The equivalence classes in Pawlak's rough set model provide the basis of "information granules" for database analysis. It is well known that there exist a one-to-one mapping between the set of equivalence relations on U and the set of partitions of U. Partitions of the set U are often identified with functions of the form $f : U \rightarrow V$ which are surjective, that is, for each $v \in V$, there exists some $u \in U$ such that $u = f(v)$.

Let U and V be two nonempty sets and consider a function $f : U \rightarrow V$, denote $f^{-1}(v) = \{u \in U | f(u) = v\}$, then we can see that, for $v_1, v_2 \in V$,

$$v_1 \neq v_2 \Longrightarrow f^{-1}(v_1) \cap f^{-1}(v_2) = \emptyset \tag{1}$$

and

$$\bigcup_{v \in V} f^{-1}(v) = U. \tag{2}$$

If f is surjective, then $\{f^{-1}(v) | v \in V\}$ forms a partition of U.

For $u \in U$, denote $[u]_f = \{x \in U | f(x) = f(u)\}$, it is easy to observe that $\{[u]_f | u \in U\}$ is a partition of U. Moreover, if $v = f(u)$, then $[u]_f = f^{-1}(v)$, in such a case, we say that $[u]_f$ is a block with the label v and we call that the pair $([u]_f, v) = (f^{-1}(v), v)$ is a labelled block induced from the function $f : U \rightarrow V$. In [4], a surjective function from U to V is called a V-labelled partition of U.

One example of class of labelled blocks are maps in the cartographic rather than the mathematical sense, in which a block is the location of a region whereas the label is the name of the same region [4]. Labelled blocks can also be employed to represent the granular information of the object-attribute values in an information system [11,24]. Consider an information system (U, A, F) in which U is a nonempty finite set of objects called the universe of discourse, A is a nonempty finite set of attribute such that $f_a : U \rightarrow V_a$ for any $a \in A$, where $V_a = \{f_a(u)|u \in U\}$ is the domain of attribute a, and $F = \{f_a|a \in A\}$ is the set of information functions. Then for each attribute $a \in A$, f_a is a surjective function from U to V_a, $([u]_{f_a}, f_a(u))$ is a labelled block in (U, A, F), the sets of all labelled blocks reflect the information granules in the information system.

3 Block-Label Approximations of Sets

In this section, we defined two pairs of lower and upper approximations of a set, one is in the sense of Pawlak which is constructed by blocks, and the other is determined by labels.

Definition 1. *Let U and V be two nonempty sets and $f : U \rightarrow V$ a function from U to V. For $X \in \mathcal{P}(U)$, a pair of lower and upper block approximations, denoted as $\underline{f}(X)$ and $\overline{f}(X)$, are subsets of U and are defined, respectively, as follows*

$$\underline{f}(X) = \{u \in U | [u]_f \subseteq X\}, \quad \overline{f}(X) = \{u \in U | [u]_f \cap X \neq \emptyset\}. \tag{3}$$

$(\underline{f}(X), \overline{f}(X))$ *is referred to as the block rough set of X with respect to (U, V, f), and $\underline{f}$ and $\overline{f} : \mathcal{P}(U) \rightarrow \mathcal{P}(U)$ are, respectively, called the lower and upper block approximation operators. The lower and upper label approximations of X with respect to (U, V, f), denoted as $\underline{L}(X)$ and $\overline{L}(X)$, are subsets of V and are, respectively, defined as follows*

$$\underline{L}(X) = \{v \in V | f^{-1}(v) \subseteq X\}, \quad \overline{L}(X) = \{v \in V | f^{-1}(v) \cap X \neq \emptyset\}. \tag{4}$$

$(\underline{L}(X), \overline{L}(X))$ *is referred to as the label rough set of X with respect to (U, V, f), and $\underline{L}$ and $\overline{L} : \mathcal{P}(U) \rightarrow \mathcal{P}(V)$ are, respectively, called the lower and upper label approximation operators. We call $((\underline{f}(X), \overline{f}(X)), (\underline{L}(X), \overline{L}(X)))$ the block-label rough set of X with respect to (U, V, f).*

According to Eq. (3), the lower block approximation, $\underline{f}(X)$, of X with respect (U, V, f) is the collection of those objects which can be classified with full certainty as elements of X in the available knowledge (U, V, f), whereas the upper block approximation $\overline{f}(X)$ is the collection of objects which can be possibly classified as elements of X using the available knowledge (U, V, f). Since the set of blocks, $\{[u]_f|u \in U\}$, is a partition of U, it can yield an equivalence binary relation R_f on U, then by the definition, we can see that the pair of block approximations are exactly set approximations in the sense of Pawlak [11], i.e.,

$$\underline{f}(X) = \cup\{[u]_f|[u]_f \subseteq X\}, \quad \overline{f}(X) = \cup\{[u]_f|[u]_f \cap X \neq \emptyset\}. \tag{5}$$

Hence, the lower block approximation $\underline{f}(X)$ is the union of blocks which are subsets of X and the upper block approximation $\overline{f}(X)$ is the union of blocks which have a nonempty intersection with X.

By Eq. (4), the lower label approximation of X with respect (U, V, f) is the collection of those labels whose corresponding blocks can be classified with full certainty as elements of X in the available knowledge (U, V, f), whereas the upper label approximation is the collection of labels whose corresponding blocks can be possibly classified as elements of X using the available knowledge (U, V, f). Therefore, the block-label approximations have more semanteme than the Pawlak approximations.

4   Properties of Block-Label Approximations

Since the block approximations are the same as the Pawlak approximations, the block approximation operators satisfy the properties of Pawlak approximation operators and we summarize as following

Theorem 1. [11] *Let U and V be two nonempty sets and $f : U \to V$ a function from U to V. Then the lower and upper block approximation operators defined in Eq. (3) satisfy the following properties:* $\forall X, Y, X_i \in \mathcal{P}(U)$, $i \in I$, I *is an index set,*

(BL0) $\underline{f}(X) = \sim \overline{f}(\sim X)$,   (BU0) $\overline{f}(X) = \sim \underline{f}(\sim X)$;
(BL1) $\underline{f}(U) = U$,   (BU1) $\overline{f}(\emptyset) = \emptyset$;
(BL2) $\underline{f}(\bigcap_{i \in I} X_i) = \bigcap_{i \in I} \underline{f}(X_i)$,   (BU2) $\overline{f}(\bigcup_{i \in I} X_i) = \bigcup_{i \in I} \overline{f}(X_i)$;
(BL3) $X \subseteq Y \Longrightarrow \underline{f}(X) \subseteq \underline{f}(Y)$, (BU3) $X \subseteq Y \Longrightarrow \overline{f}(X) \subseteq \overline{f}(Y)$;
(BL4) $\underline{f}(\bigcup_{i \in I} X_i) \supseteq \bigcup_{i \in I} \underline{f}(X_i)$,   (BU4) $\overline{f}(\bigcap_{i \in I} X_i) \subseteq \bigcap_{i \in I} \overline{f}(X_i)$;
(BL5) $\underline{f}(X) \subseteq X$,   (BU5) $X \subseteq \overline{f}(X)$;
(BL6) $\underline{f}(\emptyset) = \emptyset$,   (BU6) $\overline{f}(U) = U$;
(BL7) $\underline{f}(X) = \underline{f}(\underline{f}(X))$,   (BU7) $\overline{f}(\overline{f}(X)) = \overline{f}(X)$;
(BL8) $\overline{\overline{f}}(X) = \underline{f}(\overline{f}(X))$,   (BU8) $\overline{f}(\underline{f}(X)) = \underline{f}(X)$;
(BL9) $\overline{f}(\underline{f}(X)) \subseteq X$,   (BU9) $X \subseteq \underline{f}(\overline{f}(X))$.

The next theorem presents the basic properties of label approximation operators.

Theorem 2. *Let U and V be two nonempty sets and $f : U \to V$ a function from U to V. Then the lower and upper label approximation operators defined in Eq. (4) satisfy the following properties:* $\forall X, Y, X_i \in \mathcal{P}(U)$, $i \in I$, I *is an index set,*

(LL0) $\underline{L}(X) = \sim \overline{L}(\sim X)$,   (LU0) $\overline{L}(X) = \sim \underline{L}(\sim X)$;
(LL1) $\underline{L}(U) = V$,   (LU1) $\overline{L}(\emptyset) = \emptyset$;
(LL2) $\underline{L}(\bigcap_{i \in I} X_i) = \bigcap_{i \in I} \underline{L}(X_i)$, (LU2) $\overline{L}(\bigcup_{i \in I} X_i) = \bigcup_{i \in I} \overline{L}(X_i)$;

(LL3) $X \subseteq Y \Longrightarrow \underline{L}(X) \subseteq \underline{L}(Y)$, (LU3) $X \subseteq Y \Longrightarrow \overline{L}(X) \subseteq \overline{L}(Y)$;

(LL4) $\underline{L}(\bigcup_{i \in I} X_i) \supseteq \bigcup_{i \in I} \underline{L}(X_i)$,   (LU4) $\overline{L}(\bigcap_{i \in I} X_i) \subseteq \bigcap_{i \in I} \overline{L}(X_i)$.

Proof. It is directly follows from Eq. (4).

For $X \in \mathcal{P}(U)$ and $W \in \mathcal{P}(V)$, denote

$$f(X) = \{f(x) | x \in X\}, \quad f^{-1}(W) = \{x \in U | f(x) \in W\} = \bigcup_{w \in W} \{f^{-1}(w)\}. \quad (6)$$

It can easily be verified that

$$f^{-1}(f(X)) \supset X \quad \forall X \in \mathcal{P}(U), \quad (7)$$

and

$$f(f^{-1}(W)) = W \quad \forall W \in \mathcal{P}(V). \quad (8)$$

By Eq. (4), we can easily conclude that

$$\overline{L}(\{u\}) = \{f(u)\}, \quad u \in U. \quad (9)$$

Then, according to property (LU2), we have

$$\overline{L}(X) = \bigcup_{u \in X} \{f(u)\} = f(X), \quad X \in \mathcal{P}(U). \quad (10)$$

Theorem 3 below shows the relationship between the two types of approximations defined in Definition 1.

Theorem 3. *Let U and V be two nonempty sets and $f : U \to V$ a function from U to V. Then, for $X \in \mathcal{P}(U)$,*

$$f^{-1}(\overline{L}(X)) = \overline{f}(X), \quad f(\overline{f}(X)) = \overline{L}(X), \quad (11)$$

and

$$f^{-1}(\underline{L}(X)) = \underline{f}(X), \quad f(\underline{f}(X)) = \underline{L}(X). \quad (12)$$

Proof. For any $u \in f^{-1}(\overline{L}(X))$, let $v = f(u)$, then $f^{-1}(v) = [u]_f$. Since $f(u) = v \in \overline{F}(X)$, we have $f^{-1}(v) \cap X \neq \emptyset$, that is, $[u]_f \cap X \neq \emptyset$. Hence $u \in \overline{f}(X)$. It follows that

$$f^{-1}(\overline{L}(X)) \subseteq \overline{f}(X). \quad (13)$$

On the other hand, for any $x \in \overline{f}(X)$, let $w = f(x)$, then $[x]_f = f^{-1}(w)$. From $x \in \overline{f}(X)$, we have $[x]_f \cap X \neq \emptyset$, that is, $f^{-1}(w) \cap X \neq \emptyset$, and in turn, $w \in \overline{L}(X)$, hence $f(x) \in \overline{L}(X)$. It follows that $x \in f^{-1}(\overline{L}(X))$. Therefore,

$$\overline{f}(X) \subseteq f^{-1}(\overline{L}(X)). \quad (14)$$

Combining Eqs. (13) and (14), we conclude $f^{-1}(\overline{L}(X)) = \overline{f}(X)$. Furthermore, by Eq. (8), we then obtain $f(\overline{f}(X)) = \overline{L}(X)$.

Similarly, we can prove Eq. (12).

The following Theorems 4 and 5 show that some properties of a function can be characterized by the properties of label approximation operators.

Theorem 4. *Let U and V be two nonempty sets and $f : U \to V$ a function from U to V. Then the following statements are equivalent:*

(1) $f : U \to V$ *is surjective;*
(2) $\underline{L}(X) \subseteq \overline{L}(X)$ $\forall X \in \mathcal{P}(U)$;
(3) $\underline{L}(\emptyset) = \emptyset$;
(4) $\overline{L}(U) = V$.

Proof

"(1) $\Rightarrow$ (2)" For any $X \in \mathcal{P}(U)$ and $v \in \underline{L}(X)$, by Eq. (4), we have $f^{-1}(v) \subseteq X$. Since f is surjective, we see that $f^{-1}(v) \neq \emptyset$. Then $f^{-1}(v) \cap X \neq \emptyset$, hence $v \in \overline{L}(X)$. Thus we have proved that $\underline{L}(X) \subseteq \overline{L}(X)$.

"(2) $\Rightarrow$ (1)" If $f : U \to W$ is not surjective, there exists a $v \in V$ such that $f^{-1}(v) = \emptyset$. Then for any $X \in \mathcal{P}(U)$, we have $f^{-1}(v) \subseteq X$, by Eq. (4), $x \in \underline{L}(X)$. However, $f^{-1}(v) \cap X = \emptyset$, that is, $v \notin \overline{L}(X)$, which contradicts that $\underline{L}(X) \subseteq \overline{L}(X)$. So $f : U \to V$ is surjective.

"(2) $\Leftrightarrow$ (3) $\Leftrightarrow$ (4)" For any $X \in \mathcal{P}(U)$, in terms of Theorem 2, we have

$$
\begin{aligned}
\underline{L}(X) \subseteq \overline{L}(X) &\iff \underline{L}(X) \cap (\sim \overline{L}(X)) = \emptyset \\
&\iff \underline{L}(X) \cap \underline{L}(\sim X) = \emptyset \\
&\iff \underline{L}(X \cap (\sim X)) = \emptyset \\
&\iff \underline{L}(\emptyset) = \emptyset \\
&\iff \underline{L}(\sim U) = \sim V \\
&\iff \sim \underline{L}(\sim U) = V \\
&\iff \overline{L}(U) = V.
\end{aligned}
$$

Theorem 5. *Let U and V be two nonempty sets and $f : U \to V$ a function from U to V. Then f is injective and surjective (that is, one-to-one) if and only if*

$$\underline{L}(X) = \overline{L}(X) \quad \forall X \in \mathcal{P}(U). \tag{15}$$

Proof. "$\Rightarrow$" Assume that f is one-to-one. For any $X \in \mathcal{P}(U)$ and $v \in \underline{L}(X)$, by Eq. (4), we have $f^{-1}(v) \subseteq X$. Since f is surjective, by Theorem 4, we conclude

$$\underline{L}(X) \subseteq \overline{L}(X). \tag{16}$$

On the other hand, for any $w \in \overline{L}(X)$, by definition, $f^{-1}(w) \cap X \neq \emptyset$. Notice that f is one-to-one, then there exists a unique $u \in U$ such that $f(u) = w$, that is, $f^{-1}(w) = \{u\}$. Hence $\{u\} \cap X \neq \emptyset$, consequently, $u \in X$, and in turn, $f^{-1}(w) \subseteq X$, by the definition of lower label approximation, we have $w \in \underline{L}(X)$, therefore,

$$\overline{L}(X) \subseteq \underline{L}(X). \tag{17}$$

Combining Eqs. (16) and (17), we conclude Eq. (15).

"$\Leftarrow$" Assume that Eq. (15) holds. By property (LU1) in Theorem 2, we see that $\overline{L}(\emptyset) = \emptyset$, then, by the assumption, we have $\underline{L}(\emptyset) = \emptyset$. According to Theorem 4, we then conclude that f is surjective.

Now we are to prove that f is injective. If f is not injective, then there exists $x_1, x_2 \in U$ and $z \in V$ such that $x_1 \neq x_2$ and $f(x_1) = f(x_2) = z$. Let $X = \{x_1\}$, obviously, $x_1 \in f^{-1}(z)$, so $f^{-1}(z) \cap \{x_1\} \neq \emptyset$. Hence, by Eq. (9), $\overline{L}(X) = \{f(x_1)\} = \{z\}$. On the other hand, notice that $\{x_1, x_2\} \subseteq f^{-1}(z)$, thus $f^{-1}(z) \subseteq \{x_1\}$ does not hold, alternatively, $z \notin \underline{L}(X)$ which contradicts Eq. (15). Therefore, f is injective.

5 Conclusion

In this paper, by using labelled blocks determined by a function, we have developed a new rough set model called block-labelled rough set model. A block-labelled rough set includes two pairs of lower and upper approximations: one is the lower and upper block approximations, and the other is the lower and upper label approximations. Alternatively, a block-labelled rough set include two mechanisms of rough approximation schemes of set, one is represented by the blocks which is exactly the Pawlak rough set, the other is determined by the labels related to the blocks. Thus a block-labelled rough set has more semanteme or physical meanings than a Pawlak rough set. This model provides a new approach to describe information granules. We have also examined properties of the proposed approximation operators. We have further presented the relationship between the two pairs of lower and upper approximations. Since the labelled blocks are induced by a function, at the same time, we have employed the properties of label approximation operators to characterize properties of the function. For further study, we will investigate block-labelled rough sets as well as granular computing in complicated information systems.

Acknowledgement

This work was supported by grants from the National Natural Science Foundation of China (Nos. 60673096 and 60773174), and the Natural Science Foundation of Zhejiang Province (No. Y107262).

References

1. Bargiela, A., Pedrycz, W.: Granular Computing: An Introduction. Kluwer Academic Publishers, Boston (2002)
2. Bargiela, A., Pedrycz, W.: Toward a theory of granular computing for human-centered information processing. IEEE Transactions on Fuzzy Systems 16, 320–330 (2008)
3. Bittner, T., Smith, B.: A theory of granular partitions. In: Duckham, M., Goodchild, M.F., Worboys, M.F. (eds.) Foundations of Geographic Information Science, pp. 117–151. Taylor & Francis, Abington (2003)
4. Bittner, T., Stell, J.: Stratified rough sets and vagueness. In: Kuhn, W., Worboys, M.F., Timpf, S. (eds.) COSIT 2003. LNCS, vol. 2825, pp. 270–286. Springer, Heidelberg (2003)

5. Inuiguchi, M., Hirano, S., Tsumoto, S. (eds.): Rough Set Theory and Granular Computing. Springer, Berlin (2002)
6. Leung, Y., Li, D.: Maximal consistent block technique for rule acquisition in incomplete information systems. Information Sciences 153, 85–106 (2003)
7. Leung, Y., Wu, W.-Z., Zhang, W.-X.: Knowledge acquisition in incomplete information systems: A rough set approach. European Journal of Operational Research 168, 164–180 (2006)
8. Lin, T.Y.: Granular computing: From rough sets and neighborhood systems to information granulation and computing with words. In: European Congress on Intelligent Techniques and Soft Computing, September 8-12, pp. 602–1606 (1997)
9. Lin, T.Y., Yao, Y.Y., Zadeh, L.A. (eds.): Data Mining, Rough Sets and Granular Computing. Physica-Verlag, Heidelberg (2002)
10. Ma, J.-M., Zhang, W.-X., Leung, Y., Song, X.-X.: Granular computing and dual Galois connection. Information Sciences 177, 5365–5377 (2007)
11. Pawlak, Z.: Rough Sets: Theoretical Aspects of Reasoning about Data. Kluwer Academic Publishers, Boston (1991)
12. Pedrycz, W. (ed.): Granular Computing: An Emerging Paradigm. Physica-Verlag, Heidelberg (2001)
13. Pedrycz, W., Skowron, A., Kreinovich, V. (eds.): Handbook of Granular Computing. Wiley-Interscience, Hoboken (2008)
14. Qian, Y.H., Liang, J.Y., Dang, C.Y.: Knowledge structure, knowledge granulation and knowledge distance in a knowledge base. International Journal of Approximate Reasoning 50, 174–188 (2009)
15. Skowron, A., Stepaniuk, J.: Information granules: Towards foundations of granular computing. International Journal of Intelligent Systems 16, 57–85 (2001)
16. Wu, W.-Z., Leung, Y., Mi, J.-S.: Granular computing and knowledge reduction in formal contexts. IEEE Transactions on Knowledge and Data Engineering 21, 1461–1474 (2009)
17. Yager, R.R.: Intelligent social network analysis using granular computing. International Journal of Intelligent Systems 23, 1196–1219 (2008)
18. Yao, J.T.: Recent developments in granular computing: A bibliometrics study. In: Proceedings of IEEE International Conference on Granular Computing, Hangzhou, China, August 26–28, pp. 74–79 (2008)
19. Yao, Y.Y.: Information granulation and rough set approximation. International Journal of Intelligent Systems 16, 87–104 (2001)
20. Yao, Y.Y.: A partition model of granular computing. In: Peters, J.F., Skowron, A., Grzymała-Busse, J.W., Kostek, B.z., Świniarski, R.W., Szczuka, M.S. (eds.) Transactions on Rough Sets I. LNCS, vol. 3100, pp. 232–253. Springer, Heidelberg (2004)
21. Yao, Y.Y., Liau, C.-J., Zhong, N.: Granular computing based on rough sets, quotient space theory, and belief functions. In: Zhong, N., Raś, Z.W., Tsumoto, S., Suzuki, E. (eds.) ISMIS 2003. LNCS (LNAI), vol. 2871, pp. 152–159. Springer, Heidelberg (2003)
22. Zadeh, L.A.: Fuzzy sets and information granularity. In: Gupta, N., Ragade, R., Yager, R.R. (eds.) Advances in Fuzzy Set Theory and Applications, pp. 3–18. North-Holland, Amsterdam (1979)
23. Zadeh, L.A.: Towards a theory of fuzzy information granulation and its centrality in human reasoning and fuzzy logic. Fuzzy Sets and Systems 90, 111–127 (1997)
24. Zhang, W.-X., Leung, Y., Wu, W.-Z.: Information Systems and Knowledge Discovery (in Chinese). Science Press, Beijing (2003)

On a Criterion of Similarity between Partitions Based on Rough Set Theory

Yasuo Kudo[1] and Tetsuya Murai[2]

[1] College of Information and Systems, Muroran Institute of Technology
Mizumoto 27-1, Muroran 050-8585, Japan
kudo@csse.muroran-it.ac.jp
[2] Graduate School of Information Science and Technology, Hokkaido University
Kita 14, Nishi 9, Kita-ku, Sapporo 060-0814, Japan
murahiko@main.ist.hokudai.ac.jp

Abstract. In this paper, we introduce a criterion of similarity between partitions. The proposed similarity criterion is a generalization of an evaluation criterion of relative reducts proposed by the authors and evaluates the similarity of partitions by correctness and roughness with each other. Moreover, for comparison of similarity scores between different universes, we also propose a normalized similarity criterion.

1 Introduction

Constructing and evaluating partitions of the given universe are the most basic and important concepts in rough set theory proposed by Pawlak [5,6]. In the aspect of approximation in rough set theory, lower and upper approximations are directly based on partitions on the given universe constructed by equivalence relations. In the aspect of reasoning about data in rough set theory, calculation of relative reducts is one of the most important concepts, which corresponds indirectly to generating partitions that reproduce the positive region of decision classes, i. e., lower approximations of decision classes based on the most finest partition of the universe constructed from all condition attributes. The authors have proposed an evaluation criterion of relative reducts based on roughness of partitions constructed from the relative reducts [3,4].

In this paper, we introduce a criterion of similarity between partitions. The proposed similarity criterion is a generalization of the evaluation criterion of relative reducts [3,4] and evaluates the similarity of partitions by correctness and roughness with each other. Moreover, for comparison of similarity scores between different universes, we also propose a normalized similarity criterion.

2 Rough Set

We review the foundations of rough set theory based on mainly [8].

A decision table $DT = (U, C, d)$ is a triple, where U is a finite and non-empty set (called a universe) of objects, C is a set of condition attributes such that

H. Sakai et al. (Eds.): RSFDGrC 2009, LNAI 5908, pp. 101–108, 2009.

Table 1. An example of decision table

U	c_1	c_2	c_3	c_4	c_5	c_6	c_7	c_8	d
x_1	1	1	1	1	1	1	2	1	1
x_2	2	3	1	3	1	2	2	1	3
x_3	3	2	3	2	1	2	1	1	2
x_4	4	2	2	2	2	2	1	1	2
x_5	5	2	2	3	1	1	2	1	1
x_6	6	3	2	1	1	2	2	1	3
x_7	7	1	1	1	1	2	1	1	2
x_8	8	2	3	1	1	1	2	1	1
x_9	9	3	3	3	2	2	2	1	3
x_{10}	10	1	3	3	1	1	2	1	1

each attribute $a \in C$ is a function $a : U \to V_a$ from U to the value set V_a of a, and d is a function $d : U \to V_d$ called the decision attribute.

The indiscernibility relation R_B on U with respect to $B \subseteq C$ is defined by

$$x R_B y \iff a(x) = a(y), \ \forall a \in B. \tag{1}$$

The equivalence class $[x]_B$ of $x \in U$ by R_B is the set of objects which are not discernible with x even though using all attributes in B. Any indiscernibility relation provides a partition of U. In particular, the partition $\mathcal{D} = \{D_1, \cdots, D_m\}$ provided by the indiscernibility relation R_d based on the decision attribute d is called the set of decision classes.

For any decision class D_i $(1 \le i \le m)$C the lower approximation $\underline{B}(D_i)$ and the upper approximation $\overline{B}(D_i)$ of D_i based on R_B are defined by

$$\underline{B}(D_i) = \{x \in U \mid [x]_B \subseteq D_i\}, \tag{2}$$

$$\overline{B}(D_i) = \{x \in U \mid [x]_B \cap D_i \ne \emptyset\}. \tag{3}$$

Table 1 presents an example of a decision table that consists of a set of objects $U = \{x_1, \cdots, x_{10}\}$, a set of condition attributes $C = \{c_1, \cdots, c_8\}$, and a decision attribute d. For example, a condition attribute $c_2 \in C$ is a function $c_2 : U \to \{1, 2, 3\}$, and the value of an object $x_3 \in U$ at c_3 is 3, i. e., $c_3(x_3) = 3$. The decision attributed d provides the following three decision classes, $D_1 = \{x_1, x_5, x_8, x_{10}\}$, $D_2 = \{x_3, x_4, x_7\}$ and $D_3 = \{x_2, x_6, x_9\}$.

In this paper, we denote a decision rule constructed from a subset $B \subseteq C$ of condition attribute, the decision attribute d, and an object $x \in U$ by $(B, x) \to (d, x)$. Certainty and coverage are well known criteria for evaluating decision rules. For any decision rule $(B, x) \to (d, x)$, the certainty $Cer(\cdot)$ and the coverage $Cov(\cdot)$ of the decision rule are defined by

$$Cer((B, x) \to (d, x)) = \frac{|[x]_B \cap D_i|}{|[x]_B|}, \tag{4}$$

$$Cov((B, x) \to (d, x)) = \frac{|[x]_B \cap D_i|}{|D_i|}, \tag{5}$$

where D_i is the decision class of x and $|X|$ is the cardinality of the set X.

Relative reducts are minimal subsets of condition attributes that provide the same positive region based on the set C of all condition attributes. Formally, a relative reduct for the partition $\mathcal{D}$ is a set of condition attributes $A \subseteq C$ that satisfies the following two conditions:

1. $Pos_A(\mathcal{D}) = Pos_C(\mathcal{D})$,
2. For any proper subset $B \subset A$, $Pos_B(\mathcal{D}) \neq Pos_C(\mathcal{D})$,

where $Pos_X(\mathcal{D}) \stackrel{\text{def}}{=} \bigcup_{D_i \in \mathcal{D}} \underline{X}(D_i)$ is the positive region of decision classes based on the partition constructed from $X \subseteq C$.

For example, there are the following six relative reducts of Table 1: $\{c_1\}$, $\{c_2, c_6\}$, $\{c_2, c_7\}$, $\{c_6, c_7\}$, $\{c_3, c_4, c_6\}$, and $\{c_3, c_4, c_5, c_7\}$.

3   Evaluation of Relative Reducts Using Partitions

In this section, we review an evaluation method of relative reducts by using partitions constructed from the relative reducts [3,4].

We intend that rougher partitions constructed from relative reducts lead to better evaluation of the relative reducts. From the viewpoint of rule generation, rougher partitions constructed from relative reducts tend to generate decision rules with higher coverage values rather than finer partitions. Following this intention, we consider evaluating relative reducts by using the coverage of decision rules constructed from relative reducts.

Theorem 1 below provides a theoretical basis of our intention.

Theorem 1 ([3,4]). *For any non-empty subset $B \subseteq C$ of condition attributes, the average certainty value $ACer(B)$ and the average coverage value $ACov(B)$ of all decision rules $(B, x) \to (d, x)$ $(\forall x \in U)$ constructed from B are calculated by the following equations:*

$$ACer(B) = \frac{|U/R_B|}{\sum_{[x]_B \in U/R_B} |\{D_i \in \mathcal{D} \mid D_i \cap [x]_B \neq \emptyset\}|}, \qquad (6)$$

$$ACov(B) = \frac{|\mathcal{D}|}{\sum_{[x]_B \in U/R_B} |\{D_i \in \mathcal{D} \mid D_i \cap [x]_B \neq \emptyset\}|}. \qquad (7)$$

Note that the denominators of (6) and (7) correspond to the number of decision rules constructed from B.

In Theorem 1, if we use relative reducts as subsets of condition attributes in the given decision table, it is clear that the smaller the number of equivalence classes constructed from the relative reduct, the higher the average coverage value of decision rules generated from the relative reduct. This indicates a possibility of using the average coverage of decision rules constructed from relative reducts as an evaluation criterion for relative reducts [3,4].

Example 1. Let $\{c_2, c_6\}$ be a relative reduct of Table 1. We can construct the following five decision rules from $\{c_2, c_6\}$:

Table 2. The average certainty and the average coverage of decision rules based on relative reducts in Table 1

Relative reduct	Number of rules	Average coverage
$\{c_1\}$	10	0.3
$\{c_2, c_6\}$	5	0.6
$\{c_2, c_7\}$	5	0.6
$\{c_6, c_7\}$	3	1
$\{c_3, c_4, c_6\}$	10	0.3
$\{c_3, c_4, c_5, c_7\}$	10	0.3

- $(c_2 = 1) \wedge (c_6 = 1) \rightarrow (d = 1)$, Certainty= 1, Coverage= 1/2,
- $(c_2 = 2) \wedge (c_6 = 1) \rightarrow (d = 1)$, Certainty= 1, Coverage= 1/2,
- $(c_2 = 1) \wedge (c_6 = 2) \rightarrow (d = 2)$, Certainty= 1, Coverage= 1/3,
- $(c_2 = 2) \wedge (c_6 = 2) \rightarrow (d = 2)$, Certainty= 1, Coverage= 2/3,
- $(c_2 = 3) \wedge (c_6 = 2) \rightarrow (d = 3)$, Certainty= 1, Coverage= 1.

The average coverage is $(\frac{1}{2} + \frac{1}{2} + \frac{1}{3} + \frac{2}{3} + 1)/5 = \frac{3}{5}$ and it is equal to the value "the number of decision classes / the number of decision rules" by Theorem 1. Thus, we get the evaluation score $\frac{3}{5}$ of the relative reduct $\{c_2, c_6\}$. Table 2 shows the number of decision rules and the average coverage of the decision rules by each relative reduct. By this result, we regard the relative reduct $\{c_6, c_7\}$ as the best one that provides the roughest and most correct approximation of decision classes. Actually, the partition constructed from the relative reduct $\{c_6, c_7\}$ is identical to the set of all decision classes in Table 1.

4   A Criterion of Similarity between Partitions

Equations (6) and (7) are based on comparison of numbers of elements in two partitions. Thus, as a generalization of (6) and (7), we then introduce a criterion of similarity between partitions.

Let $U(\neq \emptyset)$ be a finite set. A partition $\mathcal{X}$ of U is a collection of subsets of U that satisfies the following properties:

1. $X_i \cap X_j = \emptyset$ for every disjoint $X_i, X_j \in \mathcal{X}$,
2. $U = \bigcup_{X \in \mathcal{X}} X$.

Let $\mathcal{X}$ and $\mathcal{Y}$ be any partitions on U. We say that $\mathcal{X}$ is a refinement of $\mathcal{Y}$ if and only if, for every $X \in \mathcal{X}$, there exists $Y \in \mathcal{Y}$ such that $X \subseteq Y$. Clearly, $|\mathcal{Y}| \leq |\mathcal{X}|$ holds if $\mathcal{X}$ is a refinement of $\mathcal{Y}$, and both $\mathcal{X}$ is a refinement of $\mathcal{Y}$ and $\mathcal{Y}$ is a refinement of $\mathcal{X}$ hold if and only if $\mathcal{X} = \mathcal{Y}$ holds.

We define the intersection $\mathcal{X} \cap \mathcal{Y}$ of $\mathcal{X}$ and $\mathcal{Y}$ by

$$\mathcal{X} \cap \mathcal{Y} \overset{\text{def}}{=} \{X \cap Y \mid X \in \mathcal{X}, Y \in \mathcal{Y}, X \cap Y \neq \emptyset\}. \tag{8}$$

It is easy to confirm that $\mathcal{X} \cap \mathcal{Y}$ is also a partition on U and $\mathcal{X} \cap \mathcal{Y}$ is a refinement of both $\mathcal{X}$ and $\mathcal{Y}$.

From the viewpoint of identifying functional dependency by using partitions, it is known that the number of elements in the intersection $\mathcal{X} \cap \mathcal{Y}$ satisfies the following inequality (e.g. [1,2]):

$$\max\left(|\mathcal{X}|, |\mathcal{Y}|\right) \leq |\mathcal{X} \cap \mathcal{Y}| \leq \min\left(|\mathcal{X}||\mathcal{Y}|, |U|\right). \tag{9}$$

This inequality indicates that $\mathcal{X}$ is a refinement of $\mathcal{Y}$ if and only if $|\mathcal{X}| - |\mathcal{X} \cap \mathcal{Y}|$.

From the viewpoint of comparison between two partitions, for each partition U/R_B constructed from a set B of condition attributes in a given decision table, we can consider that the evaluation score $ACov(B)$ of B defined by (7) compares "similarity" of the intersection $\mathcal{D} \cap U/R_B$ with respect to $\mathcal{D}$ in the sense of cardinality, and provides the highest score $ACov(B) = 1$ to B if and only if $|\mathcal{D} \cap U/R_B| = |\mathcal{D}|$, i. e., $\mathcal{D}$ is a refinement of U/R_B, and $\mathcal{D} \cap U/R_B$ is the most "similar" partition with respect to $\mathcal{D}$ in the sense of cardinality. On the other hand, the evaluation score $ACer(B)$ of B defined by (6) becomes the highest score $ACer(B) = 1$ when the partition U/R_B is a refinement of $\mathcal{D}$, that is, the intersection $\mathcal{D} \cap U/R_B$ is the most "similar" partition with respect to the partition U/R_B in the sense of cardinality.

Thus, combining and generalizing two criteria $ACer$ and $ACov$, we introduce a criterion of similarity between two partitions $\mathcal{X}$ and $\mathcal{Y}$ defined on U through comparisons of similarity between $\mathcal{X}$ and $\mathcal{X} \cap \mathcal{Y}$, and $\mathcal{Y}$ and $\mathcal{X} \cap \mathcal{Y}$ as follows.

Definition 1. *Let $\mathcal{X}$ and $\mathcal{Y}$ be any partitions on U. A criterion $Sim_U(\mathcal{X}, \mathcal{Y})$ of similarity between $\mathcal{X}$ and $\mathcal{Y}$ is defined by*

$$Sim_U(\mathcal{X}, \mathcal{Y}) = \frac{|\mathcal{X}| + |\mathcal{Y}|}{2|\mathcal{X} \cap \mathcal{Y}|}. \tag{10}$$

By this definition, it is clear that $Sim_U(\mathcal{X}, \mathcal{Y}) = Sim_U(\mathcal{Y}, \mathcal{X})$ holds. If we set $\mathcal{X} = U/R_B$ such that $B \subseteq C$ and $\mathcal{Y} = \mathcal{D}$ with respect to a given decision table (U, C, d), the similarity Sim_U by (10) is

$$Sim_U(U/R_B, \mathcal{D}) = \frac{1}{2}\left(\frac{|U/R_B|}{|U/R_B \cap \mathcal{D}|} + \frac{|\mathcal{D}|}{|U/R_B \cap \mathcal{D}|}\right)$$

$$= \frac{1}{2}\left(ACer(B) + ACov(B)\right),$$

i.e., the average of $ACer(B)$ and $ACov(B)$ with respect to B.

Proposition 1 below describes the range of scores of the similarity criterion $Sim_U(\mathcal{X}, \mathcal{Y})$ on U.

Proposition 1. *Let $\mathcal{X}$ and $\mathcal{Y}$ be any partitions on U. The similarity Sim_U defined by (10) has the following properties:*

1. $\dfrac{|\mathcal{X}| + |\mathcal{Y}|}{2\min\left(|\mathcal{X}||\mathcal{Y}|, |U|\right)} \leq Sim_U(\mathcal{X}, \mathcal{Y}) \leq 1.$

2. $Sim_U(\mathcal{X}, \mathcal{Y}) = 1$ *if and only if* $\mathcal{X} = \mathcal{Y}$.

3. $Sim_U(\mathcal{X}, \mathcal{Y}) = \dfrac{|\mathcal{X}| + |\mathcal{Y}|}{2\min\left(|\mathcal{X}||\mathcal{Y}|, |U|\right)}$ *if and only if* $|\mathcal{X} \cap \mathcal{Y}| = \min\left(|\mathcal{X}||\mathcal{Y}|, |U|\right)$.

Note that the triangle inequality with respect to Sim_U, i. e., the following inequality

$$Sim_U(\mathcal{X}, \mathcal{Y}) + Sim_U(\mathcal{Y}, \mathcal{Z}) \leq Sim_U(\mathcal{X}, \mathcal{Z}) \tag{11}$$

is not satisfied in general.

Because the range of cardinality of any partition $\mathcal{X}$ on U is $1 \leq |\mathcal{X}| \leq |U|$, Proposition 1 indicates that the minimum score of $Sim_U(\mathcal{X}, \mathcal{Y})$ between partitions $\mathcal{X}$ and $\mathcal{Y}$ on U is uniquely determined by

$$minSim(|U|) = \min_{1 \leq i \leq |U|,\ 1 \leq j \leq |U|} \frac{i + j}{2\min\left(i \times j, |U|\right)}, \tag{12}$$

where i and j are natural numbers. It is not hard to confirm that the minimum score $minSim(|U|)$ of the similarity criterion defined by (12) is monotone non-increasing with respect to the cardinality $|U|$. This causes difficulty of direct comparison of similarity scores between different universes because the minimum scores by (10) on different universes U and U' such that $|U| \neq |U'|$ may be different.

Therefore, when we need to consider comparison of similarity scores between different universes, we should consider the following *normalized similarity*.

Definition 2. *Let $\mathcal{X}$ and $\mathcal{Y}$ be any partitions on U, $Sim_U(\mathcal{X}, \mathcal{Y})$ be the criterion of similarity between $\mathcal{X}$ and $\mathcal{Y}$ defined by (10), and $minSim(|U|)$ be the minimum score of the similarity between partitions defined by (12). A criterion of normalized similarity between $\mathcal{X}$ and $\mathcal{Y}$ is defined by*

$$NS(\mathcal{X}, \mathcal{Y}) = \frac{Sim_U(\mathcal{X}, \mathcal{Y}) - minSim(|U|)}{1 - minSim(|U|)}. \tag{13}$$

From the definition of the normalized similarity by (13) and the range of the similarity criterion Sim_U by Proposition 1, it is obvious that the range of the normalized similarity NS satisfies the following properties.

Corollary 1. *Let $\mathcal{X}$ and $\mathcal{Y}$ be any partitions on U. The normalized similarity NS defined by (13) has the following properties:*

1. *$0 \leq NS(\mathcal{X}, \mathcal{Y}) \leq 1$.*
2. *$NS(\mathcal{X}, \mathcal{Y}) = 1$ if and only if $\mathcal{X} = \mathcal{Y}$.*
3. *$NS(\mathcal{X}, \mathcal{Y}) = 0$ if and only if $Sim_U(\mathcal{X}, \mathcal{Y}) = minSim(|U|)$.*

Example 2 below describes necessity of using the normalized similarity for comparing similarity scores between different universes.

Example 2. Table 3 presents another decision table which is identical to Table 1 except for absence of two elements x_9 and x_{10}. For each condition attribute c_i ($1 \leq i \leq 8$) in Table 3 and Table 1, we construct a partition U/R_{c_i} and compare with the partition U/R_d constructed from the decision attribute d.

Table 3. Another example of decision table

U	c_1	c_2	c_3	c_4	c_5	c_6	c_7	c_8	d
x_1	1	1	1	1	1	1	2	1	1
x_2	2	3	1	3	1	2	2	2	3
x_3	3	2	3	2	1	2	1	1	2
x_4	4	2	2	2	2	2	1	1	2
x_5	5	2	2	3	1	1	2	1	1
x_6	6	3	2	1	1	2	2	1	3
x_7	7	1	1	1	1	2	1	1	2
x_8	8	2	3	1	1	1	2	1	1

Table 4. The similarity and the normalized similarity of condition attributes

Attribute	Sim. in Table1	N. S. in Table1	Sim. in Table3	N. S. in Table3
c_1	0.65	0.475	0.69	0.5
c_2	0.6	0.4	0.6	0.36
c_3	0.33	0	0.375	0
c_4	0.5	0.25	0.5	0.2
c_5	0.5	0.25	0.625	0.4
c_6	0.83	0.75	0.83	0.73
c_7	0.83	0.75	0.83	0.73
c_8	0.66	0.5	0.67	0.47

Table 4 describes the similarity and the normalized similarity between partitions U/R_{c_i} and U/R_d in Table 1 ($|U| = 10$) and Table 3 ($|U| = 8$), where notations "Sim." and "N.S." in Table 4 are abbreviations of similarity and normalized similarity, respectively. The row of the condition attribute c_3 indicates that the minimum scores of similarity in Table 1 and Table 3 are different, i. e., $minSim(10) = \frac{1}{3}$ and $minSim(8) = \frac{3}{8}$. Thus, in both Table 1 and Table 3, the similarity of partitions by c_3 and d is identical to the theoretical minimum score of similarity, which concludes that the normalized similarity by c_3 and d is equal to 0 in both Table 1 and Table 3.

On the other hand, the row of c_4 indicates that comparison of the similarity scores between different universes is not appropriate, i. e. the normalized similarity scores are different between Table 1 and Table 3 even though the similarity scores are identical. Thus, we can conclude that the partitions U/R_{c_4} and U/R_d for Table 1 are relatively more similar than those partitions for Table 3.

5   Conclusion

In this paper, we introduced a criterion of similarity between partitions. The similarity criterion proposed in this paper is a generalization of an evaluation criterion of relative reducts proposed by the authors [3,4], and evaluates the similarity of partitions from the viewpoint of correctness and roughness with each other. Moreover, for comparison of similarity scores between different universes, we also proposed a normalized similarity criterion and illustrated the necessity of using the normalized similarity for comparing the similarity scores between different universes. More consideration and refinement of the proposed criteria, and comparison of the proposed criteria with other methods, for example, approximate entropy reducts [9] in the aspect of evaluation of relative reducts based on comparison of partitions, and functional dependency analysis between condition attributes and decision attributes [7] are interesting future issues.

Acknowledgments

The authors would like to thank the anonymous referees for their helpful comments. This work has been partially supported by MEXT KAKENHI (20700192).

References

1. Huhtala, Y., Karkkainen, J., Porkka, P., Toivonen, H.: Efficient Discovery of Functional and Approximate Dependencies Using Partitions. In: Proc. of the 14th International Conference on Data Engineering, pp. 392–401 (1998)
2. Kryszkiewicz, M., Lasek, P.: FUN: Fast Discovery of Minimal Sets of Attributes Functionally Determining a Decision Attribute. Transactions on Rough Sets 9, 76–95 (2008)
3. Kudo, Y.: An Evaluation Method of Relative Reducts Based on Roughness of Partitions (Extended Abstract). In: The 6th International Conferences on Rough Sets and Current Trends in Computing (RSCTC 2008): Programs & Abstracts, pp. 26–29 (2008)
4. Kudo, Y., Murai, T.: An Evaluation Method of Relative Reducts Based on Roughness of Partition. International Journal of Cognitive Informatics and Natural Intelligence (to appear)
5. Pawlak, Z.: Rough Sets. International Journal of Computer and Information Science 11, 341–356 (1982)
6. Pawlak, Z.: Rough Sets: Theoretical Aspects of Reasoning about Data. Kluwer Academic Publishers, Dordrecht (1991)
7. Pawlak, Z., Słowiński, R.: Rough Set Approach to Multi-Attribute Decision Analysis. European Journal of Operation Research 74, 443–459 (1994)
8. Polkowski, L.: Rough Sets: Mathematical Foundations. Advances in Soft Computing. Physica-Verlag, Heidelberg (2002)
9. Ślęzak, D.: Approximate Entropy Reducts. Fundamenta Informaticae 53(3-4), 365–387 (2002)

On Rough Sets with Structures and Properties

Ryszard Janicki*

Department of Computing and Software,
McMaster University,
Hamilton, ON, L8S 4K1 Canada
janicki@mcmaster.ca

Abstract. The Rough Sets paradigm is extended to the sets that have some structure (for example they are relations) and some properties (for example they are transitive relations).

1 Introduction

Consider the following problem: we have a set of data that have been obtained in an empirical manner. From the nature of the problem we know that the set should have some structure and desired properties, for example it should be partially ordered, but because the data are empirical it is not. In general case this might be just an arbitrary set without the desired structure and properties. What is the "best" approximation that have the desired structure and properties and how it can be computed? For the approximation of arbitrary relations by partial orders this problem was discussed and some solutions were proposed in [5] - within the standard theory of relations ([8,11]), and in [6] - within both the standard theory of relations and Rough Sets paradigm ([9,10]). In [6] some general Rough Sets settings for more general approximation of relations have also been proposed and analysed.

In this paper we will generalise some ideas of [6] to more sophisticated data types.

While, in general, sets are just arbitrary collections of arbitrary elements [8], when they are applied in other parts of mathematics or science, they usually have some structeres - for example they are relations, and properties - for example transitivity. They often resemble more abstract data types [1] than standard sets. Those structures and properties are essential when it comes to the problem of *approximation*.

It appears that the concept of approximation has two different intuitions in mathematics and science. The first one stem from the fact that all empirical numerical data have some errors, so in reality we never have the value x but always some interval $(x - \varepsilon, x + \varepsilon)$, i.e. the upper approximation and the lower approximation. Rough Sets exploit this idea for general sets. The second intuition can be illustrated by *least square approximation* of points in two dimensional plane (c.f. [14]). Here we know or assume that the points should be on a straight line and we are trying to find the line that fits the data best. In this case tha data have a structure (points in two dimensional plane, i.e. a relation that is a function) and should satisfied a desired property (be on the straight line). Note that even if we replace a solution $f(x) = ax + b$ by two lines $f_1(x) = ax + b + \delta$

---

* Partially supported by NSERC grant of Canada.

H. Sakai et al. (Eds.): RSFDGrC 2009, LNAI 5908, pp. 109–116, 2009.

and $f_2(x) = ax + b - \delta$, where δ is a standard error (c.f. [14]), there is no guarantee that any point resides between $f_1(x)$ and $f_2(x)$. Hence this is not the case of an upper, or lower approximation in the sense of Rough Sets. However this approach assumes that there is a well defined concept of a *metric* which allows us to minimize the distance, and this concept is not obvious, and often not even possible for non-numerical objects (see for instance [4]).

The approach presented in this paper is a mixture of both intuitions, there is no metric, but the concept of "minimal distance" is simulated by a sequence of property-driven lower and/or upper approximations.

This paper is a substantial generalisation and refinement of the ideas presented in the second half of [6].

2 Principles of Rough Sets Paradigm and Relations

To focus the intuition, in this section we will discuss only one special case of sets with structures and properties, namely, the relations.

The principles of Rough Rets [9,10] can be formulated as follows. Let U be a finite and nonepty universum of elements, and let $E \subseteq U \times U$ be an equivalence (i.e. reflexive, symmetric and transitive) relation. For each equivalence relation $E \subseteq U \times U$, $[x]_E$ will denote the equivalence class of E containing x and U/E will donote the set of all equivalence classes of E. The elements of U/E are called elementary sets and they are interpreted as basic observable, measurable, or definable sets. The pair (U, E) is referred to as a Pawlak approximation space. A set $X \subseteq U$ is approximated by two subsets of U, $\underline{A}(X)$ - called lower approximation of X, and $\overline{A}(X)$ - called upper approximation of X, where:

$$\underline{A}(X) = \bigcup \{[x]_E \mid x \in U \wedge [x]_E \subseteq X\}, \qquad \overline{A}(X) = \bigcup \{[x]_E \mid x \in U \wedge [x]_E \cap X \neq \emptyset\}.$$

Since every relation is a set of pairs the approach can be used for relations as well [12]. Unfortunately in the cases as our we want approximations to have some specific properties as irreflexivity, transitivity etc., and most of those properties are not closed under the set union operator. As it was pointed out in [16], in general one cannot expect approximations to have desired properties (see [16] for details). It is also not clear how to define the relation E for the cases as our.

However the Rought Sets can also be defined in orthogonal (sometimes called 'topological') manner [10,13,15]. For a given (U, E) we may define $\mathcal{D}(U)$ as the smallest set containing $\emptyset$, all elements of U/E and closed under set union. Clearly U/E is the set of all components generated by $\mathcal{D}(U)$ [8]. We may start with defining a space as $(U, \mathcal{D})$ where $\mathcal{D}$ is a family of sets that contains $\emptyset$ and for each $x \in U$ there is $X \in \mathcal{D}$ such that $x \in X$ (i.e. $\mathcal{D}$ is a cover of U [11]). We may now define $E_{\mathcal{D}}$ as the equivalence relation generated by the set of all components defined by $\mathcal{D}$ (see for example [8]). Hence both approaches are equivalent [10,13,16], however now for each $A \subseteq U$ we have:

$$\underline{A}(X) = \bigcup \{Y \mid Y \subseteq X \wedge Y \in \mathcal{D}\}, \qquad \overline{A}(X) = \bigcap \{Y \mid X \subseteq Y \wedge Y \in \mathcal{D}\}.$$

We can now define $\mathcal{D}$ as a set of relations having the desired properties and then calculate $\underline{R}$ and/or $\overline{R}$ with respect to a given $\mathcal{D}$. Such approach was proposed and analysed in

[16], however it seems to have only limited application. First it assumes that the set $\mathcal{D}$ is closed under both union and intersection, and few properties of relations do this. For instance transitivity is not closed under union and having a cycle is not closed under intersection. Some properties as for instance "having exactly one cycle" are preserved by neither union nor intersection. This problem was discussed in [16] and they proposed that perhaps different $\mathcal{D}$ could be used for lower and upper approximations. The approach of [16] assumes additionally that, for upper approximation there is at least one element of $\mathcal{D}$ that contains R, and, for lower approximation there exists at least one element of $\mathcal{D}$ that is uncluded in R. These are too strong assumptions for the cases like those considered in [5,6], if R contains a cycle, there is no partial order that contains R!

The problem is even bigger when we consider structures more complex than relations. Hence we need to create a new setting.

3   Sets with Structures and Properties

A *set with a structure* X is a *relational structure* (c.f. [2,8]) $X = (D_X, R_1^X, ..., R_n^X)$, where D_X is a set called the *domain* of X and each $R_i^X \subseteq \prod_{j=1}^{k_i^X} X$ is a k_i^X-ary relation on D_X. The tuple $(R_1^X, ..., R_n^X)$ is called the *structure* of X and denoted by $S(X)$. The vector $(k_1^X, ..., k_n^X)$ is called the *arity of* X. Two sets with structure $X = (D_X, S(X))$ and $Y = (D_Y, S(Y))$ are *of the same type* if they have identical arities. For example binary relations are sets with structure of arity (2) (i.e. $n = 1$, $k_1^X = 2$).

For two sets with structure of *the same type* $X = (D_X, R_1^X, ..., R_n^X)$, and $Y = (D_Y, R_1^Y, ..., R_n^Y)$, we define $X \oplus Y$, where $\oplus \in \{\cup, \cap, \setminus\}$, component-wise as $X \oplus Y = (D_X \oplus D_Y, R_1^X \oplus R_1^Y, ..., R_n^X \oplus R_n^Y)$. Similarly we define $\bigoplus_{j \in J} X_j$ for any set of indices J. We also define $X \subseteq Y \iff D_X \subseteq D_Y \wedge R_1^X \subseteq R_1^Y \wedge ... \wedge R_n^X \subseteq R_n^Y$ and the empty set with properties $\emptyset^{\{p\}} = (\emptyset, \emptyset, ..., \emptyset)$. We will usually write $\emptyset$ instead of $\emptyset^{\{p\}}$ if this will not lead to any ambiguity. We also define 2^X in the usual manner, $2^X = \{Z \mid Z \subseteq X\}$, but $\subseteq$ is as defined above for sets with a structure.

Let $X = (D_X, R_1^X, ..., R_n^X)$ be a set with a structure and let α be any first-order predicate (c.f. [3]) with the set of atomic formulae being a subset of $\{R_1^X, ..., R_n^X\}$ and all variables over D_X. Any predicate α of this kind will be called *a property over* X. The predicete α is called a *property of* X if X is a model of α, i.e. α holds for any assignment (c.f. [3]). We would like to point the difference between *a property over* R, i.e. just a statement that may or may not be true, and *a property of* R, a statement that is true for all assignments.

The question a reader might ask is "why to replace an established name as *relational structures* by a new one as *sets with structures?*" As it was already mentioned, outside pure set theory, elements of the sets usually have some structure and properties which are often used in proofs, constructions and algorithms. Even if the integers are used only as names of objects, in many algorithms the fact that they are totally ordered is utilised to increase efficiency (c.f. [1,7]). While a collection that consists of, say, a white elephant, computer mouse, empty set and a letter 'a', is a proper set (c.f. [8,11]), in most applications the sets are more homogenous, as 'sets of integers', 'vertices', 'variables', etc. In fact, when it comes to applications, the sets used resembles more *abstract data types* (c.f. [1]) than pure sets with uninterpreted elements. Clearly each abstract data type can be represented as a relational structure, however usually the set terminology is used and all the structure is used in an implicit manner.

4   Rough Approximations of Sets with Structures and Properties

Let $U = (D_U, \mathcal{S}(U))$ be a finite set with a structure called *universum* (*with a structure*) and let $\mathcal{P}$ be a set of properties *over* U. Any element $\alpha \in \mathcal{P}$ is called an *elementary property*[1]. We assume that for each $\alpha \in \mathcal{P}$ there is a non-empty family of sets $P_\alpha \subseteq 2^U$ such that $P_\alpha \neq \{\emptyset\}$ and for every $X \subseteq U$, if α is a property of X then $X \in P_\alpha$. In other words P_α is the set of all subsets of U that satisfy the property α.

Let $\mathcal{P}^\cap$ be a subset of $\mathcal{P}$ such that $\alpha \in \mathcal{P}^\cap$ iff P_α is closed under intersection, and $\mathcal{P}^\cup$ be a subset of $\mathcal{P}$ such that $\alpha \in \mathcal{P}^\cup$ iff P_α is closed under union.

We assume that $\mathcal{P} = \mathcal{P}^\cap \cup \mathcal{P}^\cup$ and the pair $(U, \mathcal{P})$ will be called an *approximation space*[2].

Let $X \subseteq U$ and $\alpha \in \mathcal{P}$. We say that:

- X has α-lower bound $\iff \exists Y \in P_\alpha . Y \subseteq X$; and $lb_\alpha(X) = \{Y \mid Y \in P_\alpha \wedge Y \subseteq X\}$,
- X has α-upper bound $\iff \exists Y \in P_\alpha . X \subseteq Y$; and $ub_\alpha(X) = \{Y \mid Y \in P_\alpha \wedge X \subseteq Y\}$.

For every family of sets $\mathcal{F} \subseteq 2^U$, we define

- $min(\mathcal{F}) = \{X \mid \forall Y \in \mathcal{F} . Y \subseteq X \Rightarrow X = Y\}$,
- $max(\mathcal{F}) = \{X \mid \forall Y \in \mathcal{F} . X \subseteq Y \Rightarrow X = Y\}$.

We are now able to provide the two main definitions of this chapter:

- If X has α-lower bound then we define its α-lower approximation as:

$$\underline{\mathbf{A}}_\alpha(X) = \bigcap \{Y \mid Y \in max(lb_\alpha(X))\}.$$

- If X has α-upper bound then we define its α-upper approximation as:

$$\overline{\mathbf{A}}_\alpha(X) = \bigcup \{Y \mid Y \in min(ub_\alpha(X))\}.$$

If X *does not have* α-lower bound (α-upper bound) then its α-lower approximation (α-upper approximation) *does not exist*. The result below shows that the above two definitions are sound when $X \in P_\alpha$.

Proposition 1. *If* $X \in P_\alpha$ *then* $\underline{\mathbf{A}}_\alpha(X) = \overline{\mathbf{A}}_\alpha(X) = X$.

Proof. If $X \in P_\alpha$ then $lb_\alpha(X) = ub_\alpha(X) = \{X\}$.   □

Directly from the definitions it follows that $\underline{\mathbf{A}}_\alpha(X)$ is well defined if $\alpha \in \mathcal{P}^\cap$ and $\overline{\mathbf{A}}_\alpha(X)$ is well defined if $\alpha \in \mathcal{P}^\cup$. The result below shows that both concepts are well defined for all $\alpha \in \mathcal{P} = \mathcal{P}^\cup \cup \mathcal{P}^\cap$.

---

[1] Even though any property can be called 'elementary', it is assumed that in any concrete case the elemetary properties are 'simple' and 'regular'. They are just atomic parts from which the real more sophisticated properties are built.

[2] This assumption is much weaker than it might appear as this is an assumption only about *elementary* properties, not about compound more sophisticated properties (see next section).

Proposition 2

1. If $\alpha \in \mathcal{P}^{\cup}$ then $\underline{\mathbf{A}}_\alpha(X) = \bigcup\{Y \mid Y \in lb_\alpha(X)\} = \bigcup\{Y \mid Y \subseteq X \wedge Y \in P_\alpha\}$.
2. If $\alpha \in \mathcal{P}^{\cap}$ then $\overline{\mathbf{A}}_\alpha(X) = \bigcap\{Y \mid Y \in ub_\alpha(X)\} = \bigcap\{Y \mid X \subseteq Y \wedge Y \in P_\alpha\}$.

Proof

(1) If $\alpha \in \mathcal{P}^{\cup}$ then $max(lb_\alpha(X)) = \{\bigcup\{Y \mid Y \in lb_\alpha(X)\}\}$.
(2) If $\alpha \in \mathcal{P}^{\cap}$ then $min(ub_\alpha(X)) = \{\bigcap\{Y \mid Y \in ub_\alpha(X)\}\}$. $\qquad\square$

The next result shows *when this model is exactly the same as the classical Rough Sets approach to relations* (the version from [15,16]).

Corollary 1

If $\alpha \in \mathcal{P}^{\cup} \cap \mathcal{P}^{\cap}$ then $\underline{\mathbf{A}}_\alpha(X) = \underline{\mathbf{A}}(X)$ and $\overline{\mathbf{A}}_\alpha(X) = \overline{\mathbf{A}}(X)$, where $\underline{\mathbf{A}}(X)$ and $\mathbf{A}(X)$ are classical upper and lower rough approximations over the space (U, P_α). $\qquad\square$

The next two results will show that our definitions of α-lower approximation and α-upper upproximation are sound, and their properties pretty close (but not identical) to those of standard rough set approximations as presented in for example [9,10]. We start with the properties of α-lower approximation.

Proposition 3. *If $X, Y \subseteq U$ have α-lower bound then:*

1. $X \subseteq Y \implies \underline{\mathbf{A}}_\alpha(X) \subseteq \underline{\mathbf{A}}_\alpha(Y)$,
2. $\underline{\mathbf{A}}_\alpha(X) \subseteq X$,
3. $\underline{\mathbf{A}}_\alpha(x) = \underline{\mathbf{A}}_\alpha(\underline{\mathbf{A}}_\alpha(X))$,
4. $\underline{\mathbf{A}}_\alpha(X \cap Y) = \underline{\mathbf{A}}_\alpha(\underline{\mathbf{A}}_\alpha(X) \cap \underline{\mathbf{A}}_\alpha(Y))$,
5. *if $\alpha \in \mathcal{P}^{\cap}$ then $\underline{\mathbf{A}}_\alpha(X \cap Y) = \underline{\mathbf{A}}_\alpha(X) \cap \underline{\mathbf{A}}_\alpha(Y)$,*
6. *if X has α-upper bound then $\overline{\mathbf{A}}_\alpha(X) = \underline{\mathbf{A}}_\alpha(\overline{\mathbf{A}}_\alpha(X))$.*

Proof
(1) Since $X \subseteq Y \implies lb_\alpha(X) \subseteq lb_\alpha(Y) \implies max(lb_\alpha(X)) \subseteq lb_\alpha(Y)$, then for each $Z \in max(lb_\alpha(X))$ there is $Z' \in max(lb_\alpha(Y))$ such that $Z \subseteq Z'$; and intersection preserves inclusion.
(2) Since $Z \in lb_\alpha(X) \implies Z \subseteq X$, and and intersection preserves inclusion.
(3) From Proposition 1 as $\underline{\mathbf{A}}_\alpha(X) \in P_\alpha$.
(4) By (1) we have $\underline{\mathbf{A}}_\alpha(X \cap Y) \subseteq \underline{\mathbf{A}}_\alpha(X)$ and $\underline{\mathbf{A}}_\alpha(X \cap Y) \subseteq \underline{\mathbf{A}}_\alpha(Y)$, so $\underline{\mathbf{A}}_\alpha(X \cap Y) \subseteq \underline{\mathbf{A}}_\alpha(X) \cap \underline{\mathbf{A}}_\alpha(Y)$. Hence by (2) and (3) $\underline{\mathbf{A}}_\alpha(X \cap Y) \subseteq \underline{\mathbf{A}}_\alpha(\underline{\mathbf{A}}_\alpha(X) \cap \underline{\mathbf{A}}_\alpha(Y))$.
By the definition we have $\underline{\mathbf{A}}_\alpha(\underline{\mathbf{A}}_\alpha(X) \cap \underline{\mathbf{A}}_\alpha(Y)) = \bigcap\{Z \mid Z \in max(lb_\alpha(\underline{\mathbf{A}}_\alpha(Y) \cap \underline{\mathbf{A}}_\alpha(Y)))\}$. Let $B \in lb_\alpha(\underline{\mathbf{A}}_\alpha(Y) \cap \underline{\mathbf{A}}_\alpha(Y)))$. This means $B \in P_\alpha \wedge B \subseteq \underline{\mathbf{A}}_\alpha(X) \cap \underline{\mathbf{A}}_\alpha(Y)$, hence $B \in P_\alpha \wedge B \subseteq X \wedge B \subseteq Y$, i.e. $B \in P_\alpha \wedge B \subseteq X \cap Y$. Therefore $B \in lb_\alpha(X \cap Y)$. In this way we proved that $lb_\alpha(\underline{\mathbf{A}}_\alpha(Y) \cap \underline{\mathbf{A}}_\alpha(Y)) \subseteq lb_\alpha(X \cap Y)$. Hence $max(lb_\alpha(\underline{\mathbf{A}}_\alpha(Y) \cap \underline{\mathbf{A}}_\alpha(Y))) \subseteq lb_\alpha(X \cap Y)$, i.e. for each $Z \in max(lb_\alpha(\underline{\mathbf{A}}_\alpha(Y) \cap \underline{\mathbf{A}}_\alpha(Y)))$ there exists $Z' \in max(lb_\alpha(X \cap Y))$, such that $Z \subseteq Z'$. Since intersection preserves inclusion this means that $\underline{\mathbf{A}}_\alpha(\underline{\mathbf{A}}_\alpha(X) \cap \underline{\mathbf{A}}_\alpha(Y)) \subseteq \underline{\mathbf{A}}_\alpha(X \cap Y)$.
(5) If $\alpha \in \mathcal{P}^{\cap}$ then $\underline{\mathbf{A}}_\alpha(X) \cap \underline{\mathbf{A}}_\alpha(Y) \in P_\alpha$ so by Proposition 1
$\underline{\mathbf{A}}_\alpha(X) \cap \underline{\mathbf{A}}_\alpha(Y) = \underline{\mathbf{A}}_\alpha(\underline{\mathbf{A}}_\alpha(X) \cap \underline{\mathbf{A}}_\alpha(Y))$.
(6) If X has α-upper bound then $\overline{\mathbf{A}}_\alpha(X) \in P_\alpha$ so by Prop. 1, $\overline{\mathbf{A}}_\alpha(X) = \underline{\mathbf{A}}_\alpha(\overline{\mathbf{A}}_\alpha(X))$. $\square$

The difference from the classical case is that intersection splits into two cases and mixing lower with upper α-approximation is conditional.

We will now present the properties of α-upper approximation.

Proposition 4. *If $X, Y \subseteq U$ have α-upper bound then:*

1. $X \subseteq Y \implies \overline{\mathbf{A}}_\alpha(X) \subseteq \overline{\mathbf{A}}_\alpha(Y)$,
2. $X \subseteq \overline{\mathbf{A}}_\alpha(X)$,
3. $\overline{\mathbf{A}}_\alpha(x) = \overline{\mathbf{A}}_\alpha(\overline{\mathbf{A}}_\alpha(X))$,
4. $\overline{\mathbf{A}}_\alpha(X \cup Y) = \overline{\mathbf{A}}_\alpha(\overline{\mathbf{A}}_\alpha(X) \cup \overline{\mathbf{A}}_\alpha(Y))$,
5. *if $\alpha \in \mathcal{P}^\cup$ then $\overline{\mathbf{A}}_\alpha(X \cup Y) = \overline{\mathbf{A}}_\alpha(X) \cup \overline{\mathbf{A}}_\alpha(Y)$,*
6. *If X has α-lower bound then $\underline{\mathbf{A}}_\alpha(X) = \overline{\mathbf{A}}_\alpha(\underline{\mathbf{A}}_\alpha(X))$.*

Proof

(1)   Since $X \subseteq Y \implies ub_\alpha(Y) \subseteq ub_\alpha(X) \implies min(ub_\alpha(Y)) \subseteq ub_\alpha(X)$, then for each $Z' \in min(ub_\alpha(Y))$ there is $Z \in min(ub_\alpha(X))$ such that $Z \subseteq Z'$; and union preserves inclusion.

(2)   Since $Z \in ub_\alpha(X) \implies X \subseteq Z$, and and union preserves inclusion.

(3)   From Proposition 1 as $\overline{\mathbf{A}}_\alpha(X) \in P_\alpha$.

(4)   By (1) we have $\overline{\mathbf{A}}_\alpha(X) \subseteq \overline{\mathbf{A}}_\alpha(X \cup Y)$ and $\overline{\mathbf{A}}_\alpha(Y) \subseteq \underline{\mathbf{A}}_\alpha(C \cup Y)$, so $\overline{\mathbf{A}}_\alpha(X) \cup \underline{\mathbf{A}}_\alpha(Y) \subseteq \underline{\mathbf{A}}_\alpha(X \cup Y)$. Hence by (2) and (3) $\underline{\mathbf{A}}_\alpha(\underline{\mathbf{A}}_\alpha(X) \cup \underline{\mathbf{A}}_\alpha(Y)) \subseteq \overline{\mathbf{A}}_\alpha(X \cup Y)$.

Since $X \subseteq \overline{\mathbf{A}}_\alpha(X)$ and $Y \subseteq \overline{\mathbf{A}}_\alpha(Y)$ then $X \cup Y \subseteq \overline{\mathbf{A}}_\alpha(X) \cup \overline{\mathbf{A}}_\alpha(Y)$, i.e. $up_\alpha(\overline{\mathbf{A}}_\alpha(X) \cup \overline{\mathbf{A}}_\alpha(Y)) \subseteq up_\alpha(X \cup Y)$, and consequently $min(up_\alpha(\overline{\mathbf{A}}_\alpha(X) \cup \overline{\mathbf{A}}_\alpha(Y))) \subseteq up_\alpha(X \cup Y)$. Hence for each $Z' \in min(up_\alpha(\overline{\mathbf{A}}_\alpha(X) \cup \overline{\mathbf{A}}_\alpha(Y)))$, there exists $Z \in min(up_\alpha(X \cup Y))$ such that $Z \subseteq Z'$. Since union preserves inclusion, we obtained $\overline{\mathbf{A}}_\alpha(X \cup Y) \subseteq \overline{\mathbf{A}}_\alpha(\overline{\mathbf{A}}_\alpha(X) \cup \overline{\mathbf{A}}_\alpha(Y))$.

(5)   If $\alpha \in \mathcal{P}^\cup$ then $\overline{\mathbf{A}}_\alpha(X) \cup \overline{\mathbf{A}}_\alpha(Y) \in P_\alpha$ so by Proposition 1 $\overline{\mathbf{A}}_\alpha(X) \cup \overline{\mathbf{A}}_\alpha(Y) = \overline{\mathbf{A}}_\alpha(\overline{\mathbf{A}}_\alpha(X) \cup \overline{\mathbf{A}}_\alpha(Y))$.

(6)   If X has α-lower bound then $\underline{\mathbf{A}}_\alpha(X) \in P_\alpha$ so by Prop. 1, $\underline{\mathbf{A}}_\alpha(X) = \overline{\mathbf{A}}_\alpha(\underline{\mathbf{A}}_\alpha(X))$. □

Here the difference from the classical case is that union splits into two cases and mixing upper with lower α-approximation is conditional.

5   Compound Properties and Mixed Approximations

Most of the interesting properties are compound properties, like for instance transitivity *and* reflexivity for relations, and they can be imposed in various orders [5,6]. In this section we will propose a framework for doing this in a systematic way.

Let $\alpha, \beta \in \mathcal{P}$. We say that β *is consitent with* α iff for every $X \in P_\alpha$:

- if X has β-lower bound then $\underline{\mathbf{A}}_\beta(X) \in P_\alpha$,
- if X has β-upper bound then $\overline{\mathbf{A}}_\beta(X) \in P_\alpha$.

We will say that α and β are *consistent* iff β *is consistent with* α and α *is consistent with* β.

We will also *assume* that for all $\alpha, \beta \in \mathcal{P}$, α and β are *consistent*.

From now on when writing a formula like $\underline{\mathbf{A}}_\beta(\overline{\mathbf{A}}_\alpha(X))$ we will assume that all necessary conditions are satisfied, i.e. in this case, X has α-upper bound and $\overline{\mathbf{A}}_\alpha(X)$ has β-lower bound.

Proposition 5

1. $\mathbf{A}_\alpha(\mathbf{A}_\beta(X)) \in P_\alpha \cap P_\beta$ *for* $\mathbf{A}_\alpha \in \{\underline{\mathbf{A}}_\alpha, \overline{\mathbf{A}}_\alpha\}, \mathbf{A}_\beta \in \{\underline{\mathbf{A}}_\beta, \overline{\mathbf{A}}_\beta\}$,
2. $\overline{\mathbf{A}}_\alpha(\underline{\mathbf{A}}_\beta(X)) \subseteq \underline{\mathbf{A}}_\beta(\overline{\mathbf{A}}_\alpha(X))$.

Proof
(1)   Because all α and β from $\mathcal{P}$ are consistent.
(2)   By Proposition 4(2), $X \subseteq \overline{\mathbf{A}}_\alpha(X)$, so $\underline{\mathbf{A}}_\beta(X) \subseteq \underline{\mathbf{A}}_\beta(\overline{\mathbf{A}}_\alpha(X))$, and $\overline{\mathbf{A}}_\alpha(\underline{\mathbf{A}}_\beta(X)) \subseteq$
$\overline{\mathbf{A}}_\alpha(\underline{\mathbf{A}}_\beta(\overline{\mathbf{A}}_\alpha(X)))$. By (1) of this proposition, $\underline{\mathbf{A}}_\beta(\overline{\mathbf{A}}_\alpha(X)) \in P_\alpha$, so by Proposition 1,
$\overline{\mathbf{A}}_\alpha(\underline{\mathbf{A}}_\beta(\overline{\mathbf{A}}_\alpha(X))) = \underline{\mathbf{A}}_\beta(\overline{\mathbf{A}}_\alpha(X))$. Therefore $\overline{\mathbf{A}}_\alpha(\underline{\mathbf{A}}_\beta(X)) \subseteq \underline{\mathbf{A}}_\beta(\overline{\mathbf{A}}_\alpha(X))$.   □

Proposition 6. *Assume that* α, β *and* $\alpha \wedge \beta$ *belong to* $\mathcal{P}$.

1. $\underline{\mathbf{A}}_{(\alpha \wedge \beta)}(X) \subseteq \underline{\mathbf{A}}_\alpha(\underline{\mathbf{A}}_\beta(X))$,
2. $\overline{\mathbf{A}}_\alpha(\overline{\mathbf{A}}_\beta(X)) \subseteq \overline{\mathbf{A}}_{(\alpha \wedge \beta)}(X)$.

Proof
(1)   Since obviously $lb_{(\alpha \wedge \beta)}(X) \subseteq lb_\beta(X)$ then $\underline{\mathbf{A}}_{(\alpha \wedge \beta)}(X) \subseteq \underline{\mathbf{A}}_\beta(X)$. Hence $\underline{\mathbf{A}}_\alpha(\underline{\mathbf{A}}_{(\alpha \wedge \beta)}$
$(X)) \subseteq \underline{\mathbf{A}}_\alpha(\underline{\mathbf{A}}_\beta(X))$. Since $\underline{\mathbf{A}}_{(\alpha \wedge \beta)}(X) \in P_\alpha$, then due to Proposition 1,
$\underline{\mathbf{A}}_\alpha(\underline{\mathbf{A}}_{(\alpha \wedge \beta)}(X)) = \underline{\mathbf{A}}_{(\alpha \wedge \beta)}(X)$, which ends the proof of (1).
(2)   Since obviously $ub_{(\alpha \wedge \beta)}(X) \subseteq ub_\beta(X)$ then $min(ub_{(\alpha \wedge \beta)}(X)) \subseteq ub_\beta(X)$. This
means $\overline{\mathbf{A}}_\beta(X) \subseteq \overline{\mathbf{A}}_{(\alpha \wedge \beta)}(X)$. Hence $\overline{\mathbf{A}}_\alpha(\overline{\mathbf{A}}_\beta(X)) \subseteq \overline{\mathbf{A}}_\alpha(\overline{\mathbf{A}}_{(\alpha \wedge \beta)}(X))$. Since $\overline{\mathbf{A}}_{(\alpha \wedge \beta)}(X) \in$
P_α, then due to Proposition 1, $\overline{\mathbf{A}}_\alpha(\overline{\mathbf{A}}_{(\alpha \wedge \beta)}(X)) = \overline{\mathbf{A}}_{(\alpha \wedge \beta)}(X)$, which ends the proof
of (2).   □

Proposition 6 suggest an important technique for the design of approximation schema. It in principle says that using a complex predicate as a property result in *worse* approximation than when the property is decomposed into simpler ones, and then we approximate all simpler properties. *This means before starting an approximation process we should think carefully how the given property could be decomposed into the simpler ones.*

Define $\widehat{\mathcal{P}} = \mathcal{P} \times \{0, 1\}$. The elements of $\widehat{\mathcal{P}}$ will be called *labelled elementary properties*. We will also write $\alpha^{(0)}$ or $\underline{\alpha}$ instead of $(\alpha, 0)$ and $\alpha^{(1)}$ or $\overline{\alpha}$ instead of $(\alpha, 1)$.

A sequence $s = \alpha_1^{(i_1)} \alpha_2^{(i_2)} .... \alpha_k^{(i_k)}$ of elements of $\widehat{\mathcal{P}}$ such that $\alpha_i \neq \alpha_{i+1}$, for $i = 1, ..., k-1$, is called a *schedule*.

For will also use $\mathbf{A}^{(0)}$ instead of $\underline{\mathbf{A}}$ and $\mathbf{A}^{(1)}$ instead of $\overline{\mathbf{A}}$.

A schedule $s = \alpha_1^{(i_1)} \alpha_2^{(i_2)} .... \alpha_k^{(i_k)}$ is *proper* if for each $X \subseteq U$ the following *mixed approximation*

$$\mathbf{A}^s(X) = \mathbf{A}_{\alpha_1}^{(i_1)}(\mathbf{A}_{\alpha_2}^{(i_2)}(...(\mathbf{A}_{\alpha_k}^{(i_k)}(X))...))$$

is well defined. Let **PS** denote the set of all proper schedules.

Each schedule $s = \alpha_1^{(i_1)} \alpha_2^{(i_2)} .... \alpha_k^{(i_k)}$ defines a *composite property*

$$\pi(s) = \alpha_1 \wedge \alpha_2 \wedge .... \wedge \alpha_k.$$

A composite property α is *approximable* if there exists a proper sequence $s \in \mathbf{PS}$ such that $\alpha = \pi(s)$. For example being a partial order [6] or pairwise comparison ranking data [4] are approximable composite properties.

The proper schedules could be interpreted as different "metrics" used for approximation purposes.

6   Final Comment

The approach presented in this paper can be called *property-driven* and it is a substantial extension of the ideas presented for relations in [15,16] and specially recently in [6]. Technically some results of [6] are just special cases of what is proven here. When thinking in terms of *properties*, very often either only lower or only upper approximation does make sense, and quite often *neither of them if the property is too sophisticated*. Due to lack of space we did not discuss this issue in details, an interested reader is referred to [6] for more on this subject. Proposition 6 might be the most useful result of this paper as it indicates how properties should be dealt with to get the best approximations. Our experience with non-numerical ranking [5] fully agrees with this result. We would like to point out that all the assumptions from Section 4 relate only to elementary properties; the requirements for compound properties are much weaker[3]. We believe the schedules can often be interpreted as "property-driven non-numerical metrics", and that finding a good schedule means finding a good approximation; which appears to be more art than science (see [5,6]). Some applications of the approach presented in this paper to the case of binary relations can be found in [6].

References

1. Aho, A.V., Hopcroft, J.E., Ullman, J.D.: Data Structures and Algorithms. Addison-Wesley, Reading (1983)
2. Burris, S., Sankappanavar, H.P.: A Course in Universal Algebra. Springer, Heidelberg (1981)
3. Huth, M.R.A., Ryan, M.D.: Logic in Computer Science. Cambridge University Press, Cambridge (2000)
4. Janicki, R.: Ranking with Partial Orders and Pairwise Comparisons. In: Wang, G., Li, T., Grzymala-Busse, J.W., Miao, D., Skowron, A., Yao, Y. (eds.) RSKT 2008. LNCS (LNAI), vol. 5009, pp. 442–451. Springer, Heidelberg (2008)
5. Janicki, R.: Some Remarks on Approximations of Arbitrary Binary Relations by Partial Orders. In: Chan, C.-C., Grzymala-Busse, J.W., Ziarko, W.P. (eds.) RSCTC 2008. LNCS (LNAI), vol. 5306, pp. 81–91. Springer, Heidelberg (2008)
6. Janicki, R.: Approximations of Arbitrary Binary Relations by Partial Orders. Classical and Rough Set Models (submitted)
7. Kleinberg, J., Tardos, E.: Algorithm Design. Addison-Wesley, Reading (2006)
8. Kuratowski, K., Mostowski, A.: Set Theory, 2nd edn. North Holland, Amsterdam (1976)
9. Pawlak, Z.: Rought Sets. International Journal of Computer and Information Sciences 34, 557–590 (1982)
10. Pawlak, Z.: Rough Sets. Kluwer, Dordrecht (1991)
11. Rosen, K.H.: Discrete Mathematics and Its Applications. McGraw-Hill, New York (1999)
12. Skowron, A., Stepaniuk, J.: Approximation of relations. In: Ziarko, W. (ed.) Rough Sets, Fuzzy Sets and Knowledge Discovery, pp. 161–166. Springer, London (1994)
13. Skowron, A., Stepaniuk, J.: Tolarence approximation spaces. Fundamenta Informaticae 27, 245–253 (1996)
14. Stewart, J.: Calculus. Concepts and Contexts. Brooks/Cole, Monterey (1997)
15. Yao, Y.Y.: Two views of the theory of rough sets in finite universes. International Journal of Approximate Reasoning 15, 291–317 (1996)
16. Yao, Y.Y., Wang, T.: On Rough Relations: An Alternative Formulations. In: Zhong, N., Skowron, A., Ohsuga, S. (eds.) RSFDGrC 1999. LNCS (LNAI), vol. 1711, pp. 82–91. Springer, Heidelberg (1999)

---

[3] However there are properties and structures that could be difficult to handle using this approach.

A Formal Concept Analysis Approach to Rough Data Tables

Bernhard Ganter and Christian Meschke*

Institut für Algebra, TU Dresden, Germany
{Bernhard.Ganter,Christian.Meschke}@tu-dresden.de

Abstract. In order to handle very large data bases efficiently, the data warehousing system ICE [5] builds so-called *rough tables* containing information that is abstracted from certain blocks of the original table. In this article we propose a formal description of such rough tables. We also investigate possibilities of mining them for implicational knowledge.

1 Introduction

Consider a large data table. It has rows, describing certain *objects*, and columns for *attributes* which these objects may have. The entry in row g and column m gives the *attribute value* that attribute m has for object g. By "large" we mean that the table has many rows, perhaps 10^9, or more. Even for a moderate number of attributes the size of such a table may be in the terabytes.

Data analysis on such a table faces complexity problems and requires a good choice of strategy. In the present paper we investigate an approach by Infobright using rough objects and granular data, and combine it with methods from Formal Concept Analysis.

Infobright Community Edition (ICE) [5,7] is an open source data warehousing system which is optimized to obtain high compression rates and to process analytic queries very quickly. ICE chops the stream of rows into so-called *rough rows*, each subsuming 65536 rows. The rough rows divide the columns into so-called *data packs*. Each data pack gets stored in a compressed form. For processing a query one does not want to decompress all data packs. Therefore ICE creates a so-called *data pack node* to every data pack. A data pack node contains meta–information about the corresponding data pack. If for instance the column contains numeric values, the data pack nodes could consist, e.g., of minimum, maximum and the sum of the data pack values. The *rough table* is the data table that has the rough rows as rows, the same attributes as the original large data table, and the data pack nodes as values.

In order to sound the possibilities of getting interesting information about the original data table from the rough table, Infobright offered a contest [6] for which they provided a rough table with 15259 rough rows (the original table has one billion rows) and 32 attributes. Furthermore, Infobright invited to propose ways to do data mining in such rough tables.

---

* Supported by the DFG research grant no. GA 216/10-1.

H. Sakai et al. (Eds.): RSFDGrC 2009, LNAI 5908, pp. 117–126, 2009.
© Springer-Verlag Berlin Heidelberg 2009

Our approach is a systematic one. Our focus is on "what can be done" rather than on "how to get quick results". Although it is likely that a large data table will contain erroneous and imprecise data, we first concentrate on the case of precise data. Approximative and fault-tolerant methods shall later be build on this basis. Please note that we had to leave out some proofs due to a lack of space. A technical report containing all proofs is available upon request.

2   Partial Formal Contexts

We assume that the reader is familiar with the basic notions of Formal Concept Analysis [4]. This theory will be used here to provide the basic data model. To encode the above mentioned granulation process we use the notion of a partial formal context. The information we are mining is in the form of implications or, more loosely, of association rules. Our aim is to infer such rules in the full data set from rules in the granulated data.

Definition 1. A partial formal context (G, M, i) consists of two sets G and M together with a mapping $i : G \times M \to \{\times, \bullet, ?\}$.   ◊

We call the elements of G the *objects* of the partial formal context, those of M the *attributes*. We read $i(g, m)$ as follows:

$$i(g, m) = \begin{cases} \times & \text{the object } g \text{ has the attribute } m, \\ \bullet & \text{the object } g \text{ does not have the attribute } m, \\ ? & \text{it is unknown if object } g \text{ has attribute } m. \end{cases}$$

Partial formal contexts have been considered under different aspects by several authors [1,2,3]. A partial formal context (G, M, j) is said to **extend** (G, M, i) if one can build it from (G, M, i) by replacing question marks "?", i.e., it holds that

$$i^{-1}(\{\times\}) \subseteq j^{-1}(\{\times\}) \quad \text{and} \quad i^{-1}(\{\bullet\}) \subseteq j^{-1}(\{\bullet\}).$$

Partial formal contexts which are maximal w.r.t. to this *extension order* are called **complete**. A formal context (G, M, I) in the usual sense, where $I \subseteq G \times M$ is a relation, is called a **completion** of a partial formal context (G, M, i) iff

$$i^{-1}(\{\times\}) \subseteq I \subseteq i^{-1}(\{\times, ?\}).$$

We say that an implication $A \to B$, where $A, B \subseteq M$, **holds** in a partial formal context (G, M, i) iff it holds in every completion. An equivalent condition is that the following holds for every object $g \in G$:

if $i(g, m) \in \{\times, ?\}$ for all $m \in A$ then $i(g, n) = \times$ for all $n \in B$.

An implication $A \to B$ **is refuted** by the partial formal context (G, M, i) if it holds in no completion. This is equivalent to the existence of an object g with

$$i(g, m) = \times \text{ for all } m \in A \text{ and } i(g, n) = \bullet \text{ for some } n \in B.$$

In order to better handle canonical formal contexts related to the partial context (G, M, i) we define for $S \subseteq \{\times, \bullet, ?\}$

$$i_S := \{(g, m) \in G \times M \mid i(g, m) \in S\} = i^{-1}(S).$$

We leave away the set brackets of S. For instance we write $i_{\times,?}$ instead of $i_{\{\times,?\}}$.

Proposition 1. Let (G, M, i) and (G, M, j) be partial formal contexts such that (G, M, j) extends (G, M, i). Then

- every implication that holds in (G, M, i) also holds in (G, M, j), and
- every implication that is refuted by (G, M, i) is also refuted by (G, M, j).

Proof. For every implication $A \to B$ that holds in (G, M, i) it follows for $g \in G$ that[1]

$$A \subseteq g^{j \times,?} \implies A^\complement \supseteq g^{j \bullet} \supseteq g^{i \bullet} \implies A \subseteq g^{i \times,?} \implies B \subseteq g^{i \times} \subseteq g^{j \times}.$$

The second item follows immediately from the observation that every object that refutes an implication in (G, M, i) also refutes this implication in (G, M, j). $\square$

3 Partial Contexts Obtained from Streams

There is a natural way how partial formal contexts arise from complete ones. Let (G, M, I) be a formal context and let $\mathcal{F}$ be a family of nonempty subsets of the object set G, i.e. $\mathcal{F} \subseteq \mathfrak{P}_{>0}(G) := \mathfrak{P}(G) \setminus \{\emptyset\}$. We obtain a partial formal context $(\mathcal{F}, M, i)$ by defining for every **block** $F \in \mathcal{F}$

$$i(F, m) := \begin{cases} \times & \text{if } F \subseteq m^I, \\ \bullet & \text{if } F \cap m^I = \emptyset, \\ ? & \text{else.} \end{cases}$$

We refer to $(\mathcal{F}, M, i)$ as the $\mathcal{F}$-**granulated** partial context to (G, M, I). Note that this reflects the situation of Infobright's rough tables from the contest [6] and is only formulated in a different language. For further details we refer the reader to the following Section 4.

Proposition 2. Let $(\mathfrak{P}_{>0}(G), M, i)$ be constructed from (G, M, I) as defined above, for the special case that $\mathcal{F} := \mathfrak{P}_{>0}(G)$. Then

- an implication that holds in $(\mathfrak{P}_{>0}(G), M, i)$ also holds in (G, M, I) and
- an implication is refuted by $(\mathfrak{P}_{>0}(G), M, i)$ iff it does not hold in (G, M, I).

---

[1] Let (G, M, R) be a formal context, i.e., $R \subseteq G \times M$. For $g \in G$ we define $g^R := \{m \in M \mid gRm\}$. For $X \subseteq G$ we define $X^R := \{m \in M \mid xRm \text{ for all } x \in X\}$. Dually one defines m^R and Y^R for $m \in M$ and $Y \subseteq M$, see [4]. Furthermore, $Y^\complement := M \setminus Y$.

Proposition 3. For every $\mathcal{F} \subseteq \mathfrak{P}_{>0}(G)$ it is true that

- no implication refuted by $(\mathcal{F}, M, i)$ holds in (G, M, I).

If $\mathcal{F}$ is a covering of G then it is true that

- every implication that holds in $(\mathcal{F}, M, i)$ also holds in (G, M, I).

Proof. Let $F \in \mathcal{F}$ be a block that refutes $A \to B$ in $(\mathcal{F}, M, i)$. Then it holds that $A \subseteq F^{i\times} = F^I$ and $B \not\subseteq F^{i\times,?}$. Hence, it follows $B \not\subseteq F^{i\times} = F^I$ which implies that $A \to B$ cannot hold in (G, M, I), since F^I is an intent containing the premise A, but not containing B.

Let $A \to B$ be an implication that holds in $(\mathcal{F}, M, i)$ and let $g \in G$. Since $\mathcal{F}$ is a covering there is a block $F \in \mathcal{F}$ containing g. Hence, it holds that $g^I \subseteq F^{i\times,?}$ which implies

$$A \subseteq g^I \implies A \subseteq F^{i\times,?} \implies B \subseteq F^{i\times} \implies B \subseteq g^I. \qquad \square$$

Now suppose that (G, M, I) is given as a stream of rows, and is chopped into data packs as described in the introduction. For each pack we take notes only if each object in the pack does have the attribute, in which case we note an "×" for the pack, or if no object in the pack has that attribute. We then note down "•". If some have and some do not, we note a question mark. This is a very strict rule, and we refer to it as **hard granulation**. Its disadvantage is that its outcome can drastically be changed by a single value in the pack. It shares this property with logical analysis: If a given logical formula does or does not hold in the original data, may be decided by a single counterexample. Proposition 3 above shows our possibilities to argue about implicational information of (G, M, I) based only on the granulated context $(\mathcal{F}, M, i)$. It is therefore necessary to investigate the circumstances under which an implication holds in or is refuted by $(\mathcal{F}, M, i)$. For both concerns it is sufficient to just take a look on implications of the form $A \to b$, where $A \subseteq M$ and $b \in M$.

Proposition 4. For $F \in \mathcal{F}$ the following three statements are equivalent:

(a) F refutes $A \to b$ in $(\mathcal{F}, M, i)$,
(b) $F \subseteq A^I \setminus b^I$,
(c) every single object $g \in F$ refutes $A \to b$ in (G, M, I).

Proof. F refutes $A \to b$ iff it holds that $A \subseteq F^{i\times} = F^I$ and $b \in F^{i\bullet} = F^{\dagger}$, which again is equivalent to $F \subseteq A^I$ and $F \subseteq b^{I\complement}$. The rest follows immediately. $\qquad \square$

The preceding propositions clarify under which conditions an implication $A \to b$ is refuted by the granulated context $(\mathcal{F}, M, i)$. If we insist on a definite answer, an answer that proves a refutation in the full data set on basis of the granulated data, these seem to be the natural conditions. But how likely is it that these conditions are satisfied? We attempt to give a first estimation. Obviously, the number $r := |A^I \setminus b^I|$ of all objects from the original data table (G, M, I) that share all attributes from A but do not have attribute b has to be large enough.

Let k be a fixed number and let $n := |G|$ be the number of objects. For the probability that a block F of cardinality k refutes $A \to b$ the following holds:

$$P(F \text{ refutes } A \to b) \;=\; \frac{\binom{r}{k}}{\binom{n}{k}} \;=\; \frac{r \cdot (r-1) \cdot \ldots \cdot (r-k+1)}{n \cdot (n-1) \cdot \ldots \cdot (n-k+1)} \;\leq\; \left(\frac{r}{n}\right)^k. \qquad (1)$$

We now assume that all $F \in \mathcal{F}$ have the same cardinality k. With the inequality from above we can conclude the following upper approximation of the probability that a partial context $(\mathcal{F}, M, i)$ refutes $A \to b$:

$$P((\mathcal{F}, M, i) \text{ refutes } A \to b) \;\leq\; \sum_{F \in \mathcal{F}} P(F \text{ refutes } A \to b) \;\leq\; |\mathcal{F}| \cdot \left(\frac{r}{n}\right)^k.$$

If we for instance assume that 95% of all objects in (G, M, I) refute $A \to b$ and that $\mathcal{F}$ contains one million blocks, i.e., $\frac{r}{n} = 95\%$ and $|\mathcal{F}| = 1.000.000$, we get that already for relatively small block sizes of $k \geq 539$ the probability that $(\mathcal{F}, M, i)$ refutes $A \to b$ is smaller than one part of a million.

Proposition 5. For $A, B \subseteq M$ the following four statements are equivalent:

(a) $A \to B$ holds in $(\mathcal{F}, M, i)$,
(b) for all $F \in \mathcal{F}$ the implication $A \subseteq F^{i \times, ?} \implies B \subseteq F^{i \times}$ holds,
(c) for all $F \in \mathcal{F}$ the implication $A \subseteq \bigcup_{g \in F} g^I \implies B \subseteq \bigcap_{g \in F} g^I$ holds,
(d) for all $F \in \mathcal{F}$ the implication $(\forall a \in A : F \not\subseteq a^t) \implies F \subseteq B^I$ holds.

If one takes a look at the third condition it becomes obvious that the bigger the block sizes $|F|$ are, the more likely it becomes that the premises are valid, and the less likely it becomes that the conclusions hold. Hence, if the number of the blocks and the sizes of the blocks are relatively large, we do not expect a lot of attribute implications to hold in $(\mathcal{F}, M, i)$: The probability that a *single* block F of cardinality k fulfills the implication from (d) is

$$P\big((\exists a \in A : F \subseteq a^t) \text{ or } F \subseteq B^I\big) \;\leq\; \sum_{a \in A} P(F \subseteq a^t) \;+\; P(F \subseteq B^I)$$

$$\leq\; |A| \cdot \left(\frac{n - |A^I|}{n}\right)^k + \left(\frac{|B^I|}{n}\right)^k.$$

Thereby the second inequation follows analogously to inequation (1). Let us assume that $A \to B$ is a *nontrivial* implication that holds in (G, M, I), i.e. $\emptyset \neq A^I \subseteq B^I \neq G$. Then for a large block size k this probability tends to be very small. Hence, for a large number of blocks it is far more improbable that $A \to B$ holds in $(\mathcal{F}, M, i)$.

4   The Contest Data Set

The Infobright data set does not come as a formal context right away, but needs some (uncritical) transformation. The formalisation of a data table which we use is that of a *many-valued context* (G, M, W, J), where G is a set of objects, M a set of many-valued attributes, W a set of attribute values and J is a ternary incidence relation satisfying

$$(g, m, v) \in J \text{ and } (g, m, w) \in J \text{ implies } v = w.$$

The standard interpretation of $(g, m, v) \in J$ is that the value of attribute m for object g is v. The value the object g has regarding to attribute m is commonly denoted with $m(g)$. To better distinguish such many-valued contexts from the formal contexts introduced first we shall refer to these sometimes as *one-valued*.

One of the standard techniques in Formal Concept Analysis expresses many-valued contexts as one-valued ones by means of *conceptual scales*. With conceptual scaling, every many-valued attribute is represented by several one-valued attributes, and the incidence to these depends on the respective attribute value. Details can be found in [4], but for the moment it suffices to know that with this technique, a data table can be transformed to a (one-valued) formal context, and this transformation can be done object-wise, one after another. As a consequence, we may transform a stream of objects with many-valued attributes into a stream of objects in a formal context. To keep things simple, we summarize: Conceptual scaling associates to each column m of the data table a set of attributes (the "scale attributes for the many-valued attribute m").

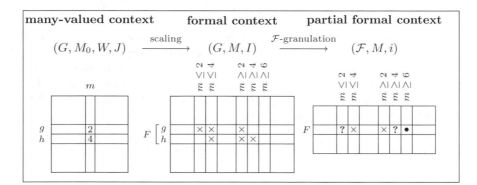

Fig. 1. A schematic illustration of interordinal scaling and $\mathcal{F}$-granulation

In the case of Infobright's contest data set we may think of the underlying, very large data table as a many-valued context (G, M_0, W, J) in which for every attribute $m \in M_0$ the set

$$W_m := m[G] := \{w \in W \mid (g, m, w) \in J \text{ for some } g \in G\}$$

of all values occurring in the column of m are ordered linearly in a canonical way. Depending on the data type of the attribute m this canonical order $\leq_m$ can for instance be the natural order of numbers or the alphabetical order of character strings. If one transforms this data table (G, M_0, W, J) into the formal context

(G, M, I) via scaling every attribute from M_0 *interordinally*, this formal context (G, M, I) directly yields to the granulated partial formal context $(\mathcal{F}, M, i)$ which contains exactly the same information as the contests rough table from [6].

We leave out the details about the interordinal scaling of the original data table (G, M_0, W, J). We refer the reader to Figure 1 to get an idea on how it works. The problem with the contest data set is that for almost every attribute $m \in M_0$ it holds that for almost every rough row $F \in \mathcal{F}$ the minimal and maximal m-values in F are exactly the overall minimal and maximal m-values, i.e.,

$$\min_{f \in F} m(f) \;=\; \min_{g \in G} m(g) \quad \text{and} \quad \max_{f \in F} m(f) \;=\; \max_{g \in G} m(g).$$

This yields to the effect that $(\mathcal{F}, M, i)$ is almost full of question marks, which minimizes the chances to verify or to refute some interesting attribute implications.

5   Soft Granulation

There is a reason why the approach of the previous section led to rather disappointing results: Our definition of the granulation process was too rigid. We defined that a block has a certain object if *all* members of a pack have the attribute, etc. As an example from the Infobright data, we mention the *minimum* parameter: It expresses that all members of the pack have a value greater or equal this one.

For a rough estimation, such parameters that can drastically be changed by a single member of the block seem inappropriate. It seems more promising to work with parameters which reflect the "tendency" of the data packs. The simplest suggestion is counting: Let us record for each data pack (F, m) the number of objects having the attribute. Formally:

$$\mathrm{supp}(F, m) \;:=\; |m^I \cap F|.$$

One calls $\mathrm{supp}(F, m)$ the **support** of the data pack (F, m). The number of objects of a block F that do not have attribute m is called its **negative support** and is defined as

$$\mathrm{nsupp}(F, m) \;:=\; |F \setminus m^I|.$$

Our granulation will now work as follows: The formal context (G, M, I) leads us to the $\mathbb{N}_0$-valued context $(\mathcal{F}, M, i)$, i.e., $i : \mathcal{F} \times M \to \mathbb{N}_0$, with

$$i(F, m) \;:=\; \mathrm{supp}(F, m).$$

What we are trying to do is mining in $(\mathcal{F}, M, i)$ for association rules that hold in (G, M, I). An **association rule** $A \to B$ consists of two attribute sets: the *premise* A and the *conclusion* B. We call

$$
\begin{aligned}
\mathrm{supp}(A) &:= & |A^I| & \qquad \text{the } \textbf{support}^2 \text{of } A, \\
\mathrm{supp}(A \to B) &:= \mathrm{supp}(A \cup B) & & \text{the } \textbf{support} \text{ of the rule } A \to B, \text{ and} \\
\mathrm{conf}(A \to B) &:= \tfrac{\mathrm{supp}(A \cup B)}{\mathrm{supp}(A)} & & \text{the } \textbf{confidence} \text{ of the rule } A \to B.
\end{aligned}
$$

Furthermore, for given thresholds $\texttt{minsupp} \in \mathbb{N}_0$ and $\texttt{minconf} \in [0,1]$ we say an association rule **holds** in (G, M, I) if its support exceeds $\texttt{minsupp}$ and its confidence exceeds $\texttt{minconf}$. Hence, association rules are a generalization of the attribute implications: The implications that hold in a formal context are exactly the association rules that hold with $\texttt{minsupp} = 0$ and $\texttt{minconf} = 1$. We say an attribute set (or a rule) is **frequent** if its support is greater or equal $\texttt{minsupp}$.

From now on we assume that $\mathcal{F}$ is a partition of the object set G and that the size of every block is known. Hence, for every data pack (F, m) we know its support and its negative support. We define approximations of the above mentioned measures just using these information:

$$\underline{\mathrm{supp}}(A) := \sum_{F \in \mathcal{F}} \max\Big\{0, |F| - \sum_{a \in A} \mathrm{nsupp}(F, a)\Big\},$$

$$\overline{\mathrm{supp}}(A) := \sum_{F \in \mathcal{F}} \min_{a \in A} \mathrm{supp}(F, a),$$

$$\underline{\mathrm{conf}}(A \to B) := \frac{\sum_{F \in \mathcal{F}} \max\Big\{0, \min_{a \in A} \mathrm{supp}(F, a) - \sum_{b \in B \setminus A} \mathrm{nsupp}(F, b)\Big\}}{\overline{\mathrm{supp}}(A)}.$$

Proposition 6. For $A, B \subseteq M$ it holds that:

$$\underline{\mathrm{supp}}(A) \leq \mathrm{supp}(A) \leq \overline{\mathrm{supp}}(A).$$

Furthermore, the inequality $\underline{\mathrm{conf}}(A \to B) \leq \mathrm{conf}(A \to B)$ holds.

Even though these approximations are very coarse in most cases, they are tight in the sense that there are cases where equality holds. We say an association rule **holds** in the granulated partial context $(\mathcal{F}, M, i)$ if

$$\texttt{minsupp} \leq \underline{\mathrm{supp}}(A \cup B) \quad \text{and} \quad \texttt{minconf} \leq \underline{\mathrm{conf}}(A \to B).$$

Corollary 1. Every association rule that holds in $(\mathcal{F}, M, i)$ also holds in (G, M, I).

For a singleton conclusion $B = \{b\}$ we can further approximate the lower approximation $\underline{\mathrm{conf}}(A \to b)$ of the confidence of rule $A \to b$ (with $b \notin A$) in the following way[3]:

$$\underline{\mathrm{conf}}(A \to b) \geq 1 - \frac{|b^I|}{\underline{\mathrm{supp}}(A)}.$$

The right side of this inequality exceeds $\texttt{minconf}$ iff the following holds:

$$\frac{|b^I|}{|G|} \geq 1 - (1 - \texttt{minconf}) \cdot \frac{\overline{\mathrm{supp}}(A)}{|G|}.$$

---

[2] It is more common to define the *support* of A as the quotient $\frac{|A^I|}{|G|}$. We choose to define it the *absolute* way since it makes the following formulas more readable.

[3] By applying the inequality $\max\{0, x\} \geq x$ to every summand in the numerator in the definition of $\underline{\mathrm{conf}}$.

Let us take for instance `minconf` $= 70\%$ and $\overline{\text{supp}}(A) = 0.6 \cdot |G|$. If in this case $82\% (= 1 - 0.3 \cdot 0.6)$ of all objects have attribute b, we can for sure read from the granulated context $(\mathcal{F}, M, i)$ that the rule is frequent. By the way, the support

$$|b^I| = \sum_{F \in \mathcal{F}} \text{supp}(F, b)$$

of the attribute b can be read from the granulated context $(\mathcal{F}, M, i)$. Hence, we get that at least for association rules with very high support and with a conclusion containing very frequent attributes, the chances that its $\underline{\text{conf}}$ value exceeds the threshold `minconf` are not too bad. But when is the rule a frequent rule? Let C be an attribute set (for instance $C = A \cup \{b\}$). It holds that

$$\underline{\text{supp}}(C) \geq |G| - \sum_{m \in C} |m^I|$$
$$\geq |G| - |C| \cdot (|G| - \underline{\text{supp}}(C))$$

The right side exceeds `minsupp` iff

$$\underline{\text{supp}}(C) \geq \frac{(|C| - 1) \cdot |G| + \text{minsupp}}{|C|}.$$

If we take for instance $|C| = 4$ and `minsupp` $= 0.2 \cdot |G|$, we get that C can be detected as frequent by just using the granulated context $(\mathcal{F}, M, i)$ if its actual support (in (G, M, I)) is at least 80% of $|G|$. Note that the approximations of $\underline{\text{supp}}(C)$ we made above were quite rigid. Hence, in practice we expect that $\underline{\text{supp}}$ gives a much better lower approximation of the actual support supp in (G, M, I) than our example may suggest.

Due to a lack of space we have to leave out the details on how to calculate a *basis* of the association rules that hold in $(\mathcal{F}, M, i)$. We will do this in a future paper. In summary our procedure will use the fact that $\underline{\text{supp}}$ yields to a closure system on M. The frequent closed attribute sets will be used to build a Luxenburger-type basis [8]. Furthermore, the following paper should investigate how to improve the approximations $\underline{\text{conf}}$ and $\underline{\text{supp}}$ if one considers background knowledge that can for instance be given by the scales used in the scaling process.

6   Conclusion

We proposed a way to describe the rough tables occurring at the data warehousing system ICE. We did that from the standpoint of Formal Concept Analysis and tried to mine these rough tables for implicational knowledge. We argued that it is very unlikely that the very rigid *minimum* and *maximum* parameters as for instance used in the contest data set [6] will yield to satisfying results. We constituted that – having in mind the data mining in rough tables – in the process of building the data pack nodes it is worth to create more sophisticated parameters that allow to give a better estimation of the distribution of the values in the data packs (like counting the number of incidences in the data packs of the scaled data table).

Ongoing work has to include the following issues: How can one efficiently calculate a basis of the association rules in Section 5? One has to explain how background knowledge can be used to improve data mining in the granulated contexts. Furthermore, experimental results are needed to find out whether the soft granulation described in Section 5 will lead to satisfying results in practice.

References

1. Baader, F., Ganter, B., Sattler, U., Sertkaya, B.: Completing Description Logic Knowledge Bases using Formal Concept Analysis, vol. 258. CEUR-WS.org (2007)
2. Bělohlávek, R., Vychodil, V.: What is a fuzzy concept lattice? In: CLA 2005, pp. 34–45 (2005) ISBN 80–248–0863–3
3. Burmeister, P., Holzer, R.: On the Treatment of Incomplete Knowledge in Formal Concept Analysis, vol. 1867, pp. 385–398. Springer, Heidelberg (2000)
4. Ganter, B., Wille, R.: Formal Concept Analysis – Mathematical Foundations. Springer, Heidelberg (1999)
5. http://www.infobright.org
6. http://web.iitd.ac.in/%7Epremi09/infobright.pdf
7. Infobright Community Edition, Technology White Paper, http://www.infobright.org/wiki/662270f87c77e37e879ba8f7ac2ea258/
8. Lakhal, L., Stumme, G.: Efficient Mining of Association Rules Based on Formal Concept Analysis, vol. 3626, pp. 180–195 (2005)
9. Pawlak, Z.: Rough Sets: Theoretical Aspects of Reasoning About Data. Kluwer Academic Publishers, Dordrecht (1991)

Covering Based Approaches to Rough Sets and Implication Lattices

Pulak Samanta[1] and Mihir Kumar Chakraborty[2]

[1] Department of Mathematics, Katwa College
Katwa, Burdwan, West Bengal, India
pulak_samanta06@yahoo.co.in
[2] Department of Pure Mathematics, University of Calcutta
35, Ballygunge Circular Road, Kolkata-700019, India
mihirc99@vsnl.com

Abstract. This paper deals with a survey of some aspects of covering based approaches to rough set theory and their implication lattices.

Keywords: Rough sets, Partition, Covering, Implication Lattice.

1 Introduction

Pawlak's rough set theory begins with an approximation space $< U, R >$ where U is a non empty set and R is an equivalence relation on U. So, the set U is partitioned. Given any subset A of U, the lower and upper approximations $\underline{A}$ and $\overline{A}$ are then defined by $\underline{A} = \{x| \ [x] \subseteq A\}$ and $\overline{A} = \{x| \ [x] \cap A \neq \phi\}$.

One can immediately observe that the following properties of lower and upper approximations hold.

(1a) $\underline{U} = U$ (Co-normality)   (1b) $\overline{U} = U$ (Co-normality)

(2a) $\underline{\phi} = \phi$ (Normality)   (2b) $\overline{\phi} = \phi$ (Normality)

(3a) $\underline{X} \subseteq X$ (Contraction)   (3b) $X \subseteq \overline{X}$ (Extension)

(4a) $\underline{X \cap Y} = \underline{X} \cap \underline{Y}$ (Multiplication)   (4b) $\overline{X \cup Y} = \overline{X} \cup \overline{Y}$ (Addition)

(5a) $\underline{(\underline{X})} = \underline{X}$ (Idempotency)   (5b) $\overline{(\overline{X})} = \overline{X}$ (Idempotency)

(6) $\underline{(\sim X)} =\sim (\overline{X}), \ \overline{(\sim X)} =\sim (\underline{X})$ (Duality)

(7a) $X \subseteq Y \Rightarrow \underline{X} \subseteq \underline{Y}$ (Monotone)   (7b) $X \subseteq Y \Rightarrow \overline{X} \subseteq \overline{Y}$ (Monotone)

(8a) $A \subseteq \underline{(\overline{A})}$ (8b) $\overline{(\underline{A})} \subseteq A$

Almost from inception of the theory, various generalizations took place one such being replacement of the partition of the set U by a covering. One starts with a set and a covering on it, that is a collection of subsets such that its union is the whole set U. A passage from partition to covering was natural from the point of view of applications also. The equivalence relation R in U originates from an attribute-value system $(U, \{A_i\}, V)$ where $\{A_i\}$ is a set of attributes and V is a set of values, each attribute A_i giving a unique value from V to each object in the universe U. Thus a partition emerges, elements having the same attribute-values being clustered together forms an equivalence class. Elements belonging

H. Sakai et al. (Eds.): RSFDGrC 2009, LNAI 5908, pp. 127–134, 2009.

to the same class are indiscernible with respect to the given set of attributes. Now indiscernibility relation is in general non-transitive - in attribute value systems such an indiscernibility arises if there are some gaps in the information viz. for some objects the value of some attribute may not be known. However, value-gaps are not the only reason for generation of non transitive indiscernibility. The clusters or granules are formed in this situation in various ways and the granules are not generally disjoint. The overlapping granules form a covering of the universe U. Since in the study we shall not be concerned with the process of granule formation, for our present purpose, as mentioned before, the pair $< U, \mathcal{C} >$ where $\mathcal{C} = \{C_i\}$ is a covering of U in a reasonable starting point.

Along with various methods of formation of granules, the lower and upper approximations of a subset of U are also defined in various ways. The objective of this paper is to present an account of various definitions of lower and upper approximations proposed so far and to study their consequences. Consequences will be marked in terms of implication latices, a notion first introduced in [3].

Given two sets A, B there are nine possible inclusions $P \subseteq Q$ where $P \in \{\underline{A}, A, \overline{A}\}$ and $Q \in \{\underline{B}, B, \overline{B}\}$. In case of partition on X we have the following equivalences $\{\underline{A} \subseteq \overline{B}\}$, $\{\underline{A} \subseteq \underline{B}, A \subseteq B\}$, $\{A \subseteq \overline{B}, \overline{A} \subseteq \overline{B}\}$, $\{A \subseteq B\}$ and $\{A \subseteq \underline{B}, \overline{A} \subseteq \underline{B}, \overline{A} \subseteq B\}$ in the sense that inclusions belonging to the same group are equivalent that is, each implies the other. These equivalence classes form a lattice with respect to inclusion again. In the present paper this lattice is the lattice for P_4. For more detail of these implication lattices see [3]. However, in case of covering based approximations, since all the relevant properties among (1) to (8) are not available the equivalence classes of inclusions are different and the implication lattices are different too.

The paper is divided into two broad sections. In the following section various definitions of the lower and upper approximations shall be presented. All these definitions are already present in rough-set literature; we have only compiled and categorized them. Categorization shall be done in terms of usual set theoretic properties. Many of these properties have already been mentioned in earlier works. But many properties were not investigated before. These are our own observations and marked with a '*' in the table in the next section.

Section two deals with the implication lattices and categorization with the help of them.

This paper ends with some concluding remarks.

2   Various Types of Lower and Upper Approximations

Let $\mathcal{C} = \{C_i\}$ be a covering of U. The following various types of granulation around an element $x \in U$ are used in defining lower and upper approximations.

$N_x^{\mathcal{C}} = \cup\{C_i \in \mathcal{C} : x \in C_i\} = Friends(x)$ [5,8,12]
$P_x^{\mathcal{C}} = \{y \in U : \forall C_i(x \in C_i \Leftrightarrow y \in C_i)\}$ (Partition generated by a covering) [4,7,8,12,17]
$N(x) = \cap\{C_i : C_i \in \mathcal{C}, x \in C_i\} = Neighbour(x)$ [9,16,17]
$Md(x) = \{C_i : x \in C_i \in \mathcal{C} \wedge (\forall S \in \mathcal{C} \wedge x \in S \subseteq C_i \Rightarrow C_i = S)\}$ [10],
$e.f(x) = U - Friends(x)$ [5]

Except $Md(x)$ all other constructs are subsets of U while $Md(x)$ is a subset of the power set of U.

Let X be a subset of U, where U is the universe. Then different types of lower and upper approximations are defined as follows.

We have used $\underline{P}_i, \overline{P}^i$ $i = 1, 2, 3, 4$ to recognize Pomykala, since to our knowledge he first studied the lower and upper approximations with the exception P_4 which was due to Pawlak. $\underline{C}_i, \overline{C}^i$ $i = 1, 2, 3, 4, 5$ are other covering based approximations which are essentially duals. C and $\overline{C}$ with symbols are also covering based, the symbols being taken from the respective papers straightway. This group of pairs barring $\underline{C}_{Gr}$ and $\overline{C}^{Gr}$ are non-duals.

$\underline{P}_1(X) = \{x : N_x^{\mathcal{C}} \subseteq X\}$
$\overline{P}^1(X) = \cup\{C_i : C_i \cap X \neq \phi\}$ [8,12,15]
$\underline{P}_2(X) = \cup\{N_x^{\mathcal{C}} : N_x^{\mathcal{C}} \subseteq X\}$
$\overline{P}^2(X) = \{z : \forall y(z \in N_y^{\mathcal{C}} \Rightarrow N_y^{\mathcal{C}} \cap X \neq \phi)\}$ [8,12]
$\underline{P}_3(X) = \cup\{C_i : C_i \subseteq X\}$
$\overline{P}^3(X) = \{y : \forall C_i(y \in C_i \Rightarrow C_i \cap X \neq \phi)\}$ [8,9,12,13,15,17]
$\underline{P}_4(X) = \cup\{P_x^{\mathcal{C}} : P_x^{\mathcal{C}} \subseteq X\}$
$\overline{P}^4(X) = \cup\{P_x^{\mathcal{C}} : P_x^{\mathcal{C}} \cap X \neq \phi\}$ [1,4,7,8,9,10,12,13,14,16,17]
$\underline{C}_1(X) = \cup\{C_i : C_i \in \mathcal{C}, C_i \subseteq X\}$
$\overline{C}^1(X) = \sim \underline{C}_1(\sim X) = \cap\{\sim C_i : C_i \in \mathcal{C}, C_i \cap X = \phi\}$ [10]
$\underline{C}_2(X) = \{x \in U : N(x) \subseteq X\}$
$\overline{C}^2(X) = \{x \in U : N(x) \cap X \neq \phi\}$ [9,10]
$\underline{C}_3(X) = \{x \in U : \exists u(u \in N(x) \wedge N(u) \subseteq X)\}$
$\overline{C}^3(X) = \{x \in U : \forall u(u \in N(x) \rightarrow N(u) \cap X \neq \phi)\}$ [10]
$\underline{C}_4(X) = \{x \in U : \forall u(x \in N(u) \rightarrow N(u) \subseteq X)\}$
$\overline{C}^4(X) = \cup\{N(x) : N(x) \cap X \neq \phi\}$ [10]
$\underline{C}_5(X) = \{x \in U : \forall u(x \in N(u) \rightarrow u \in X)\}$
$\overline{C}^5(X) = \cup\{N(x) : x \in X\}$ [10]

With the same lower approximation there are a few different upper approximations. In the following we have the symbols by corresponding authors.

$\underline{C}_*(X) = \underline{C}_-(X) = \underline{C}_\#(X) = \underline{C}_@(X) = \underline{C}_+(X) = \underline{C}_\%(X)$
$= \cup\{C_i \in \mathcal{C} : C_i \subseteq X\} \equiv \underline{P}_3(X)$ [5]
$\overline{C}^*(X) = \underline{C}_*(X) \cup \{Md(x) : x \in X \setminus X_*\}$ [5,17]
$\overline{C}^-(X) = \cup\{C_i : C_i \cap X \neq \phi\}$ [5]
$\overline{C}^\#(X) = \cup\{Md(x) : x \in X\}$ [5,17]
$\overline{C}^@(X) = \underline{C}_@(X) \cup \{C_i : C_i \cap (X \setminus \underline{C}_@(X)) \neq \phi\}$ [5]
$\overline{C}^+(X) = \underline{C}_+(X) \cup \{Neighbour(x) : x \in X \setminus \underline{C}_+(X)\}$ [5,16]
$\overline{C}^\%(X) = \underline{C}_\%(X) \cup \{\sim \cup\{Friends(y) : x \in X \setminus \underline{C}_\%(X), y \in e.f(x)\}\}$ [5]

Yet another type of lower and upper approximation is defined with the help of covering.

Let, $Gr_*(X) = \cup\{C_i \in \mathcal{C} : C_i \subseteq X\} \equiv \underline{P}_3(X)$.

This is taken as lower approximation of X and is denoted by $\underline{C}_{Gr}(X)$.

$Gr^*(X) = \cup\{C_i \in \mathcal{C} : C_i \cap X \neq \phi\} \equiv \overline{P}^1(X)$.

The upper approximation is defined by $\overline{C}^{Gr}(X) = Gr^*(X) \setminus NEG_{Gr}(X)$, where, $NEG_{Gr}(X) = \underline{C}_{Gr}(\sim X)$, $\sim X$ being the complement of X [13].

We split 4(a)(multiplication), 4(b)(addition) each into two components e.g. 4(a) is split as $\underline{A \cap B} \subseteq \underline{A} \cap \underline{B}$ and $\underline{A} \cap \underline{B} \subseteq \underline{A \cap B}$. Similarly the other.

The reason for this split is that it will be observed that one of the components may hold while the other may not.

Instead of 5(a) and 5(b) we have taken $\underline{A} \subseteq \underline{(A)}$ and $\overline{(A)} \subseteq \overline{A}$ for similar reasons.

The following table shows that the entire Picture.

	P_1	P_2	P_3	P_4	C_1	C_2	C_3	C_4	C_5	C_{Gr}	C_*	C_-	$C_\#$	$C_@$	C_+	$C_\%$
Dual	Y	Y	Y	Y	Y	Y	Y	Y	Y	Y	N	N	N	N	N	N
$\underline{\phi} = \phi = \overline{\phi}$	Y	Y	Y	Y	Y	Y	Y	Y	Y	Y	Y	Y	Y	Y$_*$	Y	Y
$\underline{U} = U = \overline{U}$	Y	Y	Y	Y	Y	Y	Y	Y	Y	Y	Y	Y	Y	Y$_*$	Y	Y
$\underline{A \cap B} \subseteq \underline{A} \cap \underline{B}$	Y	Y	Y	Y	Y	Y	Y	Y	Y	Y	Y	Y	Y	Y$_*$	Y	Y
$\overline{A \cap B} \subseteq \overline{A} \cap \overline{B}$	Y	Y	Y	Y	N	Y	N	Y	Y	N	N	N	N	N$_*$	N	N
$\underline{A} \cup \underline{B} \subseteq \underline{A \cup B}$	Y	Y	Y	Y	N	Y	N	Y	Y	N	Y	Y	N$_*$	Y	N	N
$\overline{A} \cup \overline{B} \subseteq \overline{A \cup B}$	Y	Y	Y	Y	Y	Y	Y	Y	Y	N	Y	Y	N$_*$	Y	Y	N
$A \subseteq B \Rightarrow \underline{A} \subseteq \underline{B}$	Y	Y	Y	Y	Y	Y	Y	Y	Y	Y	Y	Y	Y	Y$_*$	Y	Y
$A \subseteq B \Rightarrow \overline{A} \subseteq \overline{B}$	Y	Y	Y	Y	Y	Y	Y	Y	Y	Y	N	Y	Y	N$_*$	Y	Y
$\underline{A} \subseteq A$	Y	Y	Y	Y	Y	Y	N	Y	Y	Y	Y	Y	Y	Y$_*$	Y	Y
$A \subseteq \overline{A}$	Y	Y	Y	Y	Y	Y	N	Y	Y	N	Y	Y	Y	Y$_*$	Y	N
$\underline{A} \subseteq \overline{A}$	Y	Y	Y	Y	Y	Y	N$_*$	Y	Y	Y	Y	Y	Y	Y$_*$	Y	Y
$\underline{A} \subseteq \underline{(\overline{A})}$	Y$_*$	N$_*$	Y$_*$	Y$_*$	N$_*$	N$_*$	N$_*$	Y	N$_*$	N	Y$_*$	Y$_*$	Y	Y$_*$	N$_*$	N$_*$
$\overline{(\underline{A})} \subseteq A$	Y$_*$	N$_*$	Y$_*$	Y$_*$	N$_*$	N$_*$	N$_*$	Y	N$_*$	N$_*$	N$_*$	N$_*$	Y	Y$_*$	Y$_*$	N$_*$
$A \subseteq \overline{(\underline{A})}$	N	Y	Y	Y	Y	Y	N$_*$	N	Y	Y	Y	Y	Y	Y$_*$	Y	Y
$\overline{(\underline{A})} \subseteq \overline{A}$	N	Y	Y	Y	Y	Y	N$_*$	N	Y	Y	Y	Y	N	Y$_*$	Y	Y
$\underline{A} \subseteq \underline{(\underline{A})}$	N	N	N	Y	N$_*$	N$_*$	N	N	N	N	Y$_*$	Y	Y	Y$_*$	N	N
$\underline{(\underline{A})} \subseteq \underline{A}$	N	N	N	Y	N$_*$	N$_*$	N	N$_*$	N	N	N$_*$	N$_*$	Y	Y$_*$	Y	N

Y : Yes, the property holds.   N : No, the property does not hold.

As mentioned in the introduction properties that we have verified are marked $*$. The above table may be called the information system for the various approaches, the first row giving their names.

3 Implication Lattices

Implication lattices were first introduced in [3]. Their role in rough logics has been discussed in the same paper. We shall now demonstrate various implication lattices arising in the present context.

Implication Lattice with respect to P_1 and C_4

Properties used :

$A \subseteq B \Rightarrow \underline{A} \subseteq \underline{B}$ and $\overline{A} \subseteq \overline{B}$,

$\underline{A} \subseteq A$, $A \subseteq \overline{A}$, $\underline{(\overline{A})} \subseteq A$ and $A \subseteq \overline{(\underline{A})}$.

$\overline{A} \subseteq B \Rightarrow \overline{(\overline{A})} \subseteq \underline{B} \Rightarrow A \subseteq \overline{(\overline{A})} \subseteq \underline{B} \Rightarrow A \subseteq \underline{B}$

and $A \subseteq \underline{B} \Rightarrow \overline{A} \subseteq \overline{(\underline{B})} \subseteq B \Rightarrow \overline{A} \subseteq B$

So, $\overline{A} \subseteq B \Leftrightarrow A \subseteq \underline{B}$.

Implication Lattice with respect to P_3

Properties used :

$A \subseteq B \Rightarrow \underline{A} \subseteq \underline{B}$ and $\overline{A} \subseteq \overline{B}$,

$\underline{A} \subseteq A$, $A \subseteq \overline{A}$, $\underline{(\overline{A})} \subseteq A$, $A \subseteq \overline{(\underline{A})}$

$\underline{A} \subseteq \underline{(\underline{A})}$ and $\overline{(\overline{A})} \subseteq \overline{A}$.

$\overline{A} \subseteq B \Rightarrow \overline{(\overline{A})} \subseteq \underline{B} \Rightarrow A \subseteq \overline{(\overline{A})} \subseteq \underline{B} \Rightarrow A \subseteq \underline{B}$

and $A \subseteq \underline{B} \Rightarrow \overline{A} \subseteq \overline{(\underline{B})} \subseteq B \Rightarrow \overline{A} \subseteq B$.

So, $\overline{A} \subseteq B \Leftrightarrow A \subseteq \underline{B}$.

$\underline{A} \subseteq B \Rightarrow \underline{(\underline{A})} \subseteq \underline{B} \Rightarrow \underline{A} \subseteq \underline{(\underline{A})} \subseteq \underline{B} \Rightarrow \underline{A} \subseteq \underline{B}$.

Also, $\underline{A} \subseteq \underline{B} \Rightarrow \underline{A} \subseteq \underline{B} \subseteq B \Rightarrow \underline{A} \subseteq B$. So, $\underline{A} \subseteq B \Leftrightarrow \underline{A} \subseteq \underline{B}$.

Again, $A \subseteq \overline{B} \Rightarrow \overline{A} \subseteq \overline{(\overline{B})} \Rightarrow \overline{A} \subseteq \overline{(\overline{B})} \subseteq \overline{B} \Rightarrow \overline{A} \subseteq \overline{B}$.

Also, $\overline{A} \subseteq \overline{B} \Rightarrow A \subseteq \overline{A} \subseteq \overline{B} \Rightarrow A \subseteq \overline{B}$. So, $A \subseteq \overline{B} \Leftrightarrow \overline{A} \subseteq \overline{B}$.

Implication Lattice with respect to P_4

Properties used :

$A \subseteq B \Rightarrow \underline{A} \subseteq \underline{B}$ and $\overline{A} \subseteq \overline{B}$,

$\underline{A} \subseteq A$, $A \subseteq \overline{A}$, $\underline{(\overline{A})} \subseteq A$ and $A \subseteq \overline{(\underline{A})}$,

$\underline{A} \subseteq \underline{(\underline{A})}$ and $\overline{(\overline{A})} \subseteq \overline{A}$.

$\overline{A} \subseteq B \Rightarrow \overline{(\overline{A})} \subseteq \underline{B} \Rightarrow A \subseteq \overline{(\overline{A})} \subseteq \underline{B} \Rightarrow A \subseteq \underline{B}$

and $A \subseteq \underline{B} \Rightarrow \overline{A} \subseteq \overline{(\underline{B})} \subseteq B \Rightarrow \overline{A} \subseteq B$.

So, $\overline{A} \subseteq B \Leftrightarrow A \subseteq \underline{B}$.

Clearly, $\overline{A} \subseteq B \Rightarrow \overline{(\overline{A})} \subseteq \underline{B} \Rightarrow \overline{A} \subseteq \overline{(\overline{A})} \subseteq \underline{B} \Rightarrow \overline{A} \subseteq \underline{B}$.

$\overline{A} \subseteq \underline{B} \Rightarrow \overline{A} \subseteq \underline{B} \subseteq B \Rightarrow \overline{A} \subseteq B$. So $\overline{A} \subseteq B \Leftrightarrow \overline{A} \subseteq \underline{B}$.

$\underline{A} \subseteq B \Rightarrow \underline{(\underline{A})} \subseteq \underline{B} \Rightarrow \underline{A} \subseteq \underline{(\underline{A})} \subseteq \underline{B} \Rightarrow \underline{A} \subseteq \underline{B}$.

Also, $\underline{A} \subseteq \underline{B} \Rightarrow \underline{A} \subseteq \underline{B} \subseteq B \Rightarrow \underline{A} \subseteq B$. So, $\underline{A} \subseteq B \Leftrightarrow \underline{A} \subseteq \underline{B}$.

Again, $A \subseteq \overline{B} \Rightarrow \overline{A} \subseteq \overline{(\overline{B})} \Rightarrow \overline{A} \subseteq \overline{(\overline{B})} \subseteq \overline{B} \Rightarrow \overline{A} \subseteq \overline{B}$.

Also, $\overline{A} \subseteq \overline{B} \Rightarrow A \subseteq \overline{A} \subseteq \overline{B} \Rightarrow A \subseteq \overline{B}$. So, $A \subseteq \overline{B} \Leftrightarrow \overline{A} \subseteq \overline{B}$.

Implication Lattice with respect to $C_@$

Properties used :

$A \subseteq B \Rightarrow \underline{A} \subseteq B$, $\underline{A} \subseteq A$, $A \subseteq \overline{A}$,

$A \subseteq \overline{(\overline{A})}$, $\overline{(\underline{A})} \subseteq A$, and $\underline{A} \subseteq \overline{(\underline{A})}$.

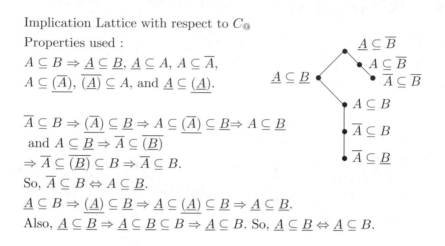

$\overline{A} \subseteq B \Rightarrow \overline{(\overline{A})} \subseteq \underline{B} \Rightarrow A \subseteq \overline{(\overline{A})} \subseteq \underline{B} \Rightarrow A \subseteq \underline{B}$

and $A \subseteq \underline{B} \Rightarrow \overline{A} \subseteq \overline{(\underline{B})}$

$\Rightarrow \overline{A} \subseteq \overline{(\underline{B})} \subseteq B \Rightarrow \overline{A} \subseteq B$.

So, $\overline{A} \subseteq B \Leftrightarrow A \subseteq \underline{B}$.

$\underline{A} \subseteq B \Rightarrow \overline{(\underline{A})} \subseteq \underline{B} \Rightarrow \underline{A} \subseteq \overline{(\underline{A})} \subseteq \underline{B} \Rightarrow \underline{A} \subseteq \underline{B}$.

Also, $\underline{A} \subseteq B \Rightarrow \underline{A} \subseteq \underline{B} \subseteq B \Rightarrow \underline{A} \subseteq B$. So, $\underline{A} \subseteq B \Leftrightarrow \underline{A} \subseteq \underline{B}$.

Implication Lattice with respect to $P_2, C_1, C_2, C_5, C_-, C_+$

Properties used :

$A \subseteq B \Rightarrow \underline{A} \subseteq \underline{B}$ and $\overline{A} \subseteq \overline{B}$,

$\underline{A} \subseteq A$, $A \subseteq \overline{A}$,

$A \subseteq \overline{(\underline{A})}$ and $\overline{(\overline{A})} \subseteq \overline{A}$.

$\underline{A} \subseteq B \Rightarrow \overline{(\underline{A})} \subseteq \overline{B}$

$\Rightarrow A \subseteq \overline{(\underline{A})} \subseteq \overline{B} \Rightarrow A \subseteq \overline{B}$.

Also, $\underline{A} \subseteq B \Rightarrow \underline{A} \subseteq \underline{B} \subseteq B \Rightarrow \underline{A} \subseteq B$. So, $\underline{A} \subseteq B \Leftrightarrow \underline{A} \subseteq B$.

$A \subseteq \overline{B} \Rightarrow \overline{A} \subseteq \overline{(\overline{B})} \Rightarrow \overline{A} \subseteq \overline{(\overline{B})} \subseteq \overline{B} \Rightarrow \overline{A} \subseteq \overline{B}$.

Also, $\overline{A} \subseteq \overline{B} \Rightarrow A \subseteq \overline{A} \subseteq \overline{B} \Rightarrow A \subseteq \overline{B}$. So, $A \subseteq \overline{B} \Leftrightarrow \overline{A} \subseteq \overline{B}$.

Implication Lattice with respect to C_*

Properties used :

$A \subseteq B \Rightarrow \underline{A} \subseteq B$, $\underline{A} \subseteq A$, $\underline{A} \subseteq \overline{(\underline{A})}$.

$\underline{A} \subseteq B \Rightarrow \overline{(\underline{A})} \subseteq \underline{B}$

$\Rightarrow \underline{A} \subseteq \overline{(\underline{A})} \subseteq \underline{B} \Rightarrow \underline{A} \subseteq \underline{B}$.

Also, $\underline{A} \subseteq B \Rightarrow \underline{A} \subseteq \underline{B} \subseteq B \Rightarrow \underline{A} \subseteq B$.

So, $\underline{A} \subseteq B \Leftrightarrow \underline{A} \subseteq B$.

Implication Lattice with respect to $C_\%, Gr$

Properties used :

$A \subseteq B \Rightarrow \underline{A} \subseteq \underline{B}$ and $\overline{A} \subseteq \overline{B}$,

$\underline{A} \subseteq A$, $\underline{A} \subseteq \underline{(A)}$ and $\overline{(\overline{A})} \subseteq \overline{A}$.

$\underline{A} \subseteq B \Rightarrow \underline{(A)} \subseteq \underline{B}$
$\Rightarrow \underline{A} \subseteq \underline{(A)} \subseteq \underline{B} \Rightarrow \underline{A} \subseteq \underline{B}$.
Also, $\underline{A} \subseteq B \Rightarrow \underline{A} \subseteq \underline{B} \subseteq B \Rightarrow \underline{A} \subseteq B$.
So, $\underline{A} \subseteq B \Leftrightarrow \underline{A} \subseteq \underline{B}$.

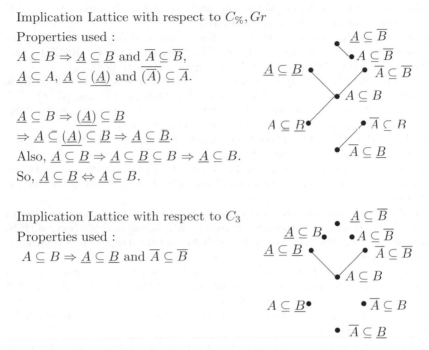

Implication Lattice with respect to C_3

Properties used :

$A \subseteq B \Rightarrow \underline{A} \subseteq \underline{B}$ and $\overline{A} \subseteq \overline{B}$

The study of implication lattices has the following significance.

• Given two Sets A and B, of the nine possible inclusions between the pairs from $\{\underline{A}, A, \overline{A}\}$ and $\{\underline{B}, B, \overline{B}\}$ how many are independent is depicted by the nodes of the diagrams.

• Which inclusion entails which one is shown.

• Any of the inclusion gives rise to a rough Modus Ponens rule [2] and a corresponding rough logic [2]. Taken with 2 the hierarchy of the logics is obtained.

Underlying modal logical systems of various rough logics are also immediately visible from the table. It may also be mentioned that it will be necessary to define and investigate modal logic systems in which necessity and possibility operations are not dual.

4   Concluding Remarks

Other issues of covering-based approaches e.g. topological and logical aspects shall be our future work.

References

1. Bonikowski, Z.: A Certain Copnception of the Calculus of Rough Sets. Notre Dame J. Formal Logic 33, 412–421 (1992)
2. Bunder, M.W., Banerjee, M., Chakraborty, M.K.: Some Rough Consequence Logics and their Interrelations. In: Peters, J.F., Skowron, A. (eds.) Transactions on Rough Sets VIII. LNCS, vol. 5084, pp. 1–20. Springer, Heidelberg (2008)

3. Chakraborty, M.K., Banerjee, M.: Rough dialogue and implication lattices. Fundamenta Informaticae 75(1-4), 123–139 (2007)
4. Chakraborty, M.K., Samanta, P.: Consistency-Degree Between Knowledges. In: Kryszkiewicz, M., Peters, J.F., Rybiński, H., Skowron, A. (eds.) RSEISP 2007. LNCS (LNAI), vol. 4585, pp. 133–141. Springer, Heidelberg (2007)
5. Liu, J., Liao, Z.: The sixth type of covering-based rough sets. In: GrC 2008 IEEE International Conference on Granular Computing 2008, pp. 438–441 (2008)
6. Pal, S.K., Polkowski, L., Skowron, A.: Rough - Neural Computing. In: Pal, S.K., Polkowski, L., Skowron, A. (eds.). Springer, Heidelberg (2004)
7. Pawlak, Z.: ROUGH SETS - Theoritical Aspects of Reasoning About Data. Kluwer Academic Publisher, Dordrecht (1991)
8. Pomykala, J.A.: Approximation, Similarity and Rough Constructions. ILLC Prepublication Series for Computation and Complexity Theory CT-93-07, University of Amsterdam
9. Li, T.-J.: Rough Approximation Operators in Covering Approximation Spaces. In: Greco, S., Hata, Y., Hirano, S., Inuiguchi, M., Miyamoto, S., Nguyen, H.S., Słowiński, R. (eds.) RSCTC 2006. LNCS (LNAI), vol. 4259, pp. 174–182. Springer, Heidelberg (2006)
10. Qin, K., Gao, Y., Pei, Z.: On Covering Rough Sets. In: Yao, J., Lingras, P., Wu, W.-Z., Szczuka, M.S., Cercone, N.J., Ślęzak, D. (eds.) RSKT 2007. LNCS (LNAI), vol. 4481, pp. 34–41. Springer, Heidelberg (2007)
11. Skowron, A., Stepaniuk, J.: Tolerance approximation spaces. Fundamenta Informaticae 27, 245–253 (1996)
12. Samanta, P., Chakraborty, M.K.: On Extension of Dependency and Consistency Degrees of Two Knowledges Represented by Covering. In: Peters, J.F., Skowron, A., Rybiński, H. (eds.) Transactions on Rough Sets IX. LNCS, vol. 5390, pp. 351–364. Springer, Heidelberg (2008)
13. Slezak, D., Wasilewski, P.: Granular Sets–Foundations and Case Study of Tolerance Spaces. In: An, A., Stefanowski, J., Ramanna, S., Butz, C.J., Pedrycz, W., Wang, G. (eds.) RSFDGrC 2007. LNCS (LNAI), vol. 4482, pp. 435–442. Springer, Heidelberg (2007)
14. Yao, Y.Y.: Constructive and algebraic methods of the theory of rough sets. Journal of Information Sciences 109, 21–47 (1998)
15. Yao, Y.Y.: On generalizing rough set theory. In: Wang, G., Liu, Q., Yao, Y., Skowron, A. (eds.) RSFDGrC 2003. LNCS (LNAI), vol. 2639, pp. 44–51. Springer, Heidelberg (2003)
16. Zhu, W.: Topological approaches to covering rough sets. ScienceDirect. Information Sciences 177, 1499–1508 (2007)
17. Zhu, W., Wang, F.-Y.: Relationship among Three Types of Covering Rough Sets. In: Proc. IEEE Int'l Conf. Grannuler Computing (GrC 2006), May 2006, pp. 43–48 (2006)

A Logical Reasoning System of Before-after Relation Based on Bf-EVALPSN

Kazumi Nakamatsu[1], Jair Minoro Abe[2], and Seiki Akama[3]

[1] University of Hyogo, Himeji, Japan
nakamatu@shse.u-hyogo.ac.jp
[2] Paulista University, Sao Paulo, Brazil
jairabe@uol.com.br
[3] University of Tsukuba, Tsukuba, Japan
sub-akama@jcom.home.ne.jp

Abstract. A paraconsistent annotated logic program called bf-EVALP-SN has been developed for dealing with before-after relations between time intervals (processes) and applied to real-time process order control. In this paper, we introduce a logical before-after relation reasoning system based on two inference rules for before-after relation with simple examples.

Keywords: Before-after relation, EVALPSN, bf-EVALPSN, annotated logic program, reasoning system.

1 Introduction

We have already developed a paraconsistent annotated logic program called Extended Vector Annotated Logic Program with Strong Negation(abbr. EVALPSN), which can deal with intelligent control and safety verification such as pipeline process control [3,4,5]. We also have developed an EVALPSN called bf(before-after)-EVALPSN to deal with before-after relations between time intervals paraconsistently, which can be applied to real-time process order control [6,7]. In this paper, we extend the result of Nakamatsu et al.[6] to a before-after relation reasoning system based on bf-EVALPSN.

Suppose that festival A starts on Feb.10th and finishes on 14th, and festival B starts on Feb.16th and finishes on 17th. Then, if we have a question, "Is festival A held before festival B ?", everyone has to answer "yes". On the other hands, if festival B starts on 11th and finishes on 12th, what about the answer for the same question ? Some people may answer "yes" and other people may do "no". There is paraconsistency in the people's knowledge. In bf-EVALPSN, a special EVALP literal $R(p_i, p_j, t) : [(m, n), \mu]$ called bf-EVALP literal whose vector annotation (m, n) paraconsistently represents the before-after relation between two processes Pr_i and Pr_j at time t is introduced. The first/second components m/n in the vector annotation (m, n) represent after/before degrees of the before-after relation, respectively. For example, the first before-after relation between

H. Sakai et al. (Eds.): RSFDGrC 2009, LNAI 5908, pp. 135–143, 2009.
© Springer-Verlag Berlin Heidelberg 2009

festivals A and B could be represented as $R(fesA, fesB, t) : [(0, 12), \alpha]$ [1], which can be intuitively interpreted that it is a fact that nobody agrees "festival A is held after festival B, and all other 12 people agree that festival A is held before festival B". Moreover, the second before-after relation of them could be represented in the EVALPSN literal $R(fesA, fesB, t) : [(4, 8), \alpha]$, which can be paraconsistently interpreted that it is a fact that 4 people agree "festival A is held after festival B" and other 8 people agree "festival A is held before festival B". We will introduce the before-after relation reasoning system to infer the vector annotation (m, n) of the bf-literal in real time according to process start/finish time information.

Suppose that we deal with n processes and their bf-relations in bf-EVALPSN, then $_nC_2$ bf-relations should be considered, which requires much more computation cost. It is not so efficient to compute directly all $_nC_2$ before-after relations based on all process start/finish time information. In order to avoid such inefficiency we also propose another before-after relation reasoning system that can reason the vector annotation of $R(p_i, p_k, t)$ from those of $R(p_i, p_j, t)$ and $R(p_j, p_k, t)$ transitively. If we use the transitive before-after relation reasoning system, only $n - 1$ before-after relations for n processes should be computed directly according to process start/finish time information and other before-after relations can be computed by the transitive reasoning system. We will also introduce the transitive before-after relation reasoning system.

This paper is organized as the following manner: first, EVALPSN and bf-EVALPSN are reviewed briefly ; next, the basic and transitive before-after relation reasoning systems are introduced with simple examples ; last, the conclution is provided.

2   Bf-EVALPSN

We review bf-EVALPSN. The details of them are refered to [3,8].

An annotation in EVALPSN has a form of $[(i, j), \mu]$ called an *extended vector annotation*. The first component (i, j) is called a *vector annotation* and the set of vector annotations constitutes the complete lattice, $\mathcal{T}_v(n) = \{(x, y) | 0 \leq x \leq n, 0 \leq y \leq n, x, y \text{ and } n \text{ are integers}\}$. The ordering($\preceq_v$) of $\mathcal{T}_v(n)$ is defined as : $(x_1, y_1) \preceq_v (x_2, y_2)$ iff $x_1 \leq x_2$ and $y_1 \leq y_2$. For each extended vector annotated literal $p : [(i, j), \mu]$, the integer i denotes the amount of positive information to support the literal p, the integer j denotes that of negative information, and the annotation $\mu \in \{\perp, \alpha, \beta, \gamma, *_1, *_2, *_3, \top\}$ is an index of deontic notions such as obligation. The set of the annotations constitutes the complete lattice, $\mathcal{T}_d$. The ordering($\preceq_d$) of $\mathcal{T}_d$ is described by the Hasse's diagram in Fig.1. Then, the complete lattice $\mathcal{T}_e(n)$ of extended vector annotations is defined as $\mathcal{T}_v(n) \times \mathcal{T}_d$. The intuitive meaning of each member of $\mathcal{T}_d$ is $\perp$ (unknown), α (fact), β (obligation), γ (non-obligation), $*_1$ (fact and obligation), $*_2$ (obligation and non-obligation), $*_3$ (fact and non-obligation), and $\top$ (inconsistency).

---

[1] α is interpreted as "it is a fact that $\cdots$".

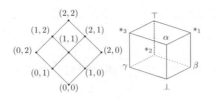
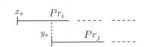

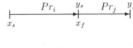

Fig. 1. Lattices $\mathcal{T}_v(2)$ and $\mathcal{T}_d$ Before/After

Fig. 2. Disjoint Before/After   Immediate Before/After

Fig. 3. Joint Before/After   S-included Before/After

There are two kinds of epistemic negation $\neg_1$ and $\neg_2$ in EVALPSN, which are defined as mappings over lattices $\mathcal{T}_v(n)$ and $\mathcal{T}_d$, respectively. There also is ontological(strong) negation($\sim$) in EVALPSN, which is defined by the epistemic negations $\neg_1$ or $\neg_2$, and it works as classical negation. Let $L_0, \cdots, L_n$ be weva-literals [2], $L_1 \wedge \cdots \wedge L_i \wedge \sim L_{i+1} \wedge \cdots \wedge \sim L_n \rightarrow L_0$ is called an *EVALPSN clause*. An *EVALPSN* is a finite set of EVALPSN clauses.

First of all, we introduce a literal $R(p_i, p_j, t)$ whose vector annotation represents the before-after relation between processes Pr_i and Pr_j at time t, which is called a *bf-literal*[3]. An extended vector annotated literal $R(p_i, p_j, t) : [\mu_1, \mu_2]$ is called a *bf-EVALP literal*, where μ_1 is a vector annotation and $\mu_2 \in \{\alpha, \beta, \gamma\}$. If an EVALPSN clause contains bf-EVALP literals, it is called a *bf-EVALPSN clause* or just a *bf-EVALP clause* if it contains no strong negation.

Now we introduce the following bf-relations represented in vector annotations called *bf-annotations*. They are described in process time charts (Fig.1-4).

Before (be)/After (af): We define the most basic bf-relations *before/after* based on the bf-relation between each start time of two processes, which are represented by bf-annotations be/af, respectively. If one process has started before/after another one, then the bf-relations are defined as "before(be)/after(af)"

Disjoint Before (db)/After (da): Bf-relations *disjoint before/after* between two processes Pr_i and Pr_j are represented by bf-annotations db/da

---

[2] $p : [(i, 0), \mu]$ and $p : [(0, j), \mu]$ are called *weva-literals*, where i, j are non-negative integers and $\mu \in \{\alpha, \beta, \gamma\}$.

[3] Hereafter, the word "**before-after**" is abbreviated as just "bf" in this paper.

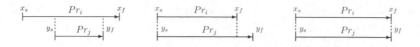

Fig. 4. Included, F-included, Paraconsistent Before/After

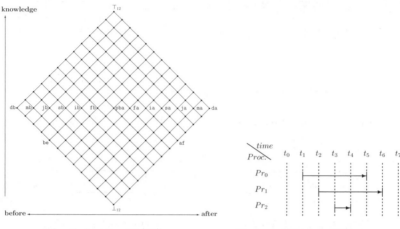

Fig. 5. Lattice $\mathcal{T}_v(12)_{bf}$ Process Schedule Chart

Immediate Before (mb)/After (ma): Bf-relations *immediate before/after* between processes Pr_i and Pr_j are represented by bf-annotations mb/ma

Joint Before (jb)/After (ja): Bf-relations, *joint before/after* between processes Pr_i and Pr_j are represented by bf-annotations jb/ja

S-included Before (sb)/After (sa): The bf-relations *s-included before/after* between processes Pr_i and Pr_j are represented by bf-annotations sb/sa

Included Before(ib)/After(ia): Bf-relations *included before/after* between processes Pr_i and Pr_j are represented by bf-annotations ib/ia

F-included Before(fb)/After(fa): The bf-relations *f-include before/after* between processes Pr_i and Pr_j are represented by bf-annotations fb/fa

Paraconsistent Before-after (pba): The bf-relation *paraconsistent before-after* between processes Pr_i and Pr_j is represented by bf-annotation pba

If we take the before-after measure over the ten bf-annotations as the horizontal order and the before-after knowledge amount of them as the vertical one, we obtain the complete bi-lattice $\mathcal{T}_v(12)_{bf}$ of bf-annotations (Fig.5). Then, there is the following correspondence between bf-annotations and vector annotations, be$(0,8)$/af$(8,0)$, db$(0,12)$/da$(12,0)$, mb$(1,11)$/ma$(11,1)$, jb$(2,10)$/ja$(10,2)$, sb$(3,9)$/sa$(9,3)$, ib$(4,8)$/ia$(8,4)$, fb$(5,7)$/fa$(7,5)$, pba$(6,6)$.

3   Bf-Relation Reasoning System

In this section, we introduce the basic and transitive bf-relation(annotation) reasoning systems. Firstly we show a simple example for reasoning bf-annotations with three processes Pr_0, Pr_1 and Pr_2 scheduled in Fig.5, and three bf-literals $R(p_0, p_1, t)$, $R(p_1, p_2, t)$ and $R(p_0, p_2, t)$.

At time t_0, as no process has started, we have no knowledge of bf-relations. Therefore, the bf-relations are unknown and the same tentative bf-annotation of all the bf-literals is $(0, 0)$.

At time t_1, as only process Pr_0 has started, it can be reasoned that both the bf-annotations of $R(p_0, p_1, t_1)$ and $R(p_0, p_2, t_1)$ are members of a set $\{\mathtt{db}(0, 12),$ $\mathtt{mb}(1, 11), \mathtt{jb}(2, 10), \mathtt{sb}(3, 9), \mathtt{ib}(4, 8)\}$. Therefore, the greatest lower bound $(0, 8)$ of the set is the tentative bf-annotation of them, which is the greatest knowledge in terms of their bf-relations, and the tentative bf-annotation of $R(p_1, p_2, t_1)$ is still unknown $(0, 0)$.

At time t_2, as process Pr_1 has started before process Pr_0 finishes, it can be reasoned that the bf-annotation of $R(p_0, p_1, t_2)$ is a member of a set $\{\mathtt{jb}(2, 10),$ $\mathtt{sb}(3, 9), \mathtt{ib}(4, 8)\}$. Therefore, the greatest lower bound $(2, 8)$ of the set is the tentative bf-annotation of it. Moreover, as process Pr_2 has not started yet, the tentative bf-annotation of others is the same $(0, 8)$.

At time t_3, as process Pr_2 has started before both processes Pr_0 and Pr_1 finish, it can be reasoned that the tentative bf-annotation of all the bf-literals is the same $(2, 8)$ as well as the case of time t_2.

At time t_4, as only process Pr_2 has finished, it can be reasoned that the tentative bf-annotation of $R(p_0, p_1, t_4)$ is still $(2, 8)$, however the final bf-annotations of other bf-literals become $\mathtt{ib}(4, 8)$.

At time t_5, as process Pr_0 has finished before process Pr_1 finishes, the bf-annotation of $R(p_0, p_1, t_5)$ becomes $\mathtt{jb}(2, 10)$. Therefore, even though process Pr_1 has not finished yet, all bf-relations between processes Pr_0, Pr_1 and Pr_2 have been determined at time t_5.

As shown in the above example, bf-relations(annotations) can be determined according to process start/finish time information. It is quite natural to adopt the following bf-relation inference rules for the basic bf-relation reasoning system.

$(0, 0)$-rule-1: If process Pr_i has started and process Pr_j has not started yet, then the tentative bf-annotation of $R(p_i, p_j, t)$ becomes $(0, 8)$ from $(0, 0)$.

$(0, 0)$-rule-2: If both processes Pr_i and Pr_j have started at the same time, the tentative bf-annotation of $R(p_i, p_j, t)$ becomes $(5, 5)$ from $(0, 0)$. They are represented by the bf-EVALPSN clause with no deontic annotation,

$$R(p_i, p_j, t):(0, 0) \wedge st(p_i, t):(1, 0) \wedge \sim st(p_j, t):(1, 0) \rightarrow R(p_i, p_j, t):(0, 8),$$
$$R(p_i, p_j, t):(0, 0) \wedge st(p_i, t):(1, 0) \wedge st(p_j, t):(1, 0) \rightarrow R(p_i, p_j, t):(5, 5),$$

where two literals $st(p_i, t)/fi(p_i, t)$ represent "process Pr_i starts/finishes at time t", and their vector annotations are members of $\{(1, 0)(\text{true}), (0, 1)(\text{false})\}$.

$(0,8)$-rule-1: If process Pr_i has finished, and process Pr_j has not started yet, then the bf-annotation of $R(p_i, p_j, t)$ becomes $\mathtt{db}(0, 12)$ from $(0, 8)$.

$(0,8)$-rule-2: If process Pr_i has finished and process Pr_j has started immediately after it, then the bf-annotation of $R(p_i, p_j, t)$ becomes $\mathtt{ib}(1, 11)$ from $(0, 8)$.

(0,8)-rule-3: If process Pr_i has started but not finished yet and process Pr_j has also started after it, then the tentative bf-annotation of $R(p_i, p_j, t)$ becomes $(2, 8)$ from $(0, 8)$. They are represented by the bf-EVALPSN clause with no deontic annotation,

$$R(p_i, p_j, t) : (0, 8) \wedge fi(p_i, t) : (1, 0) \wedge \sim st(p_j, t) : (1, 0) \rightarrow R(p_i, p_j, t) : (0, 12),$$
$$R(p_i, p_j, t) : (0, 8) \wedge fi(p_i, t) : (1, 0) \wedge st(p_j, t) : (1, 0) \rightarrow R(p_i, p_j, t) : (1, 11),$$
$$R(p_i, p_j, t) : (0, 8) \wedge \sim fi(p_i, t) : (1, 0) \wedge st(p_j, t) : (1, 0) \rightarrow R(p_i, p_j, t) : (2, 8).$$

(5,5)-rule-1: If both processes Pr_i and Pr_j have started simultaneously and only process Pr_i has finished, then the bf-annotation of $R(p_i, p_j, t)$ becomes $\mathtt{sb}(5, 7)$ from $(5, 5)$.

(5,5)-rule-2: If both processes Pr_i and Pr_j have started simultaneously and finished simultaneously, then the bf-annotation of $R(p_i, p_j, t)$ becomes $\mathtt{pba}(6, 6)$ from $(5, 5)$.

(5,5)-rule-3: If both processes Pr_i and Pr_j have started simultaneously and only process Pr_j has finished, then the bf-annotation of $R(p_i, p_j, t)$ becomes $\mathtt{sa}(7, 5)$ fom $(5, 5)$. They are represented by the bf-EVALPSN clause with no deontic annotation,

$$R(p_i, p_j, t) : (5, 5) \wedge fi(p_i, t) : (1, 0) \wedge \sim fi(p_j, t) : (1, 0) \rightarrow R(p_i, p_j, t) : (5, 7),$$
$$R(p_i, p_j, t) : (5, 5) \wedge fi(p_i, t) : (1, 0) \wedge fi(p_j, t) : (1, 0) \rightarrow R(p_i, p_j, t) : (6, 6),$$
$$R(p_i, p_j, t) : (5, 5) \wedge \sim fi(p_i, t) : (1, 0) \wedge fi(p_j, t) : (1, 0) \rightarrow R(p_i, p_j, t) : (7, 5).$$

(2,8)-rule-1: If processes Pr_i and Pr_j have started sequentially, process Pr_i has finished and process Pr_j has not finished yet, then the bf-annotation of $R(p_i, p_j, t)$ becomes $\mathtt{jb}(2, 10)$ from $(2, 8)$.

(2,8)-rule-2: If processes Pr_i and Pr_j have started sequentially and they finished at the same time, then the bf-annotation of $R(p_i, p_j, t)$ becomes $\mathtt{fb}(3, 9)$ from $(2, 8)$.

(2,8)-rule-3: If processes Pr_i and Pr_j have started sequentially and process Pr_i has not finished yet, though process Pr_j has already finished, then the bf-annotation of $R(p_i, p_j, t)$ becomes $\mathtt{ib}(4, 8)$ from $(2, 8)$. They are represented by the bf-EVALPSN clause with no deontic annotation,

$$R(p_i, p_j, t) : (2, 8) \wedge fi(p_i, t) : (1, 0) \wedge \sim fi(p_j, t) : (1, 0) \rightarrow R(p_i, p_j, t) : (2, 10),$$
$$R(p_i, p_j, t) : (2, 8) \wedge fi(p_i, t) : (1, 0) \wedge fi(p_j, t) : (1, 0) \rightarrow R(p_i, p_j, t) : (3, 9),$$
$$R(p_i, p_j, t) : (2, 8) \wedge \sim fi(p_i, t) : (1, 0) \wedge fi(p_j, t) : (1, 0) \rightarrow R(p_i, p_j, t) : (4, 8).$$

We introduce another bf-relation(annotation) inference rule called the transitive bf-relation inference rule that can infer the bf-annotation of $R(p_i, p_k, t)$ from those of $R(p_i, p_j, t)$ and $R(p_j, p_k, t)$ with three processes Pr_i, Pr_j and Pr_k starting sequentially. We show only three simple cases for describing the transitive reasoning.

Table 1. Transitive Inference Rules

rules	p_i,p_j	p_j,p_k	p_i,p_k	rules	p_i,p_j	p_j,p_k	p_i,p_k	rules	p_i,p_j	p_j,p_k	p_i,p_k
r1	(0,8)	(0,0)	(0,8)	**r11**	(0,12)	(0,0)	(0,12)	**r12**	(2,8)	(0,8)	(0,8)
r121	(2,10)	(0,8)	(0,12)	**r122**	(4,8)	(0,12)	(0,8)	**r123**	(2,8)	(2,8)	(2,8)
r1231	(2,10)	(2,8)	(2,10)	**r1232**	(4,8)	(2,10)	(2,8)	**r1233**	(2,8)	(4,8)	(4,8)
r1234	(3,9)	(2,10)	(2,10)	**r1235**	(2,10)	(4,8)	(3,9)	**r1236**	(4,8)	(3,9)	(4,8)
r1237	(3,9)	(3,9)	(3,9)	**r124**	(3,9)	(0,12)	(0,12)	**r125**	(2,10)	(2,8)	(1,11)
r126	(4,8)	(1,11)	(2,8)	**r127**	(3,9)	(1,11)	(1,11)	**r13**	(1,11)	(0,8)	(0,12)
r14	(2,8)	(5,5)	(2,8)	**r141**	(4,8)	(5,7)	(2,8)	**r142**	(2,8)	(7,5)	(4,8)
r143	(3,9)	(5,7)	(2,10)	**r144**	(2,10)	(7,5)	(3,9)	**r2**	(5,5)	(0,8)	(0,8)
r21	(5,7)	(0,8)	(0,12)	**r22**	(7,5)	(0,12)	(0,8)	**r23**	(5,5)	(2,8)	(2,8)
r231	(5,7)	(2,8)	(2,10)	**r232**	(7,5)	(2,10)	(2,8)	**r233**	(5,5)	(4,8)	(4,8)
r234	(7,5)	(3,9)	(4,8)	**r24**	(5,7)	(2,8)	(1,11)	**r25**	(7,5)	(1,11)	(2,8)
r3	(5,5)	(5,5)	(5,5)	**r31**	(7,5)	(5,7)	(5,5)	**r32**	(5,7)	(7,5)	(6,6)

Case 1. Suppose that only the first process Pr_i has started at time t, we obtain the tentative bf-annotation $(0,8)$ of $R(p_i, p_j, t)$ by $(0,0)$-rule-1 and we have the tentative bf-annotation $(0,0)$ of $R(p_j, p_k, t)$, then the vector annotation of $R(p_i, p_k, t)$ can be infered deterministically as $(0,8)$, which is formalized,

$$R(p_i, p_j, t) : (0,8) \wedge R(p_j, p_k, t) : (0,0) \rightarrow R(p_i, p_j, t) : (0,8).$$

Case 2. Suppose that processes Pr_i and Pr_j have started simultaneously at time t, we obtain the tentative bf-annotation $(5,5)$ of $R(p_i, p_j, t)$ by the $(0,0)$-rule-2 and the tentative bf-annotation $(0,8)$ of $R(p_j, p_k, t)$ by the $(0,0$-rule-1, then the vector annotation of $R(p_i, p_k, t)$ can be also reasoned deterministically as $(0,8)$, which is formalized,

$$R(p_i, p_j, t) : (5,5) \wedge R(p_j, p_k, t) : (0,8) \rightarrow R(p_i, p_j, t) : (0,8).$$

Case 3. Suppose that all processes Pr_i, Pr_j and Pr_k have started simultaneously at time t, we obtain the same tentative bf-annotation $(5,5)$ of $R(p_i, p_j, t)$ and $R(p_j, p_k, t)$ by the $(0,0)$-rule-2, then the vector annotation of $R(p_i, p_k, t)$ can be also reasoned deterministically as $(5,5)$, which is formalized,

$$R(p_i, p_j, t) : (5,5) \wedge R(p_j, p_k, t) : (5,5) \rightarrow R(p_i, p_j, t) : (5,5).$$

Only three rules have been shown though, Other transitive bf-relation inference rules are listed in Table 1. For simplicity the inference rules are represented by three vector annotations such as $(n_1, n_2)|(n_3, n_4)|(n_5, n_6)$ instead of the bf-EVALP clause with no deontic annotation,

$$R(p_i, p_j, t) : (n_1, n_2) \wedge R(p_j, p_k, t) : (n_3, n_4) \rightarrow R(p_i, p_k, t) : (n_5, n_6).$$

The transitive bf-relation inference rule name (block font) indicates the applicable order of the transitive inference rules. For example, if rule **r1** has been applied, the next applicable rules are **r11**, **r12**, **r13** or **r14**. Furthermore, if rule **r12** has been applied, one of rules **r121**,···,**r127**, can be applied at the next step ; on the other hand, if rule **r11** has been applied, there is no applicable rule that follows it and the final bf-annotation $(0, 12)$ of $R(p_i, p_k, t)$ can be derived.

Here we note that in terms of the inference rules **r122**, **r1232**, **r126**, **r141**, **r22**, **r232**, **r25** and **r31**, even though they have no following rules to be applied, they can not derive the bf-annotation of $R(p_i, p_k, t)$. For example, by rule **r1232**, even if both the bf-annotation $(4, 8)$ of $R(p_i, p_j, t)$, and the bf-annotation $(2, 10)$ of $R(p_j, p_k, t)$ are obtained, the bf-annotation of $R(p_i, p_k, t)$ can not be determined and just a tentative bf-annotation $(2, 8)$ is obtained, which implies three possibilities $\{(2, 10), (3, 9), (4, 8)\}$ as the final bf-annotation, thus $(2, 8)$-rules have to be applied at the next step for determining the final bf-annotation. Therefore, if the transitive inference rules **r122** – **r31** have been applied, $(0, 8), (2, 8), (5, 5)$-rules have to be applied by way of exception.

4   Conclusions

In this paper, we have introduced the bf-relation reasoning system based on bf-EVALPSN, which consists of the basic and transitive bf-relation inference rules.

As a related work, interval temporal logic in which bf-relations are represented in some special predicates such as *Meets*[4] has been proposed by Allen et al.[1,2] for representing knowledge of properties, actions and events. It is sure that the interval temporal logic is a logically sophisticated tool to develop practical planning or natural language understanding systems though, it does not seem to be so suitable for real-time processing because bf-relations cannot be determined until both of them have finished in the logical system. On the other hand, bf-relations(annotations) are represented by paraconsistent vector annotations more minutely in bf-EVALPSN, thus, they can be determined in real time by the basic and transitive bf-relation inference rules according to start/finish information of processes. Moreover, since Bf-EVALPSN is one of logic programs and can be implemented as both software and hardware[5], The bf-relation reasoning system can be practically applied to intelligent real-time process order control and so on.[5]

References

1. Allen, J.F.: Towards a General Theory of Action and Time. Artificial Intelligence 23, 123–154 (1984)
2. Allen, J.F., Ferguson, G.: Actions and Events in Interval Temporal Logic. J. Logic and Computation 4, 531–579 (1994)

---

[4] *Meets*(a, b) represents that time intervals a and b exist sequentially, which corresponds to bf-annotation ma/mb.

[5] This work is financially supported by Japanese Scientific Research Grant (C) No. 20500074.

3. Nakamatsu, K., Abe, J.M., Suzuki, A.: Annotated semantics for defeasible deontic reasoning. In: Ziarko, W.P., Yao, Y. (eds.) RSCTC 2000. LNCS (LNAI), vol. 2005, p. 470. Springer, Heidelberg (2001)
4. Nakamatsu, K.: Pipeline Valve Control Based on EVALPSN Safety Verification. JACIII 10, 647–656 (2006)
5. Nakamatsu, K., Mita, Y., Shibata, T.: An Intelligent Action Control System Based on Extended Vector Annotated Logic Program and its Hardware Implementation. J. Intelligent Automation and Soft Computing 13, 289–304 (2007)
6. Nakamatsu, K., Abe, J.M., Akama, S.: Paraconsistent Before-after Relation Reasoning Based on EVALPSN. Studies in Computational Intelligence, vol. 142, pp. 265–274. Springer, Heidelberg (2008)
7. Nakamatsu, K., Akama, S., Abe, J.M.: Transitive Reasoning of Before-After Relation Based on Bf-EVALPSN. In: Lovrek, I., Howlett, R.J., Jain, L.C. (eds.) KES 2008, Part II. LNCS (LNAI), vol. 5178, pp. 474–482. Springer, Heidelberg (2008)
8. Nakamatsu, K., Abe, J.M.: The development of Paraconsistent Annotated Logic Program. Intl. J. Reasoning-based Intelligent Systems 1, 92–110 (2009)

Some Proof Theoretic Results Depending on Context from the Perspective of Graded Consequence

Soma Dutta[1] and Mihir Kumar Chakraborty[2]

[1] Department of Mathematics, IBRAD School of Management and
Sustainable Development
Kolkata, West Bengal, India
somadutta77@yahoo.co.in
[2] Department of Pure Mathematics, University of Calcutta
35, Ballygunge Circular Road, Kolkata-700019, India
mihirc4@gmail.com

Abstract. In this paper we discuss some proof theoretic properties of logics generated with respect to a context from the perspective of graded consequence relation. We also indicate some implications of these results in actual application.

Keywords: Graded consequence, Proof theory, Fuzzy logic, Context dependency, Rough sets.

1 Introduction

In this paper we shall investigate some properties of graded consequence relation. These properties are in some sense counterparts of proof theoretic assertions in the classical, intuitionistic or other logics. Theory of graded consequence was introduced by Chakraborty [1,2] in 1986 as a generalization of Gentzen's notion of consequence in many-valued context. Two main features of this theory are (1) to lift many-valuedness also to the meta-level notions like consequence, consistency, tautologihood etc. and (2) to make the logic context-dependent where the context is given by a set of many-valued valuation functions $\{T_i\}_{i \in I}$. Any T_i is a mapping from the set of atomic formulae to a suitable algebraic structure, constituting the truth set (L) for the given object language. T_i is then extended over the whole set of wffs adopting truth functionality. A collection $\{T_i\}_{i \in I}$ can also be interpreted as the opinions of experts or information about the atomic sentences. As this latest interpretation, a collection $\{T_i\}_{i \in I}$ is an information system that initiates rough set theory [11] where the object set is the set of wffs, T_i is an attribute and $T_i(\alpha)$ is the value of an wff α with respect to T_i. Let us consider, for example, the following matrix where $a_{ij} = T_i(p_j) \in L$.

H. Sakai et al. (Eds.): RSFDGrC 2009, LNAI 5908, pp. 144–151, 2009.
© Springer-Verlag Berlin Heidelberg 2009

Well-formed formula	T_1	T_2	...	T_i	....	T_n
Atomic	p_1 a_{11}	a_{12}	...	a_{1i}	....	a_{1n}
	p_2 a_{21}	a_{22}	...	a_{2i}	....	a_{2n}
	... ...	...	...	...	...	...
	p_k a_{k1}	a_{k2}	...	a_{ki}	....	a_{kn}
Non-atomic	... ...	...	...	...	...	...

Any such matrix is called the context. Whatever might be the interpretation, given a set $\{T_i\}_{i\in I}$, a logic may be defined with a fuzzy or graded consequence relation that assigns a "degree" to the derivability of a wff α from a set X of premises. This degree may be interpreted as the strength or confidence in which α may be inferred from the information X given the context $\{T_i\}_{i\in I}$. We shall denote the consequence relation by $\approx_{\{T_i\}_{i\in I}}$ and the above mentioned degree or grade by $gr(X \approx_{\{T_i\}_{i\in I}} \alpha)$. One should observe the following:

- the consequence relation is a meta-logical notion and is here taken to be many-valued in general.
- an implication operator which is not necessarily the implication operator of the object language is required to define this consequence (see (Σ') below)
- although the grade $gr(X \approx_{\{T_i\}_{i\in I}} \alpha)$ is taken to be an element of L, the logical operators for the defining clause of $\approx_{\{T_i\}_{i\in I}}$ are not necessarily the same as those(if any at all) for the object language.

Thus, we need two sets of operators on the set L for computing the values of the object- level and meta-level sentences. For the present paper we shall take a complete residuated lattice (L, $*_m$, $\to_m$, 0, 1) [8]for the meta-level language and an algebraic structure (L, $*_0$,$\to_o$, 0, 1) for the object level language. Operators $*_m$ and $\to_m$ shall be needed to compute the meta-level "and" and "if-then", while $*_0$, $\to_o$ shall be used for corresponding object level-language.

The main objective of this paper is to investigate into the proof theoretic results that arise out of various conditions imposed on the object and meta-level operators and due to various interrelations among them. The rest of the paper is organized as follows. In section 2, the graded consequence relation $\approx_{\{T_i\}_{i\in I}}$ shall be defined. In section 3, general study of the proof theoretic properties of graded consequence relation will be presented. Besides, two specific cases taking definite structures for object level and meta-level will be investigated. In section 4, we will discuss the significance of proof theoretic properties in the context of graded consequence in contrast to classical situation. In section 5, there is an example of finding out actual grade of a derivation in the context of graded consequence. In the conclusion we indicate some ways of application.

2   Introduction to Graded Consequence

The notion $\approx_{\{T_i\}_{i\in I}}$ is a two-stage generalization of the notion of semantic consequence in classical two-valued logic. The latter is defined by X $\models \alpha$ if and only if for all valuations T in the truth set $\{True, False\}$ or $\{1, 0\}$, if every member of X is true under T then α is also true under T. The first stage generalization was proposed by Shoesmith and Smiley [12] through relativizing the notion $\models$ in terms of any arbitrary collection $\{T_i\}_{i \in I}$ of valuations instead of the set of all valuations. So, one gets $X \models_{\{T_i\}_{i \in I}} \alpha$ if and only if for all valuations $T \in \{T_i\}_{i \in I}$, if every member of X is true under T then α is also true under T. The second stage of generalization was accomplished by taking the valuations T_i as many-valued functions [2] the range being a lattice.

Before proceeding to the actual definition, we shall rewrite the above defining criterion of $\models_{\{T_i\}_{i \in I}}$ in a form that would be followed throughout the paper. Every valuation T_i may be identified with the set of wffs which are true under T_i. So T_i may be considered to be a subset of the set of wffs. Thus the definition of $\models_{\{T_i\}_{i \in I}}$ can be given by

(Σ) $X \models_{\{T_i\}_{i \in I}} \alpha$ iff for all valuations $T \in \{T_i\}_{i \in I}$, if $X \subseteq T$ then $\alpha \in T$.

To present the generalized version of (Σ) in many-valued context the required modifications are as follows.

(i) The valuation function T_i's are now many-valued - although the relationship between the value-set structures for the object and meta-level are not important at this stage. In fact, at this stage no particular object language is considered, nor its value set structure.

(ii) In many-valued context to evaluate the defining sentence of the right hand side of (Σ) a fuzzy implication operator for 'if-then' viz. $\rightarrow_m$ is needed and hence the value of the sentence (Σ) turns out to be

$$inf_i\{gr(X \subseteq T) \rightarrow_m gr(\alpha \in T_i)\}$$
$$= inf_i\{inf_{x \in F}((x \in X) \rightarrow_m (x \in T_i)) \rightarrow_m \alpha \in T_i\}$$
$$= inf_i\{inf_{x \in X}(1 \rightarrow_m T_i(x)) \rightarrow_m T_i(\alpha)\}.$$

'inf' is used to compute the meta-linguistic 'for all' present in (Σ).

In particular, if the $\rightarrow_m$ is taken as the residua of $*_m$ present in a residuated lattice $(L, *_m, \rightarrow_m, 0, 1)$ then the value reduces to $inf_i\{inf_{x \in X}T_i(x)) \rightarrow_m T_i(\alpha)\}$ and is considered as the grade to which α follows from the premise set X. Hence

(Σ') $gr(X \mid\approx_{\{T_i\}_{i \in I}} \alpha) = inf_i\{inf_{x \in X}T_i(x)) \rightarrow_m T_i(\alpha)\}$

$\mid\approx_{\{T_i\}_{i \in I}}$ is a graded consequence relation since it satisfies the following axioms for any general graded consequence relation $\mid\sim$, [3] viz.

GC1. If $\alpha \in X$ then $gr(X \mid\sim \alpha) = 1$,
GC2. If $X \subseteq Y$ then $gr(X \mid\sim \alpha) \leq gr(Y \mid\sim \alpha)$,
GC3. $inf_{\beta \in Y} gr(X \mid\sim \beta) * gr(X \cup Y \mid\sim \alpha) \leq gr(X \mid\sim \alpha)$.

Naturally, $\mid\sim$ is a fuzzy relation from the power set P(F) of the set of formulae F to F. These are generalizations in the many-valued context of Gentzenian axioms for consequence relation [5].

3   Proof Theoretic Properties of Graded Consequence

We divide the results of this section into three categories: properties that depend only on the conditions imposed on object level operators, properties that depend only on the interrelation of the object and meta-level operators and the third, a combination of the previous two. It is to be noted that the basic properties of $\to_m$ which constitute $|\approx_{\{T_i\}_{i\in I}}$ play significant role in asserting the following results.

Results of the first category
Theorem 1. If a $*_0$ b $\leq$ a, b then

(i) $gr(X \cup \{\alpha\} \mid\sim \gamma) \leq gr(X \cup \{\alpha\&\beta\} \mid\sim \gamma)$,
   $gr(X \cup \{\beta\} \mid\sim \gamma) \leq gr(X \cup \{\alpha\&\beta\} \mid\sim \gamma)$
(ii) $gr(X \mid\sim \alpha\&\beta) \leq gr(X \mid\sim \alpha))$, $gr(X \mid\sim \alpha\&\beta) \leq gr(X \mid\sim \beta))$
(iii) $gr(\{\alpha\&\beta\} \mid\sim \alpha) = 1$ and $gr(\{\alpha\&\beta\} \mid\sim \beta) = 1$
(iv) $gr(\{\alpha,\beta\} \mid\sim \gamma) \leq gr(\{\alpha\&\beta\} \mid\sim \gamma)$

In the object level language we have initially taken $\supset$ and $\&$ as primitive connectives. For the time being, let us add one more connective $\vee$ in the object level language and $\oplus$ as the respective operator in the corresponding algebraic structure.

Theorem2. If a, b $\leq$ a $\oplus$ b then

(i) $gr(X \mid\sim \alpha) \leq gr(X \mid\sim \alpha \vee \beta)$, $gr(X \mid\sim \beta) \leq gr(X \mid\sim \alpha \vee \beta)$.
(ii) $gr(X \cup \{\alpha \vee \beta\} \mid\sim \gamma) \leq gr(X \cup \{\alpha\} \mid\sim \gamma)$,
   $gr(X \cup \{\alpha \vee \beta\} \mid\sim \gamma) \leq gr(X \cup \{\beta\} \mid\sim \gamma)$.
(iii) $gr(\{\alpha\} \mid\sim \alpha \vee \beta) = 1$ and $gr(\{\beta\} \mid\sim \alpha \vee \beta) = 1$.

Theorem3. If $a *_0 (a \to_0 b) \leq b$ then

(i) $gr(\{\alpha\&(\alpha \supset \beta)\} \mid\sim \beta) = 1$.
(ii) $\sim \alpha \equiv \alpha \supset \underline{0}$ implies $gr(\{\alpha\& \sim \alpha\} \mid\sim \beta) = 1$.

Notes: 1. 3(ii) gives a sufficient condition for $gr(\{\alpha\& \sim \alpha\} \mid\sim \beta) = 1$. But $\sim$ needs not to be defined in terms of $\supset$ always.
   2. $gr(\{\alpha\& \sim \alpha\} \mid\sim \beta) = 1$ does not imply $gr(\{\alpha, \sim \alpha\} \mid\sim \beta) = 1$ and similarly $gr(\{\alpha\&(\alpha \supset \beta)\} \mid\sim \beta) = 1$ does not imply $gr(\{\alpha, (\alpha \supset \beta)\} \mid\sim \beta) = 1$ (see theorem 1(iv)). The converses hold for both the cases.

Proposition4. If $(a \to_o b) \wedge (b \to_o c) \leq (a \to_o c)$ then
$gr(X \mid\sim \alpha \supset \beta) *_m gr(Y \mid\sim \beta \supset \gamma) \leq gr(X \cup Y \mid\sim \alpha \supset \gamma)$.

Proposition5. If $a \wedge (a \to_o b) \leq b$ then
$gr(X \mid\sim \alpha) *_m gr(Y \mid\sim \alpha \supset \beta) \leq gr(X \cup Y \mid\sim \beta)$.

Proposition6. If $b \leq a \to_o b$ then $gr(\mid\sim \beta) \leq gr(\mid\sim \alpha \supset \beta)$.

Results of the second category
Theorem7. If $\to_m \leq \to_0$ then

(i) $gr(X \cup \{\alpha\} \mid\sim \beta) \leq gr(X \mid\sim \alpha \supset \beta)$.   (ii)$gr(\mid\sim \alpha \supset \alpha) = 1$.
(iii)$gr(\mid\sim \beta) \leq gr(\mid\sim \alpha \supset \beta)$.   (iv) $gr(\mid\sim \beta \supset (\alpha \supset \beta) = 1$.

Results of the third category

Theorem8. If $\to_m \leq \to_0$ and $\alpha \vee \beta \equiv \sim \alpha \supset \beta$ then

(i)$gr(X \cup \{\sim \alpha\} \mid \sim \beta) \leq gr(X \mid \sim \alpha \vee \beta)$.   (ii)$gr(X \mid \sim \alpha \vee \sim \alpha) = 1$.

Theorem9. If $\to_o \leq \to_m$ and $(a \to_m b) *_m (a \to_m c) \leq a \to_m (b *_m c)$ then

(i)$gr(X \mid \sim \alpha) *_m gr(Y \mid \sim \alpha \supset \beta) \leq gr(X \cup Y \mid \sim \beta)$

(ii) $\sim \alpha \equiv \alpha \supset \underline{0}$ implies $gr(\{\alpha, \sim \alpha\} \mid \sim \beta) = 1$.

Note3 Usually, for $*_m = \wedge$, $(a \to_m b) *_m (a \to_m c) \leq a \to_m (b *_m c)$ holds in Heyting algebra.

Theorem10. If $(a \to_o b) \to_o (b \to_o c) \leq (a \to_o c)$ and $\to_m \leq \to_0$ then

(i) $gr(X \mid \sim \alpha \supset \beta) *_m gr(Y \mid \sim \beta \supset \gamma) \leq gr(X \cup Y \mid \sim \alpha \supset \gamma)$

(ii)$\sim \alpha \equiv \alpha \supset \underline{0}$ implies $gr(X \mid \sim \alpha \supset \beta) *_m gr(Y \mid \sim (\sim \beta)) \leq gr(X \cup Y \mid \sim (\sim \alpha))$

(iii) $gr(X \mid \sim \alpha \supset \beta) \leq gr(X \cup \{\sim \beta\} \mid \sim (\sim \alpha))$

(iv)$gr(X \cup \{\alpha\} \mid \sim \beta) \leq gr(X \cup \{\sim \beta\} \mid \sim (\sim \alpha))$

From this general study, a picture of logics with graded notion of consequence can be assessed. But in some cases, the particular structures, taken for the algebra of object level as well as meta-level may add some new results. Let us see two such cases.

(I) Meta-level algebra: A complete pseudo Boolean algebra. Let the primitive connective of the object language be $\supset$. Let $\sim$, &,$\vee$ be defined by $\sim \alpha \equiv \alpha \supset \underline{0}$, $\alpha \& \beta \equiv \sim (\alpha \supset \sim \beta)$ and $\alpha \vee \beta \equiv \sim \alpha \supset \beta$. Let the object level algebra be an Wajsberg algebra (L,$\to_0$, 0). Then the following are obtained.

(i)$gr(X \cup \{\alpha\} \mid \sim \beta) \leq gr(X \mid \sim \alpha \supset \beta)$.

(ii) $gr(\mid \sim \alpha) = 1$ for any theorem α of Łukasiewicz logic.

(iii)$gr(\mid \sim \beta) \leq gr(\mid \sim \alpha \supset \beta)$.

(iv)$gr(X \cup \{\alpha\} \mid \sim \gamma) \leq gr(X \cup \{\alpha \& \beta\} \mid \sim \gamma)$

(v)$gr(X \mid \sim \alpha \& \beta) \leq gr(X \mid \sim \alpha)$

(vi)$gr(\{\alpha \& \beta\} \mid \sim \alpha) = 1$ and $gr(\{\alpha \& \beta\} \mid \sim \beta) = 1$

(vii) $gr(\{\alpha, \beta\} \mid \sim \gamma) \leq gr(\{\alpha \& \beta\} \mid \sim \gamma)$

(viii)$gr(X \mid \sim \alpha) \leq gr(X \mid \sim \alpha \vee \beta)$

(ix) $gr(X \cup \{\alpha \vee \beta\} \mid \sim \gamma) \leq gr(X \cup \{\alpha\} \mid \sim \gamma)$

(x) $gr(\{\alpha\} \mid \sim \alpha \vee \beta) = 1$ and $gr(\{\beta\} \mid \sim \alpha \vee \beta) = 1$.

(xi) $gr(X \mid \sim \alpha) = gr(X \mid \sim (\sim \sim \alpha))$

(xii) $gr(X \cup \{\sim \alpha\} \mid \sim \beta) \leq gr(X \mid \sim \alpha \vee \beta)$.

(xiii)$gr(X \mid \sim \alpha \vee \sim \alpha) = 1$.

(II) Meta-level algebra: An MV-algebra. Object level algebra: A complete pseudo Boolean algebra. Then the following are obtained.

(i)$gr(\mid \sim \beta) \leq gr(\mid \sim \alpha \supset \beta)$ as $b \leq a \to_o b$.

(ii)$gr(\mid \sim \alpha) = 1$ for any theorem α of Godel logic.

(iii)$gr(X \mid \sim \alpha) *_m gr(Y \mid \sim \alpha \supset \beta) \leq gr(X \cup Y \mid \sim \beta)$

(iv)$gr(\{\alpha, \sim \alpha\} \mid \sim \beta) = 1$. (from (iii))

(v)$gr(X \mid \sim \alpha \supset \beta) *_m gr(Y \mid \sim \beta \supset \gamma) \leq gr(X \cup Y \mid \sim \alpha \supset \gamma)$

(vi)gr(X $|\sim \alpha \supset \beta)*_m$ gr(Y $|\sim (\sim \beta))\leq$ gr(X $\cup$ Y $|\sim (\sim \alpha))$
(vii)gr(X $|\sim \alpha \supset \beta)\leq$ gr(X $\cup \{\sim \beta\}$ $|\sim (\sim \alpha))$
(viii)gr(X $|\sim \alpha \supset \beta)\leq$ gr(X $|\sim (\sim \beta \supset \sim \alpha))$.
(ix)gr(X $|\sim \alpha) \leq$ gr(X $|\sim (\sim\sim \alpha))$
(x)gr(X $\cup \{\alpha\}$ $|\sim \gamma) \leq$ gr(X $\cup \{\alpha \& \beta\}$ $|\sim \gamma)$
(xi) gr(X $|\sim \alpha \& \beta) \leq$ gr(X $|\sim \alpha)$
(xii)gr($\{\alpha \& \beta\}|\sim \alpha) = 1$ and gr($\{\alpha \& \beta\}|\sim \beta) = 1$
(xiii)gr(X $|\sim \alpha) \leq$ gr(X $|\sim \alpha \vee \beta)$
(xlv) gr(X $\cup \{\alpha \vee \beta\}$ $|\sim \gamma) \leq$ gr(X $\cup \{\alpha\}$ $|\sim \gamma)$
(xv) gr($\{\alpha\}|\sim \alpha \vee \beta) = 1$ and gr($\{\beta\}$ $|\sim \alpha \vee \beta) = 1$.

4   Significance of Proof Theoretic Properties

Some remarks on some of the above results may be helpful in understanding their significance.

Theorem1 contains the counterpart of the following properties of classical logical consequence $\vdash$.

$$\frac{X\cup\{\alpha\}\vdash\gamma}{X\cup\{\alpha\&\beta\}\vdash\gamma}, \quad \frac{X\vdash\alpha\&\beta}{X\vdash\alpha}, \quad \frac{X\vdash\alpha\&\beta}{X\vdash\beta}$$

Proposition 5 is a version of the rule Modus Ponens.

$$\frac{X \vdash \alpha, \; Y \vdash \alpha \supset \beta}{X \cup Y \vdash \beta}$$

While $\alpha \& (\alpha \supset \beta)\vdash \beta$, it does not necessarily imply α , $(\alpha \supset \beta)\vdash \beta$ (Theorem 3, Note) That is the meta-linguistic conjunction comma (,) and the object language conjunction & should not be treated alike. This is not so in the classical case.

Theorem 7(i) is the counterpart of deduction theorem.

$$\frac{X \cup \{\alpha\} \vdash \beta}{X \vdash \alpha \supset \beta.}$$

Theorem 10 (iii), (iv) are counterparts of

$$\frac{X \vdash \alpha \supset \beta}{X \cup \{\sim \beta\} \vdash \sim \alpha}, \qquad \frac{X \cup \{\alpha\} \vdash \beta}{X \cup \{\sim \beta\} \vdash \sim \alpha.} \quad \text{respectively.}$$

Thus, some well known logical principles (rules) hold because of certain relations hold among the operators of the truth-set algebras i.e. the semantics of the languages at the object and meta-levels. In classical two-valued logic in both the levels the two-point Boolean algebra $\{1, 0\}$ is employed. In many-valued logics the algebraic structure for the object level are varied but the algebra for the meta-level is again the two-point Boolean algebra. Fuzzy logics in the narrow [6,7,10]sense differ from the many-valued logics in the use of fuzzy premises and fuzzy conclusion - these practically mean each wff is tagged with a value from the truth set. The truth-set algebras are of wide variety but the algebra for the meta-level is two-valued Boolean. It is in the case of graded consequence that the meta-level algebra is considered non-Boolean in general.The main difference of the theory of graded consequence from other fuzzy logics lies precisely here. While in fuzzy logics a degree is calculated to the conclusion given degrees to the premises, in this theory given a set of premises X and a single formula α a "degree of derivability of α from X" is obtained.

5   An Example

This is an example of applying the principle of graded consequence to a basic rule of logic like Modus Ponens which can be repeatedly applied to obtain a derivation in the theory of graded consequence.

Example: Let $\{p_1, p_2\}$ be the propositional variables of a language and $\sim$ and $\vee$ be two connectives which are computed by $\neg a = 1 - a$ and 'max' respectively. Let $\supset$ be defined by $p_i \supset p_j \equiv \sim p_i \vee p_j (i = 1, 2)$. Let $\{T_1, T_2\}$ be a collection of fuzzy subsets over the set of formulae, generated from the above mentioned alphabet and defined by

	p_1	p_2	$p_1 \supset p_2$	$p_2 \supset p_1$	$\sim p_1 \supset p_2$	$\sim p_2 \supset p_2$	$p_1 \supset p_1$	$p_2 \supset p_2$	$p_1 \supset \sim p_2$	$p_2 \supset \sim p_1$
T_1	.7	.8	.8	.7	.8	.8	.7	.8	.3	.3
T_2	.8	.9	.9	.8	.9	.9	.8	.9	.2	.2

We have not considered the formulas $\sim p_1 \supset \sim p_1$, $\sim p_2 \supset \sim p_2$, $\sim p_2 \supset \sim p_1$, $\sim p_1 \supset \sim p_2$ in the above table as they have the same truth values as $p_1 \supset p_1$, $p_2 \supset p_2$, $p_1 \supset p_2$, $p_2 \supset p_1$ respectively.

We now calculate the grade of MP taking $([0, 1], *, \rightarrow, 0, 1)$,), an MV-algebra as the meta-level algebraic structure where $\rightarrow$ is defined by

a $\rightarrow$ b = 1 if a $\leq$ b
= (1 - a + b) , otherwise.

One can observe that the implications used at the object level and meta-level are distinct, the first one being that of Kleene-Dienes [9] and the second one that of Łukasiewicz [8] Now it can be easily shown that $inf_{\alpha,\beta}gr(\{\alpha , (\alpha \supset \beta)\} | \sim \beta) = inf_{p_i,p_j}gr(\{p'_i , (p'_i \supset p'_j)\} | \sim p'_j)$, where p'_i denotes any one of p_i and $\sim p_i$ for i = 1, 2, ...n.

    Hence $inf_{\alpha,\beta}gr(\{\alpha, (\alpha \supset \beta)\} | \sim \beta) = inf_{p_i,p_j}gr(\{p'_i, (p_i {}' \supset p'_j)\} | \sim p_{j'}) = inf_{p_i,p_j}[inf_i\{(T_i(p_{i'}) \wedge T_i(p'_i \supset p'_j)) \rightarrow T_i(p'_j)\} = .9.$

6   Conclusion

From the standpoint of use, theory of graded consequence may offer various options that is various logics at the two levels. The following diagram will give a hint.

Object Level	Lukasiewicz	Godel	Product	Kleene	...
Meta Level	Godel	Lukasiewicz	Lukasiewicz	Lukasiewicz	...

The results of section 4 will show which properties of the consequence shall follow for each choice of pairs of logics or equivalently, algebras. In the processing of data or available information any of the algebras (or logic) of object level may be used. After that while making a decision in the sense of inferencing from a set of premises some other logic (algebra) which is not necessarily classical two-valued may be used. This is the meta-level activity and depending on the necessities, the corresponding logics (algebras) may be chosen.

References

1. Chakraborty, M.K.: Use Of Fuzzy Set Theory In Introducing Graded Consequence In Multiple Valued Logic. In: Gupta, M.M., Yamakawa, T. (eds.) Fuzzy Logic in Knowledge-Based Systems, Decision and Control, pp. 247–257. Elsevier Science Publishers, B.V./North Holland, Amsterdam (1988)
2. Chakraborty, M.K.: Graded Consequence: Further Studies. Journal of Applied Non-Classical Logics 5(2), 127–137 (1995)
3. Chakraborty, M.K., Basu, S.: Graded Consequence And Some Metalogical Notions Generalized. Fundamenta Informaticae 32, 299–311 (1997)
4. Chakraborty, M.K., Dutta, S.: Grade In Meta Logical Notions: A Comparative Study of Fuzzy Logics. In: The Proceedings of Studia Logica International Conference, Trends in Logic V: Many-valued Logics and Cognition, Guangzhou, China, July 6-9 (2007)
5. Gentzen, G.: Investigations Into Logical Deductions. In: Gentzen, G., Szabo, M.E. (eds.) The collected papers, pp. 68–131. North Holland Publications, Amsterdam (1969)
6. Gerla, G.: Fuzzy Logic: Mathematical Tools For Approximate Reasoning. Kluwer Academic Publishers, Dordrecht (2001)
7. Goguen, J.A.: The Logic Of Inexact Concepts. Synthese 19, 325–373 (1968)
8. Hájek, P.: Metamathematics of Fuzzy Logic. Kluwer Academic Publishers, Dordrecht (1998)
9. Klir, J.G., Yuan, B.: Fuzzy Sets And Fuzzy Logic, Theory and Applications
10. Pavelka, J.: On Fuzzy Logic I, II, III, Zeitschrift für Math. Logik und Grundlagen d. Math 25, 45-52, 119-134, 447-464 (1979)
11. Pawlak, Z.: Rough Sets. International Journal of information and Computer Sciences 11, 341–356 (1982)
12. Shoesmith, D.J., Smiley, T.J.: Multiple Conclusion Logic. Cambridge University Press, Cambridge (1978)
13. Zadeh, L.A.: "Fuzzy Sets". Information and Control 8(3), 338–353 (1965)

IQuickReduct: An Improvement to Quick Reduct Algorithm

P.S.V.S. Sai Prasad[1] and C. Raghavendra Rao[2]

[1,2] Department of Computer and Information Sciences,
University of Hyderabad, Hyderabad
saics@uohyd.ernet.in, crrcs@uohyd.ernet.in

Abstract. Most of the Rough Sets applications are involved in conditional reduct computations. Quick Reduct Algorithm (QRA) for reduct computation is most popular since its discovery. The QRA has been modified in this paper by sequential redundancy reduction approach. The performance of this new improved Quick Reduct (IQRA) is discussed in this paper.

Keywords: Rough Sets, Feature Selection, Reduct, Quick Reduct, Redundancy, Variable Precision Rough Sets.

1 Introduction

A good amount of data is getting compiled in experimental, exploratory and interactive environments. Data pertaining to several attributes/variables on each object results in production of voluminous data. To handle this data effectively and in the light of curse of dimensionality [1] a popular technique based on statistical arguments is in the practice known as Principle Component Analysis (PCA). The PCA transforms the observable space to a hypothetical space (which is a linear combination of observable variables). This brings effective dimensional reduction by retaining good amount of information whereas suffers from interpretation aspects. With the discovery of Rough Sets by Pawlak [2] it is possible to represent the data in lower dimensional subspace of the observable space. The advantages of Rough Sets for dimensionality reduction over PCA are given in [8,9]. Identification of attributes/variables for inducing the subspace with almost the information contained in the data is an interesting and complex activity.

A minimal collection of set of attributes for meeting the above requirement is known as a Reduct which is the subject matter of several researchers. Reduct computation is one of the important activities in several Rough set based soft computing and Machine Learning systems. Reduct computation is relatively simpler whereas minimal length reduct computation is NP hard [3]. Thus getting a minimal length reduct is handled by heuristic methods.

Quick Reduct Algorithm (QRA) proposed in [4] is an efficient algorithm for finding reduct. This is widely used in several soft computing implementations using Rough Sets [12]. Some improvements to QRA are proposed in [5, 6]. But both accelerated Quick Reduct algorithm [6] and Improved Quick Reduct algorithm [5] has more time and space complexity than the QRA. QRA is aimed only at redundancy/dominancy among attributes but do not consider the redundancy/dominancy associated with the

H. Sakai et al. (Eds.): RSFDGrC 2009, LNAI 5908, pp. 152–159, 2009.

objects. The present study develops theory and algorithm for IQuickReduct (IQRA) which is an improvement to QRA in both time and space complexity.

Section 2 gives the overview of the QuickReduct algorithm. Section 3 discusses the limitations of QRA and gives the useful ness of Variable Precision Rough Sets in improving QRA. Section 4 gives the IQRA algorithm. Section 5 reports the experimental results and illustrates the advantages of the proposed algorithm and the final section is about the conclusions and future work.

2 QuickReduct Algorithm

The basic concepts of Rough sets are given in [2]. The notations used for Rough sets are described here. DT denotes the decision table comprising of U a set of objects, C set of conditional attributes and D set of decision attributes. For a given concept $X \subseteq U$, $\underline{B}X$ denotes lower approximation and $\overline{B}X$ denotes upper approximation with respect to set of attributes $B \subseteq C$. $POS_B(D)$ denotes the Positive region of B. For a set R and $R \subseteq C$, $\gamma_R(D)$ denote kappa measure which gives dependency of D on R.

For a given dataset several reducts may exist. An important application of Rough Sets to Machine Learning is in dimensionality reduction wherein the decision system is built with using only a reduct attributes. In such applications finding one reduct would be sufficient. One of the popular algorithms to find reduct is Quick Reduct algorithm proposed by A. Chouchoulas and Q. Shen [4]. Quick Reduct algorithm is a step up approach and the outline of the algorithm is given below.

Algorithm Quick Reduct(C,D)
Input: C, the set of all conditional attributes; D, the set of decision attributes.
Output: R, the attribute reduct, $R \subseteq C$.

(1) $R = \Phi$
(2) do
(3) $T = R$
(4)   for each $x \in (C - R)$
(5)   if $\gamma_{R \cup \{x\}}(D) > \gamma_T(D)$
(6)   T=RU{x}
(7)   endfor
(8)   R=T
(9) until $\gamma_R(D) = \gamma_C(D)$
(10) return R

QuickReduct algorithm initially starts with an empty set and includes an attribute in an iteration that increases the kappa in a maximum way. As QuickReduct algorithm follows a greedy based approach it has been proved in [10] that QuickReduct may not yield a reduct all times but a super reduct some times. By super reduct we mean a set of attributes which contains a reduct as a subset of it. Still QuickReduct is used widely because of fastness with which one can arrive at a set near to a reduct.

3 Limitations of Quick Reduct Algorithm and Relevance of VPRS Heuristic

The QRA is developed with an assumption that Kappa will be strictly monotonically increasing from current iteration to the next. It suffers when the Kappa is not incremental or zero at any iteration since the algorithms has no directive for the choices of an attribute. The arbiterization of selection of attribute for adding to the R leads to (i) Supper reduct and (ii) a reduct with more attributes. A knowledge driven approach is needed here to improve the performance of QRA. It is observed that QRA suffers in two cases.

Case IA: The kappa in initial iteration(s) is zero: The problem QRA more often faces is in the first iteration if kappa values for all conditional attributes become zero. Here QRA needs to take an arbitrary choice i.e. include the first conditional attribute into R, and continue to the next iteration and most often resulting in a super reduct.

Case IB: No increment in kappa in intermediate iterations: The other situation that arises while building a reduct by QRA is that the Kappa may not be strictly increasing in an intermediate iteration. This situation also leads to arbitrary choice in QRA and has the possibility of resulting in giving a super reduct.

The principles of Variable precision Rough sets (VPRS) developed by Ziarko [11] will be handy in these two cases for giving heuristic information for the selection of attribute into R. VPRS allows calculation of Kappa with a tolerance β in the range from 1 to 0.5. So by gradually reducing β and finding the conditional attribute which gives kappa gain in both Case IA and IB, a specific selection can be made for inclusion of attribute into R. The following notations are used for VPRS [7] concepts. Given $X \subseteq U$ and $B \subseteq C$, $\underline{R_B^\beta}(X)$ denotes β lower approximation and $\overline{R_B^\beta}(X)$ denotes β upper approximation. $POS_B^\beta(D)$ denote Positive region with β precision and $\gamma_B^\beta(D)$ denote kappa with β precision.

Case II: Redundancy: While studying the various instances it is noted that QRA is ignoring the redundancy prevailing about the objects. When a concept associated with an object happened to be in the POS region of a concept induced by D then that object will not contribute/add any more knowledge in further decision making i.e. in the rest of the iterations. Thus the set of objects of a POS region are redundant objects for futuristic purposes. Adoption of filtering the redundant objects in QRA algorithm would significantly influence the time and space complexity of QRA.

The following Improved Quick Reduct Algorithm (IQRA) has been developed by embedding the variable precision concept and also by taking care of the redundant object filtering/removal for finding the Reduct as improvements on QRA.

4   IQuickReduct Algorithm

Algorithm IQuickReduct:
Input: Decision table DT =(U,C $\cup$ D) where U is the set of objects and C is the set of conditional attributes and D is the set of decision attributes.

Output: Set of attributes R preserving the property $\gamma_R^1(D) = \gamma_C^1(D)$.

1. Calculate $\gamma_C^1(D)$.

2. $R = \Phi$.

3. Count=0.

4. While $\gamma_R^1(D) \neq \gamma_C^1(D)$,

5. do

6.   AvailableSet= $C - R$

7. $\beta = 1, \ \varepsilon = 0.1$.

   i. T=R

   ii. foreach $q \in AvailableSet$,

   iii. do

   iv. if $\gamma p_{R \cup \{q\}}^{\beta}(D) > \gamma p_T^{\beta}(D)$,

   v.   T= $R \cup \{q\}$

   vi. endfor

8.   if $T = R$,   /* No kappagain*/

9. $\beta = \beta - \varepsilon$

10.   if $\beta \geq 0.5$,

11.   goto 7i

12.   else

13.   R=R $\cup$ {First attribute in AvailableSet}

14.   else   /*for if in step 8*/

15.   R=T

16.   Set POSPARTIAL to $POS_R^1(D)$ for DT

/*For removal of redundant objects calculate Positive region with tolerance of 1*/

17. Reduce DT by removing the objects belonging to POSPARTIAL.

18. Count=Count+$|POSPARTIAL|$

19. $\gamma_R^1(D) = Count \div |U|$

20. endwhile

21. Return R

Note: In the above algorithm $\gamma p_R^{\beta}(D)$ is used to denote kappa value when calculated for reduced decision table and $\gamma_R^1(D)$ is used for denoting kappa value for original decision table.

The working of IQRA is similar to QRA with necessary modifications to overcome the limitations of QRA such as arbiterization of choice when maximum kappa is zero or no kapa gain and redundancy involved in calculation of kappa as discussed in the previous section. IQRA starts with R initialized to empty set and AvailableSet set to C. At the beginning of the ith iteration β is set to 1. A step as is in QRA is performed to get the attribute $x \in AvalilableSet$ which results in maximum positive value of $\gamma p_{RU\{x\}}^1(D)$. In case such 'x' is available it is included in R. If there is no 'x' which increases the current kappa indicates the situation where an arbitrary choice needs to be made in QRA. Here IQRA incorporates the VPRS heuristic. In such situation the step '7i' is repeated with reduced β until an 'x' is found or β cannot be reduced. In later case, which would be a rare occurrence, we take an arbitrary decision to include the first available attribute in R. Otherwise the attribute 'x' which is resulting in maximum $\gamma p_{RU\{x\}}^{\beta}(D)$ is included in R. After x is included in R the set of objects belonging to Positive region with tolerance of '1' are calculated. VPRS heuristic is used only for selecting an attribute to get included in R and the positive region obtained with tolerance<1, is not used for the reduction of the Decision Table. After x is included in R the set of objects belonging to Positive region with tolerance of '1' only is calculated. These set of objects are redundant for the next iterations. The rows corresponding to these objects are removed from DT before continuing for the next iteration.

$$\gamma_R^1(D) = \frac{|POSPARTIAL| + |POS_{i-1}|}{|U|} \tag{1}$$

If we denote the kappa value obtained in the reduced table as $\gamma p_{RU\{x\}}^\beta(D)$ then we need to calculate the kappa value for the original Decision table after the updating of R with 'x'. The formula in equation (1) (which can be supported by theory) is used in designing this algorithm. In the above equation (1) POS_{i-1} denotes the Positive region obtained for the original decision table till i-1 iterations and POSPARTIAL is Positive region of reduced Decision table at ith iteration. In the algorithm Count variable denotes the cardinality of Positive region for the original decision table. At the end of the iteration Count is updated by adding the cardinality of POSPARTIAL and then used for the calculation of $\gamma_R^1(D)$. The theoretical analysis of the IQRA algorithm, the validity and the influence of reduction of the decision table by removal of Positive region on the solution as well as the time and space complexities are not included in the current work here due to space constraint.

5   Experiments and Results

Both QuickReduct and IQuickReduct algorithms are implemented in Matlab [13] and are tested against standard discrete data sets available in UCI Machine learning

Table 1. Results of the Experiments for QRA and IQRA

Dataset Name	Number of Conditional attributes	Result		Time taken (in Seconds)		Time Gain in IQRA	Nature of the results
		QRA	IQRA	QRA	IQRA		
Credit	20	C1	C1	151.84	23.97	84.21	*
Australia	14	A1	A1	71.21	4.57	93.6	*
Heart	13	H1	H1	9.21	2.22	75.9	*
WDBC	9	WD	WD	6.01	2.06	66.6	*
Webtest (Website classification data)	2556	W1	W1	5902.9	230.48	96.09	*
Mofn dataset	13	[1 2 3 7 5 9 11]	[7 9 1 3 5 11]	9.66	5.75	40.5	**

A1: [14 13 5 8 2 6 1];   C1: [1 4 13 5 7 12 9 16];   H1: [11 13 5 1 4 3 6];   WD: [2 6 1 7];
W1: [1025 1288 2484 1953 2487 2146 2211 865 1623 749 1222 725 347 443 465 595 631 398 520 197 624 238 36 175 21 86 292 702 1]
* Reduct in both
** Reduct in IQRA and Super Reduct in QRA

repository [14] and in particular Web test dataset and Mofn dataset are taken from the website of Richard Jensen[15]. In IQRA algorithm ε is set to 0.1. The following Table 1 gives the detailed summary of the results obtained through the experiments.

6   Conclusions and Future Work

For all the datasets there are significant gains in time complexity for IQuickReduct algorithm over QuickReduct algorithm. Two significant advantages of IQRA over QRA are given below.

1.   ***Reduction in Time Complexity:*** This is seen in all the datasets and more significantly in large datasets. For example the webtest dataset is of the size (149 x 2557). Time taken for finding the reduct using IQuickReduct algorithm is 230.48 seconds where as for QuickReduct it turned out to be 5902.9 seconds. The time gain in IQRA is 96.09. The reduct found in both cases came out to be same. The Figure 1 gives below figure gives the estimate of the size of the space (No. of objects x No of attributes) involved in calculation of kappas in the iterations of QRA and IQRA for webtest dataset. Figure 2 illustrates the time involved in each iteration for QRA and IQRA. The large reduction in time is due to the fact that kappa values are calculated for modified DT in all iterations.

2.   ***Ability to find a better attribute set:*** As long as all the kappa gain in any iteration is not zero then the results in IQRA and QRA are same except for time gain in IQRA over QRA. In case the kappa gain is zero in any iteration for all alternatives the IQRA will search for the best candidate to be included in Minimal set using an iterative procedure using VPRS heuristic. In many experiments it is found that this happens most of the time in first iteration itself. For example for the data set Mofn, in the first iteration kappas are zero for all the conditional attributes. The QRA without any heuristics will include the first attribute in the Minimal set and continue with the remaining iterations. Hence the minimal set found in this algorithm turned out to be [1 2 3 7 5 9 11] of size 7. IQRA using VPRS heuristic selects 7 as the attribute to be included in R and in turned out that the minimal set is [7 9 1 3 5 11] of size 6. Figure 3 illustrates the use of VPRS heuristic in Mofn dataset by giving the kappa values obtained by completing iteration for both QRA and IQRA. In the first iteration for both QRA and IQRA kappa is zero. But because of the choice taken in IQRA there is a positive kappa occurring by third iteration in IQRA. But QRA needed to complete four iterations to get a positive kappa. This helped in IQRA in reaching the required kappa of '1' by 6 iterations and QRA needed to take 7 iterations. Even though extra calculations are incurred for VPRS heuristic calculations in IQRA, the reduction in data set size in iterations compensates that and there is a significant time gain of 40.5% in IQRA compared to QRA. Using RSES [16] tool on Mofn dataset we calculated the reducts for Mofn and Mofn data set has only a single reduct [1 3 5 7 9 11], which is obtained by IQuickReduct algorithm. As the result of QRA is a super set of IQRA result, the set obtained in QRA is a super reduct.

It is proposed to adopt apt heuristics for the ambiguity in the selection of the attribute which may be persistent in IQRA even reaching the lower limit of $\beta = 0.5$.

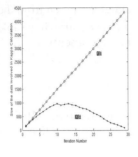 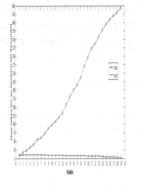 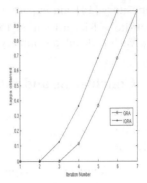

Fig. 1. Plot of Iteration Number vs. Size of the data used to calculate a Kappa Value in QRA and IQRA algorithms in webtest dataset

Fig. 2. Time taken in seconds for iteration in QRA and IQRA algorithms in webtest dataset

Fig. 3. Kappa obtained by iteration in QRA and IQRA in Mofn dataset

References

1. Bellman, R.: Adaptive Control Processes: A Guided Tour. Princeton University Press, Princeton (1961)
2. Pawlak, Z.: Rough Sets. International Journal of Computer and Information Science 11(5) (1982)
3. Nguyen, H.S., Skowron, A.: Boolean Reasoning for feature extraction problems. In: Foundations of Intelligent Systems. LNCS, pp. 117–126. Springer, Heidelberg (1997)
4. Chouchoulas, A., Shen, Q.: Rough Set-Aided Keyword Reduction for Text Categorization. Applied Artificial Intelligence 15(9), 843–873 (2001)
5. Thangavel, K., Jaganathan, P., Pethalakshmi, A., Karnan, M.: Effective Classification with Improved Quick Reduct For Medical Database Using Rough System. BIME Journal 05(1) (December 2005)
6. Pethalakshmi, A., Thangavel, K.: Performance analysis of Accelerated QuickReduct Algorithm. In: International Conference on Computational Intelligence and Multimedia Applications (2007)
7. Mi, J.-S., Wu, W.-Z., Zhang, W.-X.: Approaches to knowledge reduction based on variable precision rough set model. Information Sciences 159, 255–272 (2004)
8. Lei, C., Wan, S., Chou, T.Y.: The comparison of PCA and discrete rough set for feature extraction of remote sensing image classification – A case study on rice classification Taiwan. Comput. Geosci. 12, 1–14 (2008)
9. Zeng, A., Pan, D., Zheng, Q.-L., Peng, H.: Knowledge acquisition based on rough set theory and principal component analysis. IEEE Intelligent Systems 21(2), 78–85 (2006)
10. Chouchoulas, A., Halliwell, J., Shen, Q.: On the Implementation of Rough Set Attribute Reduction. In: Proceedings of the 2002 UK Workshop on Computational Intelligence, pp. 18–23 (2002)

11. Katzberg, J.D., Ziarko, W.: Variable precision rough sets with Asymmetric bounds. In: Ziarko, W. (ed.) Rough Sets, Fuzzy Sets and Knowledge Discovery, pp. 167–177. Springer, Heidelberg (1994)
12. Butalia, A., Dhore, M.L., Tewani, G.: Applications of Roughsets in the field of Data Mining. In: First International Conference on Emerging Trends in Engineering and Technology (July 2008)
13. Matlab, http://www.mathworks.com
14. Blake, C.L., Merz, C.J.: UCI Repository of machine learning databases, Irvine, University of California (1998), http://archive.ics.uci.edu/ml/datasets.html
15. Richard Jensen collection of datasets, http://users.aber.ac.uk/rkj/datasets/index.php
16. Rough Set Exploration System, http://alfa.mimuw.edu.pl/~rses/

Dynamic Reduct from Partially Uncertain Data Using Rough Sets

Salsabil Trabelsi[1], Zied Elouedi[1], and Pawan Lingras[2]

[1] Larodec, Institut Superier de Gestion de Tunis, Tunisia
[2] Saint Mary's University Halifax, Canada

Abstract. In this paper, we deal with the problem of attribute selection from a sample of partially uncertain data. The uncertainty exists in decision attributes and is represented by the Transferable Belief Model (TBM), one interpretation of the belief function theory. To solve this problem, we propose dynamic reduct for attribute selection to extract more relevant and stable features for classification. The reduction of the uncertain decision table using this approach yields simplified and more significant belief decision rules for unseen objects.

Keywords: Rough sets, belief function theory, uncertainty, dynamic reduct.

1 Introduction

Feature selection is an important pre-processing stage in machine learning. Rough set theory provides an attractive mechanism for feature selection [5,6,8]. The simplest approach is based on the calculation of reduct. Another issue in real world applications is the uncertain, imprecise or incomplete data. Many researches have adapted rough sets to such an uncertain environment. These extensions do not deal with partially uncertain decision attribute values. In this paper, we deal with the problem of attribute selection from partially uncertain data based on rough sets. The uncertainty exists in decision attributes and is represented by the Transferable Belief Model (TBM), one interpretation of the belief function theory. However, computing reducts from uncertain and noisy data make the results unstable, and sensitive to the sample data. All of these limit the application of rough set theory. Dynamic reducts [1] can lead to better performance in very large datasets, and also provide the ability to accommodate noisy data. The rules calculated by means of dynamic reducts are better predisposed to classify unseen cases, because these reducts are in some sense the most stable reducts, and they appear most frequently in sub-decision systems created by random samples of a given decision system. This paper is organized as follows: Section 2 provides an overview of the rough set theory. Section 3 introduces the belief function theory as understood in the TBM. In Section 4, we propose a new approach to feature selection based on dynamic reducts from partially uncertain data.

H. Sakai et al. (Eds.): RSFDGrC 2009, LNAI 5908, pp. 160–167, 2009.

2   Rough Sets

The idea of rough sets was introduced by Pawlak [6] to deal with imprecise and vague concepts. Here, we introduce only the basic notations. A decision table is an information system of the form $A = (U, C \cup \{d\})$, where $d \notin C$ is a distinguished attribute called *decision*. The value set of d is called $\Theta = \{d_1, d_2, ...d_s\}$. In this paper, the notation $c_i(o_j)$ is used to represent the value of a condition attribute $c_i \in C$ for an object $o_j \in U$. Similarly, the notation $d(o_j)$ represents the value of the decision attribute d for an object o_j. The rough sets adopt the concept of indiscernibility relation [6] to partition the object set U into disjoint subsets, denoted by U/B or IND_B. The partition that includes o_j is denoted $[o_j]_B$.

$$IND_B = U/B = \{[o_j]_B | o_j \in U\} \tag{1}$$

Where

$$[o_j]_B = \{o_i | \forall c \in B\ c(o_i) = c(o_j)\} \tag{2}$$

The equivalence classes based on the decision attribute are denoted by $U/\{d\}$.

$$IND_{\{d\}} = U/\{d\} = \{[o_j]_{\{d\}} | o_j \in U\} \tag{3}$$

Let $B \subseteq C$ and $X \subseteq U$. We can approximate X using only the information contained in B by constructing the $B - lower$ and $B - upper\ approximations$ of X, denoted $\underline{B}(X)$ and $\bar{B}(X)$, respectively, where

$$\underline{B}(X) = \{o_j | [o_j]_B \subseteq X\}\ and\ \bar{B}(X) = \{o_j | [o_j]_B \cap X \neq \emptyset\} \tag{4}$$

The objects in $\underline{B}(X)$ can be classified with certainty as members of X on the basis of knowledge in B, while the objects in $\bar{B}(X)$ can be only classified as possible members of X on the basis of knowledge in B. $Pos_C(\{d\})$, called a positive region of the partition $U/\{d\}$ with respect to C, is the set of all elements of U that can be uniquely classified to blocks of the partition $U/\{d\}$.

$$Pos_C(\{d\}) = \bigcup_{X \in U/\{d\}} \underline{C}(X) \tag{5}$$

A reduct is a minimal subset of attributes from C that preserves the positive region and the ability to perform classifications as the entire attributes set C. A subset $B \subseteq C$ is a reduct of C with respect to d, iff B is minimal and:

$$Pos_B(\{d\}) = Pos_C(\{d\}) \tag{6}$$

The core is the most important subset of attributes, it is included in every reduct.

$$Core(A,\ d) = \bigcap RED(A,\ d) \tag{7}$$

Where $RED(A, d)$ is the set of all reducts of A relative to d.

If $A = (U, C \cup \{d\})$ is a decision table, then any system $B = (U', C \cup \{d\})$ such that $U' \subseteq U$ is called a subtable of A. Let F be a family of subtables of A [1].

$$DR(A, F) = RED(A, d) \cap \bigcap_{B \in F} RED(B, d) \qquad (8)$$

Any element of $DR(A, F)$ is called an F-dynamic reduct of A. From the definition of dynamic reducts, it follows that a relative reduct of A is dynamic if it is also a reduct of all subtables from a given family F. This notation can be sometimes too restrictive so we apply a more general notion of dynamic reduct. They are called (F, ε)-dynamic reducts, where $1 \geq \varepsilon \geq 0$. The set $DR_\varepsilon(A, F)$ of all (F, ε)-dynamic reducts is defined by

$$DR_\varepsilon(A, F) = \left\{ R \in RED(A, d) : \frac{|\{B \in F : R \in RED(B, d)\}|}{|F|} \geq 1 - \varepsilon \right\} \qquad (9)$$

3   Belief Function Theory

In this section, we briefly review the main concepts underlying the belief function theory as interpreted in the Transferable Belief Model (TBM) [9,10]. Let Θ be a finite set of elementary events to a given problem, called the frame of discernment. All the subsets of Θ belong to the power set of Θ, denoted by 2^Θ. The impact of a piece of evidence on the subsets of the frame of discernment Θ is represented by a basic belief assignment (bba). The bba is a function $m : 2^\Theta \to [0, 1]$ such that:

$$\sum_{E \subseteq \Theta} m(E) = 1 \qquad (10)$$

The value $m(E)$, called a basic belief mass (bbm), represents the portion of belief committed exactly to the event E. The bba's induced from distinct pieces of evidence are combined by the rule of combination [11].

$$(m_1 \textcircled{$\cap$} m_2)(E) = \sum_{F,G \subseteq \Theta : F \cap G = E} m_1(F) \times m_2(G) \qquad (11)$$

In the TBM, beliefs to make decisions can be represented by probability functions called the pignistic probabilities denoted $BetP$ and are defined as [10]:

$$BetP(\{a\}) = \sum_{F \subseteq \Theta} \frac{|\{a\} \cap F|}{|F|} \frac{m(F)}{(1 - m(\emptyset))}, \text{ for all } a \in \Theta \qquad (12)$$

4   Dynamic Reduct under Uncertainty

Our decision system is characterized by high level of uncertain and noisy data. One of the issues with such a data is that the resulting reducts are not stable, and are sensitive to sampling. The belief decision rules generated are not suitable for classification. The solution to this problem is to redefine the concept of dynamic reduct in the new context as we have done in this paper. The rules calculated by means of dynamic reducts are better predisposed to classify unseen objects, because they are the most frequently appearing reducts in sub-decision systems created by random samples of a given decision system. In this section, we will adapt the basic concepts of rough sets such as decision system, indiscernibility relation, set approximation and positive region in order to redefine the concept of dynamic reduct in the uncertain context. The objective is to extract more stable reducts from the uncertain decision system.

4.1   Basic Concepts of Rough Sets under Uncertainty

Uncertain Decision System. Our uncertain decision system is given by $A = (U, C \cup \{ud\})$, where $U = \{o_j : 1 \leq j \leq n\}$ is characterized by a set of certain condition attributes $C=\{c_1, c_2,...,c_k\}$, and an uncertain decision attribute ud. We represent the uncertainty of each object o_j by a bba m_j expressing beliefs on decisions defined on the frame of discernment $\Theta=\{ud_1, ud_2,...,ud_s\}$ representing the possible values of ud. These bba's are given by an expert.

Example. Let us use Table 1 to describe our uncertain decision system. It contains eight objects, three certain condition attributes $C=\{a, b, c\}$ and an uncertain decision attribute $ud = e$ with two possible values $\{e_1, e_2\}$ representing Θ.

Table 1. Uncertain decision table

U	a	b	c	e
o_1	0	0	0	$m_1(\{e_1\}) = 0.95$ $m_1(\{e_2\}) = 0.05$
o_2	0	1	1	$m_2(\{e_2\}) = 1$
o_3	0	0	2	$m_3(\{e_1\}) = 0.5$ $m_3(\Theta) = 0.5$
o_4	1	0	2	$m_4(\{e_2\}) = 0.6$ $m_4(\Theta) = 0.4$
o_5	1	0	2	$m_5(\{e_2\}) = 1$
o_6	0	1	1	$m_6(\{e_2\}) = 0.9$ $m_6(\{\Theta\}) = 0.1$
o_7	1	0	0	$m_7(\{e_1\}) = 1$
o_8	1	0	1	$m_8(\{e_1\}) = 0.9$ $m_8(\{\Theta\}) = 0.1$

For the object o_3, 0.5 of beliefs are exactly committed to the decision e_1, whereas 0.5 of beliefs is assigned to the whole of frame of discernment Θ (ignorance). With bba, we can represent the certain case, like for the objects o_2, o_5 and o_7. Besides, we can represent probability case, like the bba relative to the object o_1 and possiblitic case like the consonant bba relative to the object o_3. The decision rules induced from the partially uncertain decision system are denoted belief decision rules where the decision is represented by a bba: *If a=0 and b=0 and c=2 Then $m_3(\{e_1\}) = 0.5$ $m_3(\Theta) = 0.5$.*

Indiscernibility Relation. For the condition attributes, the indiscernibility relation U/C is the same as in the certain case because their values are certain. However, the indiscernibility relation for the decision attribute $U/\{ud\}$ is not the same as in the certain case. The decision value is represented by a bba. So, we need to assign each object to the right equivalence class. The idea is to use the distance between two bba's. Many distance measures between two bba's were developed [2,3,4]. We will choose the distance measure described in [2] which satisfies properties such as non-negativity, non-degeneracy and symmetry.

For every ud_i, an uncertain decision value, we define:

$$X_i = \{o_j | dist(m(ud_i) = 1, m_j) \neq 1\} \tag{13}$$

$$IND_{\{ud\}} = U/\{ud\} = \{X_i | ud_i \in \Theta\} \tag{14}$$

Where $dist$ is a distance measure between two bba's.

$$dist(m_1, m_2) = \sqrt{\frac{1}{2}(\| \vec{m_1} \|^2 + \| \vec{m_2} \|^2 - 2 < \vec{m_1}, \vec{m_2} >)} \tag{15}$$

Where $< \vec{m_1}, \vec{m_2} >$ is the scalar product defined by:

$$< \vec{m_1}, \vec{m_2} > = \sum_{i=1}^{|2^\Theta|} \sum_{j=1}^{|2^N|} m_1(A_i) m_2(A_j) \frac{|A_i \cap A_j|}{|A_i \cup A_j|} \tag{16}$$

with A_i, $A_j \in 2^\Theta$ for $i, j = 1, 2, \cdots, |2^\Theta|$. $\| \vec{m_1} \|^2$ is then the square norm of $\vec{m_1}$.

Example. Let us continue with the same example to compute the equivalence classes based on condition attributes in the same manner as in the certain case: U/C= {{o_1}, {o_2, o_6}, {o_3}, {o_4, o_5}, {o_7}, {o_8}} and to compute the equivalence classes based on uncertain decision attribute $U/\{ud\}$ as follows:

For the uncertain decision value $ud_1 = e_1$,

$dist(m(e_1) = 1, m_1) \neq 1$
$dist(m(e_1) = 1, m_2) = 1$
$dist(m(e_1) = 1, m_3) \neq 1$
$dist(m(e_1) = 1, m_4) \neq 1$
$dist(m(e_1) = 1, m_5) \neq 1$
$dist(m(e_1) = 1, m_6) \neq 1$
$dist(m(e_1) = 1, m_7) \neq 1$
$dist(m(e_1) = 1, m_8) \neq 1$.

So, $X_1 = \{o_1, o_3, o_4, o_5, o_6, o_7, o_8\}$.

For the uncertain decision value $ud_2 = e_2$,

$dist(m(e_2) = 1, m_1) \neq 1$
$dist(m(e_2) = 1, m_2) \neq 1$
$dist(m(e_2) = 1, m_3) \neq 1$
$dist(m(e_2) = 1, m_4) \neq 1$
$dist(m(e_2) = 1, m_5) \neq 1$
$dist(m(e_2) = 1, m_6) \neq 1$

$dist(m(e_2) = 1, m_7) = 1$
$dist(m(e_2) = 1, m_8) \neq 1.$

So, $X_2 = \{o_1, o_2, o_3, o_4, o_5, o_6, o_8\}$.
$U/\{ud\} = \{\{o_1, o_3, o_4, o_5, o_6, o_7, o_8\}, \{o_1, o_2, o_3, o_4, o_5, o_6, o_8\}\}.$

Set Approximation. To compute the new *lower* and *upper* approximations for our uncertain decision table, we follow two steps:

1. For each equivalence class from U/C based on condition attributes C, combine their bba using the operator mean. The operator mean is more suitable in our case to combine these bba's than the rule of combination in eq. 11 which is proposed especially to combine different beliefs on decision for one object and not different beliefs for different objects.

2. For each equivalence class X_i from $U/\{ud\}$ based on uncertain decision attribute ud_i, we compute the new *lower* and *upper* approximations, as follows:

$$\underline{C}X_i = \{o_j | [o_j]_C \cap X_i \neq \emptyset \, and \, dist(m(ud_i) = 1, m_{[o_j]_C}) \leq threshold\} \quad (17)$$

In the *lower* approximation, we find all equivalence classes (subsets) from U/C included in X_i such that the distance between the combined bba $m_{[o_j]_C}$ and the certain bba $m(ud_i) = 1$ is less than a *threshold*. (In an uncertain context, the *threshold* is needed to give more flexibility to the set approximations). We compute the *upper* approximation in the same manner as in the certain case.

$$\bar{C}X_i = \{o_j | [o_j]_C \cap X_i \neq \emptyset\} \quad (18)$$

Example. We continue with the same example to compute the new *lower* and *upper* approximations. After the first step, we obtain the combined bba for each equivalence class from U/C using operator mean. Table 2 represents the combined bba for the equivalence classes $\{o_2, o_6\}$ and $\{o_4, o_5\}$.

Table 2. The combined bba for the subsets $\{o_2, o_6\}$ and $\{o_4, o_5\}$

Object	$m(\{e_1\})$	$m(\{e_2\})$	$m(\Theta)$
o_2	0	1	0
o_6	0	0.9	0.1
$m_{2,6}$	0	0.95	0.05
o_4	0	0.4	0.6
o_5	0	1	0
$m_{4,5}$	0	0.7	0.3

Next, we compute the *lower* and *upper* approximations for each equivalence class $U/\{ud\}$. We will use *threshold* = 0.1.

For the uncertain decision value $ud_1 = e_1$, let $X_1 = \{o_1, o_3, o_4, o_5, o_6, o_7, o_8\}$. The subsets $\{o_1\}$, $\{o_3\}$, $\{o_4, o_5\}$, $\{o_7\}$ and $\{o_8\}$ are included in X_1. We should check the distance between their bba and the certain bba $m(e_1) = 1$.

$dist(m(e_1) = 1, m_1) < 0.1$
$dist(m(e_1) = 1, m_3) > 0.1$

$dist(m(e_1) = 1, m_{4,5}) > 0.1$
$dist(m(e_1) = 1, m_7) < 0.1$
$dist(m(e_1) = 1, m_8) < 0.1$
$\underline{C}X_1=\{o_1, o_7, o_8\}$ and $\bar{C}X_1=\{o_1, o_3, o_4, o_5, o_7, o_8\}$

For uncertain decision value $ud_2=e_2$, let $X_2 =\{o_1, o_2, o_3, o_4, o_5, o_6, o_8\}$. $\underline{C}X_2=\{o_2, o_6\}$ and $\bar{C}X_2=\{o_2, o_3, o_4, o_5, o_6\}$

Positive Region. With the new *lower* approximation, we can redefine the positive region:

$$UPos_C(\{ud\}) = \bigcup_{X_i \in U/\{ud\}} \underline{C}X_i \qquad (19)$$

Example: Let us continue with the same example, to compute the positive region of A. $UPos_C(\{ud\})=\{o_1, o_2, o_6, o_7, o_8\}$

Reduct and Core. Using the new formalism of positive region, we can redefine the reduct of A as a minimal set of attributes $B \subseteq C$ such that:

$$UPos_B(\{ud\}) = UPos_C(\{ud\}) \qquad (20)$$

$$UCore(A,\ ud) = \bigcap URED(A,\ ud) \qquad (21)$$

Where $URED(A,\ ud)$ is the set of all reducts of A relative to ud.

Example. Using our example, we find that $UPos_{\{a,c\}}(\{ud\})= UPos_{\{b,c\}}(\{ud\})$ $= UPos_C(\{ud\})$. So, we have two possible reducts $\{a,c\}$ and $\{b,c\}$. The attribute c is the relative core.

4.2   Dynamic Reduct from Uncertain Data

Using the new definition of reduct in our uncertain context, we can redefine the concept of dynamic reduct as follows:

$$UDR(A,\ F) = URED(A, ud) \cap \bigcap_{B \in F} URED(B, ud) \qquad (22)$$

Where F be a family of subtables of A. This notation can be sometimes too restrictive so we apply a more general notion of dynamic reduct. They are called $(F,\ \varepsilon)$-dynamic reducts, where $1 \geq \varepsilon \geq 0$. The set $UDR_\varepsilon(A,\ F)$ of all $(F,\ \varepsilon)$-dynamic reducts is defined by:

$$UDR_\varepsilon(A,\ F)=\left\{R \in URED(A, ud) : \frac{|\{B \in F : R \in RED(B, ud)\}|}{|F|} \geq 1 - \varepsilon\right\} \qquad (23)$$

Example. To compute the dynamic reduct of the uncertain decision system A. We divide our uncertain decision system into two subtables B and B' to obtain a family F of sub-decision system. B contains the objects o_1, o_2, o_3, o_4 and B' contains the objects o_5, o_6, o_7, o_8. The two subtables have the same reducts as the whole decision system A. So, the subsets $\{a,c\}$ and $\{b,c\}$ are dynamic reducts relative to the chosen family F.

5   Conclusion and Future Work

In this paper, we have adapted the basic concepts of rough sets such as decision system, indiscernibility relation, set approximation and reduct in an uncertain context. We handle uncertainty in decision attributes using the belief function theory. We further propose dynamic reduct to address the problem of unstable reducts in uncertain decision systems. As a future work, we will experiment with many uncertain databases to evaluate the proposed feature selection based on dynamic reducts.

References

1. Bazan, J., Skowron, A., Synak, P.: Dynamic reducts as a tool for extracting laws from decision tables in: methodologies for intelligent system. In: Raś, Z.W., Zemankova, M. (eds.) ISMIS 1994. LNCS (LNAI), vol. 869, pp. 346–355. Springer, Heidelberg (1994)
2. Bosse, E., Jousseleme, A.L., Grenier, D.: A new distance between two bodies of evidence. In: Information Fusion, vol. 2, pp. 91–101 (2001)
3. Elouedi, Z., Mellouli, K., Smets, P.: Assessing sensor reliability for multisensor data fusion within the transferable belief model. IEEE Trans. Syst. Man cyben. 34(1), 782–787 (2004)
4. Fixen, D.: The modified Dempster-Shafer approach to classification. IEEE Trans. Syst. Man Cybern. A 27(1), 96–104 (1997); In: Workshop Notes, Foundations and New Directions of Data Mining, the 3rd International Conference on Data Mining, Melbourne, Florida, vol. 5663
5. Modrzejewski, M.: Feature selection using rough sets theory. In: Proceedings of the 11th International Conference on Machine Learning, pp. 213–226 (1993)
6. Pawlak, Z.: Rough Sets: Theoretical Aspects of Reasoning About Data. Kluwer Academic Publishing, Dordrecht (1991)
7. Shafer, G.: A mathematical theory of evidence. Princeton University Press, Princeton (1976)
8. Skowron, A., Rauszer, C.: The discernibility matrices and functions in information systems. In: Slowinski, R. (ed.) Intelligent Decision Support, pp. 331–362. Kluwer Academic Publishers, Boston (1992)
9. Smets, P.: The transferable belief model. Artificial Intelligence 66(2), 191–236 (1994)
10. Smets, P.: The transferable belief model for quantified belief representation. In: Gabbay, D.M., Smets, P. (eds.) Handbook of defeasible reasoning and uncertainty management systems, vol. 1, pp. 207–301. Kluwer, Dordrecht (1998)
11. Smets, P.: Application of belief transferable belief model to diagnostic problems. International journal of intelligent systems 13(2-3), 127–157 (1998)

Approximation of Loss and Risk
in Selected Granular Systems

Marcin Szczuka*

Institute of Mathematics, The University of Warsaw
Banacha 2, 02-097 Warsaw, Poland
szczuka@mimuw.edu.pl

Abstract. We discuss the notion of risk in generally understood classification support systems. We consider the situation when granularity is involved in information system we work with. We propose a method for approximating the loss function and introduce a technique for assessing the empirical risk from experimental data. We discuss the general methodology and possible directions of development in the area of constructing compound classification schemes.

1 Introduction

While constructing a decision support (classification) system for research purposes we usually rely on commonly used, convenient quality measures, such as success ratio (accuracy) on test set, coverage (support) and versatility of the classifier. While sufficient for the purposes of analysing classification methods in terms of their technical abilities, these measures sometimes fail to fit into a bigger picture.

In practical applications the classifier is usually just an element in a larger system. The decision whether to construct and then use such system is taken by the user on the of his/her assessment of the *risk* involved in making the decision.

The overall topics of risk assessment, risk management and decision making in presence of risk constitute a separate field of science. Numerous approaches have been developed so far in many areas of life, and vast literature dedicated to these issues exist (see [1], [2]). In this article we restrict ourselves to a much narrower topic of calculating (assessing) the risk associated with the use of classifier in a decision-making process.

We focus on one commonly used method for calculating a risk of (using) a classifier, which is known from the basics of statistical learning theory [3]. In this approach the risk is measured as a *summarised expectation* for creating a loss due to classifier error. It is quite common to make assessment of the involved risk by hypothesising the situations in which the gain/loss can be generated in our system, and then weighting them by the likelihood of their occurrence.

* This research was supported by the grants N N516 368334 and N N516 077837 from the Ministry of Science and Higher Education of the Republic of Poland.

H. Sakai et al. (Eds.): RSFDGrC 2009, LNAI 5908, pp. 168–175, 2009.
© Springer-Verlag Berlin Heidelberg 2009

We investigate the possibilities for approximating the risk in the situation when the standard numerical, statistical learning methods cannot be applied to full extent. The real life data is not always possible to be verified as representative, large enough or sufficiently compliant with assumptions of underlying analytical model. Also, the information we posses about the amount of loss and its probabilistic distribution may be expressed in granular rather than crisp, numerical way.

The idea of granular systems and granular computing builds on general observation, that in many real-life situations we are unable to precisely discern between similar objects. Our perception of such universe is *granular*, which means, that we are only able to observe groups of objects with limited resolution.

The existence of granularity and the necessity of dealing with it has led to formation of the *granular computing* paradigm and research on granule-based information systems (cf. [4]). The original ideas of Lotfi Zadeh (cf. [5]) has grown over time. Currently the granular computing and the notion of granularity are becoming a unifying methodologies for many branches of soft computing. Several paradigms related to rough and fuzzy sets, interval analysis, shadowed sets as well as probabilistic reasoning can be represented within granular framework, as exemplified by the contents of the handbook [4].

In the paper put forward some ideas regarding the approximate construction of two crucial components in measuring risk, i.e., the loss function and the summarisation method needed to estimate overall risk from the empirical, sample-dependant one. Our focus is on systems that support classification and decision making in the presence of vagueness, imprecision and incompleteness of information. In this paper we only address a small portion of such systems and the granules we are using are of rather basic type. We mostly address the case when a granule corresponds to an abstraction (indescernibility) class or a simple fuzzy set, without caring of its internal structure.

The paper starts with more formal introduction of risk functional, as known from statistical learning theory. Then, we discuss the possible sources of problems with such risk definition and suggest some directions, in particular an outline for a loss function approximation method. We also extend the discussion to the issue of finding the proper summarisation procedure for measuring the value of empirical risk functional. We introduce a sketch for the methods of risk calculation in case of granular systems defined using rough and fuzzy sets, which by no means represent the whole spectrum of granular systems. We conclude by pointing out several possible directions for further investigation.

2 Risk in Statistical Learning Theory

In the classical statistical learning approach, represented by seminal works of Vapnik [3,6], the risk associated with a classification method (classifier) α is defined as a functional (integral) of the *loss function* L_α calculated over an entire space with respect to probability distribution.

Formally, let X^∞ be the complete (hypothetical) universe of objects and $X \subset X^\infty$ - a finite sample.[1] In this analytical model we assume that probability distribution P is defined for entire σ-field of measurable subsets of X^∞.

Definition 1. *The risk value for a classifier α is defined as:*

$$R(\alpha) = \int_{X^\infty} L_\alpha dP$$

where $L_\alpha = L(x, f_\alpha(x))$ is the real-valued loss function defined for every point $x \in X^\infty$ where the classifier α returns the value $f_\alpha(x)$.

The classical definition of risk, as presented above, is heavily dependent on assumptions regarding the underlying analytical model of the space of discourse. While over the years several methods have been developed within the area of statistical learning in pursuit of practical means for calculating risk, there are still some important shortcomings in this approach. In particular, one has to deal with sensitivity to scarceness of the data sample, incomplete definition of loss function, and incomplete knowledge of the distribution. Even with large data sample X we may not be certain about its representativeness. The advantage of this model is that, thanks to solid mathematical grounding, it is possible to provide answers with provable quality, as long as we can assure sufficient compliance with assumptions.

In practice, the empirical risk is usually measured as an average of loss function on finite sample. For a labelled sample $z = \{x_1, \ldots, x_l\}$ of length l

$$R_{emp}(\alpha) = \frac{\sum_{i=1}^{l} L(x_i, f_\alpha(x_i))}{l}.$$

It is visible, that the ability to calculate value of loss L_α, i.e., to compare the answer of classifier with the desired one is a key element in risk calculation.

3   Approximation of Loss Function and Its Integral

The formal postulates regarding the loss function may be hard to meet, or even verify in practical situations. Nevertheless, we would like to be able to asses the loss. In this section we suggest a method for approximating the loss function from the available, finite sample.

First, we will attempt to deal with the situation when the value of loss function L_α for a classifier α is given as a set of positive real values defined for data points from a finite sample z. Let $z \in (X^\infty)^l$ be a sample consisting of l data points, by $\mathbb{R}_+$ we denote the set of non-negative reals (including 0). A function $\hat{L}_\alpha : z \mapsto \mathbb{R}_+$ is called a sample of loss function $L_\alpha : X^\infty \mapsto \mathbb{R}_+$ if L_α is an extension of $\hat{L}_\alpha$. For any $Z \subseteq X^\infty \times \mathbb{R}_+$ we introduce two projection sets:

$$\pi_1(Z) = \{x \in X^\infty : \exists y \in \mathbb{R}_+ \ (x, y) \in Z\},$$

$$\pi_2(Z) = \{y \in \mathbb{R}_+ : \exists x \in X^\infty \ (x, y) \in Z\}.$$

---

[1] Please, note that the hypothetical universe X^∞ shall not be confused with the denotation for the set of infinite sequences from the set X, that can be found in some mathematical textbooks.

3.1   The Rough Set Case

We assume that we are also given a family $\mathcal{C}$ of neighbourhoods (granules), i.e, non-empty, measurable subsets of $X^\infty \times \mathbb{R}_+$. These neighbourhoods shall be defined for a particular application. In this section we will identify these neighbourhoods with granules defined as indiscernibility classes.

Under the assumptions presented above the lower approximation of $\hat{L}_\alpha$ relative to $\mathcal{C}$ is defined by

$$\underline{\mathcal{C}}\hat{L}_\alpha = \bigcup \{c \in \mathcal{C} : \hat{L}_\alpha(\pi_1(c) \cap z) \subseteq \pi_2(c)\}. \tag{1}$$

Note, that the definition of lower approximation given by (1) is different from the traditional one, known from rough set theory [7,8]. Also, the sample z in definition of approximations (formulæ (1),(2), and (3)) is treated as a set of its elements, i.e., a subset of $(X^\infty)^l$.

One can define the upper approximation of $\hat{L}_\alpha$ relative to $\mathcal{C}$ by

$$\overline{\mathcal{C}}\hat{L}_\alpha = \bigcup \{c \in \mathcal{C} : \hat{L}_\alpha(\pi_1(c) \cap z) \cap \pi_2(c) \neq \emptyset\}. \tag{2}$$

For the moment we have defined the approximation of loss function as a pair of sets created from the elements of neighbourhood family $\mathcal{C}$. From this approximation we would like to obtain an estimation of risk. For that purpose we need to define summarisation (integration) method analogous to Def. 1. We define an integration functional based on the idea of probabilistic version of Lebesgue-Stieltjes integral [3,9].

In order to define our integral we need to make some additional assumptions. For the universe X^∞ we assume that m is a measure on a Borel σ-field of subsets of X^∞ and that $m(X^\infty) < \infty$. By m_0 we denote a measure on a σ-field of subsets of $\mathbb{R}_+$. We will also assume that $\mathcal{C}$ is a family of non-empty subsets of $X^\infty \times \mathbb{R}_+$ that are measurable relative to the product measure $\bar{m} = m \times m_0$. Finally, we assume that the value of loss function is bounded by some positive real B. Please, note that none of the above assumptions is unrealistic, and that in practical applications we are dealing with finite universes.

For the upper bound B we split the range $[0, B] \subset \mathbb{R}_+$ into $n > 0$ intervals of equal length $I_1, \dots, I_n$, where $I_i = [\frac{(i-1)B}{n}, \frac{iB}{n}]$. This is a simplification of the most general definition, where the intervals do not have to be equal. For every interval I_i we consider the sub-family $\mathcal{C}_i \subset \mathcal{C}$ of neighbourhoods such that:

$$\mathcal{C}_i = \left\{ c \in \mathcal{C} : \forall x \in (z \cap \pi_1(c)) \quad \hat{L}_\alpha(x) > \frac{(i-1)B}{n} \right\}. \tag{3}$$

With the previous notation the estimate for empirical risk is given by:

$$R_{emp}(\alpha) = \sum_{i=1}^{n} \frac{B}{n} m \left(\bigcup_{c \in \mathcal{C}_i} \pi_1(c) \right) \tag{4}$$

In theoretical setting for the formula (4) above we shall derive its limit as $n \to \infty$, but in practical situation the parameter n does not have to go to infinity. It is sufficient to find n such that for every pair of points $x_1 \neq x_2$ taken from sample z if $\hat{L}_\alpha(x_1) < \hat{L}_\alpha(x_2)$ then for some integer $i \leq n$ we have $\hat{L}_\alpha(x_1) < \frac{iB}{m} < \hat{L}_\alpha(x_2)$.

3.2 The Fuzzy Set Case

If we consider another type of neighbourhoods, one defined with use of fuzzy sets, we will find ourselves in slightly different position. In fact, we may use fuzzy granules (neighbourhoods) in two ways:

1. We can define the family of neighbourhoods $\mathcal{C}$ as a family of fuzzy sets in $X^\infty \times \mathbb{R}_+$. This approach leads to more general idea of function approximation in fuzzy granular environment.
2. We restrict fuzziness to the domain of the loss function, i.e., X^∞. The values of $L_\alpha(.)$ remain crisp. That means that we have real-valued loss function that has a set of fuzzy membership values associated with each argument.

While the former case is more general and could lead to nicer, more universal definitions of approximation (see, e.g., [10]), it is at the same time less intuitive if we want to discuss risk measures. For that reason we restrict ourselves to the latter case. The family of neighbourhoods $\mathcal{C}$ is now defined in such a way that each $c \in \mathcal{C}$ is a product of fuzzy set in X^∞ and a subset of $\mathbb{R}_+$. The family $\mathcal{C}$ directly corresponds to family of fuzzy membership functions (fuzzy sets) $\mathcal{C}_\mu$. Each $c \in \mathcal{C}$ is associated with a fuzzy membership function $\mu_c : X^\infty \mapsto [0,1]$ corresponding to the fuzzy projection of c onto X^∞. Please note that at the moment we assume nothing about the intersections of elements of $\mathcal{C}$ but, we assume that the family $\mathcal{C}$ is finite.

Again, we start with a finite sample of points in the graph of loss function $\hat{L}_\alpha :$ $z \mapsto \mathbb{R}_+$ for data points from a finite sample z. We will attempt to approximate L_α by extending its finite sample $\hat{L}_\alpha$. For $c \in \mathcal{C}$ we now introduce parameterised projections. For $0 \leq \lambda < 1$, we have:

$$\pi_1(c, \lambda) = \{x \in X^\infty : \exists y \in \mathbb{R}_+ \ ((x, y) \in c \wedge \mu_c(x) > \lambda)\},$$

$$\pi_2(c, \lambda) = \{y \in R_+ : \exists x \in X^\infty \ ((x, y) \in c \wedge \mu_c(x) > \lambda)\}.$$

The parameter λ is used to establish a cut-off value for membership. The intention behind introduction of this parameter is that in some circumstances we may want to consider only those neigbourhoods which have sufficient level of confidence (sufficiently high membership). In terms of risk approximation, we would like to consider only these situations for which the argument of loss function is sufficiently certain. Naturally, we can make projections maximally general by putting $\lambda = 0$. The result of using projection $\pi_1(c, \lambda)$ is similar to taking an *alpha-cut* known from general fuzzy set theory (see [11]).

With previous notation and under previous assumptions we now introduce an approximation of $\hat{L}_\alpha$ w.r.t. the family of neigbourhoods $\mathcal{C}$ and a threshold λ.

$$\mathcal{C}_\lambda \hat{L}_\alpha = \bigcup \{c \in \mathcal{C} : \hat{L}_\alpha(\pi_1(c, \lambda) \cap z) \subseteq \pi_2(c, \lambda)\} \tag{5}$$

It is important to notice, that while projections $\pi_1(c, \lambda)$ and $\pi_2(c, \lambda)$ are classical (crisp) sets, the resulting approximation $\mathcal{C}_\lambda \hat{L}_\alpha$ is of the same type as the original sample, i.e., it is a union of neigbourhoods (granules) which are products of fuzzy set in X^∞ and a subset of $\mathbb{R}_+$. Please also note that the union operator used in (5) is working dimension-wise, and that the resulting set $C_\lambda \hat{L}_\alpha$ does not have to be a granule (neighbourhood), as we do not assume that this union operator is a granule aggregation (fusion) operator.

With all the previously introduced notation, the empirical risk functional is introduced by first defining the building blocks (strata, neighbourhoods) as:

$$C_i^\lambda = \left\{ c \in C : \forall x \in (z \cap \pi_1(c, \lambda)) \quad \hat{L}_\alpha(x) > \frac{(i-1)B}{n} \right\}. \tag{6}$$

That leads to the estimate for empirical risk functional:

$$R_{emp}^\lambda(u) - \sum_{i=1}^n \frac{B}{n} \, m \left(\bigcup_{c \in C_i^\lambda} \pi_1(c, \lambda) \right) \tag{7}$$

The formulæ above, just as in the case of rough set risk estimates, are valid only if some assumptions can be made about the family of neighbourhoods C. Again, the assumptions that have to be met are rather reasonable, and quite possible to met if we are dealing with finite sample z and a finite family of neighbourhoods C. We have to assure that elements of C are measurable w.r.t $\bar{m} = m \times m_0$ - the product measure on $X^\infty \times \mathbb{R}_+$.

As one can see, the risk estimator (7) is parameterised by the confidence level λ. In fact the selection of proper value of λ in all steps of risk assessment in the fuzzy context is a crucial step. Depending on value of λ we may get (radically) different outcomes. This intuitively corresponds to the fact that we can get different overview of the situation depending on how specific or how general we want to be.

The notions of function approximations and risk functional that we have introduced are heavily dependent on the data sample z and decomposition of our domain into family of neighbourhoods C. It is not yet visible, how the ideas we present may help in construction of better decision support (classification) systems. In the following section we discuss these matters in some detail.

4   Classifiers, Neighbourhoods and Granulation

Insofar we have introduced the approximation of loss and the measure of risk. To show the potential use of these entities, we intend to investigate the process of creation and evaluation (scoring) of classifier-driven decision support systems.

The crucial component in all our definitions is the family of non-empty sets (neighbourhoods) C. This family represents the granular nature of the universe of discourse. We have to know this family before we can approximate loss or estimate empirical risk. In practical situations the family of neighbourhoods have to be constructed in close correlation with classifier construction. It is quite common, especially for rough sets approaches, to define these sets constructively by semantics of some formulas. An example of such formula could be the conditional part of decision rule or a template (in the sense of [12,13]). In case of fuzzy granules the neighbourhoods may be provided arbitrarily or imposed by the necessity to instantiate a set of linguistic rules (a knowledge base).

Usually the construction of proper neighbourhoods is a complicated search and optimisation task. The notions of approximation and empirical risk that we have introduced may be used to express requirements for this search/optimisation. For the purpose of making valid, low-risk decision by means of classifier α we would expect the family $\mathcal{C}$ to possess the following qualities:

1. *Precision.* In order to have really meaningful assessment of risk as well as good idea about the loss function we would like the elements of neighbourhood family to to be relatively large in terms of universe X^∞, but at the same time having possibly low variation.
2. *Relevance.* This requirement is closely connected withe previous one (precision). While attempting to precisely dissect the domain into neighbourhoods we have to keep under control the relative quality (relevance) of neighbourhoods with respect to the data sample z. We are only interested in the neighbourhoods that contain sufficient number of elements of z.
3. *Coverage and adaptability.* One of the motivations that steer the process of creating the family of neighbourhoods and the classifier is the expectation regarding its ability to generalise and adapt the solution established on the basis of finite sample to a possibly large portion of the data domain.

As discussed in points 1–3 above, the task of finding a family of neighbourhoods can be viewed as a multi-dimensional optimisation on meta-level. It is in par with the kind of procedure that has to be employed in construction of systems based on the granular computing paradigm [4,13].

Yet another is that so far we have followed the assumption made at the beginning of Section 2, that the values of loss function are given as non-negative real numbers. In real application we may face the situation when the value of loss is given to us in less precise form. One such example is the loss function expressed in relative, qualitative terms. If the value of loss is given to us by the human expert, he/she may be unable to present us with precise, numerical values. We may then be confronted with situation when the loss is expressed in qualitative terms such as "big", "negligible", "prohibitive", "acceptable". Such imprecise description of the loss function may in turn force us to introduce another training loop into our system, one that will learn how to convert the imprecise notions we have into concrete, numerical values of loss function.

5   Conclusion

In this paper we have discussed the issues that accompany the assessment of risk in classification systems on the basis of the finite set of examples. We have pointed out some sources of possible problems and outlined some directions, in which we may search for solutions that match our expectations sufficiently well.

In conclusion, we would like to go back to the more general issue of weighting the risk involved in computer-supported decision making. As we have mentioned in the introduction to this paper, in the real-life situations the human user may display various patterns in his/her risk assessment and aversion.

It is rather unrealistic to expect that it would be possible to devise and explicitly formulate a model, that sufficiently supports extensibility as well as adaptability, and at the same time compliant with human perception and applicable in many different situations. It is much more likely that in practical situation we may need to learn (or estimate) not only the parameters, but the general laws governing its dynamics, at the same time attempting preserve its flexibility and ability to adapt to new cases.

References

1. Warwick, B. (ed.): The Handbook of Risk. John Wiley & Sons, Hoboken (2003)
2. Bostrom, A., French, S., Gottlieb, S. (eds.): Risk Assessment, Modeling and Decision Support. Risk, Governance and Society. Springer, Heidelberg (2008)
3. Vapnik, V.: Statisctical Learning Theory. John Wiley & Sons, Chichester (1998)
4. Pedrycz, W., Skowron, A., Kreinovich, V. (eds.): Handbook of Granular Computing. John Wiley & Sons, New York (2007)
5. Zadeh, L.: Fuzzy sets and information granularity. In: Gupta, M., Ragade, R., Yager, R. (eds.) Advances in Fuzzy Set Theory and Application, pp. 3–18. North-Holland Publishing Co., Amsterdam (1979)
6. Vapnik, V.: Principles of risk minimization for learning theory. In: Proceedings of NIPS, pp. 831–838 (1991)
7. Pawlak, Z.: Rough Sets. Theoretical Aspects of Reasoning about Data. Kluwer Academic Publishers, Dordrecht (1991)
8. Pawlak, Z.: Rough sets, rough functions and rough calculus. In: Pal, S.K., Skowron, A. (eds.) Rough Fuzzy Hybridization: A New Trend in and Decision Making, pp. 99–109. Springer, Singapore (1999)
9. Halmos, P.: Measure Theory. Springer, Berlin (1974)
10. Höeppner, F., Klawonn, F.: Systems of information granules. In: [4], pp. 187–203
11. Yager, R.R., Filev, D.P.: Essentials of fuzzy modeling and control. John Wiley & Sons, Chichester (1994)
12. Skowron, A., Synak, P.: Complex patterns. Fundamenta Informaticae 60, 351–366 (2004)
13. Jankowski, A., Peters, J.F., Skowron, A., Stepaniuk, J.: Optimization in discovery of compound granules. Fundamenta Informaticae 85, 249–265 (2008)

An Efficient Gene Selection Algorithm Based on Tolerance Rough Set Theory

Na Jiao[1,*] and Duoqian Miao[2]

[1] Department of Computer Science and Technology, Tongji University,
Shanghai 201804, P.R. China
zdx.jn@163.com
[2] Key Laboratory of Embedded System and Service Computing,
Ministry of Education of China, Tongji University, Shanghai 201804, P.R. China

Abstract. Gene selection, a key procedure of the discriminant analysis of microarray data, is to select the most informative genes from the whole gene set. Rough set theory is a mathematical tool for further reducing redundancy. One limitation of rough set theory is the lack of effective methods for processing real-valued data. However, most of gene expression data sets are continuous. Discretization methods can result in information loss. This paper investigates an approach combining feature ranking together with feature selection based on tolerance rough set theory. Compared with gene selection algorithm based on rough set theory, the proposed method is more effective for selecting high discriminative genes in cancer classification task.

Keywords: Microarray data, gene selection, feature ranking, tolerance rough set theory, cancer classification.

1 Introduction

DNA microarray is a technology to measure the expression levels of thousands of genes, which is quite suitable for comparing the gene expression levels in tissues under different conditions, such as healthy versus diseased.

Discriminant analysis of microarray data has been widely studied to assist diagnosis. Because lots of genes in the original gene set are irrelevant or even redundant for specific discriminant problem, gene selection is usually introduced to preprocess the original gene set for further analysis.

There are two basic categories of feature selection algorithms, namely filter and wrapper models. Filter methods select feature subsets independently of any learning algorithm and rely on various measures of the general characteristics of the training data. Some statistical tests (t-test, F-test) have been shown to be effective. The idea of these methods is that features are ranked and the top ones or those that satisfy a certain criterion are selected. Wrapper methods use the predictive accuracy of a predetermined learning algorithm to determine the goodness of the selected subsets and are computationally expensive.

---

* Corresponding author.

H. Sakai et al. (Eds.): RSFDGrC 2009, LNAI 5908, pp. 176–183, 2009.

Features using existing feature selection such as filter and wrapper have redundancy because genes have similar scores in similar pathways. Rough set theory can be used to eliminate such redundancy. Rough set theory [1-6], proposed by Pawlak in 1982, is widely applied in many fields of data mining such as classification and feature selection. However, traditional rough set theory-based methods are restricted to the requirement that all data must be discrete. Existing methods [7] are to discretize the data sets and replace original data values with crisp values. This is often inadequate, as degrees of objects to the descretized values are not considered. Discretization ignores their discrimination. This may cause information loss. A better choice to solve the problem may be the use of tolerance rough set theory.

This paper presents a gene selection method based on tolerance rough set theory. By using tolerance relations, the strict requirement of complete equivalence can be relaxed, and a more flexible approach to subset selection can be developed. The proposed method is comprised two steps. In step 1, we rank all genes with the t-test and select the most promising genes. In step 2, we apply tolerance rough set theory-based method to the selected genes in step 1. The experimental results demonstrate that the proposed algorithm is more effective than gene selection approach based on rough set theory for achieving good classification performance.

2   Preliminaries

2.1   Rough Set Theory

There is a classificatory feature in gene expression data sets. We can formalize the gene expression data set into a decision system.

Definition 1. Decision table.

A decision table is defined as $T = \langle U, C \cup D, V, f \rangle$, where U is a non-empty finite set of objects; C is a set of all condition features (also called conditional attributes) and D is a set of decision features (also called decision attributes); $V = \bigcup_{a \in C \cup D} V_a$, V_a is a set of feature values of feature a; and $f : U \times (C \cup D) \to V$ is an information function for every $x \in U$, $a \in C \cup D$.

For any $B \subseteq C \cup D$, an equivalence (indiscernibility) relation induced by B on U is defined as Definition 2.

Definition 2. Equivalence relation.

$$IND(B) = \{(x, y) \in U \times U | \forall b \in B, b(x) = b(y)\}. \qquad (1)$$

The family of all equivalence classes of $IND(B)$, i.e., the partition induced by B, is given in Definition 3.

Definition 3. Partition.

$$U/IND(B) = \{[x]_B | x \in U\}, \qquad (2)$$

where $[x]_B$ is the equivalence class containing x. All the elements in $[x]_B$ are equivalent (indiscernible) with respect to B. Equivalence classes are elementary sets in rough set theory.

For any $X \subseteq U$ and $B \subseteq C$, X could be approximated by the lower and upper approximations.

Definition 4. Lower approximation and upper approximation.

$$\underline{B}X = \{x| \, [x]_B \subseteq X\}, \tag{3}$$

$$\overline{B}X = \{x| \, [x]_B \cap X \neq \emptyset\}. \tag{4}$$

Let $B \subseteq C$, the positive region of the partition $U/IND(D)$ with respect to B is defined as Definition 5.

Definition 5. Positive region.

$$POS_B(D) = \cup_{X \in U/IND(D)} \underline{B}X, \tag{5}$$

and it is the set of all samples that can be certainly classified as belonging to blocks of $U/IND(D)$ using B.

By employing the definition of the positive region it is possible to calculate the rough set degree of dependency of a set of features D on B.

Definition 6. Degree of dependency of feature.

$$\gamma_B(D) = |POS_B(D)| \, / \, |U|. \tag{6}$$

2.2 T-Test

Feature subset selection is an important step to narrowing down the feature number prior to data mining. We assume that there are two classes of samples in a gene expression data set.

Definition 7. T-test.

The t-value for gene a is expressed by:

$$t(a) = \frac{\mu_1 - \mu_2}{\sqrt{\sigma_1^2/n_1 + \sigma_2^2/n_2}}, \tag{7}$$

where μ_i and σ_i are the mean and the standard deviation of the expression levels of gene a for $i = 1, 2$. When there are multiple classes of samples, the t-value is typically computed for one class versus all the other classes. The top genes ranked by t-value can be selected for data mining. Feature set so obtained has certain redundancy because genes in similar pathways probably all have very similar score. If several pathways involved in perturbation but one has main influence it is possible to describe this pathway with fewer genes, therefore feature selection based on rough set theory is used to minimize the feature set.

2.3 Gene Selection Algorithm Based on Rough Set Theory

Gene selection algorithm based on rough set theory for gene expression data is composed of t-test and feature selection based on rough set theory. T-test

is helpful for reducing dimensionality. The algorithm without the t-test preprocessing will get worse performance. After feature ranking, top ranked n genes
are selected to form the feature set. The values of all continuous features are
discretized. Rough set theory-based feature selection method starts with the full
set and consecutively deletes one feature at a time until we obtain a reduction.

Algorithm 1. Gene selection algorithm based on rough set theory (GSRS)

> (1) Calculate t-value of each gene, select top ranked n genes to form the
>   feature set C
> (2) Discretize the feature set C
> (3) Set $P = C$
> (4) **do**
> (5)   **for each** $a \in P$
> (6)   **if** $\gamma_{P-\{a\}}(D) == \gamma_C(D)$
> (7) $P = P - \{a\}$
> (8) **until** $\gamma_{(P-\{a\})}(D) < \gamma_{(C)}(D)$
> (9) **return** P

The loop continues to evaluate in the above manner by deleting conditional
features, until the dependency value of the current reduct is less than that of
the dataset.

3   Gene Selection Algorithm Based on Tolerance Rough Set Theory

3.1   Similarity Measures

In this approach, suitable similarity measure, given in [2,3], is described in
Definition 8.

Definition 8. Similarity measure.

$$S_a(x, y) = 1 - \frac{|a(x) - a(y)|}{|a_{\max} - a_{\min}|}, \tag{8}$$

where $a \in C \cup D$, and $a_{\max}$ and $a_{\min}$ denote the maximum and minimum values
respectively for feature a. When considering more than one feature, the defined
similarities must be combined to provide a measure of the overall similarity of
objects. For a subset of features, B, the overall similarity measure is defined as
Definition 9.

Definition 9. Overall similarity measure.

$$(x, y) \in S_{B,\tau} iff \frac{\sum\limits_{a \in B} S_a(x, y)}{|B|} \geq \tau, \tag{9}$$

where τ is a global similarity threshold; it determines the required level of similarity for inclusion within tolerance classes. This framework allows for the specific

case of traditional rough set theory by defining a suitable similarity measure and threshold ($\tau = 1$). From this, for any $B \subseteq C \cup D$, $0 < \tau \leq 1$, the so-called tolerance classes that are generated by a given similarity relation for an object are defined as Definition 10.

Definition 10. Similarity relation.

$$S_{B,\tau}(x) = \{y \in U \,|\, (x, y) \in S_{B,\tau}\}. \tag{10}$$

For any $X \subseteq U$, $B \subseteq C$ and $0 < \tau \leq 1$, lower and upper approximations are then defined in a similar way to traditional rough set theory.

Definition 11. Modified lower approximation and upper approximation.

$$\underline{B_\tau} X = \{x | S_{B,\tau}(x) \subseteq X\}, \tag{11}$$

$$\overline{B_\tau} X = \{x | S_{B,\tau}(x) \cap X \neq \emptyset\}. \tag{12}$$

The tuple $\langle \underline{B_\tau} X, \overline{B_\tau} X \rangle$ is called a tolerance-based rough set. Based this, the positive region and the dependency function can be defined as follows.

Let $B \subseteq C$ and $0 < \tau \leq 1$, the positive region is defined as Definition 12.

Definition 12. Modified positive region.

$$POS_{B,\tau}(D) = \cup_{X \in U/S_{D,\tau}} \underline{B_\tau} X. \tag{13}$$

For $B \subseteq C$ and $0 < \tau \leq 1$, the tolerance rough set degree of dependency is given in Definition 13.

Definition 13. Modified degree of dependency of feature.

$$\gamma_{B,\tau}(D) = |POS_{B,\tau}(D)| \,/\, |U|. \tag{14}$$

From these definitions, a feature selection method can be formulated that uses the tolerance-based degree of dependency, $\gamma_{B,\tau}(D)$, to gauge the significance of feature subsets.

3.2 Tolerance Rough Set Theory-Based Gene Selection Method

Gene selection algorithm based on tolerance rough set theory for gene expression data combines feature ranking together with feature selection based on tolerance rough set theory. Similarly, t-test can eliminate such redundant genes. T-test is used to feature ranking as the first step and select top ranked n genes to form the feature set. Tolerance rough set theory-based feature selection method can judge every feature and delete the features that are superfluous.

Algorithm 2. Gene selection algorithm based on tolerance rough set theory (GSTRS)

    (1) Calculate t-value of each gene, select top ranked n genes to form the
        feature set C
    (2) Set $P = C$

(3) **do**
(4)   **for each** $a \in P$
(5)   **if** $\gamma_{P-\{a\},\tau}(D) == \gamma_{C,\tau}(D)$
(6) $P = P - \{a\}$
(7) **until** $\gamma_{P-\{a\},\tau}(D) < \gamma_{C,\tau}(D)$
(8) **return** P

The stopping criteria is automatically defined through the use of the dependency measure when the deletion of further features does not result in a decrease in dependency.

3.3   A Simple Example

To illustrate the operation of feature selection algorithm based on tolerance rough set theory, it is applied to a simple example dataset in Table 1, which contains three real-valued conditional features and a crisp-valued decision feature. Set $\tau = 0.8$. $C = \{a, b, c\}$. $D = \{d\}$.

Table 1. Example dataset

Objects	a	b	c	d
1	0.3	0.4	0.2	R
2	0.3	1	0.6	A
3	0.4	0.3	0.4	R
4	0.9	0.4	0.7	R
5	0.9	0.7	0.7	A
6	1	0.4	0.7	A

The following tolerance classes are generated:

$U/S_{D,\tau} = \{\{1,3,4\},\{2,5,6\}\}$,
$U/S_{C,\tau} = \{\{1\},\{2\},\{3\},\{5\},\{4,6\}\}$,
$U/S_{C-\{a\},\tau} = U/S_{\{b,c\},\tau} = \{\{1\},\{2\},\{3\},\{5\},\{4,6\}\}$,
$U/S_{C-\{b\},\tau} = U/S_{\{a,c\},\tau} = \{\{1\},\{2\},\{3\},\{5\},\{4,6\}\}$,
$U/S_{C-\{c\},\tau} = U/S_{\{a,b\},\tau} = \{\{1,3\},\{4,6\},\{2\},\{5\}\}$,
$U/S_{C-\{a,b\},\tau} = U/S_{\{c\},\tau} = \{\{1\},\{2\},\{3\},\{4,5,6\}\}$,
$U/S_{C-\{a,c\},\tau} = U/S_{\{b\},\tau} = \{\{1,3,4,6\},\{2\},\{5\}\}$,
$U/S_{C-\{b,c\},\tau} = U/S_{\{a\},\tau} = \{\{1,2,3\},\{4,5,6\}\}$.

Considering feature set , the lower approximations of the decision classes are calculated as follows:

$\underline{C_\tau}\{1,3,4\} = \underline{\{a,b,c\}_\tau}\{1,3,4\} = \{x|S_{\{a,b,c\},\tau}(x) \subseteq \{1,3,4\}\} = \{1,3\}$,
$\underline{C_\tau}\{2,5,6\} = \underline{\{a,b,c\}_\tau}\{2,5,6\} = \{x|S_{\{a,b,c\},\tau}(x) \subseteq \{2,5,6\}\} = \{2,5\}$.

Hence, the positive region can be constructed:

$POS_{C,\tau}(D) = \cup_{X \in U/S_{D,\tau}}\underline{C_\tau}X = \underline{C_\tau}\{1,3,4\} \cup \underline{C_\tau}\{2,5,6\} = \{1,2,3,5\}$.

The resulting degree of dependency is:

$$\gamma_{C,\tau}(D) = \frac{|POS_{C,\tau}(D)|}{|U|} = \frac{|\{1,2,3,5\}|}{|\{1,2,3,4,5,6\}|} = \frac{4}{6}.$$

For feature set $C - \{a\}$, the corresponding dependency degree is:

$$\gamma_{C-\{a\},\tau}(D) = \frac{|POS_{C-\{a\},\tau}(D)|}{|U|} = \frac{|\{1,2,3,5\}|}{|\{1,2,3,4,5,6\}|} = \frac{4}{6},$$

$$\gamma_{C-\{a\},\tau}(D) = \gamma_{\{b,c\},\tau}(D) = \gamma_{C,\tau}(D) = \frac{4}{6}.$$

Feature a is deleted from feature set C. Similarly, the dependency degree of feature set $\{b,c\} - \{b\}$ is:

$$\gamma_{\{b,c\}-\{b\},\tau}(D) = \frac{|POS_{\{b,c\}-\{b\},\tau}(D)|}{|U|} = \frac{|\{1,2,3\}|}{|\{1,2,3,4,5,6\}|} = \frac{3}{6},$$

$$\gamma_{\{b,c\}-\{b\},\tau}(D) = \frac{3}{6} < \gamma_{C,\tau}(D) = \frac{4}{6}.$$

Therefore, the algorithm terminates and outputs the reduct $\{b,c\}$.

4   Experiments

To evaluate the performance of the proposed algorithm, we applied it to two benchmark gene expression data sets: Lymphoma data set (http://llmpp.nih.gov/lymphoma) and Liver cancer data set (http://genome-www.stanford.edu/hcc/). The Lymphoma data set is a collection of 96 samples. There are 42 B-cell and 54 Other type samples having 4026 genes. The Liver cancer data set is a collection of gene expression measurements from 156 samples and 1648 genes. There are 82 cases of HCCs and 74 cases of nontumor livers.

GSRS and GSTRS are run on the two data sets. Firstly, t-test is employed as a filter on Lymphoma and Liver cancer. The top ranked 50 largest t-test values genes are selected. When there are missing values in data sets, these values are filled with mean values for continuous features and majority values for nominal features [8]. As two data sets are real-valued, for GSRS algorithm, discretization of every feature of the two data sets is Equal Frequency per Interval [7]. For GSTRS algorithm, set $\tau = 0.9$. The reduction results are listed in Table 2.

Two factors need to be considered for comparing GSRS and GSTRS. One is the number of selected genes. From Table 2, we can find that the number of selected genes by GSRS is equal to the number of selected genes by GSTRS.

The other considered factor is classification accuracy of the selected genes of two data sets. Two classifiers, C5.0 and KNN, are respectively adopted. As there are a relatively small number of samples, leave-one-out accuracy is adopted. The results are shown in Table 3.

Table 2. Reduction results

Data sets	Genes	Samples	GSRS	GSTRS
Lymphoma	4026	96	7	7
Liver cancer	1648	156	6	6

Table 3. The classification accuracy of two data sets

Data sets	Lymphoma		Liver cancer	
	GSRS	GSTRS	GSRS	GSTRS
KNN	93.5%	94.8%	89.6%	92.5%
C5.0	95.2%	97.4%	91.3%	94.3%

Experimental results show the selected genes by GSTRS have higher classification accuracy than the selected genes by GSRS when we take KNN classifier. While C5.0 classifier is adopted, the classification accuracy of selected genes by GSTRS is highest of all. The reason may be that GSTRS can retain the information hidden in the data.

5 Conclusions

In this paper, we address gene selection of tolerance rough set theory. By constructing an example, we show how the technique works. This paper extends the research of traditional rough set theory and establishes one direction for seeking an efficient algorithm for gene expression data. Our method is applied to the gene selection of cancer classification. Experimental results show its validity.

Acknowledgments. This paper is supported by The National Natural Science Foundation of P.R.China (no. 60475019, 60775036) and The Research Fund for the Doctoral Program of Higher Education (no. 20060247039).

References

1. Pawlak, Z.: Rough Sets. International Journal of Information Computer Science 11(5), 341–356 (1982)
2. Jensen, R., Shen, Q.: Tolerance-Based and Fuzzy-Rough Feature Selection. In: Proceedings of the 16th International Conference on Fuzzy Systems (FUZZ- IEEE 2007), pp. 877–882 (2007)
3. Parthalin, N.M., Shen, Q.: Exploring The Boundary Region of Tolerance Rough Sets for Feature Selection. Pattern Recognition 42, 655–667 (2009)
4. Miao, D.Q., Wang, J.: Information-Based Algorithm for Reduction of Knowledge. In: IEEE International Conference on Intelligent Processing Systems, pp. 1155–1158 (1997)
5. Wang, G.Y.: Rough Set Theory and Knowledge Acquisition. Xi'an Jiaotong University Press (2001) (in Chinese)
6. Li, D.F., Zhang, W.: Gene Selection Using Rough Set Theory. In: Wang, G.-Y., Peters, J.F., Skowron, A., Yao, Y. (eds.) RSKT 2006. LNCS (LNAI), vol. 4062, pp. 778–785. Springer, Heidelberg (2006)
7. Grzymala-Busse, J.W.: Discretization of Numerical Attributes. In: Klsgen, W., Zytkow, J. (eds.) Handbook of Data Mining and Knowledge Discovery, pp. 218–225. Oxford University Press, Oxford (2002)
8. Grzymala-Busse, J.W., Grzymala-Busse, W.J.: Handling Missing Attribute Values. In: Maimon, O., Rokach, L. (eds.) Handbook of Data Mining and Knowledge Discovery, pp. 37–57 (2005)

Rough Set Based Social Networking Framework to Retrieve User-Centric Information

Santosh Kumar Ray[1] and Shailendra Singh[2]

[1] Birla Institute of Technology, Mesra, International Centre, Muscat, Oman
[2] Samsung India Software Centre, Noida, India
santosh@waljat.net, shailendra.s@samsung.com

Abstract. Social networking is becoming necessity of the current generation due to its usefulness in searching the user's interest related people around the world, gathering information on different topics, and for many more purposes. In social network, there is abundant information available on different domains by means of variety of users but it is difficult to find the user preference based information. Also it is very much possible that relevant information is available in different forms at the end of other users connected in the same network. In this paper, we are proposing a computationally efficient rough set based method for ranking of the documents. The proposed method first expands the user query using WordNet and domain Ontologies and then retrieves documents containing relevant information. The distinctive point of the proposed algorithm is to give more emphasis on the concept combination based on concept presence and its position instead of term frequencies to retrieve relevant information. We have experimented over a set of standard questions collected from TREC, Wordbook, WorldFactBook and retrieved documents using Google and our proposed method. We observed significant improvement in the ranking of retrieved documents.

Keywords: Rough sets, Document Ranking, Concept Extraction, and Social Domain Networking.

1 Introduction

Today, the WorldWideWeb is growing very fast. Recently published article [1] says that the number of web pages on the internet increased tremendously and crossed 1 trillion counts in 2008 which was only 200 billion in 2006 [18]. With the growth of the WorldWideWeb based applications, an advanced Web 2.0 framework was introduced for a variety of applications such as blogging, online gaming, social networking, knowledge sharing, chat rooms etc. Social networking is related to almost every domain from general to specific domains. [17] discusses about more than 150 popular social networking websites on a variety of topics. The famous social networking websites such as Orkut [10], Facebook [5], and LinkedIn [8] are becoming essential for users ranging from school kids to qualified professionals. In a typical social networking website, Internet users are invited by the members of the social networking

H. Sakai et al. (Eds.): RSFDGrC 2009, LNAI 5908, pp. 184–191, 2009.

website to join their interest related communities, groups, and peoples. The user has freedom to explore his interest related communities and can join those communities. Also, there is no limitation on expanding social network. One can join multiple communities, groups, and peoples to get diversified information on different topics. At present, social networking websites do not have cross-website information and as a result, scattered information on different topics could be not processed together for effective use. Another important point is that sometimes the information needed by the user is not available in their network communities and it could be available in other networks as well as could be retrieved from WorldWideWeb.

To make an efficient social network, Semantic Web plays an extra-ordinary role in exchanging information conceptually. Semantic Web represents WorldWideWeb data in the form of mesh and linked in such a way so that it could be easily processed by machines on a global scale. In this paper, we are presenting a document retrieval system which will take the user question as an input and expands them to retrieve documents from the WorldWideWeb containing relevant concepts and finally ranks retrieved results as per user relevance. The research paper is organized as follows: section 2 describes related research work while section 3 explains the proposed rough set based document ranking algorithms. Section 4 shows our experiment and results. In the last section, we have stated our conclusion and future directions.

2 Related Work

Social Networking was introduced in 2003 and becoming popular very rapidly. The available social networking websites as discussed in [3], [21], and [9] are using tagging approach to improve the search mechanisms as well as for personalized recommendations. However, tagging for any kind of information, particularly for user interest, might be done by different users using different vocabularies. So tagging approach is not useful to retrieve relevant information lying at the end of other users. Therefore, conceptually expanded user input may solve the term mismatch problem in building efficient document retrieval system in social networking domain. The use of semantic web tools such as ontologies and WordNet [19] has been a preferred choice of researchers to propose input expansion methods. We have also used ontologies and WordNet combination to solve the term mismatch problem in document retrieval.

There are number of document ranking models proposed such as extended Boolean model [13], Vector space model [7], and Relevance model [4]. These models are largely dependent on the query term frequency, document length etc to rank the documents. These methods are computationally fast. However, they ignore the linguistics features and the semantics of the query as well as the documents which inversely affects their retrieval performance. [16] and [12] propose conceptual models which map a set of words and phrases to the concepts and exploit their conceptual structures for retrieval. [15] proposes an ontology hierarchy based approach for automatic topic identification which can be further extended for automatic text categorization. These models are complicated but retrieve more precise information in comparison of other statistical models. However, these methods are not able to handle imprecise information which is necessary to fulfill users need. Therefore, rough set based methods [14] [2] were proposed for document classification to handle imprecise and vague information. [6] proposes automatic classification of WWW bookmarks based on rough sets while [20] proposes extension of document frequency metric using rough sets. They have used indiscernibility relation for text categorization. In this paper, we have proposed a document ranking method which uses an extension of their research work.

3 Social Networking Based Information Retrieval System – A Rough Set Based Approach

The proposed social domain document retrieval system considers the user's interest as an input and extracts important terms then finds the semantically related concepts using its query expansion module described in [11]. These conceptually related terms along with the user input are passed to the document retrieval phase. The document retrieval phase searches for the documents relevant to the user's interests and presents a list of the document in the order of their relevancy using rough set based ranking algorithm. The proposed document ranking algorithm is not considering term frequencies for ranking of retrieved documents as the algorithms based on term frequencies tend to be more biased towards longer documents. This algorithm expands the user input, selects the relevant features from the set of documents returned by search engines and ranks extracted concept combinations according to their relevancy to the user's input. Finally, the algorithm performs re-ranking of the documents based on the position of the concept combinations in the set of documents. We are explaining algorithms in the following sections.

3.1 Concept Combination Ranking Algorithm

In this section, we are proposing an algorithm that uses the indiscernibility relation of the rough set theory to rank the concept combinations. The basic idea is based on the algorithm discussed in [20] which uses document frequency to extract the important features from a set of documents and categorizes them on the basis of their features (terms). We are extending their algorithm for ranking a concept combination. Let us assume that the user input contains concepts $C_1, C_2,...C_n$ and the input is expanded using algorithm proposed in [11]. The key concepts in the expanded set are then grouped into concept combinations using Cartesian product and ranked according to the knowledge quantity contained in them. The complete algorithm for ranking the concept combinations is described below.

Algorithm: Concept_Extraction(Q, D)
Input: User input (Q) and set of documents (D)
Output: Ranked concepts list (G_r)
Step 1: Extract key concepts $C_1, C_2, ..., C_n$ from the input.
Step 2: Expand input using expansion algorithm [11]. The resulting set is $C_1 \cup C_2 \cup ... \cup C_n$ where $C_i = C_{i1} \cup C_{i2} \cup ... \cup C_{ik}$ and C_{ij} indicates the jth semantically related word to concept C_i.
Step 3: Let $G = C_1 \times C_2 \times ... \times C_n$ where × indicates the Cartesian product.

Step 4: Define an information system I = (U, A, V, f), where $U = \{D_i \mid D_i \in D\}$, $A = \{G_i \mid G_i \in G\}$, V is the domain of values of G_i, and f is an information function (U, A) $\rightarrow$ V such that:

$$f(D_i, \ G_i) = \begin{cases} 0 & \textit{if any of the concepts in} \quad G_i \textit{ is not present in } D_i \\ 1 & \textit{if all concepts in} \quad G_i \textit{ are present in } D_i \end{cases}$$

Step 5: Determine the "Knowledge Quantity" (KQ) of G_i using the equation (1)

$$KQ_i = m(n - m) \tag{1}$$

where n and m represents cardinality of D and no. of documents in which concept group G_i occurs respectively.

Step 6: Repeat step 5 for all G_i.

Step 7: Sort G according to "Knowledge Quantity" and return G_r (Sorted G).

Step 8: END

Consider the query *"How far is Mars from our planet?"* for example. Key concepts in the query are *far, Mars, our*, and *planet*. The expanded query as explained in step 2 of algorithm *Concept_Extraction* is *"(far OR distant) AND (Mars OR "Red Planet") AND our AND (Planet OR "terrestrial planet")"*. The concept combinations, G, obtained after taking Cartesian product is:

  G =*{(far, Mars, our, planet), (far, Mars, our, terrestrial planet), (far, Red planet, our, planet), (far, Red planet, our, terrestrial planet), (distant, Mars, our, planet), (distant, Mars, our, terrestrial planet), (distant, Red planet, our, planet), (distant, Red planet, our, terrestrial planet) }.*

An information system is defined using these concept combinations, documents retrieved for the expanded query and the f-values for the concept combinations as explained in step 4 of the algorithm. In step 5 and 6, knowledge quantity of the each element of G is computed using the equation 5.4. These concept combinations are then ranked according to the knowledge quantity contained in them. The ranked concept combinations (in decreasing order) are as follows:

  G_{ranked} =*{ (distant, Mars, our, planet,) (far, Red planet, our, planet), (distant, Red planet, our, planet), (far, Mars, our, planet), (far, Mars, our, terrestrial planet), (far, Red planet, our, terrestrial planet), (distant, Red planet, our, terrestrial planet), (distant, Mars, our, terrestrial planet) }.*

3.2 Document Ranking Algorithm

The proposed document ranking algorithm considers ranked concept combination as discussed in section 4.1 and searches the document sets for these concept combinations. The underlying intuition is that a document is more relevant if it contains combination of concepts together rather than containing individual concepts. The algorithm considers the most descriptive concepts of the document which are used to define title or subtitle. Secondly, we consider those sentences more relevant which contain more number of concepts. *Algorithm Document_Ranking* describes the proposed document ranking algorithm.

Algorithm: Document_Ranking (Q, D)
Input: User query (Q) and set of documents (D)
Output: Ranked documents list (D_r)
Step 1: Run *Concept_Extraction (Q, D)* to get ranked list of concept groups.
Step 2: For each document $D_i \in D$ and concept group G_j, compute document score
(W_{i1}) using equation (2).

$$W_{i1} = \left(1 + \sum_{1 \le j \le p \text{ and } G_j \subset D_i} \left(\frac{p - r_j}{p}\right)\right) W_0 \tag{2}$$

where p is the cardinality of the set G (step 3 of *Concept_Extraction*) and r_j is the
rank of G_j obtained in step1. W_0 is the initial weight assigned to each document.
Step 3: For each document $D_i \in D$ and concept group G_j, re-compute document score
(W_{i2}) using equation (3).

$$W_{i2} = W_{i1} + k_1 W_{i1} \sum_{t \subseteq G_j \text{ and 't' is in one sentence}} \frac{a_{ts}}{b_j} + k_2 W_{i1} \sum_{t \subseteq G_j \text{ and 't' is in one subtitle}} \frac{a_{th}}{b_j}$$
$$+ k_3 W_{i1} \sum_{t \subseteq G_j \text{ and 't' is in title}} \frac{a_{tt}}{b_j} \tag{3}$$

Here k_1, k_2, and k_3 are the weights assigned for occurrence of concept combination in
sentences, sub-titles, and titles within the documents. a_{ts}, a_{th} and a_{tt} are the cardinality
of subset 't' in sentences, sub-titles, and title respectively. b_j is the cardinality of G_j.
Step 4: Rank the document set according to the scores obtained in step 3.
Step 5: END

4   Experimental Results and Discussion

To test the efficiency of the proposed algorithm, we conducted an experiment over a
set of 50 questions collected from social networking websites and further extracted
key concepts. To enhance the recall of the document retrieval, terms semantically
closer to the key concepts were determined using WordNet and domain using the
query expansion algorithm [11]. These original key concepts and semantically related
words were fed into Google in Boolean form and we downloaded 25 documents corresponding to each of these 50 questions separately. The retrieved documents were re-
ranked using *Document_Ranking* algorithm. It was observed that the average number
of documents containing correct answers in top 10 documents increased from 3.56 to
4.48. This indicates an improvement of 25% in the document retrieval.

We also observed increased number of correct answers in top ranked documents.
There were 17 questions whose answers were present in at least 5 documents out of
top 10 documents using Google but using proposed algorithm, this count increased to
25. These results reflect that the algorithm *Document_Ranking* is bringing more relevant documents to higher ranks. We have summarized our results in table 1.

Table 1. Comparative Performance Analysis

S.N	Performance Parameters	Using Google	Using Proposed Approach
1	No. of questions whose answers were present in at least 5 documents (out of first 10 documents)	17	25
2	Average no. of documents containing correct answers (out of first 10 documents)	3.56	4.48
3	Number of questions with answer in the first document	22	23
4	Average rank of the document containing first correct answer	2.78	2.44

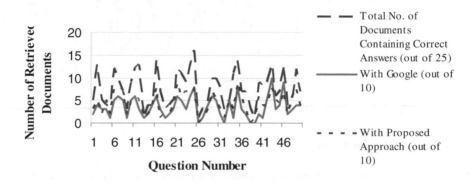

Fig. 1. Comparison of Number of Documents Containing Correct Answers

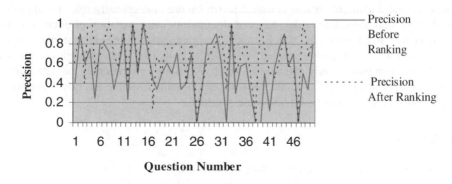

Fig. 2. Information Retrieval Precision Graph

Results of the experimental questions are shown in fig. 1. This shows the no. of documents containing correct answers is higher than original retrieval. Thus, our algorithm helps in improving the precision which is more explicitly shown in fig. 2 which is derived from fig. 1 by using the formula for precision calculation (Precision=

ratio of the number of relevant documents retrieved to the total number of relevant documents that exist for a given question). Further, we represent document rank containing first correct answer in Fig. 3. There are 28% questions for which rank of the first document with correct answer is same for Google and our proposed method. So there is no scope of improvement. While in 46% questions, the algorithm improved the ranks of the first document containing correct answer while rank of the same declined in case of 26% questions. Thus, it is clear from the fig. 3 that algorithm is improving the rank of relevant documents.

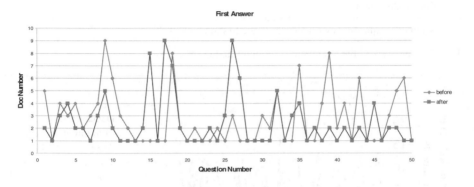

Fig. 3. Representing Documents' Rank Containing First Correct Answer

5   Conclusion and Future Scope

Social networking domain is growing rapidly and millions of users are getting benefits by sharing information on different matters. In this paper, we have presented two algorithms to rank documents conceptually. Our first algorithm ranks concept combination of the documents which is useful to find more conceptually relevant answers. Further, second algorithm ranks retrieved documents using position of concept combination which improves the precision of the information retrieval system. Though this algorithm uses modern semantic tools such as rough set and ontologies but it is a simple and computationally efficient method. We have experimented on 1250 questions collected from popular social networking domains to judge the effectiveness of the proposed method and found favorable results.

References

1. Alpert, J., Hajaj, N.: We Knew the Web was Big (2008),
   http://googleblog.blogspot.com/2008/07/
   we-knew-web-was-big.html
2. Bao, Y., Aoyama, S., Yamada, K., Ishii, N., Du, X.: A Rough Set Based Hybrid Method to Text Categorization. In: Second international conference on web information systems engineering (WISE 2001), vol. 1, pp. 254–261. IEEE Computer Society, Washington (2001)
3. Choochaiwattana, W., Spring, M.B.: Applying Social Annotations to Retrieve and Re-rank Web Resources. In: Proceedings of the International Conference on Information Management and Engineering, pp. 215–219. IEEE Computer Society, Los Alamitos (2009)

4. Crestani, F., Lalmas, M., Rijsbergen, J., Campbell, L.: Is This Document Relevant? ...Probably. A Survey of Probabilistic Models in Information Retrieval. ACM Computing Surveys 30(4), 528–552 (1998)
5. Facebook, http://www.facebook.com
6. Jensen, R., Shen, Q.: A Rough Set-Aided System for Sorting WWW Bookmarks. In: Zhong, N., Yao, Y., Ohsuga, S., Liu, J. (eds.) WI 2001. LNCS (LNAI), vol. 2198, pp. 95–105. Springer, Heidelberg (2001)
7. Lee, D.L., Chuang, H., Seamons, K.: Document Ranking and the Vector Space Model. IEEE Software 14(2), 67–75 (1997)
8. Linkedln, http://www.likedln.com
9. Marlow, C., Naaman, M., Boyd, D., Davis, A.: Position Paper, tagging, Taxonomy, Flickr, Article, To Read. In: Proceedings of the 17th ACM Conference on Hypertext and Hypermedia (HT 2006) (August 2006)
10. Orkut, http://www.orkut.com
11. Ray, S.K., Singh, S., Joshi, B.P.: Question Answering Systems Performance Evaluation – To Construct an Effective Conceptual Query Based on Ontologies and WordNet. In: Proceedings of the 5th Workshop on Semantic Web Applications and Perspectives, Rome, Italy, December 15-17. CEUR Workshop Proceedings, pp. 1613–1673 (2008)
12. Rocha, C., Schwabe, D., de Aragão, P.M.: A Hybrid Approach for Searching in the Semantic Web. In: 13th International Conference on World Wide Web, pp. 374–383. ACM, New York (2004)
13. Salton, G., Fox, E.A., Wu, H.: Extended Boolean Information Retrieval. Communications of the ACM 26(11), 1022–1036 (1983)
14. Singh, S., Dey, L.: A Rough-Fuzzy Document Grading System for Customized Text Information Retrieval. Information Processing and Management: an International Journal 41(2), 195–216 (2005)
15. Tiun, S., Abdullah, R., Kong, T.E.: Automatic Topic Identification using Ontology Hierarchy. In: Gelbukh, A. (ed.) CICLing 2001. LNCS, vol. 2004, pp. 444–453. Springer, Heidelberg (2001)
16. Vallet, D., Fernández, M., Castells, P.: An Ontology-Based Information Retrieval Model. In: Gómez-Pérez, A., Euzenat, J. (eds.) ESWC 2005. LNCS, vol. 3532, pp. 455–470. Springer, Heidelberg (2005)
17. Wikipedia List of Social Networking, http://en.wikipedia.org/wiki/List_of_social_networking_websites
18. Wirken, D.: The Google Goal Of Indexing 100 Billion Web Pages (2006), http://www.sitepronews.com/archives/2006/sep/20.html
19. WordNet, http://wordnet.princton.edu
20. Xu, Y., Wang, B., Li, J.T., Jing, H.: An Extended Document Frequency Metric for Feature Selection in Text Categorization. In: Li, H., Liu, T., Ma, W.-Y., Sakai, T., Wong, K.-F., Zhou, G. (eds.) AIRS 2008. LNCS, vol. 4993, pp. 71–82. Springer, Heidelberg (2008)
21. Zhou, D., Bian, J., Zheng, S., Zha, H., Giles, C.L.: Exploring social annotations fro information retrieval. In: Proceedings of International World Wide Web Conference, WWW 2008 (April 2008)

Use of Fuzzy Rough Set Attribute Reduction in High Scent Web Page Recommendations

Punam Bedi and Suruchi Chawla

Deptt of Computer Science, New Academic Block, University of Delhi, India
pbedi@cs.du.ac.in, sur_chawla@rediffmail.com

Abstract. Information on the web is growing at a rapid pace and to satisfy the information need of the user on the web is a big challenge. Search engines are the major breakthrough in the field of Information Retrieval on the web. Research has been done in literature to use the Information Scent in Query session mining to generate the web page recommendations. Low computational efficiency and classification accuracy are the main problems that are faced due to high dimensionality of keyword vector of query sessions used for web page recommendation. This paper presents the use of Fuzzy Rough Set Attribute Reduction to reduce the high dimensionality of keyword vectors for the improvement in classification accuracy and computational efficiency associated with processing of input queries. Experimental results confirm the improvement in the precision of search results conducted on the data extracted from the Web History of "Google" search engine.

Keywords: Fuzzy Rough Set, Information Retrieval, Information Need, Information Scent, Fuzzy Similarity.

1 Introduction

This Information on the web is growing at a rapid pace. To find the relevant documents for a specific information need of the user from a big pool of information is a big challenge. The search query of the user to the search engine is not able to fetch the sufficient relevant documents [1, 9, 11]. Work in [2, 3, 4, 5] has been done using Information Scent in Query session mining to improve the Information Retrieval precision. Query session is defined as set of clicked URLs associated with the user Query. Information scent is the subjective sense of value and cost of accessing a page based on perceptual cues with respect to the information need of the user. Users tend to click URLs with high scent associated with them [6, 10, 15, 16]. These high Scent pages uniquely satisfy the information needs of the user whereas low scent pages are less relevant to the information need of the user. Web page recommendations based on past queries can help to satisfy the information need of the current user. Each query session is represented by keywords vector weighted using Information Scent. Query sessions represented by weighted keyword vector are clustered to get the set of query sessions with similar information need. Each cluster is represented by mean weighted keyword vector. In [5] during online processing the input query vector is

H. Sakai et al. (Eds.): RSFDGrC 2009, LNAI 5908, pp. 192–200, 2009.

used to find the cluster which closely represent the information need of input query and recommend the high scent clicked web pages associated with the selected cluster. The problem that is observed is low computational efficiency and classification accuracy due to high dimensionality of keyword vector representing the clusters of query sessions. Attribute Selection is important for reducing computational cost of classification. All the keywords of query sessions keyword vector are not equally important from the point of view of identifying the different information need represented by clusters. Classification assumes that all attribute are equally important when classifying using nearest neighbour approach and is sensitive to sparse data representation which affects the classification accuracy. In [8] Research has been done for personalizing the web search using Rough Fuzzy method to personalize the web search more effectively by identifying the discerning keywords for focussed web search using Fuzzy set discretization of real valued term weight of document vector and attribute reduction using Rough set attribute reduction. In this paper Fuzzy Rough Set attribute reduction has been applied to reduce the keyword vectors representing the query sessions to those keywords of keyword vector which are all imperative to identify the different information need associated with identified clusters. The reduced set of keyword vector representing each cluster of query sessions reduces the space complexity due to reduction in memory requirement for storing the clusters mean weighted keyword vector. Time complexity will be improved in online processing in computing the clusters which best represent the information need associated with input query. Thus computational efficiency is improved with the reduction in time and space complexity. The Classification Accuracy of the input queries to the clusters of query sessions is improved with reduced relevant attribute set obtained using Fuzzy Rough Attribute Reduction which uniquely identifies the information need of the user associated with the input query after removing those attributes which were redundant and irrelevant from perspective of identifying the information need associated with the cluster.

This paper is organized as follows section 2 explains the Computation of Information Scent, section 3 explains Fuzzy Rough Approach for Attribute Reduction , section 4 explains the use of FRSAR in Information Retrieval, section 5 presents the Experimental study and section 6 concludes the paper.

2 Information Scent Computation

2.1 Information Scent Metric

The Inferring User Need by Information Scent (IUNIS) algorithm provides various combinations of parameters to quantify the Information Scent [6] [10].The factors that are taken are page access PF.IPF weight and TIME that are used to quantify the information scent associated with the clicked page in a query session. The information scent sid is calculated for each page P_{id} in a given Query session i as follows.

$$S_{id} = PF.IPF(P_{id}) * Time(P_{id}) \forall d \in 1..n \quad (1)$$

$$PF.IPF(P_{id}) = f_{Pid}/\max(f_{Pid}) * \log(M/m_{Pid}) \text{ where } d \in 1..n \quad (2)$$

where n is the number of unique clicked web pages in query session i.

PF.IPF(P_{id}) and Time(P_{id}) are defined as follows.

PF.IPF(P_{id}) : PF corresponds to the page Pid normalized frequency f_{Pid} in a given query session Qi and IPF correspond to the ratio of total number of query sessions M in the whole log to the number of query sessions m_{Pid} that contain the given page Pid.

Time(P_{id}) : It is the ratio of time spent on the page Pid in a given session Qi to the total duration of session Q_i.

3 Fuzzy Rough Approach for Attribute Reduction

3.1 Fuzzy Rough Set Attribute Reduction (FRSAR)

Rough Set Attribute Reduction is well suited to discrete information system $S1=(U,A)$. In discrete information system all the objects in the table have discrete value for their data objects. However it is often the case that data object in the information system have both real and crisp attribute. Rough set theory in RSAR fails to find the similarity of data object having real value attribute [12,13]. Fuzzy Rough approach is used in this paper for the keyword reduction of query sessions vectors representing the information needs of the query sessions on the web. The keyword vector of query sessions is weighted using Information Scent.

In order to apply Fuzzy Rough set concept to the reduction of attribute in data set containing real attribute without loss of information, Fuzzy similarity relation Rp is used to determine the extent to which the two data object are similar having real valued attribute. The crisp lower and upper approximation PX and $\overline{P}X$ becomes Fuzzy lower and upper approximation set with membership function $\mu_{PX}(x)$ and $\mu_{\overline{PX}}(x)$ respectively. The P positive region of D, i.e. $POS_P(D)$ become Fuzzy set whose membership function is defined by $\mu_{POSP}(D)$. The dependency of attribute set D on P is given by gemma'$_P$(D).

Consider class X represented by Fuzzy set X in U whose membership function is defined as μ_X, the lower approximation of X denoted by PX and the upper approximation of X denoted by $\overline{P}X$ are the fuzzy sets of U whose membership functions are defined as below.

$$\mu_{\underline{P}X}(x) = \underset{\forall x' \in U}{S} \min(\mu_{Rp}(x, x'), T_{y \in U}(S(N(\mu_{Rp}(x', y)), \mu_X(y)))) \tag{3}$$

$$\mu_{\overline{P}X}(x) = \underset{\forall x' \in U}{S} \min(\mu_{Rp}(x, x'), S_{y \in U}(T(\mu_{Rp}(x', y), \mu_X(y)))) \tag{4}$$

$$\mu_X : U \to |0,1| \tag{5}$$

$$\mu x(x) = \begin{vmatrix} 1 : x \in U \, and \, D \ (x) = X \\ 0 : x \in U \, and \, D \ (x) \neq X \end{vmatrix} \tag{6}$$

where

$$T(a,b) = \min(a,b)$$
$$S(a_1, a_2, \ldots, a_n) = \max(a_1, a_2, a_3, \ldots, a_n)$$
$$N(a) = 1 - a$$
$$\forall(a, b, a_1, a_2, \ldots, a_n) \in [0,1]$$

The membership of an object $x \in U$ to the fuzzy positive region can be defined

$$\mu_{POS\,P(D)}(x) = S(\mu_{PX}(x)) \tag{7}$$
$$X \in U/\bar{D}$$

Using the definition of fuzzy positive region, the new dependency function can be defined as follows.

$$\text{gemma'}_P(D) = (\sum_{x \in U} \mu_{POSP(D)}(x))/|U| \tag{8}$$

4   Use of FRSAR in Information Retrieval

4.1   Query Sessions Representation

Information need associated with the query session is modeled using Information Scent and content of clicked URLs. Each query session is represented by keyword vector weighted by information scent as given by the equation below.

$$Q_i = \sum_{d=1}^{n} s_{id} * P_{id} \tag{9}$$

In above formula n is the number of distinct clicked pages in the session Q_i and s_{id} (information scent) is calculated for each page P_{id} in a given session Q_i. P_{id} is a keyword vector describing the content of the page P_{id} using tf.idf (where tf represents term frequency and idf represents inverse document frequency).

The query sessions vector are clustered using k-means Algorithm in [17,18] to generate clusters of query sessions optimized by criterion function. A score or criterion function measures the quality of resulting clusters. This is used by common vector space implementation of k-means algorithm [18]. The function measures the average similarity between vectors and the centroid of clusters that are assigned to. Let C_p be a cluster found in a k-way clustering process ($p \in 1..k$) and let c_p be the centroid of pth cluster. The criterion function I is defined as follows:

$$I = 1/M \sum_{p=1}^{k} \sum_{v_i \in C_p} sim(v_i, c_p) \tag{10}$$

where M is the total number of query sessions in all clusters and vi is the vector representing some query session belonging to the cluster Cp and centroid cp of the cluster Cp is defined as given below.

$$c_p = (\sum_{v_i \in C_p} v_i)/|C_p| \tag{11}$$

where $|C_p|$ denotes the number of query sessions in cluster C_p. $sim(v_i, c_p)$ is calculated using cosine measure.

Let U be a finite set of query sessions keyword vector called Universe and A be the set of attribute describing the query sessions where $A=(C \cup D)$ and (U,A) is an information system S1. Decision column D represents the label of cluster to which query session belongs and each cluster uniquely represents the specific information need. The query sessions keyword vectors are stored in (U,A). Each query session in clusters is stored as row in (U,A) and C columns represent the keywords of weighted keyword vector of query sessions. A particular cell S1(row,col) represent the weight of keyword represented by column col of query session vector which is labelled by row. The membership function of Fuzzy Similarity Relation R_p to find the extent of similarity of query sessions keyword vector is defined below.

$\mu_{Rp}(x,y) = \{cos(x,y): x,y \in U$ and x,y are weighted keyword vector of query session with attributes in set P , cos(x,y) calculates the cosine similarity of x and y vector such that μ_{Rp} satisfies the following properties

$$\mu_{Rp}(x,x)=1 \ \forall \ x \in U \quad \mu_{Rp}(x,y)=\mu_{Rp}(y,x) \forall \ x,y \in U \quad \mu_{Rp}(x,z) \geq \mu_{Rp}(x,y) \wedge \mu_{Rp}(y,z)$$

Fuzzy Rough Set Attribute Reduction algorithm operates on the (U,A) information system to reduce the size of keyword vectors without loss of information represented by query sessions in (U,A).

4.2 Fuzzy Rough Set Attribute Reduction of Clustered Query Sessions Keyword Vector

The following algorithm is used to generate the reduct R which is a subset of C using dependency function gemma'$_R$(D). gemma'$_R$(D) is measure of dependency of decision attribute set D on R condition attribute set .

1. R= C where $C=\{k_1,k_2,k_3,..,k_{|Keywords|}\}$, D={1,2,..,|Clusters|}, gemma'prev=0, gemma'best = gemma'$_R$(D),Y={ }
2. T=R
3. gemma'prev=gemma'best
4. gemma'best = -1
5. for all x ∈ C and x ∉ Y
        if gemma' $_{R-\{x\}}$ (D) > gemma'best
                T=R - {x}
                Temp={x}
            gemma'best=gemma'$_T$(D)
        end if
6. end for

7.   if(round(gemma'best) = round(gemma'prev))
        R=T
            Y=Y ∪ Temp
8.   end if
9.   if round(gemma'best) = round(gemma'prev))
10.   goto step 2
    else
11.   return R

The above algorithm generates R by incrementally removing the least informative attribute from it till there is no change in the value of dependency function. The round function used in the algorithm returns the value rounded to the nearest integer. |Keywords| represent the count of all distinct keywords of clicked URLs present in the data set after all stopword removal and stemming using Porter Stemming Algorithm. |Clusters| represent the count of clusters obtained in Query sessions mining.

4.3 High Scent Web Page Recommendations in Information Retrieval Using FRSAR

The proposed method of High Scent Web page recommendations using Query sessions keyword vector reduction with FRSAR in Query session mining is given below.

Offline Processing
1. Clustered Query sessions are represented in the form of information system $S1=(U,A)$ where $A=(C \cup D)$ where C are set of keywords of keyword vector representing all query sessions and D is the class label of the cluster to which query session belongs.
2. Apply the Fuzzy Rough Set Attribute Reduction to reduce the dimensionality of information system using Fuzzy similarity relation for query sessions keyword vector in FRSAR approach given in section 4.2.
3. Use reduced set of attribute R to define each cluster mean keyword vector.

Online processing
1. The input query is represented in the keyword vector scaled to the dimension of reduced set R.
2. The input query similarity to each cluster mean term vector is calculated to classify the input query to the nearest cluster which best represent the information need similar to that of input query.
3. The High Scent web pages associated with the selected cluster will be recommended for a given input query.

5 Experimental Study

Experiment was performed on the data collected from Web history of "Google" search engine. The data set was generated by users who had expertise in specific domains mainly entertainment, academics and sports. The Web history of "Google" search engine contains the following fields for each clicked URLs.

1. Time of the Day   2. Query terms   3. Clicked URLs

On submission of input query, "Google" search engine returns a result page consists of URLs retrieved for a given query along with the content information about URLs. In the experiment only those query sessions in the data set were selected which had at least one click in their answer. Query sessions considered consists of query terms along with clicked URLs. The numbers of distinct URLs in the collected data set were found to be 3145. The data set was pre-processed to get 400 query sessions. The data set generated from web history was loaded into database format to be processed further.

The experiment was performed on Pentium IV PC with 1 GB RAM under Windows XP using JADE (Java Agent Development Environment) and Oracle database. Web Sphinx crawler was used to fetch the clicked documents of query sessions in the data set. Each query session was transformed into the vector representation using Information Scent and content of clicked URLs and stored in the database. The k-Means algorithm was executed several times for different values of k and criterion function was computed for each value of k. The criterion function was found to have maximum value at k=8 where k is the number of clusters. The similarity of vectors was measured using cosine formula for weighted term vector. Clusters of 400 query sessions were stored in the form of information system $S1=(U,A)$ where $A=(C \cup D)$.

The initial dimensionality of C was 1429 that is 1429 keyword attributes were representing each clustered query sessions vector in Information System S1. D was the class label of the clusters to which query sessions vector belongs. The dimensionality of reduced set of attributes obtained using Fuzzy Rough Attribute Reduction Algorithm was 314 which is 21% of original set of attributes.

In order to analyze the effectiveness of keyword reduction in query sessions mining in satisfying the information need of the users in information retrieval, the performance of both the approaches with and without using FRSAR was evaluated using randomly selected test input queries which were categorized as untrained queries set. The untrained queries were those queries which did not have sessions associated with them in data set and are categorised as unseen queries. Some of the queries in each of the category are given below in Table 1.

Table 1. Sample List of Untrained Queries

Category	Queries
Untrained Set	Movie song, Space food, novels, magazine, movies ,Numbness, Nature, family play Games, movie pictures, software download, online tutorial, Free download mp3, skies of arcadia pictures.

The experiment was performed using 46 untrained queries distributed in each of the domain. The precision was evaluated on untrained set of queries belonging to each of the domain considered for both without FRSAR and with FRSAR. The average precision was calculated for first 2 result pages and users mark the relevant documents within the list of URLs retrieved using "Google" search engine along with web page recommendation for a given query using both with and without FRSAR. The Fig 1 shows the average precision calculated for untrained queries .The average precision is improved for untrained queries using keyword reduction with FRSAR. It is apparent that dimensionality reduction helps to identify those attributes which uniquely identify the different information need represented by clusters of query sessions.

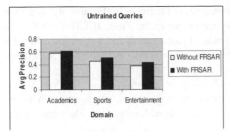

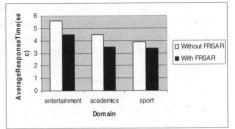

Fig. 1. The Average precision of without FRSAR and with FRSAR on untrained set of input queries

Fig. 2. The Average response time of search results with web page recommendation with FRSAR and without FRSAR on untrained set of input queries

Fig 2 shows the dimensionality reduction also has the significant impact on the time complexity of online processing phase. The time to classify the input query is reduced significantly. This effect is reflected in the average response time of web page recommendations for input queries with FRSAR on Google Search engine. The online processing time decreases tremendously which is significant for a system like search engine which require quick response time. The storage requirement for clustered data set has been reduced due to reduced keyword vector obtained using FRSAR. Fuzzy rough set attribute reduction is used for dimensionality reduction before online processing phase use the attributes belonging to the resultant reduct. The computational cost of Fuzzy Rough set Attribute Reduction has no impact on the run time efficiency of online processing phase of High Scent web page recommendation in Information Retrieval.

6 Conclusion

This paper presented an approach to improve the information retrieval precision by improving the identification of the past query sessions similar in information need to that of input queries coupled with the improvement in computational efficiency using Fuzzy Rough set Attribute Reduction (FRSAR). FRSAR reduce the large dimensionality of weighted keyword vector of clusters of query sessions using Fuzzy Similarity Relation without loss of information. Experiment was conducted on the data extracted from the Web History of "Google" search engine. Experiments used Fuzzy Rough Set Attribute Reduction in offline processing before online processing phase use the attributes belonging to the resultant reduct. Experiments show the improvement in the Information Retrieval precision confirming that FRSAR reduces the dimensionality of clusters of query sessions vector without loss of information.

References

1. Baeza-Yates, R., Ribeiro-Neto, B.: Modern Information Retrieval. Addison Wesley, Reading (1999)
2. Bedi, P., Chawla, S.: Improving Information Retrieval Precision using Query log mining and Information Scent. Information Technology Journal, Asian Network for Scientific Information 6(4), 584–588 (2007)

3. Chawla, S., Bedi, P.: Improving Information Retrieval Precision by Finding Related Queries with similar Information need using Information Scent. In: Proc., ICETET 2008 – The 1st International Conference on Emerging Trends in Engineering and Technology, July 16-18, pp. 486–491. IEEE Computer Society Press and Papers also available in IEEE Xplore (2008)

4. Chawla, S., Bedi, P.: Personalized Web Search using Information Scent. In: Proc. CISSE 2007 - International Joint Conferences on Computer, Information and Systems Sciences, and Engineering, Technically Co-Sponsored by: Institute of Electrical & Electronics Engineers (2007)

5. Chawla, S., Bedi, P.: Finding Hubs and authorities using Information scent to improve the Information Retrieval precision. In: Proc. ICAI 2008 -The 2008 International Conference on Artificial Intelligence, WORLDCOMP 2008, July 14-17 (2008)

6. Chi, E.H., Pirolli, P., Chen, K., Pitkow, J.: Using Information Scent to model User Information Needs and Actions on the Web. In: Proc. ACM CHI 2001 Conference on Human Factors in Computing Systems, pp. 490–497 (2001)

7. Chouchoulas, A., Shen, Q.: Rough Set Aided Keyword Reduction for Text Categorization. Applied Artificial Intelligence 15(9), 843–873 (2001)

8. Duan, Q., Miao, D., Zhang, H., Zheng, J.: Personalized web retrieval based on Rough Fuzzy method. Journal of Computational Information Systems 3(2), 203–208 (2007)

9. Gudivada, V.N., Raghavan, V.V., Grosky, W., KasanaGottu, R.: Information Retrieval on World Wide Web, pp. 58–68. IEEE expert, Los Alamitos (1997)

10. Heer, J., Chi, E.H.: Identification of Web User Traffic Composition using Multi-Modal clustering and Information Scent. In: Proc. of Workshop on Web Mining, SIAM Conference on Data Mining, pp. 51–58. SIAM, Philadelphia (2001)

11. Jansen, M., Spink, A., Bateman, J., Saracevic, T.: Real life Information retrieval: a Study of user queries on the Web. ACM SIGIR Forum 32(1), 5–17 (1998)

12. Jansen, R., Shen, Q.: Semantic –Preserving Dimensionality Reduction:Rough and Fuzzy – Rough-Based Approach. IEEE Transactions on Knowledge and Data Engineering 16(12) (2004)

13. Jansen, R., Shen, Q.: Fuzzy-Rough Attribute Reduction with Application to Web Categorization. Fuzzy Set and Systems 141(3), 469–485 (2004)

14. Pawlak, Z., Grzymala-Busse, J., Slowinski, R., Ziarko, W.: Rough Sets. Communications of the ACM 38(11), 88–95 (1995)

15. Pirolli, P.: Computational models of information scent-following in a very large browsable text collection. In: Proc. ACM CHI 1997 Conference on Human Factors in Computing Systems, pp. 3–10 (1997)

16. Pirolli, P.: The use of proximal information scent to forage for distal content on the world wide web. In: Working with Technology in Mind: Brunswikian, Resources for Cognitive Science and Engineering. Oxford University Press, Oxford (2004)

17. Zhao, Y., Karypis, G.: Comparison of agglomerative and partitional document clustering Algorithms. In: SIAM Workshop on Clustering High-dimensional Data and its Applications (2002)

18. Zhao, Y., Karypis, Y.: Criterion functions for document clustering. Technical report, University of Minnesota, Minneapolis, MN (2002)

19. Zhong, N., Dong, J., Ohsuga, S.: Using Rough Set with Heuristics for Feature Selection. Journal of Intelligent Information System 16, 199–214 (2001)

A Rough Sets Approach for
Personalized Support of Face Recognition

Daryl H. Hepting[1], Timothy Maciag[1], Richard Spring[1],
Katherine Arbuthnott[2], and Dominik Ślęzak[3,4]

[1] Department of Computer Science, University of Regina
3737 Wascana Parkway, Regina, SK, S4S 0A2 Canada
dhh@cs.uregina.ca, maciagt@cs.uregina.ca, spring1r@cs.uregina.ca
[2] Campion College, University of Regina
3737 Wascana Parkway, Regina, SK, S4S 0A2 Canada
katherine.arbuthnott@uregina.ca
[3] Institute of Mathematics, University of Warsaw
Banacha 2, 02-097 Warsaw, Poland
[4] Infobright Inc.
Krzywickiego 34 pok. 219, 02-078 Warsaw, Poland
slezak@infobright.com

Abstract. The activity of facial recognition is routine for most people; yet describing the process of recognition, or describing a face to be recognized reveals a great deal of complexity inherent in the activity. Eyewitness identification remains an important element in judicial proceedings. It is very convincing, yet it is not very accurate. We studied how people sorted a collection of facial photographs and found that individuals may have different strategies for similarity recognition. In our analysis of the data, we have identified two possible strategies. We apply rough set based attribute reduction methodology to this data in order to develop a test to identify which of these strategies an individual is likely to prefer. We hypothesize that by providing a personalized search and filter environment, individuals would be more adequately equipped to handle the complexity of the task, thereby increasing the accuracy of identifications. Furthermore, the rough set based analysis may help to more clearly identify the different strategies that individuals use for this task. This paper provides a description of the preliminary study, our computational approach that includes an important pre-processing step, discusses results from our evaluation, and provides a list of opportunities for future work.

1 Introduction

Eyewitness identification holds a prominent role in many judicial settings, yet it is generally not accurate. Verbal overshadowing [1] is an effect that can obscure a witness's recollection of face when she is asked to describe the face to create a composite sketch. Alternatively, if the witness is asked to examine a large collection of photos, her memory may become saturated and she may mistakenly judge

H. Sakai et al. (Eds.): RSFDGrC 2009, LNAI 5908, pp. 201–208, 2009.

the current face similar to another she has examined examined (i.e., inaccurate source monitoring) and not to the one she is trying to recall [2]. We hypothesize that if the presentation of images can be personalized, the eyewitness may have to deal with fewer images, minimizing both of the negative effects discussed. This research takes the first steps along that path.

This paper discusses an analysis of data from a sorting study, which avoided verbalization completely while sorting. Each participant was asked to group a stack of 356 photos according to perceived similarity. As a participant encountered a photo, she could only place that photo and not disturb any existing piles. Indirectly, each participant made 63,190 pairwise similarity judgements. This quantity of data made it a good candidate for rough sets attribute reduction methodology.

Section 2 describes in more detail the initial study and data analysis that occurred. Section 3 describes the pre-processing developed to limit the number of pairs (or objects) needed to apply the attribute reduction methodology. Section 4 presents results from an exploration of the selected pairs. Section 5 presents some preliminary conclusions and avenues for future work.

2   Sorting Facial Photographs for Similarity

The stimulus photo set comprised equal numbers ($n = 178$) of Caucasian and First Nation faces. Cross-race identification of faces is an important topic of ongoing research [3], and our sorting study seeks to contribute to this body of work. We have focused on similarity judgements as a way to understand the way people perceive structure in the stimuli set. It may be that not everyone perceives the same structure. Therefore, if a person's preferred structure could be ascertained easily, it could be used to improve identification accuracy.

Figure 1 shows two photos from the stimulus set. Photographs were of the head and shoulders of each individual in a front facing pose wearing casual clothing. Subjects for these photographs were positioned 5 feet from the camera and 2 feet from the background wall. All distinguishing materials (e.g., glasses, piercing) were removed for the purposes of the photograph. All photographs were edited using Adobe Photoshop 7.0. Photographs for the facial recognition task were cropped to include only the subject's head and shoulders, while the background colour was changed from white to grey. The photographs were laminated on 5 by 4 inch cards. Participants were asked to view photos one at a time and place each photo on a pile with similar photos, without disturbing existing piles. Therefore, not all participants would make this direct comparison only if one photo was visible when the other was being placed. The number of piles was not constrained. Within the 25 participants, the number of piles made ranged between 4 and 38.

From a record of which photos each participant placed together, two things were done:

- a list of all possible pairs (356 choose 2 = 63,190) was constructed along with the judgement of similar (in the same pile) or dissimilar (in different piles)

Fig. 1. 2 photos from the collection which participants were asked to sort according to similarity

– the classification of photos was summarized in the following way. The number of Caucasian (C) and First Nations (F) photos in each pile was expressed as a percentage. The CHITEST function in Microsoft Excel was used to compare the ratio of C to F against an expected equal distribution. If $p < 0.05$ the pile was classified as C, if C >F or as F if F > C. The pile was classified as U (for undecided) if $p \geq 0.05$. All pictures in that pile were then labelled as C, F, or U. The numbers for all piles were totalled and expressed as a percentage (shown in Figure 2).

Figure 2 shows all participants plotted according to their percentages of photos classifed as C, F, and U. Many participants made only piles that could be identified as either C or F. These are found along the bottom line between vertices 1 and 3 in Figure 2. Other participants therefore had some number of undecided piles (and photos), labelled U according to the procedure outlined above. A threshold of 5% was set for the percentage of U and two groups were formed. We hypothesize that these groups correspond to different strategies for facial recognition, which we have labelled as "uses-race" (U < 5%, $n = 14$) and "uses-not-race" (U $\geq$ 5%, $n = 11$), because we hypothesize that race is being used by former group but not by the latter.

We seek to find a simple way to classify participants according to these groups, which we hypothesize will allow for personalization of the eyewitness identification process.

The strategy (uses-race or uses-not-race) then becomes the decision variable as we begin to apply the rough set attribute reduction methodology [4]. The objective is to reduce the number of pairs required as input to discriminate between the two strategies, as the original number of pairs is impractical.

3   Pre-processing

If the two groups identified in the last section are meaningfully different, then we hypothesize that we should be able to distinguish them with the help of pairs for

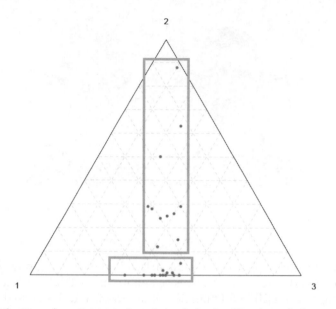

Fig. 2. Distribution of participants based on their classification of photos. Each point reflects the mix of C, F, and U photos identified by a participant. Vertex 1 is Caucasian, Vertex 3 is First Nations, and Vertex 2 is Undecided. The lower rectangle identifies participants in the "uses-race" group (with no or very few photos classified as undecided). The upper rectangle identifies participants in the "uses-not-race" group (with many photos classified as undecided).

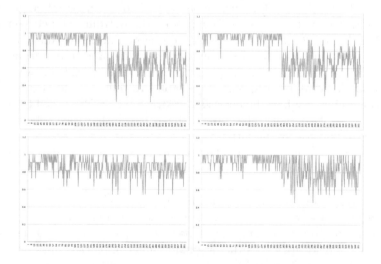

Fig. 3. The two photos from Figure 1 are compared against all other photos. The top graphs show results for the "uses-race" group and the bottom graphs show results for the "uses-not-race" group. For both the left and right photos from Figure 1, the two groups behave differently.

which the similarity ratings differed between the groups. Figure 3, which shows similarity ratings between the uses-race and uses-not-race groups, indicates that this may be a fruitful course of action.

Following an approach similar to the feature extraction/selection phase in knowledge discovery and data mining, we choose pairs with the following condition: that one group rated the pair very similar and the other group did not. We parameterize this in the following way. The difference between ratings must be greater than or equal to X (shown on the horizontal axis in Table 1) and that one group's rating of distance for the pair must be less than or equal to Y (show on the vertical axis in Table 1).

Table 1. The results for varying parameters for pair selection. The horizontal axis represents the absolute difference between the distances for a pair between the two groups. It is most strict at the left. The vertical axis represents the maximum rated distance for a pair by one of the groups. It is most strict at the top. For our study, we considered values in the upper left quadrant of this table, and used RSES on each of those sets of pairs. Results are shown in Table 2.

	≥ 0.9	≥ 0.8	≥ 0.7	≥ 0.6	≥ 0.5	≥ 0.4	≥ 0.3	≥ 0.2	≥ 0.1	≥ 0.0
≤ 0.1	0	0	0	1	2	3	5	6	6	6
≤ 0.2	0	0	0	7	11	28	35	37	38	38
≤ 0.3	0	0	17	82	197	350	467	556	585	605
≤ 0.4	0	0	17	130	401	840	1253	1584	1775	1881
≤ 0.5	0	0	17	130	798	2393	4536	6737	7825	8297
≤ 0.6	0	0	17	130	798	2925	6592	10450	12925	14488
≤ 0.7	0	0	17	130	798	2925	7480	13634	18241	22156
≤ 0.8	0	0	17	130	798	2925	7480	17371	28260	35903
≤ 0.9	0	0	17	130	798	2925	7480	17371	35398	47589
≤ 1.0	0	0	17	130	798	2925	7480	17371	35398	63190

4   Rough Set Attribute Reduction Methodology

Each of the parameter combinations in the upper left quadrant of Table 1 led to a number of photo pairs being selected for processing. The number of pairs processed in RSES (Rough Sets Exploration System) ranged from 2 to 798. For each set of pairs, the following steps were undertaken.

1. Preprocessing: Split input file (50/50): Each file in the analysis was split with approximately 50% of participants in a training set (data from 12 participants) and 50% of participants (data from 13 participants) in a testing set. The files comprised objects each representing the result of a pairwise comparison of facial photographs (0 if similar, 1 if dissimilar). The decision class was the group (either uses-race or uses-not-race), illustrated in Figure 2.

2. Training: Calculate the reducts in training file using genetic algorithms in RSES. The genetic algorithms procedure calculates the top N reducts possible for a given analysis. For the purposes of our analysis, we chose N = 10 in order to pick the top 10 reducts possible (if indeed 10 top reducts could be found).
3. Testing: Using the reducts generated in step 2, test the results on the data in the testing set.
4. Classification: Observe the classification accuracy of the train and test procedure in steps 1-3 and report the results.

All results had 100 percent coverage, which means that the classifier based on the reducts generated from an ensemble of reducts was able to recognize everything, which is valuable in itself.

Table 2. Results from running RSES on the pairs from the upper left quadrant of Table 1. In the case of 1 object, accuracy was computed by direct comparison of judgement on pair to decision variable.

Condition	Objects	Total Reducts	Reducts by Size 1	2	3	4	Global Accuracy
0.1v0.6	1	-	-	-	-	-	84*
0.1v0.5	2	1	1				62
0.2v0.6	7	3	1		2		85
0.2v0.5	11	10		3	7		69
0.3v0.7	17	10		8	2		92
0.3v0.6	82	10	3	4	3		92
0.3v0.5	197	8		1	4	3	92
0.4v0.7	17	10		10			100
0.4v0.6	131	10		6	4		85
0.4v0.5	401	8		3	2	3	85
0.5v0.7	17	10	1	9			92
0.5v0.6	131	9		8	1		92
0.5v0.5	798	8		4	3	1	92

The table for rough sets analysis is constructed in the following way: each row in the table represents an individual participant each column in the table (object) represents a pairwise photo distance. If two photos were said to be similar (placed in the same pile) then the object value is 0 (the distance between them is 0). If two photos were said to be dissimilar (placed in different piles) then the object value is 1 (the distance between them is 1). The decision variable is the value determined in Section 2, which indicates the decision of whether the participant belongs to the "uses-race" group or the "uses-not-race" group. It was anticipated that rough sets analysis could reduce the necessary pairwise comparisons to classify a participant, and thereby aid personalization efforts.

Table 3. A closer look at the solution with 100% global accuracy. Pair 1859a-1907a is the one shown in Figure 1.

	Unique Pairs	
Reducts	Pair	Frequency
032-129, 1859a-1907a	1859a-1907a	6
0062a-4488a, 032-130	032-129	3
0003a-8230a, 032-129	059-128	2
023-116, 059-128	032-130	2
023-116, 1859a-1907a	023-116	2
032-129, 059-128	8230a-9265a	1
1859a-1907a, 4488a-9622a	4488a-9622a	1
046-087, 1859a-1907a	046-087	1
032-130, 1859a-1907a	0062a-4488a	1
1859a-1907a, 8230a-9265a	0003a-8230a	1

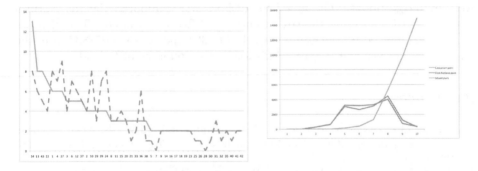

Fig. 4. These figures illustrate some interesting questions that remain unexplored. On the left, are reducts that use popular photo pairs more useful than others? These popular pairs are those where the dashed line is higher than the solid line. On the right, even though we hypothesize that not all people are using race for similarity ratings, very few mixed race pairs are rated as similar.

Amongst the results obtained from the RSES analysis, 117 unique pairs were identified in various reducts. The most frequently occurring pair was present in 13 different reducts. 44 pairs appeared more than once. When examining the solution with 100% global accuracy (condition 0.4v0.7), laid out in Table 3, one also finds a number of repeated pairs.

5 Conclusion and Future Work

Through this effort, we have found a very succinct test to classify people into one of two proposed strategy groups. Namely, we proved that rough sets can help in accuracy and clarity of the results. Our plan is to examine the validity of this test with more participants. This test will also help to clarify the strategies being used. It remains to be tested whether the 10 unique pairs listed in Table 3 are the minimum needed to achieve 100% global accuracy. Figure 4 illustrates some other open questions. The amount of overlap of attributes in reducts is particularly interesting. More than simply counting occurences, a more detailed analysis of the correlations between attributes is also warranted.

We have explored this data with RSES. However, there are more advanced rough set approaches and tools, with their own parameters, that can be used in this situation. It may be indeed useful to experiment with ensembles of reducts, approximate reducts, and so on, both within and outside RSES.

Also, although we have labelled the groups as uses-race/uses-not-race, it is interesting to note that only 1 of the pairs used in the reducts is a mixed pair. More work to understand the uses-not-race strategy especially is required. As we improve our understanding of this data, we may also be able to find success using other decision classes.

Acknowledgements. The first four authors were supported by the Natural Sciences and Engineering Research Council (NSERC) of Canada. The fifth author was supported by the grants N N516 368334 and N N516 077837 from the Ministry of Science and Higher Education of the Republic of Poland.

References

1. Schooler, J.W., Ohlsson, S., Brooks, K.: Thoughts beyond words: when language overshadows insight. Journal of Experimental Psychology: General 122, 166–183 (1993)
2. Dysart, J., Lindsay, R., Hammond, R., Dupuis, P.: Mug shot exposure prior to lineup identification: Interference, transference, and commitment effects. Journal of Applied Psychology 86(6), 1280–1284 (2002)
3. Platz, S., Hosch, H.: Cross-racial/ethnic eyewitness identification: A field study 1. Journal of Applied Social Psychology (January 1988)
4. Pawlak, Z.: Rough set approach to knowledge-based decision support. European Journal of Operational Research 99, 48–57 (1997)

Taking Fuzzy-Rough Application to Mars[*]

Fuzzy-Rough Feature Selection for Mars Terrain Image Classification

Changjing Shang, Dave Barnes, and Qiang Shen

Dept. of Computer Science, Aberystwyth University, Wales, UK
{cns,dpb,qqs}@aber.ac.uk

Abstract. This paper presents a novel application of fuzzy-rough set-based feature selection (FRFS) for Mars terrain image classification. The work allows the induction of low-dimensionality feature sets from sample descriptions of feature patterns of a much higher dimensionality. In particular, FRFS is applied in conjunction with multi-layer perceptron and K-nearest neighbor based classifiers. Supported with comparative studies, the paper demonstrates that FRFS helps to enhance the effectiveness and efficiency of conventional classification systems, by minimizing redundant and noisy features. This is of particular significance for on-board image classification in future Mars rover missions.

1 Introduction

The panoramic camera instruments on the Mars Exploration Rovers have acquired a large volume of high-resolution images, which provides substantial information to characterize the Mars environment [1,4]. Automated analysis of such images has since become an important task, especially for surveying places (e.g. for geologic cues) in Mars [8,12]. Any progress towards automated detection and recognition of objects within Mars images, including different types of rocks and their surroundings, will make a significant contribution to the accomplishment of this task.

Mars terrain images vary significantly in terms of intensity, scale and rotation, and are blurred with noise. These factors make Mars image classification a challenging problem. One critical step to successfully build an image classifier is to extract and use informative features from given images [3,7,9]. To capture the essential characteristics of such images, many features may have to be extracted without explicit prior knowledge of what properties might best represent the underlying scene reflected by the original image. Yet, generating more features increases computational complexity and measurement noise, and not all such features may be useful to perform classification. Thus, it is desirable to employ a technique that can determine the most significant features, based on sample measurements, to simplify the classification process, while ensuring high classification performance.

This paper presents an approach for performing large-scale Mars terrain image classification, by exploiting the recent advances in fuzzy-rough set-based

---

[*] Work funded by the Daphne Jackson Trust and the Royal Academy of Engineering.

H. Sakai et al. (Eds.): RSFDGrC 2009, LNAI 5908, pp. 209–216, 2009.

feature selection techniques [6]. As such, fuzzy-rough sets are, for the first time, applied to tasks relevant to space engineering. Experimental results show that this application ensures rapid and accurate learning of classifiers. This is of great importance to on-board image classification in future Mars rover missions. The rest of this paper is organized as follows. Section 2 introduces the Mars terrain images under investigation. Sections 3, 4 and 5 outline the key component techniques used in this work, including feature extraction, (fuzzy-rough) feature selection and feature pattern classification. Section 6 shows the experimental results, supported by comparative studies. The paper is concluded in Section 7.

2   *McMurdo* Panorama Image

This work concentrates on the classification of the *McMurdo* panorama image, which is obtained from the panoramic camera on NASA's Mars Exploration Rover Spirit and presented in approximately true color [4]. Such an image reveals a tremendous amount of detail in part of Spirit's surroundings, including many dark, porous-textured volcanic, brighter and smoother-looking rocks, sand ripple, and gravel (mixture of small stones and sand). Fig. 1 shows the most part of the original *McMurdo* image (of a size 20480×4124). This image, excluding the areas occupied by the instruments and their black shadows, is used for the work here, involving five major image types (i.e. classes) which are of particular interest. These image types are: textured or smoothed dark rock (C1), orange colored bedding rock (C2), light gray rock (C3), sand (C4), and gravel (C5), which are illustrated in Fig. 2. The ultimate task of this research is to detect and recognize these five types of image over a given region.

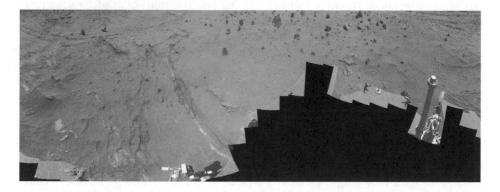

Fig. 1. Mars McMurdo panorama image

3   Feature Extraction

Many techniques may be used to capture and represent the underlying characteristics of a given image [3,10]. In this work, local grey level histograms and

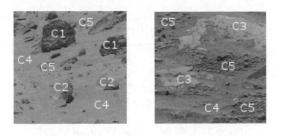

Fig. 2. Image classes (C1: rock1, C2: rock2, C3: rock3, C4: sand, C5: gravel)

the first and second order color statistics are exploited to produce a feature pattern for each individual pixel. This is due to the recognition that such features are effective in depicting the underlying image characteristics and are efficient to compute. Also, the resulting features are robust to image translation and rotation, thereby potentially suitable for classification of Mars images.

3.1 Color Statistics-Based Features

Color images originally given in the RGB (Red, Green and Blue) space are first transformed to those in the HSV (Hue, Saturation and Value) color space [10]. These spaces are in bijection with one another, and the HSV space is widely used in the literature. Six features are then generated per pixel, by computing the first order (mean) and the second order (standard deviation, denoted by STD) color statistics with respect to each of the H, S and V channels, from a neighborhood of the pixel. The size of such neighborhoods is pre-selected by trial and error (which trades off between the computational efficiency in measuring the features and the representative potential of the measured features).

3.2 Local Histogram-Based Features

To reduce computational complexity, in extracting histogram-based features, given color images are first transformed to grey-level (GL) images. For a certain pixel, a set of histogram features $H_i, i = 1, 2, ..., B$, are then calculated within a predefined neighborhood, with respect to a certain bin size B. Here, the neighborhood size is for convenience, set to the same as that used in the above color feature extraction, and H_i denotes the normalized frequency of the GL histogram in bin i. To balance between effectiveness and efficiency, B is empirically set to 16 in this work. In addition, two further GL statistic features are also generated, namely, the mean and STD (which are different from their color statistics-based counterparts of course).

4   Fuzzy-Rough Set-Based Feature Selection

Let U be the set of pixels within a given image, P be a subset of features, and D be the set of possible image classes. The concept of fuzzy-rough dependency measure [6], of D upon P, is defined by

$$\gamma_P(D) = \frac{\sum\limits_{x \in U} \mu_{POS_{R_P}(D)}(x)}{|U|} \tag{1}$$

where

$$\mu_{POS_{R_P}(D)}(x) = \sup_{X \in U/D} \mu_{\underline{R_P}X}(x) \tag{2}$$

$$\mu_{\underline{R_P}X}(x) = \inf_{y \in U} I(\mu_{R_P}(x, y), \mu_X(y)) \tag{3}$$

and U/D denotes the (equivalence class) partition of the image (i.e. pixel set) with respect to D, and I is a fuzzy implicator and T a t-norm. R_P is a fuzzy similarity relation induced by the feature subset P:

$$\mu_{R_P}(x, y) = T_{A \in P}\{\mu_{R_{\{A\}}}(x, y)\} \tag{4}$$

That is, $\mu_{R_{\{A\}}}(x, y)$ is the degree to which pixels x and y are deemed similar with regard to feature A. It may be defined in many ways, but in this work, the following commonly used similarity relation [5] is adopted:

$$\mu_{R_{\{A\}}}(x, y) = 1 - \frac{|A(x) - A(y)|}{A_{max} - A_{min}} \tag{5}$$

where $A(x)$ and $A(y)$ stand for the value of feature $A \in P$ of pixel x and that of y, respectively, and A_{max} and A_{min} are the maximum and minimum values of feature A. The fuzzy-rough set-based feature selection (FRFS) method works by greedy hill-climbing. It employs the above dependency measure to choose which features to add to the subset of the current best features and terminates when the addition of any remaining feature does not increase the dependency.

5   Image Classifiers

Multi-layer perceptron neural networks [11] and K-nearest neighbors (KNN) [2] are used here to accomplish image classification, by mapping input feature patterns onto the underlying image class labels. For learning such classifiers, a set of training data is selected from the typical parts (see Fig. 2) of the *McMurdo* image, with each pixel represented by a feature pattern which is manually assigned an underlying class label.

6   Experimental Results

From the *McMurdo* image of Fig. 1, a set of 270 subdivided non-overlap images with a size of 512×512 each are used to perform this experiment. 816 pixel points are selected from 28 of them, which are each labeled with an identified class index (i.e. one of the five image types: rock1, rock2, rock3, sand and gravel) for training and verification. The rest of all these images are used as unseen data for classification. Each training pixel is represented by a pattern of 24

Table 1. Feature meaning and reference

No.	Meaning	No.	Meaning	No.	Meaning	No.	Meaning	No.	Meaning
1	Mean(GL)	2	STD(GL)	3	Mean(H)	4	STD(H)	5	Mean(S)
6	STD(S)	7	Mean(V)	8	STD(V)	9-24	H_i		

features (see Section 3). Of course, the actual classification process only uses subsets of selected features. The performance of each classifier is measured using classification accuracy, with ten-fold cross validation.

For easy cross-referencing, Table 1 lists the reference numbers of the original features that may be extracted, where $i = 1, 2, ..., 16$. In the following, for KNN classification, the results are first obtained with K set to 1, 3, 5, 8, and 10. For the MLP classifiers, to limit simulation cost, only those of one hidden layer are considered here with the number of hidden nodes set to 8, 12, 16, 20, or 24. Those classifiers which have the highest accuracy, with respect to a given feature pattern dimensionality and a certain number of nearest neighbors or hidden nodes, are then taken for performance comparison.

6.1 Comparison with the Use of All Original Features

This subsection shows that, at least, the use of a selected subset of features does not significantly reduce the classification accuracy as compared to the use of the full set of original features. For this problem, FRFS returns 8 features, namely, STD(GL), Mean(H), STD(H), Mean(S), STD(S), Mean(V), H_4, H_{15} (i.e. features $2, 3, 4, 5, 6, 7, 12$ and 23 in Table 1), out of the original twenty-four. That is, a reduction rate of two-third. Table 2 lists the correct classification rates produced by the MLP and KNN classifiers with 10-fold-cross-validation, where the number (N) of hidden nodes and that (K) of the nearest neighbors used by these MLP and KNN classifiers are also provided (in the first column).

Table 2. FRFS-selected vs. full set of original features

Classifier	Set	Dim.	Feature No	Rate
MLP(N=20)	FRFS	8	$2, 3, 4, 5, 6, 7, 12, 23$	94.0%
MLP(N=20)	Full	24	$1, 2, ..., 23, 24$	94.0%
KNN(K=8)	FRFS	9	$2, 3, 4, 5, 6, 7, 12, 23$	89.1%
KNN(K=5)	Full	24	$1, 2, ..., 23, 24$	89.2%

The results demonstrate that the classification accuracy of using the eight FRFS-selected features is the same as that of using the twenty-four original features for MLP classifiers (94.0%), and is very close to that for KNN classifiers (89.1% vs. 89.2%). This is indicative of the potential of FRFS in reducing not only redundant feature measurements but also the noise associated with such measurements. Clearly, the use of FRFS helps to improve both effectiveness and efficiency of the classification process. Note that although the number of original features is not large, for on-board Martian application, especially in relation to the task of classifying large-scale images, any reduction of feature measurements is of great practical significance.

6.2   Comparison with the Use of PCA-Returned Features

Principal component analysis (PCA) [2] is arguably one of the most popular methods for dimensionality reduction, it is adopted here as the benchmark for comparison. Fig. 3 shows the classification results of the KNN and MLP classifiers using a different number of principal features. For easy comparison, the results of the KNN and MLP that use 8 FRFS-selected features are also included in the figure, which are represented by ∗ and ∘, respectively.

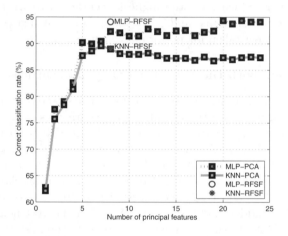

Fig. 3. Performance of KNN and MLP vs. the number of principal components

These results show that the MLP classifier which uses FRFS-selected features has a substantially higher classification accuracy amongst all those classifiers using a subset of features of the same dimensionality (i.e. 8). This is achieved via a considerably simpler computation, due to the substantial reduction of the complexity in input patterns. The results also show the cases where PCA-aided (MLP or KNN) classifiers each employ a feature subset of a different dimensionality. However, these classifiers still generally underperform than the corresponding FRFS-aided ones, whether they are implemented using MLP or KNN. This situation only changes when almost the full set of PCA-returned features is used where the MLP classifiers may perform similarly or slightly better (if 20 or 22 principal components are used). Yet, this is at the expense of requiring many more feature measurements and much more complex classifier structures. Besides, PCA-returned features lose the underlying meaning of the original.

6.3   Classified and Segmented Images

The ultimate task of this research is to classify Mars panoramic camera images and to detect different objects or regions in such images. The MLP which employs the 8 FRFS-selected features, and which was trained by the given 816 labeled feature patterns, is taken to accomplish this task: the classification of the entire image of Fig. 1 (excluding those regions as indicated previously). As an illustration, three classified images are shown in Fig. 4, numbered by (a), (b) and (c) respectively, where five different colors represent the five image types (rock1, rock2, rock3, sand and gravel). From this, boundaries between different class regions can be identified and marked with white lines, resulting in the segmented images also given in Fig. 4, numbered by (d), (e) and (f) correspondingly.

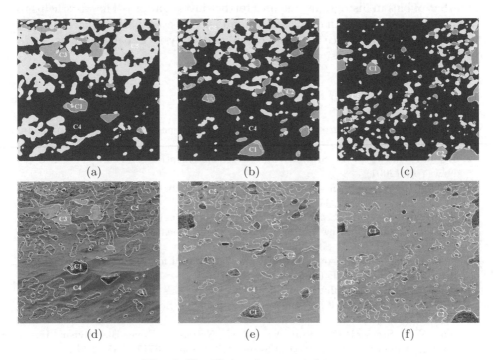

(a)   (b)   (c)

(d)   (e)   (f)

Fig. 4. Classified and segmented image

From these classified images, it can be seen that the five image types vary in terms of their size, rotation, color, contrast, shapes, and texture. For human eyes it can be difficult to identify boundaries between certain image regions, such as those between sand and gravel, and those between rock2 and sand. However, the classifier is able to perform under such circumstances, showing its robustness to image variations. This indicates that the small subset of features selected by FRFS indeed convey the most useful information of the original. Note that classification errors mainly occur within regions representing sand and gravel. This may be expected since gravel is itself a mixture of sand and small stones.

7   Conclusion

This paper has presented a study on Mars terrain image classification, supported by advanced fuzzy-rough set-based feature selection techniques. For the first time, fuzzy-rough sets have been adopted to help solving problems in space engineering. Although the real-world images encountered are large-scale and complex, the resulting feature pattern dimensionality of selected features is manageable. Conventional classifiers such as MLP and KNN that are built using such selected features generally outperform those using more features or an equal number of features obtained by classical approaches represented by PCA. This is confirmed by systematic experimental investigations (though the influence of parameter set-up for feature extraction, e.g. the number of pixels in neighbors and that of bins in histograms, requires further investigation). The work helps to accomplish challenging image classification tasks effectively and efficiently. This is of particular significance for classification and analysis of real images on board in future Mars rover missions.

References

1. Castano, R., et al.: Current results from a rover science data analysis system. In: Proc. of IEEE Aerospace Conf. (2006)
2. Duda, R.O., Hart, P.E., Stork, D.G.: Pattern classification, 2nd edn. Wiley & Sons, New York (2001)
3. Huang, K., Aviyente, S.: Wavelet feature selection for image classification. IEEE Trans. Image Proc. 17, 1709–1720 (2008)
4. http://marswatch.astro.cornell.edu/pancam_instrument/mcmurdo_v2.html
5. Jensen, R., Shen, Q.: New approaches to fuzzy-rough feature selection. IEEE Trans. Fuzzy Syst. 17(4), 824–838 (2009)
6. Jensen, R., Shen, Q.: Computational intelligence and feature selection: rough and fuzzy approaches. IEEE Press/Wiley (2008)
7. Kachanubal, T., Udomhunsakul, S.: Rock textures classification based on textural and spectral features. Proc. of World Academy of Science, Eng. and Tech. 29, 110–116 (2008)
8. Kim, W.S., Steele, R.D., Ansar, A.I., Al, K., Nesnas, I.: Rover-Based visual target tracking validation and mission infusion. AIAA Space. 6717-6735 (2005)
9. Lepisto, L., Kunttu, I., Visa, A.: Rock image classification based on k-nearest neighbour voting. Vis. Im. and Sig. Proc., IEE Proc. 153(4), 475–482 (2006)
10. Martin, D.R., Fowlkes, C.C., Malik, J.: Learning to detect natural image boundaries using local brightness, color, and texture cues. IEEE Trans. Patt. Anal. and Mach. Inte. 26, 530–549 (2004)
11. Rumelhart, D., Hinton, E., Williams, R.: Learning internal representations by error propagating. In: Rumelhart, D., McClell, J. (eds.) Parallel Distributed Processing. MIT Press, Cambridge (1986)
12. Thompson, D.R., Castano, R.: Performance comparison of rock detection algorithms for autonomous planetary geology. In: Proc. of IEEE Aerospace Conf. paper no. 1251 (2007)

Rough Image Colour Quantisation

Gerald Schaefer[1], Huiyu Zhou[2], Qinghua Hu[3], and Aboul Ella Hassanien[4]

[1] Department of Computer Science, Loughborough University, Loughborough, U.K.
[2] Queen's University Belfast, Belfast, U.K.
[3] Harbin Institute of Technology, China
[4] Information Technology Department, Cairo University, Giza, Egypt

Abstract. Colour quantisation algorithms are essential for displaying true colour images using a limited palette of distinct colours. The choice of a good colour palette is crucial as it directly determines the quality of the resulting image. Colour quantisation can also be seen as a clustering problem where the task is to identify those clusters that best represent the colours in an image. In this paper, we use a rough c-means clustering algorithm for colour quantisation of images. Experimental results on a standard set of images show that this rough image quantisation approach performs significantly better than other, purpose built colour quantisation algorithms.

1 Introduction

Colour quantisation is a common image processing technique that allows the representation of true colour images using only a small number of colours. True colour images typically use 24 bits per pixel resulting overall in 2^{24}, i.e. more than 16 million different colours. Colour quantisation uses a colour palette that contains only a small number of distinct colours (usually between 8 and 256) and pixel data are then stored as indices to this palette. Clearly the choice of the colours that make up the palette is of crucial importance for the quality of the quantised image. However, the selection of the optimal colour palette is known to be an np-hard problem [1]. In the image processing literature many different algorithms have been introduced that aim to find a palette that allows for good image quality of the quantised image [1,2,3]. Soft computing techniques such as genetic algorithms have also been employed to extract a suitable palette [4,5].

Colour quantisation can also be seen as a clustering problem where the task is to identify those clusters that best represent the colours in an image. In this paper, we use a rough c-means clustering algorithm for colour quantisation of images. The rough c-means clustering algorithm utilises two sets for each cluster, a lower and an upper approximation. Through iterative adjustment of the cluster centres, the algorithm converges towards a good colour palette. Experimental results on a standard set of images show that this rough image quantisation performs significantly better than other, purpose built colour quantisation algorithms.

H. Sakai et al. (Eds.): RSFDGrC 2009, LNAI 5908, pp. 217–222, 2009.

2 Rough Colour Quantisation

Colour quantisation can be seen as a clustering problem where the task is to identify those clusters that best represent the colours in an image. In this paper we employ a rough c-means clustering algorithm for this purpose.

Lingras *et al.* [6] introduced a rough set inspired clustering algorithm based on the well known c-means algorithm. In their rough c-means approach, each cluster c_k is described not only by its centre m_k, but also contains additional information, in particular its lower approximation $\underline{c_k}$, its upper approximation $\overline{c_k}$, and its boundary area $c_k^b = \overline{c_k} - \underline{c_k}$. Lingras *et al.*'s algorithm proceeds in the following steps:

Step 1: *Initialisation:* Each data sample is randomly assigned to one lower approximation. As the lower approximation of a cluster is a subset of its upper approximation, this also automatically assigns the sample to the upper approximation of the same cluster.

Step 2: *Cluster centre calculation:* The cluster centres are updated as

$$m_k = \begin{cases} \omega_l \sum_{x_i \in \underline{c_k}} \frac{x_i}{|\underline{c_k}|} + \omega_b \sum_{x_i \in c_k^b} \frac{x_i}{|c_k^b|} & \text{if } c_k^b \neq \{\} \\ \omega_l \sum_{x_i \in \underline{c_k}} \frac{x_i}{|\underline{c_k}|} & \text{otherwise} \end{cases} \quad (1)$$

The cluster centres are hence determined as a weighted average of the samples belonging to the lower approximation and the boundary area, where the weights ω_l and ω_b define the relative importance of the two sets.

Step 3: *Sample assignment:* For each data sample the closest cluster centre is determined and the sample assigned to its upper approximation. Then, all clusters that are at most ϵ further away than the closest cluster are determined. If such clusters exist, the sample will also be assigned to their upper approximations. If no such cluster exist, the sample is assigned also to the lower approximation of the closest cluster.

Step 4: *Termination:* If the algorithm has converged (i.e., if the cluster centres do not change any more, or after a pre-set number of iterations), terminate, otherwise go to Step 2.

Strictly speaking, this algorithm does not implement all properties set out for rough sets [7], and hence belongs to the reduced interpretation of rough sets as lower and upper approximations of data [8].

Peters [9] noticed some potential pitfalls of the algorithm as proposed by Lingras *et al.* in terms of objective function and numerical stability, and suggested some improvements to overcome these. Equation 1 is revised to

$$m_k = \omega_l \sum_{x_i \in \underline{c_k}} \frac{x_i}{|\underline{c_k}|} + \omega_u \sum_{x_i \in \overline{c_k}} \frac{x_i}{|\overline{c_k}|} \quad (2)$$

with $\omega_l + \omega_u = 1$, i.e. as a convex combination of lower and upper approximation means. In order to overcome the possibility of situations with empty lower approximations, Peters suggests two possible ways of addressing this, either by modifying the calculation of cluster centres so that for empty lower approximations the cluster centre is calculated as the average of samples in the upper approximation, or by ensuring that each lower approximation has at least one member. In our approach we choose the latter by assigning the data sample closest to the cluster centre to its lower approximation.

In addition, we perform a different initialisation procedure than Lingras *et al.* and Peters. Rather than randomly assigning samples to clusters, we generate random cluster centres first and then proceed with Steps 3, 2 and 4 (i.e., steps 2 and 3 reversed) of the algorithm.

3   Experimental Results

For our experiments we used six standard images commonly used in the colour quanti-sation literature (*Lenna, Peppers, Mandrill, Sailboat, Airplane, and Pool* - see Figure 1) and applied the our rough c-means colour quantisation algorithm to generate quantised images with a palette of 16 colours.

Fig. 1. The six test images used in the experiments: (*Lenna, Peppers, Mandrill, Sailboat, Pool, and Airplane.* (from left to right, top to bottom).

To put the results obtained into context, we have also implemented four popular colour quantisation algorithms to generate corresponding quantised images with a palette size of 16, namely Popularity algorithm [1], Median cut quantisation [1], Octree quantisation [2], and Neuquant [3]. For our rough c-means approach, we adopt the changes proposed by Peter's with parameters $\omega_l = 0.7, \omega_u = 0.3, \epsilon = 0.001$ (image pixel values are

Table 1. Quantisation results, given in terms of PSNR [dB]

	Lenna	Peppers	Mandrill	Sailboat	Pool	Airplane	average
Popularity algorithm [1]	22.24	18.56	18.00	8.73	19.87	15.91	17.22
Median cut [1]	23.79	24.10	21.52	22.01	24.57	24.32	23.39
Octree [2]	27.45	25.80	24.21	26.04	29.39	28.77	26.94
Neuquant [3]	27.82	26.04	24.59	26.81	27.08	28.24	26.73
Rough c-means (mean)	28.63	26.67	25.02	27.62	29.40	30.50	27.98
Rough c-means (max)	28.77	26.81	25.10	27.82	30.17	31.03	28.28

normalised to $[0; 1]^3$). For all algorithms, pixels in the quantised images were assigned to their nearest neighbours in the colour palette to provide the best possible image quality.

The results are listed in Table 1, expressed in terms of peak signal to noise ratio (PSNR) defined as

$$\text{PSNR}(I_1, I_2) = 10 \log_{10} \frac{255^2}{MSE(I_1, I_2)} \tag{3}$$

with MSE (the mean-squared error) calculated as

$$\text{MSE}(I_1, I_2) = \frac{1}{3nm} \sum_{i=1}^{n} \sum_{j=1}^{m} [(R_1(i, j) - R_2(i, j))^2 + \tag{4}$$
$$(G_1(i, j) - G_2(i, j))^2 + (B_1(i, j) - B_2(i, j))^2]$$

where $R(i, j)$, $G(i, j)$, and $B(i, j)$ are the red, green, and blue pixel values at location (i, j), and n and m are the dimensions of the images.

From Table 1 we can see that of the dedicated colour quantisation algorithms Octree and Neuquant clearly outperform the Popularity and Median Cut methods. For our rough c-means approach we ran the algorithm 10 times (randomly initialising the cluster centres) on each image and report both the average and the highest PSNR of these 10 runs in Table 1. Looking at the results, it is obvious that the rough c-means approach achieves significantly better image quality than any of the other algorithms, including Octree and Neuquant. In fact, on average, our colour quantisation approach provides an increase in PSNR of about 1 (mean)/1.5 (max) which is quite remarkable.

An example of this performance is given in Figure 2 which shows the *Airplane* image together with the images colour quantised by all algorithms. Error images (or image distortion maps) are commonly employed for judging the difference between images or the performance of competing algorithms [10]. For each quantised image, we therefore also provide an error image that represents the difference between the original and the palettised image (the squared error at each pixel location is calculated, the resulting image then inverted and a gamma function applied to increase the contrast). It can be seen that popularity based colour quantisation does not work very well. Median cut performs better but not as well as Octree and Neuquant which provide much improved image quality. However, our proposed rough c-means algorithm outperforms all other approaches and clearly produces the image with the highest image fidelity.

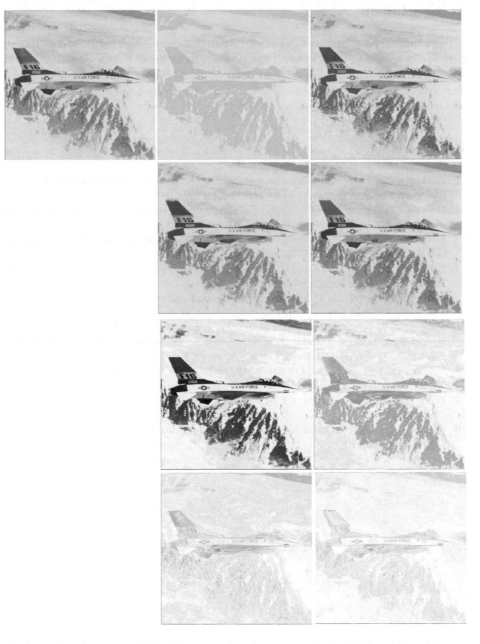

Fig. 2. *Airplane* image (top-left) and corresponding images quantised with (from left to right, top to bottom): Popularity, Median cut, Octree, Neuquant, rough c-means. Also shown are the error images of the quantised images compared to the original.

4   Conclusions

In this paper we proposed a rough c-means based colour quantisation algorithm. Rough c-means is applied to extract cluster centres corresponding to palette entries of colour quantised images. Experimental results obtained on a set of common test images have demonstrated that this approach can not only be effectively employed but clearly outperforms dedicated colour quantisation algorithms.

References

1. Heckbert, P.S.: Color image quantization for frame buffer display. ACM Computer Graphics (ACM SIGGRAPH 1982 Proceedings) 16(3), 297–307 (1982)
2. Gervautz, M., Purgathofer, W.: A simple method for color quantization: Octree quantization. In: Glassner, A. (ed.) Graphics Gems, pp. 287–293 (1990)
3. Dekker, A.: Kohonen neural networks for optimal colour quantization. Network: Computation in Neural Systems 5, 351–367 (1994)
4. Scheunders, P.: A genetic c-means clustering algorithm applied to color image quantization. Pattern Recognition 30(6), 859–866 (1997)
5. Nolle, L., Schaefer, G.: Color map design through optimization. Engineering Optimization 39(3), 327–343 (2007)
6. Lingras, P., West, C.: Interval set clustering of web users with rough k-means. Journal Intell. Inform. Syst. 23, 5–16 (2004)
7. Pawlak, Z.: Rough sets. Int. Journal Inform. Comput. Sci. 11, 145–172 (1982)
8. Yao, Y., Li, X., Lin, T., Liu, Q.: Representation and classification of rough set models. In: 3rd Int. Workshop on Rough Sets and Soft Computing, pp. 630–637 (1994)
9. Peters, G.: Some refinements of rough k-means clustering. Pattern Recognition 39, 1481–1491 (2006)
10. Zhang, X., Wandell, B.: Color image fidelity metrics evaluated using image distortion maps. Signal Processing 70(3), 201–214 (1998)

Part of Speech (POS) Tag Sets Reduction and Analysis Using Rough Set Techniques

Mohamed Elhadi and Amjd Al-Tobi

Department of Computer Science
Sultan Qaboos University, Oman
elhadi@squ.edu.om

Abstract. The motivation behind this work stems from an earlier work where text was transformed into strings of syntactical structures and used in similarity calculations using sequence algorithm on a string generated by a POS tagger. The performance of computations was greatly affected by the size of the string which in itself is the result of the type of tags used. Generated tags range from several (minimum of nine) general ones to many more (hundreds) detailed tags. Figuring out which tags and what combination of tags affect the realization of meanings, dependencies or relationships that exist in the text is an important issue. The resulting tag set reduction using rough sets and consequently string reduction has resulted in an improved efficiency in similarity calculations between documents while maintaining the same level of accuracy. Such finding was very encouraging.

Keywords: Rough sets, POS tagging, Data reduction, String comparison, Similarity calculations.

1 Introduction

Work done and presented in this paper was motivated by previously conducted experiments on the use of Part of Speech (POS) tags and Longest Common Sequence (LCS) algorithm in similarity calculations for Copy Detection [1,2]. In those experiments, POS tagging was used to extract the document's syntactical structures and represent the whole text as an ordered string of tags before it was passed for comparison by LCS algorithm. For large collections, accuracy and performance of computations using the produced string were greatly affected by the size of the string.

Selection of different combinations of tags can have an inherent impact on the realization of meanings manifested as dependencies and relationships that exist within the text. Data reduction through the discovery and analysis of such dependencies and relationships can be useful for similarity calculations and its applications.

Techniques for discovering dependencies and relationships within data are generally categorized into statistical or machine learning fields [3,4]. Rough set based techniques [5,6] are one of such methods that are useful in understanding the importance of different factors that contribute to dependencies and relationships in data as well data reduction.

H. Sakai et al. (Eds.): RSFDGrC 2009, LNAI 5908, pp. 223–230, 2009.
© Springer-Verlag Berlin Heidelberg 2009

It is the aim of this work to investigate the use of rough sets techniques in the discovery and analysis of relationships that exit within POS tags and the reduction of factors (tags) that contribute to the overall build of text as a representation of meaning.

The rest of the paper is organized as follows. Section 2 describes tagging and similarity calculation. Section 3 contains a description of rough sets and the Rosetta [20,15] tool used in the analysis. Section 4 contains descriptions of the experiments and discussions. Finally, section 5 contains the conclusion.

2 Similarity Calculation Using LCS and POS Tagging

Document management and text analysis including copy detection, near-copy detection, and similarity-based techniques in general have become very important with the growth of the web and the emergence of digital libraries [7,8]. String representations and processing tasks are widely used in text analysis in many fields to help find and cluster similar documents [9, 10, 13, 16]. Genomic and protein sequence alignments heavily relied on string manipulation techniques to calculate similarity between new and existing sequences. Many of those techniques can be readily utilized in text similarity calculation if suitable sequences could be produced.

A major impediment to treating text sequences in a similar fashion as to biological sequences, is to do with the nature of the strings themselves, lack of an appropriate alphabet, and the lack of a theory like that of evolution for biological sequences, which can be used to explain them.

A chunk of text, however, can be viewed as a string made of some meaningful numerable units (alphabets) allowing modified text to be handled and thought of as being a result of some intervention or application of some edit operations. Thus text can be considered as strings of syntactical unites derived from POS tagging instead of using actual characters or words as is commonly done [1]. The created strings capture syntactic and some of the semantics contained in the writing style of authors and the relationships defined by the grammar and order of POS tags.

The use of syntactical structures instead of actual text greatly reduces the dimensionality of documents while less information is lost. Still, however, among other things, the issue of selection of better tags to use for the creation of string for further reduction of those strings warrants more investigation.

2.1 Similarity Calculation

It is more common to use syntactical approaches when calculating similarities of text through fingerprinting [17], information retrieval [9] and hybrid techniques [21,25].

Fingerprinting uses chunking of text into small chunks where each chunk is hashed using a hashing algorithm to produce a list of values representing the document. These values are then used to compare one document's hash values to other documents' values to detect similarities [13]. Information retrieval, on the other hand, focuses on representing documents based on their words and frequencies using indexes with an appropriate model/technique to evaluate similarities between documents [9]. Attempts have been made to combine some of the above techniques [21,25].

2.2 POS Tagging and Tag-Sets

There are basic POS tags that are used by all taggers when annotating text. These basic parts are the verb, noun, pronoun, adjective, adverb, preposition, conjunction and interjection [12] with some taggers adding the article [14]. Many tagging systems extend these basic tags to describe additional grammatical features, such as, singular/plural, number, tense, gender and even punctuations [19].

The TreeTagger [2] is one such tagger and is used in this experiment. It has a tag set that contains about 55 tags. TreeTagger is able to cope with ungrammaticalities in the input. This tagger is reported to have achieved the highest accuracy in comparison to other taggers. Its accuracy reached up to 96.36%. [23,24].

3   Rough Sets

Rough set theory was proposed by Pawlak [5,6] as a mathematical framework to deal with incomplete and uncertain data.

Given a finite non-empty set U of objects, called a universe. The limits of discernibility of objects are formally expressed by an equivalence relation over a set of objects. Each object of U is characterized by a description, in the form of a set of attribute-values represented in a tabular format referred to as Information System. Table 1 is an example from current work where objects are documents and attributes are the tags with tag frequencies as values.

3.1 Approximation Space

The equivalence relation is called an *indiscernibility relation $R(C)$*, where C is a set of condition attributes used to represent objects belonging to the domain of interest U. The attributes are discrete and finite-valued properties of objects. Each attribute a belonging to C is a mapping $a: U \rightarrow V_a$, where V_a is a finite set of values called the *domain* of the attribute a.

The indiscernibility relation represents prior knowledge about the universe U expressed in terms of identity of values of the condition attributes C on objects.

The pair (U, R) is called an approximation space and the equivalence classes of R are called elementary sets.

3.2 Rough Approximations

If we let R* be a collection of all elementary sets, then any definable subset of the universe U is a set union of some elementary sets. All other subsets are undefinable or rough. For any definable set X there exits an uncertainty free criterion for determining the membership status in the set of any object belonging to the universe U. The criterion is referred to as a description of the set X, denoted as des(X). If the set X is rough, the defining description does not exist and the membership status of some objects with respect to the set X cannot be determined with certainty. Rough sets can be approximately characterized by two definable sets called lower and upper approximations respectively.

— The lower approximation of X is a union of elementary classes totally included in X, that is, this is the largest definable set contained in the rough set X. Objects belonging to the lower approximation with certainty belong to the set X.
— The upper approximation is the smallest definable set containing the rough set X. Objects belonging to the upper approximation possibly belong to the set X.

In addition, based on the upper and lower approximations, the boundary area consists of objects whose membership status with respect to the set X is uncertain.

That is, the boundary area is a union of such elementary classes which have only partial overlap with the set X. The union of all elementary classes which are completely disjoint from the set X is called the negative region of X. The negative region is a largest definable collection of objects which with certainty do not belong to X. Readers are referred to [6,22] for more details.

Table 1. Sample frequency table with few documents and tags

Docs/Tags	Articles	Adjectives	Verbs	Adverbs	Conjunctions	Preposition	Interjection	Pronoun	Noun
42650	19	27	57	7	5	33	0	23	83
42651	44	43	92	12	7	51	0	20	160
95037	26	26	73	14	10	32	0	24	141

Rosetta, a general-purpose tool for discernibility-based modeling [20,15], was used in this work. It is a toolkit for analyzing tabular data within the framework of rough set theory.

4   Experiments and Discussions

In the approach adopted in this work we used three tags-sets of 9, 19 and 55 tags, where a given document is processed to produce a string of tokens using TreeTagger. These tags are then mapped into single characters to reduce costs.

Step 1: Convert each news document from XML format to text files.

Step 2: Tag each document in the set using 9, 19 and 55 tag sets.

Step 3: Map tags into single character tags.

Step 4: Produce tables of tag frequencies of occurrence in documents for each set.

Step 5: Divide the results, based on the writer and the general topics into tables that use the tags as condition attributes where Writer-based set used the respective writer of the article as decision attribute,   General topic-based set used the respective general topic category of the article as decision attribute and Specific topic based set used the respective specific topic of the article as decision attribute.

Step 6: Run rough sets on total of 6 produced tables of frequencies. The results of each of the subsets are analyzed looking for meaningful relationships and dependencies. Johnson's algorithm which invokes a variation of a simple greedy algorithm to compute a single reduct only was used.

Fig. 1. The adopted procedure

The used documents were taken from Reuter's corpus [18]. Reuter has defined a number of categories using writer, region, industry and topic as a bases for their manually classified documents. For purpose of validation, ease of analysis and consistency with work already done [1], a total of 1512 pre-selected set of documents from the Reuter's collection were used. The set was further divided into three data sets of 333 documents each. Figure 1 illustrates the procedure used.

4.1   Reuters's Collection: 9-Tags Tag Set: Resulting Reducts and Validations

The aim here was to see if what is considered by tagger as the minimum tag set is really a minimum and whether there can be any smaller sets that have the same data dependencies and relationships. The relationship of the 9 tags set relative to the three decisions of writer, general topic and specific topic using produced frequencies tables was carried out for the whole set of 1512 documents and was validated on the three pre-selected data sets documents.

As is shown in Table 2, and through the use of Johnson algorithm [13], rough sets were able to suggest a much smaller subset (made of 4 tags) as an alternative to the 9 tags set. The suggested subset is made of **Articles, Adjectives, Verbs, and Nouns**. This is an overall reduction of 56% as suggested by the data using writer, general and specific data collections. Same tags were suggested in all cases except for one difference where a **conjunction** was suggested instead of noun.

Validations of results confirmed these suggestions when the suggested reduced subsets of tags were used along with the original 9 set, to detect duplication in Reuter's collection. The results based on new suggested reducts were compared to those of original 9-tag set looking top 90% score using the normalized Longest Common Subsequence (LCS) [1]. The two new reducts were able to get exact duplicate documents (23 in total) as original 9-tags tag set. As a matter of fact, even lower scores of 80% contained same hits that are duplicates.

4.2   Reuters's Collection: 19-Tags Tag Set: Resulting Reducts and Validations

The aim here was to see if a larger set of tags can contain smaller subsets that would be used to realize the dependencies in the data. This was done by analyzing the relationship of the 19-tags set relative to the three decisions of writer, general and specific topics using produced frequencies tables.

As is shown in Table 2, rough sets, using the Johnson algorithm, suggested a much smaller subset (made of 4 tags) as an alternative to the original 19-tags. The suggested subset is made of **Punctuation, Article, Adjective, Noun and Conjunction Subordinating.** As the table suggests all data sets have produced same reduct except for one where the noun is replaced by a Conjunction Subordinating tag.

The proposed reducts constitute a reduction of 79%. It is worth noting that three of the suggested tags in this case are the same as those of the 9 tag set obtained in the previous experiment.

To validate results, we looked the top 90% range of the normalized LCS score in the copy detection tested sets. The two new reducts were able to get exact duplicate documents (23 in total) as original 19-tags tag set. Lower scores of 70% and higher contained same results with only those in high end (80%) range being exact duplicates.

4.3 Reuters's Collection: 55-Tags Tag Set: Resulting Reducts and Validations

In this third experiment the TreeTagger's largest possible tag set was used with the aim of confirming previous results obtained in duplicate detection experiments [1] and analyzing the effect of using an expended set of tags. This tag set is made of 55 tags compromising of groups of tags ranging from usual variable and noun categories to punctuations and ordinals. The relationship of a 55 tags tag-set were analyzed using rough sets relative to the three decisions of writer, general and specific topic using frequencies tables of the sets of documents.

As is shown in Table 2, rough sets was able to suggest a much smaller subset (made of 3 tags) as an alternative to the 55 tags set we have used as the bases for this experiment. The suggested subset is made of **Conjunction Subordinating, Common noun singular, and Proper noun singular**. As the table suggests all data sets have produced same reduct.

The proposed reducts constitute a huge reduction of 95%. It is worth noting again that three of the suggested tags in this case do relate to the 9 and 19 tags sets reducts.

Validations results confirmed these suggestions. Looking at the top 90% range of the normalized LCS score in the tested sets, with the new reduct was able to get exact documents (23 in total) as original 55-tags set with more matches in the lower scores of 80%.

Table 2. List of Reducts from the 9, 19 and 55 tag sets

	Reducts			Reducts
9-1	Articles, Adjective, Verb, Noun		19-2	Punctuation, Article ,Adjective, Conjunction Subordinating
9-2	Articles, Adjective, Verb, conjunction		55-1	Conjunction Subordinating, Common noun singular, Proper noun singular
19-1	Punctuation, Article, Adjective, Noun			

There was a total match of the top 90% results across all tag sets. These results compare well to the largest 55 set original results obtained in [1].

5 Performance

One of the motivations behind the above attempts of tag set reduction was to improve the performance of LCS time while still maintaining same or approximant accuracy. Above validation results did confirm the accuracy of the much reduced sets.

Hence as a consequence of the huge string reductions; the system's performance was improved. For example there was more than 42% and more than 87% savings in time comparing the 9 tags set to one of its reducts and the 19 tags set to one of its reducts respectively. This is considered to be a very favorable gain in performance.

6 Conclusion

Motivated by the need to improve performance while still maintaining the same accuracy when calculating similarity using POS tags and LCS algorithm for use in Copy

Detection [1,2], experiments were conducted to reduce the size of the string representative of documents. POS tagging was used to extract the document's syntactical structures and represent the whole text as an ordered string of tags. Size of the string that is produced using the newly reduced set for large collection of document as well as the different combination of tags and their sizes can have an inherent impact on the realization of the text's meaning. Discovery, analysis of dependencies and relationships can lead to reduction of strings.

Rough set based techniques were applied to tag set reduction. Results were investigated and analyzed using three representative tag sets.

There was a total match of the top 90% results not just when comparing the reducts' results to individual original tag sets but also across all tag sets. Results compared well to original results obtained in pervious experiments [1]. The results were consistent across tags sets and on all the three data sets used. Results confirmed the accuracy of the reduced sets and improved system's performance was improved.

References

1. Elhadi, M., Al-Tobi, A.: Use of Text Syntactical Structures in Detection of Document Duplicates. In: Third IEEE International Conference on Digital Information Management, University of East London, London, UK (2008)
2. Elhadi, M., Al-Tobi, A.: Webpage Duplicate Detection Using Combined POS and Sequence Alignment Algorithm. In: World Congress on Computer Science and Information Engineering, Los Angeles/Anaheim, USA (2009)
3. Koppel, M., Argamon, S., Schler, J.: Computational Methods in Authorship Attribution. Journal of the American Society for Information Science and Technology 60, 9–26 (2009)
4. Zheng, R., Li, J., Chen, H., Huang, Z.: A framework for authorship identification of on line messeges: writing style features and classification techniques. Journal of American society of Information Sciences and technology 57, 378–393 (2006)
5. Pawlak, Z.: Rough sets. International Journal of Computer and Information Sciences 11, 341–356 (1982)
6. Pawlak, Z.: Rough Sets - Theoretical Aspects of Reasoning: About Data. Kluwer Academic Publishers, Dordrecht (1991)
7. Maguitman, A.G., Menczer, F., Roinestad, H., Vespignani, A.: Algorithmic Detection of Semantic Similarity. In: Proceedings of the 14th international conference on World Wide Web, pp. 107–116 (2005)
8. Mihalcea, R., Corley, C., Strapparava, C.: Corpus-based and Knowledge-based Measures of Text Semantic Similarity. In: Proceedings of The Twenty-First National Conference on Artificial Intelligence and the Eighteenth Innovative Applications of Artificial Intelligence Conference (2006)
9. Steinberger, R., Pouliquen, B., Hagman, J.: Cross-lingual Document Similarity Calculation Using the Multilingual Thesaurus EUROVOC. In: Gelbukh, A. (ed.) CICLing 2002. LNCS, vol. 2276, pp. 415–424. Springer, Heidelberg (2002)
10. Campbell, D.M., Chen, W.R., Smith, R.D.: Copy Detection Systems for Digital Documents. In: Proceedings of Advances in Digital Libraries, pp. 78–88. IEEE, Los Alamitos (2000)
11. Shivakumar, N., Garcia-Molina, H.: SCAM: A Copy Detection Mechanism for Digital Documents. In: Proceedings of 2nd International Conference in Theory and Practice of Digital Libraries (1995)

12. MacFadyen, H.: The Parts of Speech (2007),
    http://www.arts.uottawa.ca/writcent/hypergrammar/partsp.html
13. Johnson, D.S.: Approximation algorithms for combinatorial problems. Journal of Computer and System Sciences 9, 256–278 (1974)
14. ELC Courses: Parts of Speech: English Language Centre, University of Victoria (1997),
    http://web2.uvcs.uvic.ca/elc/StudyZone/330/grammar/parts.htm
15. Øhrn, A.: Discernibility and Rough Sets in Medicine: Tools and Applications, PhD thesis, Department of Computer and Information Science, Norwegian University of Science and Technology, Trondheim, Norway. NTNU report 1999:133, IDI report 1999:14, 239 pages (1999) ISBN 82-7984-014-1
16. Bull, J., Collins, C., Coughlin, E., Sharp, D.: Technical Review of Plagiarism Detection Software Report: Computer Assisted Assessment Centre, University of Luton, Luton, UK (2003)
17. Schleimer, S., Wilkerson, D.S., Aiken, A.: Winnowing: Local Algorithms for Document Fingerprinting. In: Proceedings of the 2003 ACM SIGMOD International Conference on Management of Data, pp. 76–85 (2003)
18. REUTERS, Reuters Corpus (Volume 1: English Language, 1996-08-20 to 1997-08-19), Released date: November 3, 2000, NIST (2000)
19. Hu, X.R., Atwell, E.: A survey of machine learning approaches to analysis of large corpora. In: Proceedings of the Workshop on Shallow Processing of Large Corpora, Lancaster University, UK, pp. 45–52 (2003)
20. Komorowski, J., Øhrn, A., Skowron, A.: The ROSETTA Rough Set Software System. In: Klösgen, W., Zytkow, J. (eds.) Handbook of Data Mining and Knowledge Discovery, ch. D.2.3. Oxford University Press, Oxford (2002),
    http://www.lcb.uu.se/tools/rosetta/downloads.php
21. Clough, P.: Old and new challenges in automatic plagiarism detection: Department of Information Studies, University of Sheffield (2003)
22. Wong, S.K.M., Ziarko, W.: On learning and evaluation of decision rules in the context of rough sets. In: Proceedings of the International Symposium on Methodologies for Intelligent Systems, Knoxville, Tennessee, pp. 224–308 (1986)
23. Schmid, H.: Probabilistic Part-of-Speech Tagging Using Decision Trees. In: Proceedings of International Conference on New Methods in Language Processing, Manchester, UK, pp. 44–49 (1994)
24. Schmid, H.: Improvements in Part-of-Speech Tagging With an Application To German. In: EACL SIGDAT workshop, Dubai, UAE (1995)
25. Liu, Y., Liang, L.: A Dual-method Model for Copy Detection. In: Proceedings of the IEEE/WIC/ACM international conference on Web Intelligence and Intelligent Agent Technology, Hong Kong Convention and Exhibition Centre, Hong Kong, pp. 634–637. IEEE, Los Alamitos (2006)
26. Lexicon and Textcorpora Group: TreeTagger - a language independent part-of-speech tagger: Institut für Maschinelle Sprachverarbeitung, Universität Stuttgart, Germany (2003),
    http://www.ims.uni-stuttgart.de/projekte/corplex/TreeTagger/

More on Intuitionistic Fuzzy Soft Sets

Pabitra Kumar Maji[*]

Department of Pure Mathematics
University of Calcutta
35, Ballygunge Circular Road
Kolkata-19
West Bengal, India
pabitra_maji@yahoo.com

Abstract. New operations on intuitionistic fuzzy soft sets have been introduced in this paper. Some results relating to the properties of these operations have been established. An example has also been introduced as an application of these operations.

Keywords: Soft Sets, Fuzzy Soft Sets, Intuitionistic Fuzzy Soft Sets, Similarity Measurements.

1 Introduction

Most of the real life problems in social sciences, engineering, medical sciences, economics etc. the data involved are imprecise in nature. The solutions of such problems involve the use of mathematical principles based on uncertainty and imprecision. Some of these problems are essentially humanistic and thus subjective in nature (e.g. human understanding and vision system), while others are objective, yet they are firmly embedded in an imprecise environment. A number of theories have been proposed for dealing with uncertainties in an efficient way. Some of these are the theory of probability, fuzzy set theory [12], intuitionistic fuzzy sets [1, 2], vague sets [5], theory of interval mathematics [2], rough set theory [11] etc. All these theories, however, are associated with an inherent limitation, which is possibly due to the inadequacy of the parameterization tool associated with these theories. Molodtsov [10] initiated a novel concept of soft sets theory as a new mathematical tool for dealing with uncertainties which is free from the above limitations. In many of real life problems it is very often observed that the parameters involved in the system are uncertain fuzzy in nature. On the basis of this point, we have some extensions of soft sets theory in [7, 8, 9]. In [6] we can find the further work of soft sets theory.

In the present work, we have extended the intuitionistic fuzzy soft sets defining new operations on it. Some properties of these operations have also been studied.

---

[*] The work is supported by the University Grants Commission (U G C) as UGC Research Award No. F. 30-1/2009(SA- I I) dt. 2nd July, 2009.

H. Sakai et al. (Eds.): RSFDGrC 2009, LNAI 5908, pp. 231–240, 2009.

2 Theory of Soft Sets

Molodtsov [10] defined the soft set in the following way:

Let U be a universe set and E be a set of parameters. Let P (U) denotes the power set of U and A $\subset$ E.

Definition 2.1 [10]. A pair (F, A) is called a soft set over U, where F is a mapping given by F : A $\rightarrow$ P (U). In other words, a soft set over U is a parameterized family of subsets of the universe U. For $\epsilon \in$ A, F (ϵ) may be considered as the set of ϵ- approximate elements of the soft set (F,A).

3 Intuitionistic Fuzzy Sets

We recollect some relevant basic preliminaries, in particular, the works of Atanassov [1]. Let a set E be fixed. An intuitionistic fuzzy set or IFS in E is an object having the form A = {<x, $\mu_A(x)$, $\nu_A(x)$>| x $\in$ E } where, the function

$\mu_A : E \rightarrow [0, 1]$ and $\nu_A : E \rightarrow [0, 1]$ define the degree of membership and the degree of non-membership respectively of the element x ($\in$ E) to the set A. For any x $\in$ E , $0 \leq \mu_A(x) + \nu_A(x) \leq 1$. The indeterministic part for x denoted by $\pi_A(x)$, where $\pi_A(x) = 1 - \mu_A(x) - \nu_A(x)$. Clearly, $0 \leq \pi_A(x) \leq 1$.

If A and B are two IFSs of the set E, then

A $\subset$ B iff $\forall$ x $\in$ E, $\mu_A(x) \leq \mu_B(x)$ and $\nu_A(x) \geq \nu_B(x)$

A $\subset$ B iff B $\supset$ A; A = B iff $\forall$ x $\in$ E, $\mu_A(x) = \mu_B(x)$ and $\nu_A(x) = \nu_B(x)$

A^c= { < x, $\nu_A(x)$, $\mu_A(x)$ >| x $\in$ E }; A $\cup$ B = {<x, max ($\mu_A(x)$, $\mu_B(x)$),min ($\nu_A(x)$, $\nu_B(x)$)>| x $\in$ E}

A $\cap$ B = {<x, min ($\mu_A(x)$, $\mu_B(x)$), max ($\nu_A(x)$, $\nu_B(x)$)>| x $\in$ E }

A . B = {<x, $\mu_A(x) . \mu_B(x)$, $\nu_A(x) + \nu_B(x) - \nu_A(x) . \nu_B(x)$>| x $\in$ E }.

Now we recollect some preliminaries from [4, 7, 9].

4 Intuitionistic Fuzzy Soft Sets

Definition 4.1 [7]. Consider U and E as a universe set and a set of parameters respectively. Let $P(U)$ denotes the set of all intuitionistic fuzzy sets of U. Let A $\subset$ E. A pair (F, A) is an intuitionistic fuzzy soft set over U, where F is a mapping given by F: A $\rightarrow$ P (U).

Definition 4.2 [7]. For two intuitionistic fuzzy soft sets (F, A) and (G, B) over a common universe U, we say that (F, A) is an intuitionistic fuzzy soft subset of (G, B) if

    i.   A $\subset$ B, and

    ii. $\forall$ $\epsilon \in$ A, F (ϵ) is an intuitionistic fuzzy subset of G (ϵ).

Definition 4.3 [7]. Two intuitionistic fuzzy soft sets (F, A) and (G, B) over a common universe U are said to be fuzzy soft equal if (F, A) is an intuitionistic fuzzy soft subset of (G, B) and (G, B) is an intuitionistic fuzzy soft subset of (F, A).

Definition 4.4 [7]. If (F, A) and (G, B) be two intuitionistic fuzzy soft sets then, "(F, A) AND (G, B)" is an intuitionistic fuzzy soft set denoted by (F, A) $\wedge$ (G, B) is defined by

$$(F, A) \wedge (G, B) = (H, A \times B), \text{ where } H(\alpha, \beta) = F(\alpha) \cap G(\beta), \forall (\alpha, \beta) \in A \times B.$$

Definition 4.5 [7]. If (F, A) and (G, B) be two intuitionistic fuzzy soft sets then, "(F, A) OR (G, B)" is an intuitionistic fuzzy soft set denoted by (F, A) $\vee$ (G, B) is defined by (F, A) $\vee$ (G, B) = (O, A $\times$ B),

where $O(\alpha, \beta) = F(\alpha) \cup G(\beta), \forall (\alpha, \beta) \in A \times B.$

Definition 4.6 [9]. Dot of two intuitionistic fuzzy soft sets (F, A) and (G, B) over the common universe U is the intuitionistic fuzzy soft sets denoted by '(F, A).(G, B)' and is defined as

(F, A).(G, B) = (H, A$\cap$B), where,
H(e) = {<m, $\mu_{F(e)}$(m). $\mu_{G(e)}$(m), $\nu_{F(e)}$(m)+ $\nu_{G(e)}$(m) - $\nu_{F(e)}$(m). $\nu_{G(e)}$(m)>|m$\in$ U}, if e $\in$ A $\cup$ B,
  = {<m, $\mu_{F(e)}$(m). $\mu_{F(e)}$(m), $\nu_{F(e)}$(m)+ $\nu_{F(e)}$(m) - $\nu_{F(e)}$(m). $\nu_{F(e)}$(m)>|m$\in$ U}, if e $\in$ A - B,
  = {<m, $\mu_{G(e)}$(m). $\mu_{G(e)}$(m), $\nu_{G(e)}$(m)+ $\nu_{G(e)}$(m) - $\nu_{G(e)}$(m). $\nu_{G(e)}$(m)>|m$\in$ U}, if e $\in$ B - A.

Definition 4.7 [4]. For any two intuitionistic fuzzy sets A and B of E , similarity measure $S(A, B)$ between A and B is defined by $S(A, B)$ and is defined as,

$$S(A, B) = \frac{\sum_y \vec{A}_y . \vec{B}_y}{\sum_y \vec{A}_y^2 \vee \sum_y \vec{B}_y^2} \quad , \text{ where } \vec{A}_x \text{ is the vector } (\mu_A(x), \pi_A(x)) \text{ and}$$

$\vec{B}_x$ is the vector $(\mu_B(x), \pi_B(x))$, $\forall$ x $\in$ E.

Now we define new operations on intuitionistic fuzzy soft sets.

Let U be a universal set. E be a set of parameters and A be a subset of E. Let the intuitionistic fuzzy soft set (F, A) = {<m, $\mu_{F(e)}$(m), $\nu_{F(e)}$(m)>|m$\in$ U and e$\in$ A}, where $\mu_{F(e)}$(m), $\nu_{F(e)}$(m) be the membership and non-membership functions respectively.

5 The Necessity Operation on Intuitionistic Fuzzy Soft Set

Definition 5.1. The necessity operation on an intuitionistic fuzzy soft set (F, A) is denoted by $\square$(F, A) and is defined as $\square$(F, A) = {<m, $\mu_{F(e)}$(m), 1 - $\mu_{F(e)}$(m)>|m$\in$ U and e$\in$ A}. Here $\mu_{F(e)}$(m) is the membership function of m for the parameter e and F is a mapping F : A $\rightarrow$ P(U), P(U) is the set of all intuitionistic fuzzy sets of U.

Example 5.1. Let there are five objects as the universal set where U = { m_1, m_2, m_3, m_4, m_5 }and the set of parameters as E = { beautiful, moderate, wooden, muddy, cheap, costly }and
Let A = {beautiful, moderate, wooden}. Let the attractiveness of the objects represented by the intuitionistic fuzzy soft sets (F, A) is given as

F(beautiful) ={ $m_{1/(.6,.4)}$, $m_{2/(.7,.2)}$, $m_{3/(.5,.4)}$, $m_{4/(.6,.3)}$, $m_{5/(.8,.1)}$}, F(moderate) ={ $m_{1/(.7,.2)}$, $m_{2/(.8,.1)}$, $m_{3/(.7,.2)}$, $m_{4/(.8,.1)}$, $m_{5/(1,0)}$}and F(wooden) ={ $m_{1/(.8,.1)}$, $m_{2/(.6,0)}$, $m_{3/(.6,.2)}$, $m_{4/(.2,.4)}$, $m_{5/(.3,.5)}$}.

Then the intuitionistic fuzzy soft sets ¤(F, A) becomes as

F(beautiful) ={ $m_{1/(.6,.4)}$, $m_{2/(.7,.3)}$, $m_{3/(.5,.5)}$, $m_{4/(.6,.4)}$, $m_{5/(.8,.2)}$}, F(moderate) ={ $m_{1/(.7,.3)}$, $m_{2/(.8,.2)}$, $m_{3/(.7,.3)}$, $m_{4/(.8,.2)}$, $m_{5/(1,0)}$}and F(wooden) ={ $m_{1/(.8,.2)}$, $m_{2/(.6,.4)}$, $m_{3/(.6,.4)}$, $m_{4/(.2,.8)}$, $m_{5/(.3,.7)}$}.

Let (F, A) and (G, B) be two intuitionistic fuzzy soft sets over the same universe U and A, B be two sets of parameters. Then we have the following propositions:

Proposition 5.1

i.   ¤[(**F, A**)∪(**G, B**)] = ¤(**F, A**) ∪ ¤(**G, B**).
ii.   ¤[(**F, A**)∩(**G, B**)] = ¤(**F, A**) ∩ ¤(**G, B**).
iii.   ¤¤ (**F, A**) = ¤ (**F, A**).
iv.   ¤[(**F, A**)]^n = [¤(**F, A**)]^n, for any finite positive integer n.
v.   ¤[(**F, A**)∪(**G, B**)]^n = [¤(**F, A**) ∪ ¤ (**G, B**)]^n.
vi.   ¤[(**F, A**)∩(**G, B**)]^n = [¤(**F, A**) ∩ ¤(**G, B**)]^n.

Proof

i. ¤[(F, A)∪(G, B)]

$$= ¤\{<m, \max (\mu_{F(e)}(m), \mu_{G(e)}(m)), \min(\vee_{F(e)}(m), \vee_{G(e)}(m))>|m\in U\}$$
$$= \{<m, \max (\mu_{F(e)}(m), \mu_{G(e)}(m)), 1- \max (\mu_{F(e)}(m), \mu_{G(e)}(m)) >|m \in U\}$$
$$= \{<m, \max (\mu_{F(e)}(m), \mu_{G(e)}(m)), \min(1- \mu_{F(e)}(m), 1-\mu_{G(e)}(m)) >|m \in U\}$$
$$=\{<m, \mu_{F(e)}(m), 1- \mu_{F(e)}(m)>|m\in U\}\cup\{<m,\mu_{G(e)}(m), 1- \mu_{G(e)}(m)>|m\in U\}$$
$$= ¤(F, A) \cup ¤ (G, B).$$

Hence the result is proved.

ii.   ¤ [(F, A)∩(G, B)]

$$= ¤\{<m, \min (\mu_{F(e)}(m), \mu_{G(e)}(m)), \max(\vee_{F(e)}(m), \vee_{G(e)}(m))>|m\in U\}$$
$$= \{<m, \min (\mu_{F(e)}(m), \mu_{G(e)}(m)), 1- \min (\mu_{F(e)}(m), \mu_{G(e)}(m)) >|m \in U\}$$
$$= \{<m, \min (\mu_{F(e)}(m), \mu_{G(e)}(m)), \max(1- \mu_{F(e)}(m), 1-\mu_{G(e)}(m)) >|m \in U\}$$
$$=\{<m, \mu_{F(e)}(m), 1- \mu_{F(e)}(m)>|m\in U\}\cap\{<m,\mu_{G(e)}(m), 1- \mu_{G(e)}(m)>|m\in U\}$$
$$= ¤(F, A) \cap ¤ (G, B).$$

Hence the result.

iii. Let (F, A) = {<m, $\mu_{F(e)}(m)$, $\vee_{F(e)}(m)$>|m∈ U and e∈ A}.

Then ¤(F, A) = {<m, $\mu_{F(e)}(m)$, 1 - $\mu_{F(e)}(m)$>|m∈ U and e∈ A}.

So ¤¤(F, A) = {<m, $\mu_{F(e)}(m)$, 1 - $\mu_{F(e)}(m)$>|m∈ U and e∈ A}.
Hence the result follows.

iv.   Let the intuitionistic fuzzy soft set (F, A) = {<m, $\mu_{F(e)}(m)$, $\vee_{F(e)}(m)$>|m∈ U and e∈ A}.

Then for any finite positive integer n
$$(F, A)^n = \{<m, [\mu_{F(e)}(m)]^n, 1-[1 - \vee_{F(e)}(m)]^n>|m\in U \text{ and } e\in A\} \quad [8]$$
So, ¤(F, A)^n = {<m, $[\mu_{F(e)}(m)]^n$, 1- $[\mu_{F(e)}(m)]^n$>|m∈ U and e∈ A}.
Again,   [¤(F, A)]^n = {<m, $[\mu_{F(e)}(m)]^n$, 1- $[\mu_{F(e)}(m)]^n$>|m∈ U and e∈ A} as
$$¤(F, A) = \{<m, \mu_{F(e)}(m), 1 - \mu_{F(e)}(m)>|m\in U \text{ and } e\in A\}.$$

Hence the result.

v. As $(F, A)^n \cup (G, B)^n = [(F, A) \cup (G, B)]^n$ [9]

$\qquad \square[(F, A) \cup (G, B)]^n = [\square[(F, A) \cup (G, B)]]^n$,   by the proposition 5.1.iv

$\qquad \qquad \qquad \qquad = [\square(F, A) \cup \square(G, B)]^n$,   by the proposition 5.1.i

vi. As $(F, A)^n \cap (G, B)^n = [(F, A) \cap (G, B)]^n$ [9]

$\qquad$ So, $\square[(F, A) \cap (G, B)]^n = [\square[(F, A) \cap (G, B)]]^n$,   by the proposition 5.1.iv

$\qquad \qquad \qquad \qquad = [\square(F, A) \cap \square(G, B)]^n$,   by the proposition 5.1.ii

The result is proved.

Now we shall define another operation, the possibility operation on intuitionistic fuzzy soft sets.

Let U be a universal set. E be a set of parameters and Λ be a subset of E. Let the intuitionistic fuzzy soft set

$(F, A) = \{<m, \mu_{F(e)}(m), \nu_{F(e)}(m)>|m \in U \text{ and } e \in A\}$, where $\mu_{F(e)}(m)$, $\nu_{F(e)}(m)$ be the membership and non-membership functions respectively.

Definition 5.2. Let U be the universal set and E be the set of parameters. The possibility operation on the intuitionistic fuzzy soft set (F, A) is denoted by $\Diamond(F, A)$ and is defined as

$$\Diamond(F, A) = \{<m, 1 - \nu_{F(e)}(m), \nu_{F(e)}(m)>|m \in U \text{ and } e \in A\}.$$

Example 5.2. Let there are five objects as the universal set where $U = \{m_1, m_2, m_3, m_4, m_5\}$. Also let the set of parameters as E = { beautiful, costly, cheap, moderate, wooden, muddy } and A = { costly, cheap, moderate}.

The cost of the objects represented by the intuitionistic fuzzy soft sets (F, A) is given as

F(costly) = { $m_{1/(.7,.2)}$, $m_{2/(.8,0)}$, $m_{3/(.8,.1)}$, $m_{4/(.9,0)}$, $m_{5/(.6,.2)}$}, F(cheap) = { $m_{1/(.5,.2)}$, $m_{2/(.7,.1)}$, $m_{3/(.4,.2)}$, $m_{4/(.8,.1)}$, $m_{5/(.4,.2)}$}and F(moderate) = { $m_{1/(.8,.2)}$, $m_{2/(.6,.3)}$, $m_{3/(.5,.1)}$, $m_{4/(.9,0)}$, $m_{5/(.7,.1)}$}.

Then the intuitionistic fuzzy soft set $\Diamond(F, A)$ is as

F(costly) = { $m_{1/(.8,.2)}$, $m_{2/(1,0)}$, $m_{3/(.9,.1)}$, $m_{4/(1,0)}$, $m_{5/(.8,.2)}$}, F(cheap) = { $m_{1/(.8,.2)}$, $m_{2/(.9,.1)}$, $m_{3/(.8,.2)}$, $m_{4/(.9,.1)}$, $m_{5/(.8,.2)}$}and F(moderate) = { $m_{1/(.8,.2)}$, $m_{2/(.7,.3)}$, $m_{3/(.9,.1)}$, $m_{4/(1,0)}$, $m_{5/(.9,.1)}$}.

Let (F, A) and (G, B) be two intuitionistic fuzzy soft sets over the same universe U and A, B be two sets of parameters. Then we have the propositions:

Proposition 5.2

i. $\qquad \Diamond[(F, A) \cup (G, B)] = \Diamond(F, A) \cup \Diamond(G, B)$.

ii. $\qquad \Diamond[(F, A) \cap (G, B)] = \Diamond(F, A) \cap \Diamond(G, B)$.

iii. $\qquad \Diamond\Diamond(F, A) = \Diamond(F, A)$.

iv. $\qquad \Diamond[(F, A)]^n = [\Diamond(F, A)]^n$, for any finite positive integer n.

v. $\qquad \Diamond[(F, A) \cup (G, B)]^n = [\Diamond(F, A) \cup \Diamond(G, B)]^n$.

vi. $\qquad \Diamond[(F, A) \cap (G, B)]^n = [\Diamond(F, A) \cap \Diamond(G, B)]^n$.

Proof

i. $\Diamond[(\,F,A\,)\cup(\,G,B\,)]$

$= \Diamond\{<m, \max(\mu_{F(e)}(m), \mu_{G(e)}(m)),\ \min(\nu_{F(e)}(m), \nu_{G(e)}(m))>|m\in U\}$

$= \{<m, 1- \min(\nu_{F(e)}(m), \nu_{G(e)}(m)),\ \min(\nu_{F(e)}(m), \nu_{G(e)}(m))>|m \in U\}$

$= \{<m, \max(1- \nu_{F(e)}(m), 1-\nu_{G(e)}(m)),\ \min(\nu_{F(e)}(m), \nu_{G(e)}(m))>|m \in U\}$

$= \{<m, 1-\nu_{F(e)}(m), \nu_{F(e)}(m)>|m\in U\}\cup\{<m, 1-\nu_{G(e)}(m), \nu_{G(e)}(m)>|m\in U\}$

$= \Diamond(\,F,A\,)\cup\Diamond(\,G,B\,).$

Hence the result is proved.

ii. $\Diamond[(\,F,A\,)\cap(\,G,B\,)]$

$= \Diamond\{<m, \min(\mu_{F(e)}(m), \mu_{G(e)}(m)),\ \max(\nu_{F(e)}(m), \nu_{G(e)}(m))>|m\in U\}$

$= \{<m, 1- \max(\nu_{F(e)}(m), \nu_{G(e)}(m)), \max(\nu_{F(e)}(m), \nu_{G(e)}(m))>|m \in U\}$

$= \{<m, \min(1-\nu_{F(e)}(m), 1-\nu_{G(e)}(m)),\ \max(\nu_{F(e)}(m), \nu_{G(e)}(m))>|m \in U\}$

$= \{<m, 1-\nu_{F(e)}(m), \nu_{F(e)}(m)>|m\in U\}\cap\{<m, 1-\nu_{G(e)}(m), \nu_{G(e)}(m)>|m\in U\}$

$= \Diamond(\,F,A\,)\cap\Diamond(\,G,B\,).$

Hence the result is proved.

iii. $\Diamond(\,F,A\,) = \{<m, 1 - \nu_{F(e)}(m),\ \nu_{F(e)}(m)>|m\in U$ and $e\in A\}.$

So $\Diamond\Diamond(\,F,A\,) = \{<m, 1 - \nu_{F(e)}(m),\ \nu_{F(e)}(m)>|m\in U$ and $e\in A\}.$

Hence the result.

iv. For any positive finite integer n, $(\,F,A\,)^n = \{<m, [\mu_{F(e)}(m)]^n,\ 1-[1-\nu_{F(e)}(m)]^n>|m\in U\}\ \forall\ e\in A$, by [9]

So, $\Diamond(\,F,A\,)^n = \{<m, 1-[1-[1-\nu_{F(e)}(m)]^n],\ 1-[1-\nu_{F(e)}(m)]^n>|m\in U\}$

$= \{<m, [1-\nu_{F(e)}(m)]^n,\ 1-[1-\nu_{F(e)}(m)]^n>|m\in U\}\ \forall\ e\in A.$

Again $[\Diamond(\,F,A\,)]^n = \{<m, [1-\nu_{F(e)}(m)]^n,\ 1-[1-\nu_{F(e)}(m)]^n>|m\in U\}\ \forall\ e\in A.$

Hence the result follows.

v. As $[(\,F,A\,)\cup(\,G,B\,)]^n = (\,F,A\,)^n\cup(\,G,B\,)^n$, by [9]

$\Diamond[(\,F,A\,)\cup(\,G,B\,)]^n = \Diamond(\,F,A\,)^n\cup\Diamond(\,G,B\,)^n.$

The result is proved.

vi. As $[(\,F,A\,)\cap(\,G,B\,)]^n = (\,F,A\,)^n\cap(\,G,B\,)^n$, by [9]

$\Diamond[(\,F,A\,)\cap(\,G,B\,)]^n = \Diamond(\,F,A\,)^n\cap\Diamond(\,G,B\,)^n.$

Hence the result follows.

For any intuitionistic fuzzy soft set $(\,F,A\,)$ we have the following propositions.

Proposition 5.3

  i. $\Box(\mathbf{F},\mathbf{A}) \subset (\mathbf{F},\mathbf{A}) \subset \Diamond(\mathbf{F},\mathbf{A})$

  ii. $\Diamond\Box(\mathbf{F},\mathbf{A}) = \Box(\mathbf{F},\mathbf{A})$

  iii. $\Box\Diamond(\mathbf{F},\mathbf{A}) = \Diamond(\mathbf{F},\mathbf{A})$

Proof

i. Let $(\,F,A\,)$ be an intuitionistic fuzzy soft set over the universe U.

Then $(\,F,A\,) = \{<m, \mu_{F(e)}(m), \nu_{F(e)}(m)>|m\in U\}$ where $e \in A$.

So, $\Box(\,F,A\,) = \{<m, \mu_{F(e)}(m), 1-\mu_{F(e)}(m)>|m\in U\}$, and

$\Diamond(\,F,A\,) = \{<m, 1-\nu_{F(e)}(m), \nu_{F(e)}(m)>|m\in U\}.$

Since $1 - \mu_{F(e)}(m) > \nu_{F(e)}(m)$ □(F, A) $\subset$ (F, A). [as $\mu_{F(e)}(m) + \nu_{F(e)}(m) + \pi_{F(e)}(m) = 1$]
Again since $1 - \nu_{F(e)}(m) > \mu_{F(e)}(m)$, (F, A) $\subset \lozenge$ (F, A).
Hence the result follows.

ii. For the intuitionistic fuzzy soft set (F, A) = { <m, $\mu_{F(e)}(m)$, $\nu_{F(e)}(m)$>|m∈ U},
where e ∈ A.

   We have □(F, A) = { <m, $\mu_{F(e)}(m)$, $1-\mu_{F(e)}(m)$ >|m∈ U}.
So $\lozenge$□ (F, A) = { <m, $1-(1-\mu_{F(e)}(m))$, $1-\mu_{F(e)}(m)$ >|m∈ U}.
       = { <m, $\mu_{F(e)}(m)$, $1-\mu_{F(e)}(m)$ >|m∈ U}.
       = □(F, A).

iii. The proof is similar to the proof of the proposition 5.3.ii.

Let (F, A) and (G, B) be two intuitionistic fuzzy soft sets over the common universe U, then we have the following propositions:

Proposition 5.4

i.	□ [(**F, A**) ∧ (**G, B**)] = □ (**F, A**) ∧ □ (**G, B**).
ii.	□ [(**F, A**) ∨ (**G, B**)] = □ (**F, A**) ∨ □ (**G, B**).
iii.	$\lozenge$ [(**F, A**) ∧ (**G, B**)] = $\lozenge$ (**F, A**) ∧ $\lozenge$ (**G, B**).
iv.	$\lozenge$ [(**F, A**) ∨ (**G, B**)] = $\lozenge$ (**F, A**) ∨ $\lozenge$ (**G, B**).

Proof

i. Let (H, A × B) = (F, A) ∧ (G, B).
Hence, (H, A × B) = { <m, $\mu_{H(\alpha,\beta)}(m)$, $\nu_{H(\alpha,\beta)}(m)$>|m∈ U },
where $\mu_{H(\alpha,\beta)}(m)$ = min { $\mu_{F(\alpha)}(m)$, $\mu_{G(\beta)}(m)$ } and $\nu_{H(\alpha,\beta)}(m)$ = max { $\nu_{F(\alpha)}(m)$, $\nu_{G(\beta)}(m)$ }.
   So, □ (H, A × B) = { <m, $\mu_{H(\alpha,\beta)}(m)$, $1 - \mu_{H(\alpha,\beta)}(m)$>|m∈ U }, for (α, β) ∈ A × B

   = { < m, min ($\mu_{F(\alpha)}(m)$, $\mu_{G(\beta)}(m)$), $1 -$ min ($\mu_{F(\alpha)}(m)$, $\mu_{G(\beta)}(m)$) > |m∈ U }
   = { < m, min ($\mu_{F(\alpha)}(m)$, $\mu_{G(\beta)}(m)$), max ($1 - \mu_{F(\alpha)}(m)$, $1 - \mu_{G(\beta)}(m)$) > |m∈ U }
   = { < m, $\mu_{F(\alpha)}(m)$, $1 - \mu_{F(\alpha)}(m)$ > |m∈ U} AND {<m, $\mu_{G(\beta)}(m)$, $1 - \mu_{G(\beta)}(m)$>|m∈ U}
   = □ (F, A) ∧ □ (G, B).
Hence the result is proved.

ii. Let (O, A × B) = (F, A) ∨ (G, B).
Hence, (O, A × B) = { <m, $\mu_{O(\alpha,\beta)}(m)$, $\nu_{O(\alpha,\beta)}(m)$>|m∈ U },
where $\mu_{O(\alpha,\beta)}(m)$ = max { $\mu_{F(\alpha)}(m)$, $\mu_{G(\beta)}(m)$ } and $\nu_{O(\alpha,\beta)}(m)$ = min { $\nu_{F(\alpha)}(m)$, $\nu_{G(\beta)}(m)$ }.
   So, □ (O, A × B) = { <m, $\mu_{O(\alpha,\beta)}(m)$, $1 - \mu_{O(\alpha,\beta)}(m)$>|m∈ U }, for (α, β) ∈ A × B

   = { < m, max ($\mu_{F(\alpha)}(m)$, $\mu_{G(\beta)}(m)$), $1 -$ max ($\mu_{F(\alpha)}(m)$, $\mu_{G(\beta)}(m)$) > |m∈ U }
   = { < m, max ($\mu_{F(\alpha)}(m)$, $\mu_{G(\beta)}(m)$), min ($1 - \mu_{F(\alpha)}(m)$, $1 - \mu_{G(\beta)}(m)$) > |m∈ U }
   = { < m, $\mu_{F(\alpha)}(m)$, $1 - \mu_{F(\alpha)}(m)$ > |m∈ U} OR {<m, $\mu_{G(\beta)}(m)$, $1 - {G(\beta)}(m)$> |m∈ U}
   = □ (F, A) ∨ □ (G, B).
Hence the result is proved.

iii. Let (H, A × B) = (F, A) ∧ (G, B).

Hence, (H, A × B) = { <m, $\mu_{H(\alpha,\beta)}$(m), $v_{H(\alpha,\beta)}$ (m)>|m∈ U },

where $\mu_{H(\alpha,\beta)}$(m) = min {$\mu_{F(\alpha)}$(m), $\mu_{G(\beta)}$(m)} and $v_{H(\alpha,\beta)}$ (m) = max {$v_{F(\alpha)}$(m), $v_{G(\beta)}$(m)}.

So, ◊ (H, A × B) = { <m, 1 - $v_{H(\alpha,\beta)}$ (m), $v_{H(\alpha,\beta)}$ (m) >|m∈ U }, for (α, β) ∈ A × B

= { < m, 1 - max ($v_{F(\alpha)}$(m), $v_{G(\beta)}$(m)), max ($v_{F(\alpha)}$(m), $v_{G(\beta)}$(m)) > |m∈ U }

= { < m, min (1- $v_{F(\alpha)}$(m), 1- $v_{G(\beta)}$(m)), max ($v_{F(\alpha)}$(m), $v_{G(\beta)}$(m)) > |m∈ U }

= {< m, 1- $v_{F(\alpha)}$(m), $v_{F(\alpha)}$(m)> |m∈ U} AND {<m, 1- $v_{G(\beta)}$(m), $v_{G(\beta)}$(m)> |m∈ U}

= ◊ (F, A) ∧ ◊ (G, B).

Hence the result is proved.

iv. The proof is similar to the proof of the proposition 5.4.iii.

6   An Application of Newly Defined Operation on IFSS

Here we consider the problem of selecting the most suitable object out of n alternatives based on m parameters where information available are intuitionistic fuzzy soft in nature. Suppose a person wants to select an object with certain characteristics. If all the objects are of similar quality, then it is very difficult to choose the appropriate object. For the sake of completeness, let the characteristics in terms of parameter as E = {long, very long, short, costly, very costly, moderate}. Suppose that the person wants to select an object from the view point of its possible 'size' and 'cost'. Let there are four objects with almost same quality as the universe, U = {o_1, o_2, o_3, o_4}. These four objects are chosen for the parameter long, very long and moderate. The 'cost' of these four objects are considered for the parameters costly and moderate. Since the data present are not crisp but intuitionistic fuzzy soft, it is difficult to select the appropriate object as usual. The decision for selection will be made with the help of 'similarity measurement method'. In this method we obtain that particular object o_k dominates all the objects if

S (S, F_k) ≥ S (S, F_i), ∀ i, where F_k is the criteria value of o_k. In case a tie occurs, we select that object corresponding to which the total indeterministic part is maximum.

For a particular problem let the intuitionistic fuzzy soft sets (F, A) which represents the size of the objects is as

F(long) ={ $o_{1/(.6,.2)}$, $o_{2/(.7,.2)}$, $o_{3/(.8,.1)}$, $o_{4/(.6,.2)}$}, F(very long) ={ $o_{1/(.7,.2)}$, $o_{2/(.8,.2)}$, $o_{3/(.6,.3)}$, $o_{4/(.7,.2)}$}and

F(moderate) ={ $o_{1/(.4,.3)}$, $o_{2/(.6,.2)}$, $o_{3/(.7,.1)}$, $o_{4/(.8,.2)}$}.

Also let the cost of the objects represented by the intuitionistic fuzzy soft sets (G, B) is as

G(moderate) ={ $o_{1/(.5,.2)}$, $o_{2/(.4,.4)}$, $o_{3/(.6,.3)}$, $o_{4/(.5,.4)}$}, F(costly) ={ $o_{1/(.6,.2)}$, $o_{2/(.7,.1)}$, $o_{3/(.8,.2)}$, $o_{4/(.5,.1)}$}.

Then the tabular representation of the IFSS (F, A) ∧ (G, B) is as below:

U	(long, moderate)	(long, costly)	(very long, moderate)	(very long, costly)	(moderate, moderate)	(moderate, costly)
o_1	(.5,.2)	(.6,.2)	(.5,.2)	(.6,.2)	(.4,.3)	(.4,.3)
o_2	(.4,.4)	(.7,.2)	(.4,.4)	(.7,.2)	(.4,.4)	(.6,.2)
o_3	(.6,.3)	(.8,.2)	(.6,.3)	(.6,.2)	(.6,.3)	(.7,.2)
o_4	(.5,.4)	(.5,.2)	(.5,.4)	(.5,.2)	(.5,.4)	(.5,.2)

Hence, the tabular representation of the IFSS ◊[(F, A) ∧ (G, B)] is as below:

U	(long, moderate)	(long, costly)	(very long, moderate)	(very long, costly)	(moderate, moderate)	(moderate, costly)
o_1	(.8,.2)	(.8,.2)	(.8,.2)	(.8,.2)	(.7,.3)	(.7,.3)
o_2	(.6,.4)	(.8,.2)	(.6,.4)	(.8,.2)	(.6,.4)	(.8,.2)
o_3	(.7,.3)	(.8,.2)	(.7,.3)	(.8,.2)	(.7,.3)	(.8,.2)
o_4	(.6,.4)	(.8,.2)	(.6,.4)	(.8,.2)	(.6,.4)	(.8,.2)

By similarity measurement we have

$S(S, F_1) = 0.77,$
$S(S, F_2) = 0.70,$
$S(S, F_3) = 0.75,$
$S(S, F_4) = 0.70.$

Here, the maximum value is $S(S, F_1)$. Therefore, according to his choice parameter the object o_1 will be the appropriate option for the person.

7 Conclusion

We have introduced new operations on intuitionistic fuzzy soft set and some properties of these operations have also been established. A simple example has been presented as an application of this mathematical tool.

References

1. Atanassov, K.: Intuitionistic Fuzzy Sets. Fuzzy Sets and Systems 20, 87–96 (1986)
2. Atanassov, K.: Operators Over Interval Valued Intuitionistic Fuzzy Sets. Fuzzy Sets and Systems 64, 159–174 (1994)
3. Biswas, R.: On Fuzzy Sets and Intuitionistic Fuzzy Sets. NIFS 3, 3–11 (1997)
4. De, S.K., Biswas, R., Roy, A.R.: Multicriteria Based Decision Making Using Intuitionistic Fuzzy Set Approach. The Journal of Fuzzy Mathematics 6(4), 837–842 (1998)
5. Gau, W.L., Buchrer, D.J.: Vague Sets. IEEE Trans. Systems Man Cybernet 23(2), 610–614 (1993)
6. Haci, A., Naim, C.: Soft Sets and Soft Groups. Information Sciences 177, 2726–2735 (2007)
7. Maji, P.K., Biswas, R., Roy, A.R.: Intuitionistic Fuzzy Soft Sets. The Journal of Fuzzy Mathematics 9(3), 677–692 (2001)
8. Maji, P.K., Biswas, R., Roy, A.R.: Soft Set Theory. Computers and Mathematics with Applications 45, 555–562 (2003)

9. Maji, P.K., Roy, A.R., Biswas, R.: On Intuitionistic Fuzzy Soft Sets. The Journal of Fuzzy Mathematics 12(3), 669–683 (2004)
10. Molodtsov, D.: Soft Set Theory- First Result. Computers and Mathematics with Applications 37, 19–31 (1999)
11. Pawlak, Z.: Rough Sets. International Journal of Information and Computer Sciences 11, 341–356 (1982)
12. Zadeh, L.A.: Fuzzy Sets. Infor. and Control. 8, 338–353 (1965)
13. Atanassov, K.T.: Intuitionistic Fuzzy Sets Theory and Applications. Physica – Verlag, Heidelberg (1999)
14. Prade, Z., Dubois, D.: Fuzzy Sets and Systems: Theory and Applixcations. Academic Press, London (1980)
15. Zimmerman, H.J.: Fuzzy Set Theory and its Applications. Kluwer Academic Publishers, Boston (1996)

On Fuzzy Contra α-Irresolute Maps

M. Parimala[1] and R. Devi[2]

[1] Lecturer, Department of Mathematics
Bannari Amman Institute of Technology,
Sathyamangalam – 638 401
Tamilnadu, India
rishwanthpari@gmail.com
[2] Reader, Department of Mathematics
Kongunadu Arts and Science College
Coimbatore – 641 029
Tamilnadu, India
rdevicbe@yahoo.com

Abstract. In this paper, we consider a new weak and strong forms of *fuzzy* α*-irresolute* and *fuzzy* α*-closure* via the concept of $F g\alpha$*-closed* sets, which we call *fuzzy approximately* α*-irresolute* maps, *fuzzy approximately*α*-closed* maps and *fuzzy contra* α*-irresolute* maps. Moreover, it turns out that we can use these notions to obtain a new characterization of *fuzzy* α-$T_{1/2}$ spaces.

Keywords: *Fuzzy topological spaces*, $F g\alpha$*-closed* set, $F\alpha$*-open* sets, $F\alpha$*-open* maps, *Fuzzy* α*-irresolute* maps and *Fuzzy* α-$T_{1/2}$ spaces.

AMS(2000) Subject classification: 54A40.

1 Introduction

The concept of fuzzy set and fuzzy set operations were first introduced by Zadeh in his classical paper [6]. Subsequently several authors have applied various basic concepts from general topology to fuzzy sets and developed the theory of fuzzy topological spaces. The concept of *fuzzy generalized* α*-closed* sets was introduced by R. K. Saraf and S. Mishra [5]. In 2004 M.Caldas [2] defined and studied weak and strong forms of irresolute maps in general topology.

In this paper we introduce the concept of *irresoluteness* called *Fap* α-*irresolute* maps and *Fap* α-*closed* maps by using $F g\alpha$-*closed* sets and study some of their basic properties , this definition enables us to obtain conditions under which maps and inverse maps preserves $g\alpha$-*closed* sets. Also, in this paper we present a new generalization of *irresoluteness* called *fuzzy contra* α-*irresolute* map. Finally, we also characteraize the class of *fuzzy* α-$T_{1/2}$ spaces in terms of *Fap* α-*irresolute* and *Fap* α-*closed* maps.

H. Sakai et al. (Eds.): RSFDGrC 2009, LNAI 5908, pp. 241–246, 2009.
© Springer-Verlag Berlin Heidelberg 2009

2   Preliminaries

Throughout this paper (X, τ), (Y, σ) and (Z, η) denote fuzzy topological spaces (briefly, fts) in Chang's [3] sense, on which no separation axioms are assumed unless explicitly stated. For a fuzzy set A of a fuzzy topological space X, the notion $cl(A)$, $int(A)$ and $1 - A$ denote the closure, the interior and the complement of A respectively. In order to make the concepts of the paper as self contained as possible, we briefly describe certain definitions, notions and some properties. A fuzzy set A of a space (X, τ) is called a *fuzzy α-open* (breifly, $F\alpha$-open) set [3] if $A \leq int(cl(int(A)))$ and a *fuzzy α-closed* (breifly, $F\alpha$-closed) set if $cl(int(cl(A))) \leq A$. By $F\alpha O(X, \tau)$, we denote the family of all $F\alpha$-open sets of fts X.

Definition 2.1. A subset A of a topological space (X, τ) is called a *fuzzy generalizedα-closed* (briefly $Fg\alpha$-closed) set [5] if $\alpha cl(A) \leq H$ whenever $A \leq H$ and H is $F\alpha$-open in (X, τ). The complement of $Fg\alpha$-closed set is called $Fg\alpha$-open set.

Definition 2.2. A map $f : (X, \tau) \rightarrow (Y, \sigma)$ is called a *Fuzzy pre α-closed* (resp. *fuzzy pre α-open*) if for every $F\alpha$-closed (resp. $F\alpha$-open) set B in (X, τ), $f(B)$ is $F\alpha$-closed (resp. $F\alpha$-open) in (Y, σ).

Definition 2.3. A map $f : (X, \tau) \rightarrow (Y, \sigma)$ is called a *fuzzy α-irresolute* [4] if for each $V \in F\alpha O(Y, \sigma)$, $f^{-1}(V) \in F\alpha O(X, \tau)$.

3   On *Fap α-Irresolute*, *Fap α-Closed* and *Fc α-Irresolute* Maps

Definition 3.1. A map $f : (X, \tau) \rightarrow (Y, \sigma)$ is called a *fuzzy approximately α-irresolute* (briefly, Fa α-irresolute) map if, $\alpha cl(A) \leq f^{-1}(H)$ whenever H is a $F\alpha$-open subset of (Y, σ), A is $Fg\alpha$-closed subset of (X, τ) and $A \leq f^{-1}(H)$.

Definition 3.2. A map $f : (X, \tau) \rightarrow (Y, \sigma)$ is called a *fuzzy approximately α-closed* (briefly, Fa α-closed) map if, $f(H) \leq \alpha int(A)$ whenever H is a $Fg\alpha$-open subset of (Y, σ), A is $F\alpha$-closed subset of (X, τ) and $f(H) \leq A$.

Theorem 3.3

(i) A map $f : (X, \tau) \rightarrow (Y, \sigma)$ is called a *Fap α-irresolute* if $f^{-1}(A)$ is $F\alpha$-closed in (X, τ) for every $A \in F\alpha O(Y, \sigma)$.
(ii) A map $f : (X, \tau) \rightarrow (Y, \sigma)$ is called a *Fap α-closed* if $f(B) \in F\alpha O(Y, \sigma)$ for every $F\alpha$-closed subset B of (X, τ).

Proof

(i) Let $F \leq f^{-1}(A)$, where $A \in F\alpha O(Y, \sigma)$ and F is a $Fg\alpha$-closed subset of (X, τ). Therefore, $\alpha cl(F) \leq \alpha cl(f^{-1}(A)) = f^{-1}(A)$. Thus f is *Fap α-irresolute*.

(ii) Let $f(B) \leq A$, where B is $F\alpha$-closed subset of (X, τ) and A is a $Fg\alpha$-open subset of (Y, σ). Therefore $\alpha int(f(B)) \leq \alpha int(A)$. Then $f(B) \leq \alpha int(A)$. Thus f is Fap α-closed.

Theorem 3.4. Let $f : (X, \tau) \to (Y, \sigma)$ be a map from a space (X, τ) to a space (Y, σ).

(i) If the $F\alpha$-open and $F\alpha$-closed sets of (X, τ) coincide, then f is Fap α-irresolute if and only if $f^{-1}(A)$ is $F\alpha$-closed in (X, τ) for every $A \in F\alpha O(Y, \sigma)$.

(ii) If the $F\alpha$-open and $F\alpha$-closed sets of (Y, σ) coincide, then f is Fap α-closed if and only if $f(B) \in F\alpha O(Y, \sigma)$ for every $F\alpha$-closed subset B of (X, τ).

Proof

(i) The sufficiency is stated in theorem 3.3.
Necessity. Assume that f is Fap α-irresolute. Let A be an arbitrary subset of (X, τ) such that $A \leq Q$ where $Q \in F\alpha O(X, \tau)$. Then by hypothesis $\alpha cl(A) \leq \alpha cl(Q) = Q$. Therefore all subsets of (X, τ) are $Fg\alpha$-closed (and hence all are $Fg\alpha$-open). So for any $A \in F\alpha O(Y, \sigma)$, $f^{-1}(A)$ is $Fg\alpha$-closed in (X, τ). Since f is Fap α-irresolute, $\alpha cl(f^{-1}(A)) \leq f^{-1}(A)$. Therefore $\alpha cl(f^{-1}(A)) = f^{-1}(A)$, i.e. $f^{-1}(A)$ is $F\alpha$-closed in (X, τ).

(ii) The sufficiency is clear by theorem 3.3.
Necessity. Assume that f is Fap α-closed. As in (i), we obtain that all subsets of (Y, σ) are $Fg\alpha$-open. Therefore for any $F\alpha$-closed subset B of (X, τ), $f(B)$ is Fap α-closed $f(B) \leq \alpha int(f(B))$. Hence $f(B) = \alpha int(f(B))$, i.e. $f(B)$ is $F\alpha$-open.

Corollary 3.5. Let $f : (X, \tau) \to (Y, \sigma)$ be a map such that:

(i) If the $F\alpha$-open and $F\alpha$-closed sets of (X, τ) coincide, then f is Fap α-irresolute if and only if f is $F\alpha$-irresolute.

(ii) If the $F\alpha$-open and $F\alpha$-closed sets of (Y, σ) coincide, then f is Fap α-closed if and only if f is $Fpre$ α-closed.

Theorem 3.6. If a map $f : (X, \tau) \to (Y, \sigma)$ is surjective $F\alpha$-irresolute and Fap α-open, then $f^{-1}(A)$ is $Fg\alpha$-open whenever A is $Fg\alpha$-open subset of (Y, σ).

Proof. Let A be $Fg\alpha$-open subset of (Y, σ). Suppose that $F \leq f^{-1}(A)$, where $F \in F\alpha O(X, \tau)$. Taking compliments, we obtain $f^{-1}(A^c) \leq F^c$ or $A^c \leq f^{-1}(F^c)$. Since f is an Fap α-open and $\alpha int(A) = A \wedge cl(int(cl(A)))$ and $\alpha cl(A) = A \vee cl(int(cl(A)))$, then $(\alpha int(A))^c = \alpha cl(A^c) \leq f(F^c)$. It follows that $(f^{-1}(\alpha int(A)))^c \leq F^c$ and hence $F \leq f^{-1}(\alpha int(A))$. Since f is $F\alpha$-irresolute $f^{-1}(\alpha int(A))$ is $F\alpha$-open. Thus, we have $F \leq f^{-1}(\alpha int(A)) = \alpha int(f^{-1}(\alpha int(A))) \leq \alpha int(f^{-1}(A))$. This implies that $f^{-1}(A)$ is $Fg\alpha$-open in (X, τ).

Theorem 3.7. If a map $f : (X, \tau) \to (Y, \sigma)$ and $g : (Y, \sigma) \to (Z, \eta)$ be two maps such that $gof : (X, \tau) \to (Z, \eta)$, then

(i) gof is Fap α-*closed*, if f is *Fuzzy pre* α-*closed* and g is Fap α-*closed*.
(ii) gof is Fap α-*closed*, if f is Fap α-*closed* and g is *Fuzzy pre* α-*open* and g^{-1} preserves $Fg\alpha$-*open* sets.
(iii) gof is Fap α-*irresolute* if f is Fap α-*irresolute* and g is $F\alpha$-*irresolute*.

Proof

(i) Suppose B is $F\alpha$-*closed* set in (X, τ) and A is $Fg\alpha$-*open* subset of (Z, η) for which $(gof)(B) \leq A$. Then $f(B)$ is $F\alpha$-*closed* in Y because f is *Fuzzy pre* α-*closed*. Since g is Fap α-*closed*, $g(f(B)) \leq \alpha int(A)$. This implies that gof is Fap α-*closed*.

(ii) Suppose B is $F\alpha$-*closed* set in (X, τ) and A is $Fg\alpha$-*open* subset of (Z, η) for which $(gof)(B) \leq A$. Hence $f(B) \leq g^{-1}(A)$. Then $f(B) \leq \alpha int(g^{-1}(A))$, because $g^{-1}(A)$ is $Fg\alpha$-*open* and f is Fap α-*closed*. Thus $(gof)(B) = g(f(B)) \leq g(\alpha int(g^{-1}(A))) \leq \alpha int(g(g^{-1}(A))) = g(f(B)) \leq \alpha int(A)$. This implies that gof is Fap α-*closed*.

(iii) Suppose E is $Fg\alpha$-*closed* subset of (X, τ) and $H \in F\alpha O(Z, \eta)$, for which $E \leq (gof)^{-1}(H)$. Then $g^{-1}(H) \in F\alpha O(Y)$ because g is $F\alpha$-*irresolute*. Since f is Fap α-*irresolute*, $\alpha cl(E) \leq f^{-1}(g^{-1}(H)) = (gof)^{-1}(H)$. This proves that gof is Fap α-*irresolute*.

Definition 3.8. A map $f : (X, \tau) \to (Y, \sigma)$ is called a *Fuzzy contra-α-irresolute* (briefly, Fc α-*irresolute*)if $f^{-1}(A)$ is $F\alpha$-*closed* in (X, τ) for each $A \in F\alpha O(Y, \sigma)$.

Definition 3.9. A map $f : (X, \tau) \to (Y, \sigma)$ is called a *Fuzzy contra-pre-α-closed* (briefly, Fcp α-*closed*) if $f^{-1}(B) \in F\alpha O(Y, \sigma)$ for each $F\alpha$-*closed* set B of (X, τ).

Definition 3.10. A map $f : (X, \tau) \to (Y, \sigma)$ is called a *perfectly fuzzy contra-α-irresolute* (briefly, Pfc α-*irresolute*)if the inverse image of every $F\alpha$-*open* set in Y is $F\alpha$-*clopen* in X.

Every *pfc-α-irresolute* map is Fc-α-*irresolute*. Clearly, the following diagram holds and none of its immplicatins are reversible.

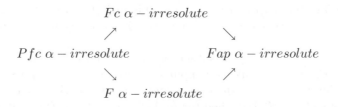

The next two theorems establish conditions under which maps and inverse maps preserve *fuzzy generalized α-closed* sets.

Theorem 3.11. If a map $f : (X, \tau) \to (Y, \sigma)$ is Fap α-*irresolute* and *Fuzzy pre-α-closed* then for every $Fg\alpha$-*closed* set E of (X, τ), $f(A)$ is $Fg\alpha$-*closed* set of (Y, σ).

Proof. Let A be $Fg\alpha$-closed set of (X,τ), let $f(A) \le B$ where $B \in F\alpha O(Y,\sigma)$. Then $A \le f^{-1}(B)$ holds. Since f is Fap α-irresolute, $\alpha cl(A) \le f^{-1}(B)$ and hence $f(\alpha cl(A)) \le B$. Therefore, we have $\alpha cl(A) \le \alpha cl(f(\alpha cl(A))) = f(\alpha cl(A)) \le B$. Hence $f(A)$ is $Fg\alpha$-closed in Y.

Theorem 3.12. Let $f : (X,\tau) \to (Y,\sigma)$ and $g : (Y,\sigma) \to (Z,\eta)$ be two maps such that $gof : (X,\tau) \to (Z,\eta)$, then

(i) gof is Fc α-irresolute, if g is *fuzzy irresolute* and f is Fc α-irresolute
(ii) gof is Fc α-irresolute, if g is Fc α-irresolute and f is *fuzzy irresolute*.

Proof. It is straight forward. Now we state the following theorems whose proof is straight forward and hence omitted.

Theorem 3.13. Let $f : (X,\tau) \to (Y,\sigma)$ be a map. Then the following conditionds are equivalent.

(i) f is Fc α-irresolute.
(ii) The inverse image of each $F\alpha$-closed set in Y is $F\alpha$-open in X.

Theorem 3.14. Let $f : (X,\tau) \to (Y,\sigma)$ and $g : (Y,\sigma) \to (Z,\eta)$ be two maps and $gof : (X,\tau) \to (Z,\eta)$, then

(i) gof is Fcp α-open, if f is *Fuzzy pre-α-open* and g is Fcp α-open.
(ii) gof is Fcp α-open, if f is Fcp α-open and g is *Fuzzy pre-α-open*.

Theorem 3.15. Let $f : (X,\tau) \to (Y,\sigma)$ and $g : (Y,\sigma) \to (Z,\eta)$ be two maps such that $gof : (X,\tau) \to (Z,\eta)$ is a Fcp α-open map. Then

(i) gof is $F\alpha$-irresolute surjection, then g is Fcp α-open.
(ii) gof is $F\alpha$-irresolute injection, then f is Fcp α-open.

Proof.

(i) Suppose A is any arbitrary $F\alpha$-open set in Y. Since f is $F\alpha$-irresolute, $f^{-1}(A)$ is $F\alpha$-open in X. Moreover, gof is Fcp α-open and f is surjective, then $(gof)(f^{-1}(A)) = g(A)$ is $F\alpha$-closed in Z. This implies that g is a Fcp α-open map.
(ii) Suppose A is any arbitrary $F\alpha$-open set in X. Since gof is Fcp α-open, $(gof)(A)$ is $F\alpha$-closed in Z. Since g is a $F\alpha$-irresolute injection, $g^{-1}(gof)(A) = f(A)$ is $F\alpha$-closed in Y. This implies that f is a Fcp α-open map.

4   A Characterization of $F\alpha$-$T_{1/2}$ Spaces

In the following result, we offer a characterization of the class of $F\alpha$-$T_{1/2}$ spaces by using the concepts of Fap α-irresolute and Fap α-closed maps.

Definition 4.1. A space (X,τ) is said to be *Fuzzy α-$T_{1/2}$* (briefly, $F\alpha$-$T_{1/2}$) space, if every $Fg\alpha$-closed set is $F\alpha$-closed.

Theorem 4.2. Let (X,τ) be a fuzzy topological space, then the following statements are equivalent:

(i) (X, τ) is $F\alpha$-$T_{1/2}$ space.
(ii) For every fuzzy topological space (Y, σ) and every map $f : (X, \tau) \to (Y, \sigma)$, f is $Fap\ \alpha$-irresolute.

Proof. $(i) \Rightarrow (ii)$ Let A be a $Fg\alpha$-closed subset of (X, τ) and suppose that $A \leq f^{-1}(B)$, where $B \in F\alpha O(Y)$. Since, (X, τ) is $F\alpha$-$T_{1/2}$ space, A is $F\alpha$-closed (i.e. $A = \alpha cl(A)$). Therefore, $\alpha cl(A) \leq f^{-1}(B)$, then f is $Fap\ \alpha$-irresolute.
$(ii) \Rightarrow (i)$ Let B be a $Fg\alpha$-closed subset of (X, τ) and let Y be the set X with topology $\sigma = \{1, B, 0\}$. Finally let $ff : (X, \tau) \to (Y, \sigma)$ be the identity map. By assumption, f is $Fap\ \alpha$-irresolute, since B is $Fg\alpha$-closed in (X, τ) and $F\alpha$-open in (Y, σ) and $B \leq f^{-1}(B)$. It follows that $\alpha cl(B) \leq B$. Hence B is $F\alpha$-closed in (X, τ) and therefore it is $F\alpha$-$T_{1/2}$.

Theorem 4.3. Let (Y, σ) be a fuzzy topological space. Then the following statements are equivalent:

(i) (Y, σ) is $F\alpha$-$T_{1/2}$ space.
(ii) For every space (Y, σ) and every map $f : (X, \tau) \to (Y, \sigma)$, f is $Fap\ \alpha$-closed.

Proof. Analogous to theorem 4.2 making the obvious changes.

References

[1] Bin Shahna, A.S.: On *fuzzy strongly semi continuity* and *fuzzy pre continuity*. Fuzzy sets and systems 44(2), 303–308 (1991)
[2] Caldas, M.: *Weak* and *strong* forms of irresolute map. internal. J. Math. Math. Sciences 23, 253–259 (2004)
[3] Chang, C.L.: *Fuzzy topological spaces*. J. Math. Anal. Appl. 24, 182–190 (1968)
[4] Prasad, R., Thakur, S.S., Saraf, R.K.: *Fuzzy α-irresolute mapping*. Jour. Fuzzy Math. 2, 235–239 (1994)
[5] Saraf, R.K., Mishra, S.: *Fgα-closed* sets. J. Tripura Math. Soc. 2, 27–32 (2000)
[6] Zadeh, L.A.: *Fuzzy sets*. Information control 8, 338–353 (1965)

On Fuzzy Strongly α-I-Open Sets

A. Selvakumar and R. Devi

Department of Mathematics
Kongunadu Arts and Science College
Coimbatore – 641 029
Tamilnadu, India
selvam_mphil@yahoo.com, rdevicbe@yahoo.com

Abstract. In this paper, we introduce the notion of Fuzzy Strongly α-I-open sets in fuzzy ideal topological spaces and investigate some of their properties. Further we study the continuous functions for the above set and derive the some of their properties.

Keywords: fuzzy α-I-open set, fuzzy Strongly α-I-open set and B-I set.

2000 Mathematics Subject Classification: 54A05, 54D10, 54F65, 54G05.

1 Introduction

The notion of fuzzy α-open sets in fuzzy topological spaces was introduced and investigated by A.S. Bin Shahna [1]. Ideals in fuzzy topological spaces have been considered since 1997. In 2007, E.Hatir et al. [3] have introduced the concept of fuzzy semi-I-open set via fuzzy ideals. The notion of fuzzy pre-I-open sets was introduced and investigated by A. Nasef et al. [5]. Recently in [8], fuzzy α-I-open sets has defined and using this sets proved decomposition of continuity in fuzzy ideal topological spaces. In this paper, we introduce fuzzy strongly α-I-open sets and establish a decomposition of continuity.

For a space A of a fuzzy topological space (X, τ), $cl(A), int(A)$ is denoted by closure of A and interior of A respectively. A non-empty collection of fuzzy sets I of a sets X is called a fuzzy ideal [4,7] if and only if (i) if $A \in I$ and $A \subseteq B$, then $B \in I$ (heredity) and (ii) if $A, B \in I$ then $A \cup B \in I$ (finite additivity). The triple (X, τ, I) means a fuzzy topological space with a fuzzy ideal I and fuzzy topology τ. For (X, τ, I), the fuzzy local function $A \leq X$ with respect to τ and I is denoted by $A^*(\tau, I)$ (briefly A^*) and is defined as $A^*(\tau, I) = \vee\{x \in X : A \wedge U \notin I$ for every $U \in \tau\}$. Fuzzy closure operator of a fuzzy set A in (X, τ, I) is defined as $cl^*(A) = A \vee A^*$. In (X, τ, I), the collection $\tau^*(I)$ means an extension of fuzzy topological space than τ via fuzzy ideal which is constructed by considering the class $\beta = \{U - E : U \in \tau, E \in I\}$ as a base. A fuzzy subset A in (X, τ, I) is called fuzzy ideal open [6] if $A \leq int(A^*)$. The collection of all fuzzy ideal open sets in (X, τ, I) will be denoted by $FIO(X)$.

H. Sakai et al. (Eds.): RSFDGrC 2009, LNAI 5908, pp. 247–252, 2009.

2   Preliminaries

First we will recall some definitions used in sequel.

Definition 2.1. A subset A of a fuzzy ideal topological space (X, τ, I) is said to be

1. fuzzy α-I-open [8] if $A \leq int(cl^*(int(A)))$,
2. fuzzy pre-I-open [5] if $A \leq int(cl^*(A))$ and
3. fuzzy semi-I-open [3] if $A \leq cl^*(int(A))$.

Definition 2.2. A function $f : (X, \tau, I) \to (Y, \sigma)$ is said to be

1. fuzzy continuous [2] if for every $V \in \sigma$, $f^{-1}(V)$ is fuzzy open,
2. fuzzy semi-I-continuous [3] if for every $V \in \sigma$, $f^{-1}(V)$ is fuzzy semi-I-open and
3. fuzzy pre-I-continuous [5] if for every $V \in \sigma$, $f^{-1}(V)$ is fuzzy pre-I-open.

3   Fuzzy Strongly α-I-Open Sets

Definition 3.1. A subset A of a fuzzy ideal topological space (X, τ, I) is said to be

1. t-I-set if $int(cl^*(A)) = int(A)$,
2. B-I-set if $A = U \wedge V$, where $U \in \tau$ and V is a t-I-set,
3. C-I-set if $A = U \wedge V$, where $U \in \tau$ and $int(cl^*(int(V))) = int(V)$ and
4. A-I-set if $A = U \wedge V$, where $U \in \tau$ and $V = (int(V))^*$.

Definition 3.2. A subset A of a fuzzy ideal topological space (X, τ, I) is said to be fuzzy strongly α-I-open set if A is fuzzy α-I-open as well as a B-I-set.

The family of all fuzzy strongly α-I-open sets (resp. fuzzy α-I-open sets)in (X, τ, I) is denoted by $FS\alpha IO(X, \tau)$ or $FS\alpha IO(X)$ (resp. $F\alpha IO(X, \tau)$ or $F\alpha IO(X)$).

Theorem 3.3. *Every fuzzy strongly α-I-open set is fuzzy α-I-open.*

Proof. It is obvious.

Theorem 3.4. *Every fuzzy strongly α-I-open set is fuzzy pre-I-open.*

Proof. Let A be fuzzy strongly α-I-open set in (X, τ, I). Then A is fuzzy α-I-open set, $A \leq int(cl^*(int(A))) \leq int(cl^*(A))$. Therefore A is fuzzy pre-I-open.

Definition 3.5. A subset A of a fuzzy ideal space (X, τ, I) is fuzzy $*$-dense in itself if $A \leq A^*$.

Theorem 3.6. *Let (X, τ, I) be a fuzzy ideal topological space. If A is fuzzy $*$-dense in itself, then $A^* = cl^*(A)$.*

Proof. Since A is fuzzy $*$-dense in itself, $A \leq A^*$. Then $cl^*(A) = A \vee A^* = A^*$.

Theorem 3.7. *In (X, τ, I), A is fuzzy strongly α-I-open if and only if there exists $U \in \tau$ such that $U \leq A \leq int(cl^*(U))$.*

Proof. Let $A \in FS\alpha IO(X)$. Then we have $A \leq int(cl^*(int(A)))$. Take $int(A) = U$. Then $U \leq A \leq int(cl^*(U))$.

Conversely, let U be a fuzzy open set such that $U \leq A \leq int(cl^*(U))$. Since $U \leq A$, $U \leq int(A)$ and hence $int(cl^*(U)) \leq int(cl^*(int(A)))$. Thus, we obtain $A \leq int(cl^*(int(A)))$.

Theorem 3.8. *If A is a fuzzy strongly α-I-open set in (X, τ, I) and $A \leq B \leq int(cl^*(A))$, then B is fuzzy strongly α-I-open in (X, τ, I).*

Proof. Since $A \in FS\alpha IO(X)$, there exists a fuzzy open set U such that $U \leq A \leq int(cl^*(A))$. By Theorem 3.7., we obtain $B \in FS\alpha IO(X)$.

Lemma 3.9. [6] *For any (X, τ) if $U \leq X$ and $V \in \tau$ then*

(1) $cl(U^*(I)) \leq cl(U)$
(2) $V \wedge U^*(I) \leq (V \wedge U)^*(I)$.

Theorem 3.10. *The intersection of any fuzzy open set and fuzzy strongly α-I-open set is fuzzy strongly α-I-open set.*

Proof. Let $U \in \tau$ and $V \in FS\alpha IO(X)$ in a fuzzy ideal topological space (X, τ, I), then

$$U \wedge V \leq U \wedge int(cl^*(int(V)))$$
$$\leq U \wedge int(int(V) \vee (int(V))^*)$$
$$= int(U \wedge (int(V) \vee (int(V))^*))$$
$$= int(int(U \wedge V) \vee (int(U \wedge V))^*) \quad \text{(by Lemma 3.9.)}$$
$$= int(cl^*(int(U \wedge V)))$$
$$U \wedge V \in F\alpha IO(X).$$

Theorem 3.11. *Every B-I-set is a C-I-set.*

Proof. Let A be a B-I set. Then $A = U \wedge V$, where $U \in \tau$ and V is a t-I-set. Then $int(V) = int(cl^*(V)) \geq int(cl^*(int(V))) \geq int(V)$ and hence $int(V) = int(cl^*(int(V)))$. This shows that A is a C-I-set.

Definition 3.12. A subset A of a fuzzy ideal topological space (X, τ, I) is said to be FI-locally closed set if $A = U \wedge V$, where $U \in \tau$ and $V = V^*$.

Theorem 3.13. *Let (X, τ, I) be a fuzzy ideal topological space. A subset A of X is FI-locally closed set if A is both fuzzy open and A-I-set.*

Proof. Let A be a fuzzy open and A-I-set, then $A = G \wedge V$, where $G \in \tau$ and $V = (int(V))^* = V^*$. This shows that A is FI-locally closed set.

The following Theorem gives a characterization of fuzzy open sets in terms of fuzzy strongly α-I-open sets and A-I-sets.

Theorem 3.14. *Let (X, τ, I) be a fuzzy ideal topological space. A subset A of (X, τ, I) is fuzzy pre-I-open and B-I-set if A is fuzzy strongly α-I-open.*

Proof. Let A be fuzzy strongly α-I-open set. Since every fuzzy α-I-open set is fuzzy pre-I-open, then A is fuzzy pre-I-open and B-I-set.

Theorem 3.15. *Let (X, τ, I) be a fuzzy ideal topological space. A subset A of (X, τ, I) is fuzzy strongly α-I-open if and only if it is fuzzy semi-I-open, fuzzy pre-I-open and B-I-set.*

Proof

Necessity. It follows from the fact that every fuzzy α-I-open set is fuzzy semi-I-open and fuzzy pre-I-open.

Sufficiency. Let A be a fuzzy semi-I-open, fuzzy pre-I-open and B-I-set. Then, we have $A \leq int(cl^*(A)) \leq int(cl^*(cl^*(int(A)))) = int(cl^*(int(A)))$. This shows that A is fuzzy α-I-open set and also A is B-I-set. Therefore A is fuzzy strongly α-I-open set.

4   Fuzzy Strongly α-I-Continuous Maps

Definition 4.1. A function $f : (X, \tau, I) \to (Y, \sigma)$ is said to be

(1) A-I-continuous if for every $V \in \sigma$, $f^{-1}(V)$ is A-I-set,
(2) B-I-continuous if for every $V \in \sigma$, $f^{-1}(V)$ is B-I-set,
(3) C-I-continuous if for every $V \in \sigma$, $f^{-1}(V)$ is C-I-set and
(4) FI-locally continuous if for every $V \in \sigma$, $f^{-1}(V)$ is FI-locally *closed*.

Definition 4.2. A mapping $f : (X, \tau, I) \to (Y, \sigma)$ is said to be fuzzy strongly α-I-continuous if for every $V \in \sigma$, $f^{-1}(V)$ is fuzzy strongly α-I-open.

Theorem 4.3. *Every fuzzy strongly α-I-continuous map is fuzzy pre-I-continuous.*

Proof. It follows from Theorem 3.4.

Theorem 4.4. *Let $f : (X, \tau, I) \to (Y, \sigma)$ be any mapping. Then f is FI-locally continuous map if f is both fuzzy continuous and A-I-continuous.*

Proof. It follows from Theorem 3.13.

Theorem 4.6. *Let $f : (X, \tau, I) \to (Y, \sigma)$ be any mapping. Then f is fuzzy pre-I-continuous and B-I-continuous if f is fuzzy strongly α-I-continuous.*

Proof. It follows from Theorem 3.14.

Theorem 4.7. *Let $f : (X, \tau, I) \to (Y, \sigma)$ be any mapping. Then f is fuzzy strongly α-I-continuous if and only if it is fuzzy semi-I-continuous, fuzzy pre-I-continuous and B-I-continuous.*

Proof. It follows from Theorem 3.15.

Definition 4.8. A mapping $f : (X, \tau, I) \to (Y, \sigma, I)$ is said to be fuzzy strongly α-I-irresolute if $f^{-1}(V)$ is fuzzy strongly α-I open in X for every fuzzy strongly α-I-open set V of Y.

Theorem 4.9. *Let $f : (X, \tau, I) \to (Y, \sigma)$ and $g : (Y, \sigma) \to (Z, \eta)$ be mappings. Then the composition $g \circ f : X \to Z$ is fuzzy strongly α-I-continuous if g is fuzzy continuous and f is fuzzy strongly α-I-continuous.*

Proof. Let W be any fuzzy open subset of Z. Since g is fuzzy continuous, $g^{-1}(W)$ is fuzzy open in Y. Since f is fuzzy strongly α-I-continuous, then $(g \circ f)^{-1}(W) = f^{-1}(g^{-1}(W))$ is fuzzy strongly α-I-open in X and hence $g \circ f$ is fuzzy strongly α-I-continuous.

Theorem 4.10. *Let $f : (X, \tau, I_1) \to (Y, \sigma, I_2)$ and $g : (Y, \sigma, I_2) \to (Z, \eta, I_3)$ be mappings. Then the composition $g \circ f : X \to Z$ is fuzyy strongly α-I-continuous if g is fuzzy strongly α-I-continuous and f is fuzzy strongly α-I-irresolute.*

Proof. Let W be any fuzzy open subset of Z. Since g is fuzzy strongly α-I-continuous, $g^{-1}(W)$ is fuzzy strongly α-I-open in Y. Since f is fuzzy strongly α-I-irresolute, then $(g \circ f)^{-1}(W) = f^{-1}(g^{-1}(W))$ is fuzzy strongly α-I-open in X and hence $g \circ f$ is fuzzy strongly α-I-continuous.

Theorem 4.11. *Let $f : (X, \tau, I_1) \to (Y, \sigma, I_2)$ and $g : (Y, \sigma, I_2) \to (Z, \eta, I_3)$ be mappings. Then the composition $g \circ f : X \to Z$ is fuzzy strongly α-I-irresolute if both f and g are fuzzy strongly α-I-irresolute.*

Proof. Let W be any fuzzy strongly α-I-open subset of Z. Since g is fuzzy strongly α-I-irresolute, $g^{-1}(W)$ is fuzzy strongly α-I-open in Y. Since f is fuzzy strongly α-I-irresolute, then $(g \circ f)^{-1}(W) = f^{-1}(g^{-1}(W))$ is fuzzy strongly α-I-open in X and hence $g \circ f$ is fuzzy strongly α-I-irresolute.

References

[1] Bin Shahna, A.S.: On fuzzy strong semi-continuity and fuzzy pre-continuity. Fuzzy sets and systems 44, 303–308 (1991)
[2] Chang, C.L.: Fuzzy topological spaces. J. Math. Annl. Appl. 24, 182–190 (1968)

[3] Hatir, E., Jafari, S.: Fuzzy semi-I-open sets and fuzzy semi-I-continuity via fuzzy idealization. Chaos, Solitons and Fractals 34, 1220–1224 (2007)

[4] Mahmoud, R.: Fuzzy ideals, fuzzy local functions and fuzzy topology. J. Fuzzy Math., Los Angels 5(1), 165–172 (1997)

[5] Nasef, A.A., Hatir, E.: On fuzzy pre-I-open sets and decomposition of fuzzy I-continuity. In: Chaos, Solitons and Fractals (Article in press)

[6] Nasef, A.A., Mahmoud, R.A.: Some topological applications via fuzzy ideals. Chaos, Solitons and Fractals 13, 825–831 (2002)

[7] Sarkar, D.: Fuzzy ideal theory, fuzzy local function and generated fuzzy topology. Fuzzy Sets Syst. 87, 117–123 (1997)

[8] Yuksel, S., Gursel Caylak, E., Acikgoz, A.: On fuzzy α-I-continuous and fuzzy α-I-open functions. Chaos, Solitons and Fractals (Article in press)

Lattice Structures of Rough Fuzzy Sets

Guilong Liu

School of Information Science,
Beijing Language and Culture University,
Beijing 100083, China

Abstract. This paper studies lattice structure of the family of rough fuzzy sets in a given approximation space. Our result is an extension of standard rough sets. Starting with the definition of rough fuzzy sets, the union, intersection and pseudocomplementation operations are generalized. Then it is proved that the above family with these operations is a distributive lattice. This paper also gives the characterization of borderline region of rough fuzzy sets.

Keywords: Rough sets, Fuzzy sets, Rough fuzzy sets, Lower approximations, upper approximations, Lattices.

1 Introduction

A more general concept than that of Boolean algebra is that of a lattice. A lattice is an important mathematical structure, lattice structures often appear in computing and mathematical application such as models of information flow and computer circuit.

Rough set theory was firstly proposed by Pawlak [10] in 1982. It is an extension of set theory for the study of intelligent systems characterized by uncertain information. Since then Pawlak rough sets are extended by many authors [1,4,6,7,8,9,13,14,15]. Work on lattice structure of rough sets is important, much research has been done in lattice structure of rough sets, For example, Dai [3] discussed rough 3-valued algebra. J.Pomykala and J.A.Pomykala [12] studied the lattice structure of the family of rough sets in a given approximation space and proved that the above family is a Stone algebra. In this paper, we consider the similar problem for rough fuzzy sets, we give a similar lattice structure in fuzzy environment, our result is an extension of [12] in fuzzy environment. Finally, we gives the characterization of borderline region of rough fuzzy sets.

The paper is organized as follows. Section 2 introduces relevant definition of rough fuzzy sets. Section 3 gives some basic properties of rough fuzzy sets. Section 4 considers a lattice from rough fuzzy sets. Section 5 characterizes the borderline region of rough fuzzy sets. Finally, Section 6 concludes the paper.

2 Rough Fuzzy Sets

Dubois and Prade [4] have proposed fuzzy generalizations of rough approximations. For any given universal set U, let $X : U \rightarrow [0,1]$ be a fuzzy set [16],

H. Sakai et al. (Eds.): RSFDGrC 2009, LNAI 5908, pp. 253–260, 2009.

$X(x), x \in U$, giving the degree of membership of x in X. Power set, fuzzy power set of U is denoted by $P(U)$, $F(U)$, respectively. If $A, B \in F(U)$, then $A \cap B$, $A \cup B$ are the two fuzzy sets pointwise defined as $(A \cap B)(x) = A(x) \wedge B(x)$ and $(A \cup B) = A(x) \vee B(x)$ for each $x \in U$. where $\wedge$ denotes minimum and $\vee$ maximum. If $A \in P(U)$ or $A \in F(U)$ we denote by $-A$ the complement of A. That is, for crisp subset A of U, $-A = U - A$, and for fuzzy set A in U, $(-A)(x) = 1 - A(x)$. Let R be an equivalence relation on U, the quotient set is denoted by U/R. We use $[x]$ to denote an equivalence class in R containing an element $x \in U$. For $x, y \in U, R(x, y)$ is defined by $R(x, y) = \begin{cases} 1, (x, y) \in R \\ 0, (x, y) \notin R \end{cases}$.

Rough fuzzy set [4] can be rewritten as following:

Definition 2.1. Let U be a universal set and R be an equivalence relation on U. The lower and upper approximations of the fuzzy set $X \in F(U)$, denoted $\underline{R}X$ and $\overline{R}X$, respectively, are defined as fuzzy sets in U such that

$$(\overline{R}X)(x) = \vee_{y \in U}(R(x, y) \wedge X(y)), x \in U$$

and

$$(\underline{R}X)(x) = \wedge_{y \in U}((1 - R(x, y)) \vee X(y)), x \in U$$

The pair $RX = (\underline{R}X, \overline{R}X)$ is referred to as a rough fuzzy set. $\underline{R}, \overline{R} : F(U) \to F(U)$ are respectively referred to as lower and upper rough fuzzy approximation operators.

The following property is obvious:

Proposition 2.1. Let U be a universal set and R be an equivalence relation on U. Then for $x \in U$

$$(\overline{R}X)(x) = \vee_{y \in [x]} X(y) \text{ and } (\underline{R}X)(x) = \wedge_{y \in [x]} X(y)$$

Note that if X is a crisp subset of U, then Definition 2.1 coincides with the definition of Pawlak lower and upper approximations.

3   Properties of Rough Fuzzy Sets

In this section, we give some basic properties of rough fuzzy sets. Recall that for fuzzy set X in U and $\lambda \in [0, 1]$, the unit interval, the λ-cut of X is defined as $X_\lambda = \{x \in U | X(x) \geq \lambda\}$. We need a technical lemma.

Lemma 3.1. Let $X, Y \in F(U)$, then $X = Y$ if and only if $X_\alpha = Y_\alpha$ for all $\alpha \in [0, 1]$.

Proof. By the decomposition theorem of fuzzy sets, $X = \bigcup_{\alpha \in [0,1]} \alpha X_\alpha = \bigcup_{\alpha \in [0,1]} \alpha Y_\alpha = Y$.

Using Lemma 3.1, the following theorem can be easily derived.

Theorem 3.1. Let U be a universal set and R be an equivalence relation on U. Then for $X, Y \in F(U)$

(1) $(\underline{R}X)_\alpha = \underline{R}X_\alpha, (\overline{R}X)_\alpha = \overline{R}X_\alpha$ for all $\alpha \in [0,1]$;

(2) if $X \in P(U)$, then $\underline{R}X, \overline{R}X \in P(U)$;

(3) $\underline{R}X \subseteq X \subseteq \overline{R}X$ for all $X \in F(U)$;

(4) $\underline{R}\emptyset = \overline{R}\emptyset = \emptyset, \underline{R}U = \overline{R}U = U$;

(5) $\overline{R}(X \cup Y) = \overline{R}(X) \cup \overline{R}(Y), \underline{R}(X \cap Y) = \underline{R}X \cap \underline{R}Y$;

(6) $X \subseteq Y$ implies $\underline{R}X \subseteq \underline{R}Y, \overline{R}X \subset \overline{R}Y$;

(7) $\overline{R}(X \cap Y) \subseteq \overline{R}(X) \cap \overline{R}(Y), \underline{R}(X \cup Y) \supseteq \underline{R}X \cup \underline{R}Y$;

(8) $\underline{R}(-X) = -\overline{R}X, \overline{R}(-X) = -\underline{R}X$;

(9) $\underline{R}\underline{R}X = \overline{R}\underline{R}X = \underline{R}X, \overline{R}\overline{R}X = \underline{R}\overline{R}X = \overline{R}X$;

(10) for $y, z \in [x], \underline{R}X(y) = \underline{R}X(z), \overline{R}X(y) = \overline{R}X(z)$, hence $\underline{R}X, \overline{R}X$ are fuzzy sets in quotient set U/R.

Proof. (1) By Proposition 2.1, $x \in (\overline{R}X)_\alpha \Leftrightarrow \overline{R}X(x) \geq \alpha \Leftrightarrow \vee_{y\in[x]}X(y) \geq \alpha$ $\Leftrightarrow \exists y \in [x], X(y) \geq \alpha \Leftrightarrow \exists y \in [x], y \in X_\alpha \Leftrightarrow \exists y \in [x] \cap X_\alpha \Leftrightarrow [x] \cap X_\alpha \neq \emptyset \Leftrightarrow$ $x \in \overline{R}X_\alpha$. $(\overline{R}X)_\alpha = \overline{R}X_\alpha$ can be proved in a similar way.

(9) By Lemma 3.1, we only need to prove $(\underline{R}\underline{R}X)_\alpha = (\overline{R}\underline{R}X)_\alpha = (\underline{R}X)_\alpha$ for all $\alpha \in [0,1]$. By Proposition 2.1(7) and (1), $(\underline{R}\underline{R}X)_\alpha = \underline{R}\underline{R}X_\alpha = \overline{R}\underline{R}X_\alpha = (\overline{R}\underline{R}X)_\alpha = \underline{R}X_\alpha = (\underline{R}X)_\alpha$.

(10) Since $y, z \in [x]$, we have $[x] = [y] = [z]$, hence $\overline{R}X(y) = \vee_{t\in[y]}X(t) = \vee_{t\in[z]}X(t) = \overline{R}X(z)$ and $\underline{R}X(y) = \wedge_{t\in[y]}X(t) = \wedge_{t\in[z]}X(t) = \underline{R}X(z)$.

The proofs of remaining parts can be found in [4] or [10,11].

4   Lattice from Rough Fuzzy Sets

Let U be a universal set and let R be an equivalence relation on U, $[x]$ will stand for the equivalence class of the relation R determined by $x \in U$. For each equivalence class $[x]$ we choice an element in $[x]$ called a representative element, the set of the all representative elements is denoted by I.

The rough equality between sets is defined in the following way [11,12]: $X \approx Y$ if and only if $\underline{R}X = \underline{R}Y$ and $\overline{R}X = \overline{R}Y$ for any $X, Y \in P(U)$.

Obviously $\approx$ is an equivalence relation on $P(U)$ [12] and the induced quotient set will be denoted by $P(U)/\approx$. Any equivalence class $[X]$ of relation $\approx$ can be written as follows:

$$[X] = \{Y \in P(U) | \underline{R}X = \underline{R}Y, \overline{R}X = \overline{R}Y\} \equiv (\underline{R}X, \overline{R}X) = RX$$

Rough set intersection $\sqcap$, union $\sqcup$, and complement $-$ are defined by set operations as follows: for two rough sets $RX = (\underline{R}X, \overline{R}X), RY = (\underline{R}Y, \overline{R}Y)$,

(1) $(\underline{R}X, \overline{R}X) \sqcap (\underline{R}Y, \overline{R}Y) = (\underline{R}X \cap \underline{R}Y, \overline{R}X \cap \overline{R}Y)$;

(2) $(\underline{R}X, \overline{R}X) \sqcup (\underline{R}Y, \overline{R}Y) = (\underline{R}X \cup \underline{R}Y, \overline{R}X \cup \overline{R}Y)$;

(3) $-(\underline{R}X, \overline{R}X) = (-\overline{R}X, -\underline{R}X)$.

The system $(P(U)/\approx, \sqcap, \sqcup, -, R\emptyset, RU)$ is a complete distributive pseudocomplemented lattice [12]. We will show that this result can be extended to the case of rough fuzzy sets. The rough equality between fuzzy sets can also be defined as follows:

$X \approx Y$ if and only if $\underline{R}X = \underline{R}Y$ and $\overline{R}X = \overline{R}Y$ for any $X, Y \in F(U)$.

It is easy to verify that rough equality $\approx$ of fuzzy sets is also an equivalence relation on $F(U)$, and the induced quotient set will be denoted by $F(U)/\approx$. For $X \in F(U)$, any equivalence class $[X]$ of relation $\approx$ on $F(U)$ can be written as follows:

$$[X] = \{Y \in F(U) | \underline{R}X = \underline{R}Y, \overline{R}X = \overline{R}Y\} \equiv (\underline{R}X, \overline{R}X) = RX.$$

Now, our main goal in this section is to show that the $F(U)/\approx$, the family of all rough fuzzy sets, is also a distributive lattice. We need a lemma.

Lemma 4.1. Let U be a universal set and R be an equivalence relation on U. I is the set of the all representative elements in equivalence classes of R. Then for all $X, Y \in P(U)$

(1) $\underline{R}(\underline{R}X \cup Y) = \underline{R}X \cup \underline{R}Y$;
(2) $\overline{R}(\overline{R}X \cap Y) = \overline{R}X \cap \overline{R}Y$;
(3) $\overline{R}(\overline{R}X \cap I) = \overline{R}X$;
(4) $\underline{R}((\overline{R}X \cup \overline{R}Y) \cap I) \subseteq \underline{R}X \cup \underline{R}Y$;
(5) $\underline{R}(\underline{R}X \cup \underline{R}Y \cup ((\overline{R}X \cup \overline{R}Y) \cap I)) = \underline{R}X \cup \underline{R}Y$;
(6) $\overline{R}(\underline{R}X \cup \underline{R}Y \cup ((\overline{R}X \cup \overline{R}Y) \cap I)) = \overline{R}X \cup \overline{R}Y$;
(7) $\underline{R}((\underline{R}X \cap \underline{R}Y) \cup (\overline{R}X \cap \overline{R}Y \cap I)) = \underline{R}X \cap \underline{R}Y$;
(8) $\overline{R}((\underline{R}X \cap \underline{R}Y) \cup (\overline{R}X \cap \overline{R}Y \cap I)) = \overline{R}X \cap \overline{R}Y$.

Proof. (1) If $x \in \underline{R}(\underline{R}X \cup Y)$, then $[x] \subseteq \underline{R}X \cup Y$, hence $[x] \subseteq \underline{R}X$ or $[x] \subseteq Y$, that is, $x \in \underline{R}X$ or $x \in \underline{R}Y$ which is to mean that $\underline{R}(\underline{R}X \cup Y) \subseteq \underline{R}X \cup \underline{R}Y$.

By Theorem 3.1(6),$\underline{R}(\underline{R}X \cup Y) \supseteq \underline{R}X \cup \underline{R}Y$. Thus $\underline{R}(\underline{R}X \cup Y) = \underline{R}X \cup \underline{R}Y$.

(3) If $x \in \overline{R}(\overline{R}X \cap I)$, then $[x] \cap \overline{R}X \cap I \neq \emptyset$, hence $[x] \cap \overline{R}X \neq \emptyset$, thus $x \in \overline{R}X$, and $\overline{R}(\overline{R}X \cap I) \subseteq \overline{R}X$. Conversely, if $x \in \overline{R}X$, then $[x] \subseteq \overline{R}X \neq \emptyset$, but $\emptyset \neq [x] \cap I = [x] \cap \overline{R}X \cap I$, and $x \in \overline{R}(\overline{R}X \cap I)$, i.e., $\overline{R}(\overline{R}X \cap I) \supseteq \overline{R}X$.

(5) By (4), $\underline{R}(\underline{R}X \cup \underline{R}Y \cup ((\overline{R}X \cup \overline{R}Y) \cap I)) = \underline{R}X \cup \underline{R}Y \cup \underline{R}((\overline{R}X \cup \overline{R}Y) \cap I) = \underline{R}X \cup \underline{R}Y$.

(6) Using (3), $\overline{R}(\underline{R}X \cup \underline{R}Y \cup ((\overline{R}X \cup \overline{R}Y) \cap I)) = \underline{R}X \cup \underline{R}Y \cup \overline{R}((\overline{R}X \cup \overline{R}Y) \cap I) = \overline{R}X \cup \overline{R}Y$. The proofs of remaining parts are similar.

Lemma 4.1 can be extended to the case of rough fuzzy set. That is, we can obtain the following lemma.

Lemma 4.2. Let U be a universal set and R be an equivalence relation on U. I is the set of the all representative elements in equivalence classes of R. Then for all $X, Y \in F(U)$

(1) $\underline{R}(\underline{R}X \cup Y) = \underline{R}X \cup \underline{R}Y$;
(2) $\overline{R}(\overline{R}X \cap Y) = \overline{R}X \cap \overline{R}Y$;
(3) $\overline{R}(\overline{R}X \cap I) = \overline{R}X$;
(4) $\underline{R}((\overline{R}X \cup \overline{R}Y) \cap I) \subseteq \underline{R}X \cup \underline{R}Y$;
(5) $\underline{R}(\underline{R}X \cup \underline{R}Y \cup ((\overline{R}X \cup \overline{R}Y) \cap I)) = \underline{R}X \cup \underline{R}Y$;
(6) $\overline{R}(\underline{R}X \cup \underline{R}Y \cup ((\overline{R}X \cup \overline{R}Y) \cap I)) = \overline{R}X \cup \overline{R}Y$;

(7) $\underline{R}((\underline{R}X \cap \underline{R}Y) \cup (\overline{R}X \cap \overline{R}Y \cap I)) = \underline{R}X \cap \underline{R}Y$;
(8) $\overline{R}(\underline{R}X \cap \underline{R}Y \cup ((\overline{R}X \cap \overline{R}Y) \cap I)) = \overline{R}X \cap \overline{R}Y$.

Proof. We only prove Property (3) here. Continue using Lemma 3.1 and Lemma 4.1, for each $\lambda \in [0,1]$, $(\overline{R}(\overline{R}X \cap I))_\lambda = \overline{R}(\overline{R}X \cap I)_\lambda = \overline{R}(\overline{R}X_\lambda \cap I) = \overline{R}X_\lambda$. This proves Property (3). The proofs of remaining Properties are similar.

In order to prove that the $F(U)/\approx$ is a distributive lattice, we introduce the intersection $\sqcap$, union $\sqcup$, and complement - operations as follows: for two rough fuzzy sets $RX = (\underline{R}X, \overline{R}X), RY - (\underline{R}Y, \overline{R}Y), X, Y \in F(U)$,

(1) $(\underline{R}X, \overline{R}X) \sqcap (\underline{R}Y, \overline{R}Y) = (\underline{R}X \cap \underline{R}Y, \overline{R}X \cap \overline{R}Y)$;
(2) $(\underline{R}X, \overline{R}X) \sqcup (\underline{R}Y, \overline{R}Y) = (\underline{R}X \cup \underline{R}Y, \overline{R}X \cup \overline{R}Y)$;
(3) $-(\underline{R}X, \overline{R}X) = (-\overline{R}X, -\underline{R}X) = (\underline{R}(-X), \overline{R}(-X))$.

Theorem 4.1. Let U be a universal set and R be an equivalence relation on U. I is the set of the all representative elements in equivalence classes of R. Then for all $X, Y \in F(U)$

(1) $\underline{R}X \cup \underline{R}Y \cup ((\overline{R}X \cup \overline{R}Y) \cap I) \in (\underline{R}X \cup \underline{R}Y, \overline{R}X \cup \overline{R}Y)$;
(2) $(\underline{R}X \cap \underline{R}Y) \cup (\overline{R}X \cap \overline{R}Y \cap I) \in (\underline{R}X \cap \underline{R}Y, \overline{R}X \cap \overline{R}Y)$.

Proof. By Lemma 4.2(1),(4), $\underline{R}(\underline{R}X \cup \underline{R}Y \cup ((\overline{R}X \cup \overline{R}Y) \cap I)) = \underline{R}X \cup \underline{R}Y \cup \underline{R}((\overline{R}X \cup \overline{R}Y) \cap I)) = \underline{R}X \cup \underline{R}Y$. Similarly, $\overline{R}(\underline{R}X \cup \underline{R}Y \cup ((\overline{R}X \cup \overline{R}Y) \cap I)) = \overline{R}X \cup \overline{R}Y$, this prove Part (1). The proof of Part (2) is analogous to that of Part (1) and we omit it.

By Theorem 4.1, since the conclusions are independence on the choice of representative element of the equivalence classes $[x]$, the intersection, union, and complement operations are well-defined.

Theorem 4.2. Let U be a universal set and R be an equivalence relation on U. Then the $(F(U)/\approx, \sqcap, \sqcup, -, R\emptyset, RU)$ is a distributive pseudocomplemented lattice with the least element $R\emptyset$ and the greatest element RU.

Proof. For all $X, Y \in F(U), (\underline{R}X \cup \underline{R}Y, \overline{R}X \cup \overline{R}Y) \neq \emptyset$, and $(\underline{R}X \cap \underline{R}Y, \overline{R}X \cap \overline{R}Y) \neq \emptyset$.

It is obvious that $\sqcap, \sqcup$ are commutative, associative, and idempotent operations. We show one of distributive identities, e.g. $RX \sqcap (RY \sqcup RZ) = (RX \sqcap RY) \sqcup (RX \sqcap RZ)$, for all $X, Y, Z \in F(U)$. Indeed

$$RX \sqcap (RY \sqcup RZ) = (\underline{R}X, \overline{R}X) \sqcap ((\underline{R}Y, \overline{R}Y) \sqcup (\underline{R}Z, \overline{R}Z))$$
$$= (\underline{R}X, \overline{R}X) \sqcap (\underline{R}Y \cup \underline{R}Z, \overline{R}Y \cup \overline{R}Z)$$
$$= (\underline{R}X \cap (\underline{R}Y \cup \underline{R}Z), \overline{R}X \cap (\overline{R}Y \cup \overline{R}Z))$$
$$= ((\underline{R}X \cap \underline{R}Y) \cup (\underline{R}X \cap \underline{R}Z), (\overline{R}X \cap \overline{R}Y) \cup (\overline{R}X \cap \overline{R}Z))$$
$$= (\underline{R}X \cap \underline{R}Y, \overline{R}X \cap \overline{R}Y) \sqcup (\underline{R}X \cap \underline{R}Z, \overline{R}X \cap \overline{R}Z)$$
$$= (RX \sqcap RY) \sqcup (RX \sqcap RZ).$$

As for the pseudocomplement of $RX = (\underline{R}X, \overline{R}X)$, using Lemma 7 of [7], there exists a pseudocomplement $((RX)_\alpha)^* = (\underline{R}X_\alpha^*, \overline{R}X_\alpha^*)$ of $(RX)_\alpha = RX_\alpha$ for all

$\alpha \in [0,1]$. Thus $Y = \cup_{\alpha \in [0,1]} \alpha((RX)_\alpha)^* = (\cup_{\alpha \in [0,1]} \alpha(\underline{R}X_\alpha)^*, \cup_{\alpha \in [0,1]} \alpha(\overline{R}X_\alpha^*))$ is the pseudocomplement of RX.

The following property is straightforward:

Lemma 4.3. Let U be a universal set and R be an equivalence relation on U. for all $X, Y \in F(U), \alpha \in [0,1]$

(1) $(RX \sqcap RY)_\alpha = RX_\alpha \sqcap RY_\alpha$;
(2) $(RX \sqcup RY)_\alpha = RX_\alpha \sqcup RY_\alpha$.

Theorem 4.3. Let U be a universal set and R be an equivalence relation on U. Then the $(F(U)/\approx, \sqcap, \sqcup, -, R\emptyset, RU)$ is a Stone algebra.

Proof. Using Lemma 8 of [12] and Lemma 4.3, the Stone identity is valid, hence $F(U)/\approx$ is a Stone algebra.

Because of theorem 4.1, the intersection, union, and complement operations can be also rewritten as follows:

(1) $RX \sqcap RY = R(\underline{R}X \cup \underline{R}Y \cup ((\overline{R}X \cup \overline{R}Y) \cap I))$,
(2) $RX \sqcup RY = R(\underline{R}X \cap \underline{R}Y \cup ((\overline{R}X \cap \overline{R}Y) \cap I))$, and
(3) $-RX = R(-X)$.

From the Theorem 4.2, it is obvious that lattice $(P(U)/\approx, \sqcap, \sqcup, -, R\emptyset, RU)$ is a sublattice of the distributive lattice $(F(U)/\approx, \sqcap, \sqcup, -, R\emptyset, RU)$.

5 Characterization of Borderline Region

Banerjee and Pal [1] provided a measure of roughness of a fuzzy set. For $X \in U$, the borderline region $BN_R(X) = \overline{R}X - \underline{R}X$ of X is [11], in a sense, the undecidable area of the universal set U, i.e., none of the objects belonging to the boundary can be classified with certainty into X or $-X$ as far as knowledge R is concerned. In this section, we consider the borderline region in the complete distributive lattice $(P(U)/\approx, \sqcap, \sqcup, -, R\emptyset, RU)$. For class $[X] \in P(U)/\approx$, the borderline region $BN_R[X]$ of class $[X]$ is defined as follows:

$$BN_R[X] = \overline{R}X - \underline{R}X$$

It is obvious that this definition is independence on the choice of representatives of the equivalence class $[X]$. Now we prove a relation between the borderline region of classes $[X], [Y], [X] \sqcap [Y]$, and $[X] \sqcup [Y]$.

Theorem 5.1. Let R be an equivalence relation on finite universal set U, $(P(U)/\approx, \sqcap, \sqcup, -, R\emptyset, RU)$ the corresponding complete distributive lattice. Then for all $[X], [Y] \in P(U)/\approx$

$$|BN_R([X] \sqcup [Y])| = |BN_R[X]| + |BN_R[Y]| - |BN_R([X] \sqcap [Y])|$$

Where $|X|$ denote the cardinality of a set X.

Proof. $|BN_R([X] \sqcup [Y])| = |\overline{R}X \cup \overline{R}Y| - |\underline{R}X \cup \underline{R}Y|$

$= (|\overline{R}X| - |\underline{R}X|) + (|\overline{R}Y| - |\underline{R}Y|) - (|\overline{R}X \cap \overline{R}Y| - |\underline{R}X \cap \underline{R}Y|)$

$= |BN_R[X]| + |BN_R[Y]| - |BN_R([X] \sqcap [Y])|.$

Analogous to the crisp situation, we extend the Theorem 5.1 to the case of fuzzy sets. We introduce the borderline region in the distributive lattice $(F(U)/ \approx , \sqcap, \sqcup, -, R\emptyset, RU)$.

Definition 5.1. The borderline region of class $[X] \in F(U)/ \approx$, denoted by $BN_R^{\alpha,\beta}[X]$, with respect to parameters α, β, where $0 < \alpha \le \beta \le 1$, is defined as follows:

$$BN_R^{\alpha,\beta}[X] = \overline{R}X_\beta - \underline{R}X_\alpha$$

By Theorem 3.1, we have

$$BN_R^{\alpha,\beta}[X] = (\overline{R}X)_\beta - (\underline{R}X)_\alpha$$

If $X \approx Y$, then $BN_R^{\alpha,\beta}[X] = BN_R^{\alpha,\beta}[Y]$. Similar to the crisp situation, we prove a relation between the borderline regions of fuzzy set classes $[X], [Y], [X] \sqcap [Y]$, and $[X] \sqcup [Y]$.

Theorem 5.2. Let R be an equivalence relation on finite universal set U, $(F(U)/ \approx, \sqcap, \sqcup, -, R\emptyset, RU)$ the corresponding distributive lattice. Then for all $[X], [Y] \in F(U)/ \approx$ and parameters $\alpha, \beta, 0 < \alpha \le \beta \le 1$

$$|BN_R^{\alpha,\beta}([X] \sqcup [Y])| = |BN_R^{\alpha,\beta}[X]| + |BN_R^{\alpha,\beta}[Y]| - |BN_R^{\alpha,\beta}([X] \sqcap [Y])|$$

Proof. By Definition 5.1, we have

$$|BN_R^{\alpha,\beta}([X] \sqcup [Y])| = |\overline{R}X_\beta \cup \overline{R}Y_\beta| - |\underline{R}X_\alpha \cup \underline{R}Y_\alpha|$$

$$= (|\overline{R}X_\beta| - |\underline{R}X_\alpha|) + (|\overline{R}Y_\beta| - |\underline{R}Y_\alpha|) - (|\overline{R}X_\beta \cap \overline{R}Y_\beta| - |\underline{R}X_\alpha \cap \underline{R}Y_\alpha|)$$

$$= |BN_R^{\alpha,\beta}[X]| + |BN_R^{\alpha,\beta}[Y]| - |BN_R^{\alpha,\beta}([X] \sqcap [Y])|.$$

This completes the proof.

6   Conclusions

Rough fuzzy sets is the generalization of Pawlak rough sets, we study lattice properties of rough fuzzy sets. J.Pomykala and J.A.Pomykala [12] proved that the family of rough sets is a Stone algebra. We extend this result to the case of rough fuzzy sets. That is, analogous to standard rough set theory, we proved that the family of rough fuzzy sets with the union, intersection and pseudocomplementation operations is also a Stone algebra. We also give the characterization of borderline region of rough fuzzy sets.

Acknowledgements

This work is partially supported by the National Natural Science Foundation of China (No. 60973148) and the Key Project of the Chinese Ministry of Education (No. 108133).

References

1. Banerjee, M., Pal, S.K.: roughness of a fuzzy set. Information Sciences 93, 235–246 (1996)
2. Bonikowski, Z., Bryniarski, E., Mybraniec, S.U.: Extensions and intentions in the rough set theory. Information Sciences 107, 149–167 (1998)
3. Dai, J.H.: Rough 3-valued algebras. Information Sciences 178, 1986–1996 (2008)
4. Dubois, D., Prade, H.: Rough fuzzy sets and fuzzy rough sets. International Journal of General System 17, 191–208 (1990)
5. Huynh, V.N., Nakamori, Y.: A roughness measure for fuzzy sets. Information Sciences 173, 255–275 (2005)
6. Liu, G.L.: The transitive closures of matrices over distributive lattices. In: The Proceedings of the 2006 IEEE International Conference on Granular Computing, pp. 63–66 (2006)
7. Liu, G.L.: Axiomatic Systems for Rough Sets and Fuzzy Rough Sets. International Journal of Approximate Reasoning 48, 857–867 (2008)
8. Liu, G.L.: Generalized rough sets over fuzzy lattices. Information Sciences 178, 1651–1662 (2008)
9. Liu, G.L., Zhu, W.: The algebraic structures of generalized rough set theory. Information Sciences 178(21), 4105–4113 (2008)
10. Pawlak, Z.: Rough sets. International Journal of Computer and Information Sciences 11, 341–356 (1982)
11. Pawlak, Z.: Rough sets-theoretical aspects of reasoning about data. Kluwer Academic Publishers, Boston (1991)
12. Pomykala, J., Pomykala, J.A.: The stone algebra of rough sets. Bulletin of the Polish Academy of Sciences Mathematics 36, 495–508 (1988)
13. Yao, Y.Y., Lin, T.Y.: Generalization of rough sets using modal logic. Intelligent Automation and Soft Computing, an International Journal 2, 103–120 (1996)
14. Yao, Y.Y.: A comparative study of fuzzy sets and rough sets. Information Sciences 109, 227–242 (1998)
15. Yao, Y.Y.: Constructive and algebraic methods of the theory of rough sets. Information Sciences 109, 21–47 (1998)
16. Zadeh, L.A.: Fuzzy sets. Information and Control 8, 8338–8353 (1965)

Analysis of Ergodicity of a Fuzzy Possibilistic Markov Model

B. Praba, R. Sujatha, and V. Hilda Christy Gnanam

Department of Mathematics, SSN College of Engineering, Kalavakkam,
Chennai - 603 110, India
brprasuja@yahoo.co.in,
hildagnanam@gmail.com

Abstract. Performing ergodicity analysis is essential to study the long realization of a model. In this paper we analyze the ergodicity, i.e.,the existence of the limiting fuzzy transition possibility matrix with identical rows for the fuzzy transition possibility matrix $\tilde{H}$ of a fuzzy possibilistic Markov model which contains a state j such that the transition from every state to the state j is a sure event.

Keywords: Triangular fuzzy number, fuzzy possibilistic Markov model, max-min composition, fuzzy graph, equivalence class, ergodicity.

1 Introduction

Fuzzy Markov chains are the frequently used Mathematical models in fuzzy reliability theory. There are considerable amount of works have been done on fuzzy Markov Models [1],[2], [3], [5]. In [1], Avrachenkov and Sanchez defined ergodicity of a fuzzy Markov model and left out the problem of finding the more general conditions ensuring the ergodicity of fuzzy Markov model as an open problem. In [6], we have defined the fuzzy possibilistic Markov model whose fuzzy transition possibilities are triangular fuzzy numbers on [0,1] and the rows of its fuzzy transition possibility matrix are possibility distributions. We have studied the properties of the fuzzy possibilistic Markov model with n states. And we have given the necessary conditions for the ergodicity (existence of the limiting fuzzy transition possibility matrix with identical rows) of a fuzzy possibilistic Markov model with three states among which there is a state j such that the fuzzy transition possibility of reaching the j from every state i is (1,1,1). In this paper, we are extending the above mentioned work for n states.

Consider a system with various components in which the control transfers between the components follow Markovian property. If its behavior, the control transfer is characterized using fuzzy possibility measures, then this system can be modeled as FPMM. Suppose the system consists a component such that the event of reaching that component from every component is an sure event, then the long term behavior of the system can be obtained through the ergodicity analysis as given in the preceding section.

H. Sakai et al. (Eds.): RSFDGrC 2009, LNAI 5908, pp. 261–268, 2009.

The formation of this paper is as follows. In the next section, we see some already defined definitions which are used in this work. Section three contains a few theorems and corollaries about the existence of the limiting fuzzy transition possibility matrix with identical rows. And we end up with some conclusions.

2 Preliminaries

Definition 2.1 (Fuzzy Possibilistic Markov model - FPMM [6]). Consider a fuzzy possibilistic stochastic process $\{X(t), t = 0, 1, \ldots\}$ with discrete state space $\mathfrak{S} = \{1,2,3, \ldots\}$. If it possesses the Markov property (the given present and future states are independent of the past states),
$$\tilde{\pi}(X(t+1) = j \mid X(t) = i, X(t-1) = k, \ldots, X(0) = m)$$
$$= \tilde{\pi}(X(t+1) = j \mid X(t) = i) = \tilde{\pi}_{ij}$$
where $\tilde{\pi}_{ij}$ is the triangular fuzzy number on $[0,1]$ and it represents the fuzzy possibility of control transfers from state 'i' at t^{th} step to state 'j' at $(t+1)^{th}$ step, then this fuzzy possibilistic stochastic process is called the fuzzy possibilistic Markov Model (FPMM).

These fuzzy transition possibilities form a square fuzzy matrix which is called the fuzzy transition possibility matrix. And its rows are fuzzy possibility distributions. i.e each row maximum is $(1,1,1)$.

Definition 2.2 (Limiting fuzzy transition possibility matrix). We can say that the powers of the fuzzy transition possibility matrix $\tilde{H}_{n \times n}$ will converge if there exists a positive integer t so that $\tilde{H}^m = \tilde{H}^{m+1}$ for $m \geq t$. Then $\tilde{H}^m$ is called the limiting fuzzy transition possibility matrix of $\tilde{H}$ as follows.

Definition 2.3 (Fuzzy Graph by Fuzzy Matrix). A fuzzy matrix $\tilde{A} = (a_{ij})$ can be regarded as a fuzzy graph $\tilde{G}_{\tilde{A}}$, where the edge (i, j), $i, j \in \{1,2,\ldots,n\}$, has a weight $a_{ij} \in [0,1]$

Hence our fuzzy matrix $\tilde{H} = (\tilde{h}_{ij})$ can be viewed as a fuzzy graph $\tilde{G}_{\tilde{H}}$ whose edges (i,j), i, j belongs to the state space of the corresponding FPMM, have the weight $\tilde{h}_{ij}$ which are the triangular fuzzy numbers on $[0,1]$. And the definitions, results given for $\tilde{G}_{\tilde{A}}$ [4], can also be true for $\tilde{G}_{\tilde{H}}$.

Definition 2.4 (Cycle, Length and Strength). A cycle C in $\tilde{G}_{\tilde{H}}$ is a sequence of distinct vertices $i_0, i_1, \ldots, i_{q-1}, i_0$ where $i_0, i_1, \ldots, i_{q-1}$ belong to the state space. The length of the cycle $|C|$ is q. The strength of C is $\mathrm{M}(C) = \min\{\tilde{h}_{i_0 i_1}, \tilde{h}_{i_1 i_2}, \ldots, \tilde{h}_{i_{q-1} i_0}\}$. And a loop is a cycle with length 1.

Definition 2.5 (Mutually Linked Cycles). Let C_1 and C_2 be two cycles in $\tilde{G}_{\tilde{H}}$. If $C_1 \cap C_2 \neq \emptyset$, the the cycles C_1 and C_2 are mutually linked.

Definition 2.6 (Strongest Cycle). A strongest cycle is a cycle C_1, if there exist no sequence of cycles $C_1, \ldots, C_r$, $r \geq 2$, satisfying
   1. C_i and C_{i+1} are mutually independent
   2. $\mathrm{M}(C_1) = \mathrm{M}(C_2) = \ldots = \mathrm{M}(C_{r-1}) < \mathrm{M}(C_r)$

Definition 2.7 (Equivalent Class). Let Ω be the set of all strongest cycles in a fuzzy graph. The equivalence relation $(\Omega \sim)$ is defined as follows:

$C \sim C' \Leftrightarrow$ there exist a sequence $C_1(=C)$, ..., $C_r(=C')$, such that C_i and C_{i+1} are mutually linked. And $[C]$ denotes the equivalent class by the equivalence relation $(\Omega \sim)$. $\sigma_{[C]}$ denotes the greatest common divisor of the length of all strongest cycles belonging to $[C]$.

Theorem 1. Let $\tilde{H} = (\tilde{h}_{ij})$ be a fuzzy matrix and $[C_1]$, $[C_1]$, ..., $[C_m]$ be all equivalent classes where C_i is a strongest cycle of the fuzzy graph $\tilde{G}_{\tilde{H}}$. Thus the period of a fuzzy matrix is the least common multiple of $\sigma_{[C_1]}$, $\sigma_{[C_2]}$, ..., $\sigma_{[C_m]}$.

Corollary 1. A fuzzy matrix $\tilde{H}$ converges if and only if $\sigma_{[C]} = 1$ for all strongest cycles $\tilde{G}_{\tilde{H}}$.

3 Ergodicity of a FPMM

In classical Markov model, the tth step transition probabilities $p_{ij}(t)$ of finite, irreducible, aperiodic Markov chains become independent of i as $t \to \infty$. That is all the entries in the jth column of the tth step transition probability matrix become equal as $t \to \infty$. Hence, the limiting transition probability matrix will have identical rows. For FPMM, the limiting fuzzy transition possibility matrix $\tilde{H}$ need not have identical rows, even though it is recurrent. Hence using the ergodicity definition given in [1], we define the same for FPMM as follows. *A FPMM is said to be ergodic if it is aperiodic and has limiting fuzzy transition possibility matrix with identical rows.*

Let us discuss the existence of the limiting fuzzy transition possibility matrix with identical rows for a FPMM with n states among which there is a state j such that the events of reaching the state j from every other states $1, 2, ...,n$ are sure events. That is its fuzzy transition possibility matrix $\tilde{H}$ of order n contains a column j such that its entries are equal to $(1, 1, 1)$.

Let us consider the column maximum of each column of $\tilde{H}^t$ by excluding the jth row and the jth column of $\tilde{H}^t$, $t = 1,2,...,n$. And $\tilde{H}^1 = \tilde{H}$. Suppose $\tilde{h}_{jk}$ is greater than or equal to k^{th} column maximum in $\tilde{H}$. Since each rows j^{th} entry is $(1,1,1)$, while doing the max min composition between each row of $\tilde{H}$ and k^{th} column of $\tilde{H}^t$, $t = 1,2, ..., n$, the resultant k^{th} column entries are equal to $\tilde{h}_{jk}$. i.e ., for $z = 1,2, ..., n$ and $t = 1, 2, ... ,n$,

$$\tilde{h}_{zk}^{t+1} = \max\{\min(\tilde{h}_{z1}, \tilde{h}_{1k}^t), \min(\tilde{h}_{z2}, \tilde{h}_{2k}^t), ..., \min(\tilde{h}_{zj}, \tilde{h}_{jk}^t), ..., \min(\tilde{h}_{zn}, \tilde{h}_{nk}^t)\}.$$

$$= \min(\tilde{h}_{zj}, \tilde{h}_{jk}^t) \ [\because \tilde{h}_{jk} \geq \text{the } k^{th} \text{ column maximum in } \tilde{H} \Rightarrow \tilde{h}_{jk}^t = \tilde{h}_{jk}].$$

$$= \tilde{h}_{jk}^t = \tilde{h}_{jk} \ [\because \tilde{h}_{zj} = (1,1,1)].$$

Hence to get the entries of a column k in $\tilde{H}^t$ equal, k^{th} entry of jth row$(\tilde{h}_{jk})$ should be greater than or equal to the kth column maximum in $\tilde{H}$.

Thus, the existence of the identical rows in the limiting fuzzy transition possibility matrix of $\tilde{H}$ depends on the entries of the jth row of $\tilde{H}$ which is discussed in the preceding lemma, theorem and corollaries. Let $a_0 = \min$ {column maximums of $\tilde{H}$} and consider the column maximum of each column of $\tilde{H}^t$ by excluding the jth row and the jth column of $\tilde{H}^t$, $t = 1,2,\ldots,n$.

Lemma 1. *Let $\tilde{G}_{\tilde{H}}$ be the fuzzy graph defined by $\tilde{H}$ and $\tilde{H}$ be contain a column j with all its entries equal to (1,1,1). Suppose its jth row entries are less than or equal to a_0 except $\tilde{h}_{jj}$ and $\tilde{G}_{\tilde{H}}$ contains a strongest cycle C with $M(C) \geq a_0$. Then in $\tilde{H}^t$, $t \geq |C|$, the column maximum of each column $a \in C$ will be equal to $M(C)$.*

Proof. Let $C = (e,p,l,k,e)$ be the strongest cycle $\tilde{G}_{\tilde{H}}$. Then
$$\tilde{h}^2_{kp} = \max \{\min(\tilde{h}_{k1}, \tilde{h}_{1p}), \min(\tilde{h}_{k2}, \tilde{h}_{2p}), \ldots, \min(\tilde{h}_{ke}, \tilde{h}_{ep}), \ldots,$$
$$\min(\tilde{h}_{kn}, \tilde{h}_{np})\}.$$
If there is a min term $\min(\tilde{h}_{kt}, \tilde{h}_{tp})$ such that $\min(\tilde{h}_{kt}, \tilde{h}_{tp}) > \min(\tilde{h}_{ke}, \tilde{h}_{ep})$, then $\tilde{G}_{\tilde{H}}$ will have a cycle $C_2 = (k,t,p,l,k)$ and $M(C_2) \geq M(C)$. If $M(C_2) = M(C)$, then C_2 is an element of the equivalent class $[C]$. If $M(C_2) > M(C)$, then there exist a sequence C, C_2 where C, C_2 are mutually linked and $M(C) < M(C_2)$. And C_2 is the strongest cycle. But we have considered that $\tilde{G}_{\tilde{H}}$ contains the strongest cycle C. Hence, there is no other min terms which are greater than $\min(\tilde{h}_{ke}, \tilde{h}_{ep})$ and
$$\tilde{h}^2_{kp} = \min(\tilde{h}_{ke}, \tilde{h}_{ep}) \geq M(C) [\because M(C) = \min\{\tilde{h}_{ep}, \tilde{h}_{pl}, \tilde{h}_{lk}, \tilde{h}_{ke}\}]. \tag{1}$$
Similarly
$$\tilde{h}^3_{lp} = \min(\tilde{h}_{lk}, \tilde{h}^2_{kp}) = \min(\tilde{h}_{lk}, \tilde{h}_{ke}, \tilde{h}_{ep}) \geq M(C), \tag{2}$$
$$\tilde{h}^4_{pp} = \min(\tilde{h}_{lk}, \tilde{h}_{ke}, \tilde{h}_{ep}, \tilde{h}_{pl}) = M(C). \tag{3}$$
And in higher powers,
$$\left.\begin{aligned}
\tilde{h}^4_{pp} = \tilde{h}^4_{ee} = \tilde{h}^4_{kk} = \tilde{h}^4_{ll} = \tilde{h}^8_{pp} = \tilde{h}^8_{ee} = \tilde{h}^8_{kk} = \tilde{h}^8_{ll} = \ldots = M(C) \\
\tilde{h}^5_{ep} = \tilde{h}^5_{ke} = \tilde{h}^5_{lk} = \tilde{h}^5_{pl} = \tilde{h}^9_{ep} = \tilde{h}^9_{ke} = \tilde{h}^9_{lk} = \tilde{h}^9_{pl} = \ldots = M(C) \\
\tilde{h}^6_{kp} = \tilde{h}^6_{le} = \tilde{h}^6_{pk} = \tilde{h}^6_{el} = \tilde{h}^{10}_{kp} = \tilde{h}^{10}_{le} = \tilde{h}^{10}_{pk} = \tilde{h}^{10}_{el} = \ldots = M(C) \\
\tilde{h}^7_{lp} = \tilde{h}^7_{pe} = \tilde{h}^7_{ek} = \tilde{h}^7_{kl} = \tilde{h}^{11}_{lp} = \tilde{h}^{11}_{pe} = \tilde{h}^{11}_{ek} = \tilde{h}^{11}_{kl} = \ldots = M(C)
\end{aligned}\right\}. \tag{4}$$
Now let us see about the entries of the column p other than its entries kp, lp, ep, pp in higher powers of $\tilde{H}$.

Let $f \in \mathfrak{S} - C$ and the entries of the columns k, e, p, l in $\tilde{H}$ other than lk, ke, ep, pl be less than $M(C)$. For $t \geq 2$,

$$\tilde{h}^t_{fp} = \max\{\min(\tilde{h}_{f1}, \tilde{h}^{t-1}_{1p}), \min(\tilde{h}_{f2}, \tilde{h}^{t-1}_{2p}), \ldots, \min(\tilde{h}_{fb}, \tilde{h}^{t-1}_{bp}), \ldots,$$

$$\min(\tilde{h}_{fn}, \tilde{h}^{t-1}_{np})\}.$$

$\tilde{h}^t_{fp}$ will be greater than $M(C)$, only if there is a min term $\min(\tilde{h}_{fa}, \tilde{h}^{t-1}_{ap})$ such that its value is greater than $M(C)$. From equations (1),(2),(3),(4), it is clear that $\tilde{h}^{t-1}_{ap} \geq M(C)$ only if $a \in C$ and for other a, $\tilde{h}^{t-1}_{ap}$ is less than $M(C)$. $\tilde{h}_{fa}$ cannot be greater than $M(C)$, because the entries of the columns k, e, p, l in $\tilde{H}$ other than lk, ke, ep, pl are less than $M(C)$. Hence there is no min term such that $\min(\tilde{h}_{fa}, \tilde{h}^{t-1}_{ap}) > M(C)$ and
$$\tilde{h}^t_{fp} < M(C), \ t \geq 2. \tag{5}$$

Similarly in $\tilde{H}^t$, $t \geq 2$ the entries of the columns e, k, l other than the entries occurred in (4) will become less than $M(C)$. Then in $\tilde{H}^t$, $t \geq 2$, the column maximum of the columns k, e, p, l will be $M(C)$.

Suppose at least one of the pth column entries say $\tilde{h}_{dp} \geq M(C)$ where $d \in \mathfrak{S} - C$. Then

$$\tilde{h}^2_{dl} = \max\{\min(\tilde{h}_{d1}, \tilde{h}_{1l}), \min(\tilde{h}_{d2}, \tilde{h}_{2l}), \ldots, \min(\tilde{h}_{dp}, \tilde{h}_{pl}), \ldots,$$
$$\min(\tilde{h}_{dn}, \tilde{h}_{nl})\}.$$
$$\tilde{h}^2_{dl} = \min(\tilde{h}_{dp}, \tilde{h}_{pl}) \geq M(C) \; [\because \text{ in } \tilde{H} \text{ the entries of the } l^{th} \text{ column other}$$
$$\text{than } \tilde{h}_{pl} \text{ are less than } M(C) \text{ and } \tilde{h}_{dp} \geq M(C)].$$

Similarly, $\tilde{h}^3_{dk}$ and $\tilde{h}^4_{de}$ are greater than or equal to $M(C)$. The preceding steps imply that the value of $\tilde{h}^t_{da}$ depends only $\min(\tilde{h}_{dp}, \tilde{h}^{t-1}_{pa})$ where $a \in C$. Hence, from equation (4), it is clear that for $a \in C$,

$$\tilde{h}^t_{da} = M(C), t \geq 4. \tag{6}$$

By equations (5, 6), in $\tilde{H}^t$, $t \geq 4$, the entries of the columns $a \in C$ other than the entries occurred in (4) become less than or equal to $M(C)$. Hence in $\tilde{H}^t$, $t \geq |C|$ the column maximum of each column $a \in C$ is $M(C)$. □

Remark 1. *From the above lemma, it is observed that since the columns $f_v \in \mathfrak{S} - C$, (v - the index number), should not form any strongest cycle C_1 with $M(C_1) \geq a_0$, in $\tilde{H}$ the columns corresponding to $f_v \in \{\mathfrak{S} - C\}$ should be either $\tilde{h}_{af_v} \geq a_0$ or both $\tilde{h}_{af_v}$, $\tilde{h}_{f_v f_w}$ are greater than or equal to a_0 where $v \neq w$ and $a \in C$. And in $\tilde{H}^t$, $t \geq |C|$ the column maximum of each column corresponding to the states in $\{\mathfrak{S} - C\}$ will be either $\min(\tilde{h}_{af_v}, M(C))$ or $\min(\tilde{h}_{af_v}, \tilde{h}_{f_v f_w}, M(C))$.*

Example 1. Let $\tilde{H}$ be

$$\begin{pmatrix} (1,1,1) & (0,0.1,0.3) & (0.1,0.2,0.3) & (0.1,0.3,0.5) & (0.2,0.4,0.6) & (0,0.1,0.3) \\ (1,1,1) & (0.1,0.2,0.3) & (0.1,0.3,0.5) & (0.1,0.2,0.3) & (0.1,0.3,0.5) & (0.1,0.2,0.3) \\ (1,1,1) & (0.1,0.3,0.5) & (0,0.1,0.3) & (0.2,0.7,0.8) & (0.1,0.3,0.5) & (0,0,0) \\ (1,1,1) & (0,0,0) & (0.3,0.6,0.7) & (0.1,0.3,0.5) & (0,0.1,0.3) & (0.2,0.7,0.8) \\ (1,1,1) & (0.2,0.4,0.6) & (0.2,0.4,0.6) & (0.3,0.6,0.7) & (0.1,0.2,0.3) & (0.1,0.2,0.3) \\ (1,1,1) & (0.3,0.6,0.7) & (0,0,0) & (0.2,0.4,0.6) & (0.1,0.5,0.6) & (0.2,0.4,0.6) \end{pmatrix}$$

$a_0 = (0.1, 0.5, 0.6)$ and it has the strongest cycle $C = (4, 3, 4)$ with $M(C) = (0.3, 0.6, 0.7)$. In $\tilde{H}^t$, $t \geq 2$, the 3rd, 4th column maximum will be $(0.3, 0.6, 0.7)$. Since the 6th column maximum is $\tilde{h}_{46} = (0.2, 0.7, 0.8)$, $4 \in C$ and $\tilde{h}_{62} = (0.3, 0.6, 0.7)$, $\tilde{h}_{65} = (0.1, 0.5, 0.6)$, in $\tilde{H}^t$, $t \geq 2$, the 6th column maximum will be $\min\{(0.2, 0.7, 0.8), M(C)\}$, the 2nd column maximum will be $\min\{(0.3, 0.6, 0.7), M(C)\}$ and the 5th column maximum will be $\min\{(0.1, 0.5, 0.6), M(C)\}$. It oscillates with period two for $t \geq 6$.

Theorem 2. *Let the fuzzy graph $\tilde{G}_{\tilde{H}}$ defined by $\tilde{H}$ contain a strongest cycle C with $M(C) \geq a_0$ and $F = \{b \mid \tilde{h}_{ba} \geq M(C), a \in C\}$. If at least one of $\{\tilde{h}_{jd} \mid$*

$d \in (C \cup F)$ } $\geq M(C)$, then in higher powers of $\tilde{H}$ the entries of each column $a \in C$ will become equal and they are greater than or equal to $M(C)$. And the limiting fuzzy transition possibility matrix with identical rows will exist.

Proof. Let $C = (e, p, l, k, e)$ be a strongest cycle with $M(C) \geq a_0$ and $F = \{b \mid \tilde{h}_{ba} \geq M(C), a \in C\}$. Suppose at least one of $\{\tilde{h}_{jd} \mid d \in (C \cup F)\}$ is greater than or equal to $M(C)$ say $\tilde{h}_{jl} \geq M(C)$. Then there exist cycle $C_1 = (j, l, j)$ with $M(C_1) \geq M(C)$, because $\tilde{h}_{lj}$ is (1,1,1) and $\tilde{h}_{jl} \geq M(C)$. Since the cycle $C_2 = (j, j)$ has strength $M(C_2) = (1,1,1)$ which is greater than $M(C)$, there exist a sequence C, C_1, C_2 such that each pair $(C, C_1), (C_1, C_2)$ is mutually linked and $M(C) \leq M(C_1) < M(C_2)$. Hence C_2 is the strongest cycle with $\sigma_{[C_2]} = 1$. By corollary (1), $\tilde{H}$ converges. Thus the limiting fuzzy transition possibility matrix of $\tilde{H}$ exists.

Since $\tilde{h}_{jl} \geq M(C)$ and each rows j^{th} entry in $\tilde{H}$ is (1 1 1), there is an integer $m \geq t$, $t = 2$ such that for $z = 1, 2, \ldots, n$

$$\tilde{h}_{zl}^m = \max \{ \min(\tilde{h}_{z1}, \tilde{h}_{1l}^{m-1}), \min(\tilde{h}_{z2}, \tilde{h}_{2l}^{m-1}), \ldots, \min(\tilde{h}_{zj}, \tilde{h}_{jl}^{m-1}), \ldots,$$
$$\min(\tilde{h}_{zn}, \tilde{h}_{nl}^{m-1})\}.$$

Suppose $\min(\tilde{h}_{zg}, \tilde{h}_{gl}^{m-1}) > M(C)$. That is both $\tilde{h}_{zg}, \tilde{h}_{gl}^{m-1}$ should be greater than $M(C)$. By lemma (1), in $\tilde{H}^t$, $t \geq |C|$, the column maximum of each column $a \in C$ will be $M(C)$. Hence $\tilde{h}_{gl}^{m-1}$ cannot be greater than $M(C)$ and there is no min term other than $\min(\tilde{h}_{zj}, \tilde{h}_{jl}^{m-1})$ whose value is greater than $M(C)$. And $\tilde{h}_{zl}^m = \min(\tilde{h}_{zj}, \tilde{h}_{jl}^{m-1}) = \tilde{h}_{jl}^{m-1} = \tilde{h}_{jl}$ [$\because \tilde{h}_{zj} = \tilde{h}_{jj} = (1,1,1)$ and $\tilde{h}_{jl} \geq M(C)$]. i.e., in $\tilde{H}^m$, $m \geq t$, $t = 2$ all the lth column entries become equal to $\tilde{h}_{jl}$. Then,

$$\tilde{h}_{jk}^2 = \max \{\min(\tilde{h}_{j1}, \tilde{h}_{1k}), \min(\tilde{h}_{j2}, \tilde{h}_{2k}), \ldots, \min(\tilde{h}_{jl}, \tilde{h}_{lk}), \ldots,$$
$$\min(\tilde{h}_{jn}, \tilde{h}_{nk})\}.$$
$$\tilde{h}_{jk}^2 \geq M(C) \; [\because M(C) = \min\{\tilde{h}_{ep}, \tilde{h}_{pl}, \tilde{h}_{lk}, \tilde{h}_{ke}\} \Rightarrow \tilde{h}_{lk} \geq M(C) \Rightarrow$$
$$\min(\tilde{h}_{jl}, \tilde{h}_{lk}) \geq M(C)\;].$$

Since $\tilde{h}_{jk}^2 \geq M(C)$ and in $\tilde{H}$ each rows j^{th} entry is (1 1 1), there exist $m \geq t$, $t = 3$ such that $\tilde{h}_{zk}^m = \tilde{h}_{jk}^2$ for $z = 1, 2, \ldots, n$. Also $\tilde{h}_{le}^2, \tilde{h}_{je}^3, \tilde{h}_{jp}^4$ become greater than or equal to $M(C)$. Consequently, $\tilde{h}_{ze}^t = \tilde{h}_{je}^3$ for $t \geq 4$ and $\tilde{h}_{zp}^t = \tilde{h}_{jp}^4$ $t \geq 5$, $z = 1, 2, \ldots, n$. Thus the entries of each column $a \in C$ become equal in $\tilde{H}^t$, $t \geq |C|$.

Let $f_v \in \{\mathfrak{S} - C\}$, $v = 1, 2, \ldots$, the index number. Then the remark (1) implies that in $\tilde{H}$, at least one entry of the columns corresponding to the states of $\{\mathfrak{S} - C\}$ should be in the form either $\tilde{h}_{af_v} \geq a_0$ or both $\tilde{h}_{af_v}, \tilde{h}_{f_v f_w}$ are greater than or equal to a_0 and their column maximum in $\tilde{H}^t$, $t \geq |C|$ will be less than or equal to $M(C)$. If $\tilde{h}_{af_v} \geq a_0$, then for some integer m where $2 \leq m \leq |C|$, in $\tilde{H}^m$, $\tilde{h}_{lf_v}^m$ will be greater than or equal to $\min\{\tilde{h}_{af_v}, M(C)\}$. In $\tilde{H}^{m+1}$, $\tilde{h}_{jf_v}^{m+1}$ will be equal to $\tilde{h}_{lf_v}^m$, since $\tilde{h}_{jl}$ is greater than or equal to $M(C)$ and $\min\{\tilde{h}_{af_v}, M(C)\}$ is less than or equal to $M(C)$. Then in $\tilde{H}^t$, $t \geq |C|$, the entries of the column f_v will be equal to $\tilde{h}_{lf_v}^m$, because the column maximum of f_v is $\min\{\tilde{h}_{af_v}, M(C)\}$, $\tilde{h}_{lf_v}^m \geq \min\{\tilde{h}_{af_v}, M(C)\}$ and in $\tilde{H}$ each rows j^{th} entry is (1,1,1).

If $\tilde{h}_{af_v}$ and $\tilde{h}_{f_v f_w} \geq a_0$, then $\tilde{h}^2_{af_w} = \min\{\tilde{h}_{af_v}, \tilde{h}_{f_v f_w}\}$ which is greater than or equal to a_0. As explained above, there exist an integer m where $3 \leq m \leq |C|$, in $\tilde{H}^m$, $\tilde{h}^m_{lf_w}$ will be greater than or equal to $\min\{\tilde{h}_{af_v}, \tilde{h}_{f_v f_w}, M(C)\}$. And $\tilde{h}^{m+1}_{jf_w}$ will be equal to $\tilde{h}^m_{lf_w}$, since $\tilde{h}_{jl} \geq M(C)$ and $\min\{\tilde{h}_{af_v}, \tilde{h}_{f_v f_w}, M(C)\} \leq M(C)$. Then in $\tilde{H}^t$, $t \geq |C|$, the entries of the column f_w will become equal to $\tilde{h}^m_{lf_w}$, because the column maximum of f_w is $\min\{\tilde{h}_{af_v}, \tilde{h}_{f_v f_w}, M(C)\}$, $\tilde{h}^m_{lf_v}$ is greater than or equal to $\min\{\tilde{h}_{af_v}, M(C)\}$ and in $\tilde{H}$ each rows j^{th} entry is $(1,1,1)$. Hence in $\tilde{H}^t$, $t \geq |C|$, the entries of the columns $f_v \in \{\mathfrak{S} - C\}$ becomes equal and the limiting fuzzy transition possibility matrix exists with identical rows. □

Example 2. Let

$$\tilde{H} = \begin{pmatrix} (0.1,0.2,0.3) & (1,1,1) & (0.3,0.6,0.7) & (0.1,0.2,0.3) & (0.2,0.7,0.8) \\ (0.1,0.3,0.5) & (1,1,1) & (0,0.1,0.3) & (0.1,0.2,0.3) & (0.1,0.2,0.3) \\ (0.1,0.2,0.3) & (1,1,1) & (0.1,0.2,0.3) & (0.1,0.3,0.5) & (0.1,0.3,0.5) \\ (0.2,0.4,0.6) & (1,1,1) & (0.1,0.3,0.5) & (0,0.1,0.3) & (0.1,0.5,0.6) \\ (0,0.1,0.3) & (1,1,1) & (0.2,0.4,0.6) & (0.1,0.5,0.6) & (0.1,0.2,0.3) \end{pmatrix}$$

$a_0 = (0.2, 0.4, 0.6)$ and $\tilde{G}_{\tilde{H}}$ contains strongest cycle $C = (4,5,4)$. And its strength $M(C) = (0.1, 0.5, 0.6)$. Since $\sigma_{[C]} = 2$, it oscillates with period 2 at $t \geq 5$. Suppose $\tilde{h}_{24} = (0.3, 0.6, 0.7)$. Then by theorem 2, the limiting fuzzy transition possibility matrix will exist with identical rows equal to $[(0.2, 0.4, 0.6), (1,1,1), (0.2, 0.4, 0.6), (0.3, 0.6, 0.7), (0.1, 0.5, 0.6)]$ at $t = 4$.

Corollary 2. *Let the fuzzy graph $\tilde{G}$ defined by $\tilde{H}$ contain a strongest cycle C_1 and its equivalent class be $[C_1]$. Let $S(C_1) = \bigcup_{i=1}^r C_i$ and $F = \{b \mid \tilde{h}_{ba} \geq M(C_1), a \in S(C_1)\}$. If at least one of $\{\tilde{h}_{jd} \mid d \in [S(C_1) \cup F]\} \geq M(C_1)$, then in higher powers of $\tilde{H}$ the entries of each column corresponding to the states of S will become equal and they are greater than or equal to $M(C_1)$. Also the limiting fuzzy transition possibility matrix exists with identical rows.*

Corollary 3. *Let the fuzzy graph $\tilde{G}$ defined by $\tilde{H}$ contain the equivalence classes $[C_1], [C_2], \ldots$ and $F_i = \{b \mid \tilde{h}_{ba} \geq M(C_i), a \in S(C_i)\}$.*

- *If at least one of $\{\tilde{h}_{jd} \mid d \in [S(C_i) \cup F_i]\} \geq M(C_i)$, then in higher powers of $\tilde{H}$ the entries of each column corresponding to the states of $S(C_i)$ will become equal and they are $\geq M(C_i)$.*
- *If $F_i \cap S(C_f) \neq \phi$ and at least one of $\tilde{h}_{jx} \geq \max\{M(C_i), M(C_f)\}$, $x \in [F_i \cap S(C_f)]$, then in higher powers of $\tilde{H}$ the entries of the columns corresponding to the states of $S(C_i)$, $S(C_f)$ will become equal.*

Example 3. Let

$$\tilde{H} = \begin{pmatrix} (0.1,0.3,0.5) & (1,1,1) & (0.1,0.3,0.5) & (0.1,0.2,0.3) & (0.2,0.4,0.6) \\ (0.1,0.2,0.3) & (1,1,1) & (0,0.1,0.3) & (0.1,0.3,0.5) & (0.1,0.2,0.3) \\ (0.1,0.3,0.5) & (1,1,1) & (0.1,0.2,0.3) & (0.3,0.6,0.7) & (0.1,0.3,0.5) \\ (0,0.1,0.3) & (1,1,1) & (0.3,0.6,0.7) & (0,0.1,0.3) & (0,0.1,0.3) \\ (0.2,0.4,0.6) & (1,1,1) & (0.1,0.3,0.5) & (0.2,0.7,0.8) & (0.1,0.2,0.3) \end{pmatrix}$$

$a_0 = (0.2,\ 0.4,\ 0.6)$ and $\tilde{G}_{\tilde{H}}$ contains strongest cycles $C_1 = (3,\ 4,\ 3)$ with $M(C_1) = (0.3,\ 0.6,\ 0.7)$ and $C_2 = (1,\ 5,\ 1)$ with $M(C_2) = (0.2,\ 0.4,\ 0.6)$. And $F_1 = \{b\,/\,\tilde{h}_{ba} \geq M(C_1), a \in C_1\} = \{5\}$ has intersection with C_2. If $\tilde{h}_{25} = (0.3,0.6,0.7)$, then the limiting fuzzy transition possibility matrix exists with identical rows $[(0.2, 0.4, 0.6), (1,1,1), (0.3, 0.6, 0.7), (0.3, 0.6, 0.7), (0.3, 0.6, 0.7)]$ at $t = 4$.

4 Conclusions

In crisp case, a Markov chain is said to be ergodic if it is irreducible and aperiodic. But a FPMM is ergodic if it is aperiodic and the limiting fuzzy transition possibility matrix exist with identical rows. In this paper, we have analyzed the existence of the limiting fuzzy transition possibility matrix with identical rows for a FPMM which contains a state j such that the fuzzy transition possibility of control transfers from every state i to state j is $(1,1,1)$ ($\tilde{h}_{ij} = (1,1,1)$ for $i = 1,2,\ldots,n$) which ensures its ergodicity.

By the theorem 2 and corollaries (2, 3), it is concluded that the ergodicity of the FPMM depends on the entries of the jth row of $\tilde{H}$. The long term behavior of a system which could be modeled as a FPMM and contains a component such that the event of reaching that component from every component is an sure event, can be obtained by analyzing its ergodicity using the above mentioned theorem and corollaries. Our future work is to extend the above results for a general FPMM.

Acknowledgments. We thank the management of SSN Institutions and Principal for providing necessary facilities to carry out this work.

References

1. Avrachekov, K.E., Sanchez, E.: Fuzzy Markov Chain: Specificities and Properties. Fuzzy Optimization and Decision Making 1, 143–159 (2002)
2. Buckly, J.J., Eslami, E.: Fuzzy Markov Chains:Uncetain Probabilities. Math Ware and Soft Computing 9, 33–41 (2002)
3. Buckly, J.J., Feuring, T., Hayashi, Y.: Fuzzy Markov Chains. In: Proc. 9th IFSA World Congress and 20th NAFIPS International Conference, pp. 2708–2711 (2001)
4. Imai, H., et al.: The Period of Powers of a Fuzzy Matrix. Fuzzy Sets and Systems 109, 405–408 (2000)
5. Kruse, R., Emden, R.B., Cordes, R.: Processor Power Considerations- An Application of Fuzzy Markov Chains. Fuzzy Sets and Systems 21, 289–299 (1987)
6. Praba, B., Sujatha, R., Hilda Christy Gnanam, V.: Classification and Steady State Analysis for Fuzzy Possibilistic Markov Chain (Communicated)
7. Trivedi, K.S.: Probability and Statistics with Reliability, Queuing and Computer Science Applications. Prentice Hall, Englewood Cliffs (2002)

A Fuzzy Random Variable Approach to Restructuring of Rough Sets through Statistical Test

Junzo Watada, Lee-Chuan Lin, Minji Qiang, and Pei-Chun Lin

Graduate School of Information, Production and Systems, Waseda University,
2-7 Hibikino, Wakamatsu, Kitakyushu 808-0135, Fukuoka, Japan
junzow@osb.att.ne.jp, leechuanlin@gmail.com, peichunpclin@gmail.com

Abstract. Usually it is hard to classify the situation where randomness and fuzziness exist simultaneously. This paper presents a method based on fuzzy random variables and statistical t-test to restructure a rough set. The algorithms of rough set and statistical t-test are used to distinguish whether a subset can be classified in the object set or not. The expected-value-approach is also applied to calculate the fuzzy value with probability into a scalar value.

Keywords: Rough Sets, Fuzzy t-test, Restructuring, Fuzzy Random Variable.

1 Introduction

We often have problems in classifying data under hydride uncertainty of both randomness and fuzziness. For example, linguistic data always have these features. However, as the meaning of each linguistic datum can be interpreted by a fuzzy set and the variability of the individual meaning may be understood as a random event, fuzzy random variable is a concept which can be applied to such a situation. In this research linguistic data are obtained randomly as a fuzzy random variable and after the fuzzy random variables are defined, the expected-value-approach will be applied to calculate them into some scalar values. Finally the subset, using its expectation values of ransom samples, will be distinguished whether to be classified into the object set or not by applying the method of rough set and statistical t-test.

Fuzzy random variable has been a basic tool in constructing the framework of decision making models under fuzzy random environment, and a number of practical optimization problems have been studied based on fuzzy random variables, such as inventory, risk management, portfolio selection, renewal process, and regression analysis (see [9,10]). Nevertheless, the reliability and redundancy optimization models under fuzzy random environment have not been well established in the literature.

The remainder of this paper is organized as follows. We give an overview of rough set theory and fuzzy random variables in Sections 2 and 3, respectively.

H. Sakai et al. (Eds.): RSFDGrC 2009, LNAI 5908, pp. 269–277, 2009.

The expected-value-approach, which can calculate the fuzzy random variables into scalar values, will also be explained in Section 3. In Section 4, the statistical t-test model which is built according to the rough sets theory will be explained including its principles and features. In the end, we will summarize this paper in conclusions.

2 Preliminaries

In this section, we recall some basic concepts on fuzzy variable and fuzzy random variable which make it easier to follow further discussions on the models. Assume that $(\Gamma, \mathcal{P}(\Gamma), \mathrm{Pos})$ is a possibility space, where $\mathcal{P}(\Gamma)$ is the power set of Γ, X is a fuzzy variable defined on $(\Gamma, \mathcal{P}(\Gamma), \mathrm{Pos})$ with membership function μ_X, and r is a real number. As a well-known fuzzy measure, possibility measure of a fuzzy event $X \le r$ is defined as

$$\mathrm{Pos}\{X \le r\} = \sup_{t \le r} \mu_X(t). \tag{1}$$

A self-dual set function, named credibility measure, is formed by [3] as follows

$$\mathrm{Cr}\{X \le r\} = \frac{1}{2}\left(1 + \sup_{t \le r} \mu_X(t) - \sup_{t > r} \mu_X(t)\right). \tag{2}$$

A fuzzy variable X is said to be positive if the credibility of $X \le 0$ is zero, i.e., $\mathrm{Cr}\{X \le 0\} = 0$. Furthermore, fuzzy variable X is said to be convex if all the α-cut sets of X are convex sets on $\Re$. In addition, for an n-ary fuzzy vector $\boldsymbol{X} = (X_1, X_2, \cdots, X_n)$, where each individual coordinate X_k is a fuzzy variable for $k = 1, 2, \cdots, n$, the membership function of $\boldsymbol{X}$ is given by taking the minimum of the individual coordinates as follows

$$\mu_{\boldsymbol{X}}(\boldsymbol{t}) = \bigwedge_{i=1}^{n} \mu_{X_i}(t_i), \tag{3}$$

where $\boldsymbol{t} = (t_1, \cdots, t_n) \in \Re^n$.

Definition 1 ([4]). *Suppose that $(\Omega, \Sigma, \mathrm{Pr})$ is a probability space, $\mathcal{F}_v$ is a collection of fuzzy variables defined on possibility space $(\Gamma, \mathcal{P}(\Gamma), \mathrm{Pos})$. A fuzzy random variable is a map $\xi : \Omega \to \mathcal{F}_v$ such that for any Borel subset B of $\Re$, $\mathrm{Pos}\{\xi(\omega) \in B\}$ is a measurable function of ω.*

Example 1. Let X be a random variable defined on probability space $(\Omega, \Sigma, \mathrm{Pr})$. We call ξ a triangular fuzzy random variable, if for every $\omega \in \Omega$, $\xi(\omega)$ is a triangular fuzzy variable defined on some possibility space $(\Gamma, \mathcal{P}(\Gamma), \mathrm{Pos})$, e.g.,

$$\xi(\omega) = \Big(X(\omega) - 1, X(\omega), X(\omega) + 1\Big).$$

We say ξ is a normal fuzzy random variable, denoted by $\mathcal{N}_{\mathcal{F}}(X, b), b > 0$, if for every $\omega \in \Omega$, the membership function of $\xi(\omega)$ is

$$\mu_{\xi(\omega)}(r) = \exp\left(\frac{-(r - X(\omega))^2}{b}\right).$$

In addition, a fuzzy random variable ξ is said to be positive if for almost every $\omega \in \Omega$, $\xi(\omega)$ is a positive fuzzy variable. For example, we can construct a positive normal fuzzy random variable ξ as

$$\mu_{\xi(\omega)}(r) = \begin{cases} \exp\left(-(r - X(\omega))^2/b\right), & r \geq 0 \\ 0, & r < 0. \end{cases} \quad (4)$$

In this paper, the above positive normal fuzzy random variable ξ is denoted by $\mathcal{N}_{\mathcal{F}}^{+}(X, b)$.

In order to measure an event $\xi \in B$ induced by fuzzy random variable ξ, where B is any Borel subset of $\Re$, the mean chance measure (see [5]) is defined as

$$\text{Ch}\left\{\xi \in B\right\} = \int_{\Omega} \text{Cr}\left\{\xi(\omega) \in B\right\} \text{Pr}(\mathrm{d}\omega). \quad (5)$$

Example 2. Consider a triangular fuzzy random variable ξ with $\xi(\omega) = (X(\omega) + 2, X(\omega) + 3, X(\omega) + 4)$, where X is a discrete random variable, which takes on values $X_1 = 2$ with probability 0.4, and $X_2 = 4$ with probability 0.6. Now we calculate the mean chance of event $\xi \leq 7$.

3 Fuzzy Random Variables and the Expected-Value Approach

Given some universe Γ, let Pos is a possibility measure defined on the power set $P(\Gamma)$ of Γ. Let $\Re$ be the set of real numbers. A function $Y : \Gamma \to \Re$ is said to be a fuzzy variable defined on Γ [6]. The possibility distribution μ_Y of Y is defined by $\mu_Y(t) = Pos\{Y = t\}$. For fuzzy variable Y with possibility distribution μ_Y, the possibility, necessity and credibility of event $\{Y \geq r\}$ are given, as follows:

$$Pos\{Y \leq r\} = \sup_{t \leq r} \mu_Y(t), \quad (6)$$

$$Nec\{Y \leq r\} = 1 - \sup_{t > r} \mu_Y(t), \quad (7)$$

$$Cr\{Y \leq r\} = \frac{1}{2}(1 + \sup_{t \leq r} \mu_Y(t) \quad (8)$$

It should be noted that the credibility measure is an average of the possibility and the necessity measure, i.e. $Cr\{\cdot\} = (Pos\{\cdot\} + Nec\{\cdot\})/2$, and it is a self-dual set function for any A in $P(\Gamma)$. The motivation behind the introduction of the credibility measure is to develop a certain measure which is a sound aggregate of the two extreme cases such as the possibility (expressing a level of overlap and being highly optimistic in this sense) and necessity. Based on credibility measure, the expected value of a fuzzy variable is presented as follows.

Definition 2 ([3]). *Let Y be a fuzzy variable. The expected value of Y is defined as:*

$$E[Y] = \int_0^{\infty} Cr\{Y \geq r\}dr - \int_{-\infty}^0 Cr\{Y \leq r\}dr \quad (9)$$

provided that the two integrals are finite.

Example 3. Assume that $Y = (c, a^l, a^r)_T$ is a triangular fuzzy variable whose possibility distribution is

$$\mu_Y(x) = \begin{cases} \dfrac{x - a^l}{c - a^l}, & a^l \leq x \leq c \\ \dfrac{a^r - x}{a^r - c}, & c \leq x \leq a^r \\ 0, & otherwise. \end{cases}$$

Making use of (9), we determine the expected value of Y to be

$$E[Y] = \frac{a^l + 2c + a^r}{4}. \tag{10}$$

Next the definitions of fuzzy random variable and its expected value and variance operators will be explained. For more theoretical results on fuzzy random variables, one may refer to Liu and Liu [1], and Wang and Watada [7], [8].

Definition 3 ([1]). *Suppose that (Ω, Σ, Pr) is a probability space, F_ν is a collection of fuzzy variables defined by possibility space $(\Gamma, P(T), Pos)$. A fuzzy random variable is a mapping $X : \Omega \to F_\nu$ such that for any Borel subset B of $\Re$, $Pos\{X(\omega) \in B\}$ is a measurable function of ω.*

Let X be fuzzy random variable on Ω. From the above definition, we can know for each $\omega \in \Omega$, $X(\omega)$ is a fuzzy variable. Furthermore, a fuzzy random variable $X(\omega)$ is said to be positive if for almost every ω, fuzzy variable $X(\omega)$ is positive almost surely.

Example 4. Let V be a random variable defined on probability space (Ω, Σ, Pr). Define that for every $\omega \in \Omega$, $X(\omega) = (V(\omega) + 2, V(\omega) - 2, V(\omega) + 6)_T$ which is a triangular fuzzy variable defined on some possibility space $(\Gamma, P(T), Pos)$. Then X is a triangular fuzzy random variable.

For any fuzzy random variable X on Ω, for each $\omega \in \Omega$, the expected value of the fuzzy variable $X(\omega)$ is denoted by $E[X(\omega)]$, which has been proved to be a measurable function of ω. Given this, the expected value of the fuzzy random variable X is defined as the mathematical expectation of the random variable $E[X(\omega)]$.

Definition 4. *Let X be fuzzy random variable defined on a probability space (Ω, Σ, Pr). The expected value of X is defined as:*

$$E[\xi] = \int_\Omega [\int_0^\infty Cr\{\xi(\omega) \geq r\}dr - \int_{-\infty}^0 Cr\{\xi(\omega) \leq r\}dr]Pr(d\omega) \tag{11}$$

Example 5. Consider the triangular fuzzy random variable X as defined in Example 4. Suppose that V is a discrete random variable, which takes values $V_1 = 3$ with probability 0.2, and $V_2 = 6$ with probability 0.8. The expected value of X can be calculated.

From the distribution of random variable V, we know that the fuzzy random variable X takes fuzzy variables $X(V_1) = (5, 1, 9)_T$ with probability 0.2 and $X(V_2) = (8, 4, 12)_T$ with probability 0.8. We need to compute the expected values of fuzzy random variables $X(V_1)$ and $X(V_2)$. That is , $E[X(V_1)] = (1 + 2 \times 5 + 9) \div 4 = 5$ and $E[X(V_2)] = (4 + 2 \times 8 + 12) \div 4 = 8$. Finally, by Definition 3, the expected value of X is $E[X] = 0.2 \times E[X(V_1)] + 0.8 \times E[V_2] = 7.4$.

Definition 5 ([1]). *Let X be a fuzzy random variable defined on a probability space $(\Omega, \Sigma, \Pr)$ with expected value e. The variance of X is defined as*

$$Var[X] = E[(X - e)^2] \tag{12}$$

where $e = E[X]$ is given by Definition 4.

Example 6. Consider the triangular fuzzy random variable X defined in Example 4. Let us calculate the variance of X. From the distribution of random variable V, we know that the fuzzy random variable X takes fuzzy variables $X(V_1) = (5, 1, 9)_T$ with probability 0.2, and $X(V_2) = (8, 4, 12)_T$ with probability 0.8. From Example 4, $E(X) = 7.4$. Then $Var(X) = E[(X(V_1) - 7.4)^2] \cdot 0.2 + E[(X(V_2) - 7.4)^2] \cdot 0.8$.

Therefore, from Definition 1, we obtain $E[(X(V_1) - 7.4)^2] = E[Y_1^2]$ as $E[Y_1^2] = 12.08$.

Similarly, we obtain $E[(X(V_2) - 7.4)^2] = E[Y_2^2] = 4.25$. Thus, $Var(X) = 0.2 \cdot E(X(V_1) - 7.4)^2 + 0.8 \cdot E(X(V_2) - 7.4)^2 = 0.2 \times 12.08 + 0.8 \times 4.25 = 5.81$.

4   Confidence Intervals

Table 1 illustrates a format of data to be dealt with here, where independent attributes X_{ik} and decision attribute Y_i, for all $i = 1. \cdots , N$ and $k = 1, \cdots , K$ are fuzzy random variables, which are defined as

$$Y_i = \bigcup_{t=1}^{M_{Y_i}} \left\{ \left(Y_i^t, Y_i^{t,l}, Y_i^{t,r} \right)_T, p_i^t \right\}, \tag{13}$$

$$X_{ik} = \bigcup_{t=1}^{M_{X_{ik}}} \left\{ \left(X_{ik}^t, X_{ik}^{t,l}, X_{ik}^{t,r} \right)_T, q_{ik}^t \right\}, \tag{14}$$

respectively. This means that all values are given as fuzzy numbers with probabilities, where fuzzy variables $(Y_i^t, Y_i^{t,l}, Y_i^{t,r})_T$ and $(X_{ik}^t, X_{ik}^{t,l}, X_{ik}^{t,r})_T$ are associated with probability p_i^t and q_{ik}^t for $i = 1, 2, \cdots , N$, $k = 1, 2, \cdots , K$ and $t = 1, 2, \cdots , M_{Y_i}$ or $t = 1, 2, \cdots , M_{X_{ik}}$, respectively.

Before discussing the restructuring of rough sets using fuzzy random attributes with confidence interval, we define the confidence interval which is induced by the expectation and variance of a fuzzy random variable. When we consider the one sigma confidence $(1 \times \sigma)$ interval of each fuzzy random variable, we can express it as the following interval

$$I[e_X, \sigma_X] \triangleq \left[E(X) - \sqrt{Var(X)}, \ E(X) + \sqrt{Var(X)} \right], \tag{15}$$

Table 1. Data given for fuzzy random attributes

Sample	Decision Attribute	Independent Attributes				
i	Y	X_1	X_2	$\cdots$	X_k	$\cdots$, X_K
1	Y_1	X_{11}	X_{12}	$\cdots$	X_{1k}	$\cdots$, X_{1K}
2	Y_2	X_{21}	X_{22}	$\cdots$	X_{2k}	$\cdots$, X_{2K}
$\vdots$	$\vdots$	$\vdots$	$\vdots$		$\vdots$	$\vdots$
i	Y_i	X_{i1}	X_{i2}	$\cdots$	X_{ik}	$\cdots$, X_{iK}
$\vdots$	$\vdots$	$\vdots$	$\vdots$		$\vdots$	$\vdots$
N	Y_N	X_{N1}	X_{N2}	$\cdots$	X_{Nk}	$\cdots$ X_{NK}

Table 2. Confidence intervals calculated for attributes

Sample	Decision Attribute	Independent attributes		
i	$I[e_Y, \sigma_Y]$	$I[e_{X_1}, \sigma_{X_1}]$	$\cdots$	$I[e_{X_K}, \sigma_{X_K}]$
1	$I[e_{Y_1}, \sigma_{Y_1}]$	$I[e_{X_{11}}, \sigma_{X_{11}}]$	$\cdots$	$I[e_{X_{1K}}, \sigma_{X_{1K}}]$
2	$I[e_{Y_2}, \sigma_{Y_2}]$	$I[e_{X_{21}}, \sigma_{X_{21}}]$	$\cdots$	$I[e_{X_{2K}}, \sigma_{X_{2K}}]$
$\vdots$	$\vdots$	$\vdots$		$\vdots$
i	$I[e_{Y_i}, \sigma_{Y_i}]$	$I[e_{X_{i1}}, \sigma_{X_{i1}}]$	$\cdots$	$I[e_{X_{iK}}, \sigma_{X_{iK}}]$
$\vdots$	$\vdots$	$\vdots$		$\vdots$
N	$I[e_{Y_N}, \sigma_{Y_N}]$	$I[e_{X_{N1}}, \sigma_{X_{N1}}]$	$\cdots$	$I[e_{X_{NK}}, \sigma_{X_{NK}}]$

which is called a one-sigma confidence interval. Similarly, we can define two-sigma and three-sigma confidence intervals. All of these confidence intervals are called σ-confidence intervals. Table 3 shows the data with one-sigma confidence interval.

5   An Example

In this section, we present a simple example to visualize how to use the proposed CI-FRRM. Assume that the data of fuzzy random independent and decision attributes (4 samples and 2 attributes) are given in the Tables 3 and 4, respectively.

First of all, we need to calculate all the $I[e_{X_{ik}}, \sigma_{X_{ik}}]$ and $I[e_{Y_k}, \sigma_{Y_k}]$ for $i = 1, 2, 3, 4, k = 1, 2$. By using the calculation in Example 6, we obtain all the pairs $\left(e_{X_{ik}}, \sigma_{X_{ik}}\right)$ and $\left(e_{Y_k}, \sigma_{Y_k}\right)$ as shown in Table 5.

Hence, the confidence intervals for the input data and output data can be calculated in the form

$$I[e_{X_{ki}}, \sigma_{X_{ki}}] = [e_{X_{ki}} - \sigma_{X_{ki}}, e_{X_{ki}} + \sigma_{X_{ki}}] \tag{16}$$

Table 3. Independent attributes

No.	X_1
1	$X_{11} = \left((3, 2, 4)_T, 0.5; (4, 3, 5)_T, 0.5\right)$
2	$X_{21} = \left((6, 4, 8)_T, 0.5; (8, 6, 10)_T, 0.5\right)$
3	$X_{31} = \left((12, 10, 14)_T, 0.25; (14, 12, 16)_T, 0.75\right)$
4	$X_{41} = \left((14, 12, 16)_T, 0.5; (16, 14, 18)_T, 0.5\right)$

No.	X_2
1	$X_{12} = \left((2, 1, 3)_T, 0.1; (4, 3, 5)_T, 0.9\right)$
2	$X_{22} = \left((3, 2, 4)_T, 0.5; (4, 3, 5)_T, 0.5\right)$
3	$X_{32} = \left((12, 10, 16)_T, 0.2; (14, 12, 16)_T, 0.8\right)$
4	$X_{42} = \left((18, 16, 20)_T, 0.2; (21, 20, 22)_T, 0.8\right)$

Table 4. Decision attribute

No.	Y
1	$Y_1 = \big((14, 10, 16)_T, 0.4; (18, 16, 20)_T, 0.6\big)$
2	$Y_2 = \big((17, 16, 18)_T, 0.8; (20, 18, 22)_T, 0.2\big)$
3	$Y_3 = \big((22, 20, 24)_T, 0.3; (26, 24, 28)_T, 0.7\big)$
4	$Y_4 = \big((32, 30, 34)_T, 0.4; (36, 32, 40)_T, 0.6\big)$

Table 5. Expectation and standard deviation of the data

i	$\big(e_{X_{i1}}, \upsilon_{X_{i1}}\big)$	$\big(e_{X_{i2}}, \upsilon_{X_{i2}}\big)$	$\big(e_{Y_i}, \sigma_{Y_i}\big)$
1	$(3.5, 0.56)$	$(3.8, 0.75)$	$(16.2, 7.68)$
2	$(7.0, 2.25)$	$(3.5, 0.56)$	$(17.6, 2.41)$
3	$(13.5, 1.87)$	$(13.7, 4.20)$	$(24.8, 4.68)$
4	$(15.0, 2.25)$	$(20.4, 2.00)$	$(34.4, 8.24)$

Table 6. Confidence intervals of the input data

i	$I[e_{X_{i1}}, \sigma_{X_{i1}}]$	$I[e_{X_{i2}}, \sigma_{X_{i2}}]$
1	$[2.94, 4.06]$	$[3.05, 4.75]$
2	$[4.75, 9.25]$	$[2.94, 4.06]$
3	$[11.63, 15.37]$	$[9.50, 17.90]$
4	$[12.75, 17.25]$	$[18.40, 22.40]$

Table 7. Confidence intervals of the output data

i	$I[e_{Y_i}, \sigma_{Y_i}]$
1	$[8.52, 23.88]$
2	$[15.19, 20.01]$
3	$[20.12, 29.48]$
4	$[26.16, 42.64]$

Table 8. Restructuring of fuzzy random data

i	X_{i1}	X_{i1}	Y_i
1	0	0	0
2	1	0	0
3	1	1	1
4	1	1	1

and

$$I[e_{Y_i}, \sigma_{Y_i}] = [e_{Y_i} - \sigma_{Y_i}, e_{Y_i} + \sigma_{Y_i}], \tag{17}$$

respectively, for $i = 1, 2$ and $k = 1, 2, 3, 4$. They are listed in the Tables 6 and 7, respectively.

6   T-Testing Procedure

After we obtained the expected value of the fuzzy random variables through the mathematical approach mentioned in Section 3, a new algorithm applying statistical t-test is designed to identify the inclusion efficiency of the upper approximation in rough set theory.

Suppose 30 subsets are included in the collectivity set and each subset has 50 sample values. A discrimination ratio for the collectivity set also exists which means when the classification result of each subset is larger than the discrimination ratio, the subset will be included into the object set. The statistical t-test is applied to verify the result of each subset in order to make out whether the result of samples accords with the result of the whole subset.

In this study, one-side testing is applied and the testing equation is shown as follows:

$$T = \frac{\overline{X} - \mu_0}{\frac{S_n^0}{\sqrt{n}}} \tag{18}$$

where $\overline{X}$ is the average of each subset μ_0 is the threshold set for each subset. S_n^* is the modified sample variance of each subset, and n is the number of the subset values. The significance level used in this study is 5%. The sequence of the test is shown as follows:

- Calculate the modified sample variance and the value T.
- Comparing the value T and $t_{1-\frac{\alpha}{2}}(n-1)$.
- If the value $T < t_{1-\frac{\alpha}{2}}(n-1)$, then accept the result of sampling.
- If the value $T >= t_{1-\frac{\alpha}{2}}(n-1)$, then reject the result of sampling.

7   Conclusions

This research aims to solve the problem of classification when the object contains vagueness, randomness and fuzziness. At first, we proposed a rough set approach because rough set deals well with the vagueness. Secondly we apply the concepts of fuzzy random variable as well as the method of expected-value-approach to handle the problem of randomness and fuzziness. After we obtained the expected value of fuzzy random variables through the above method, the algorithm of statistical t-test is adopted to reach the goal of classification. Furthermore, to apply this approach with some real data is also considered in the near future.

References

1. Liu, Y.-K., Liu, B.: Fuzzy random variable: A scalar expected value operator. Fuzzy Optimization and Decision Making 2(2), 143–160 (2003)
2. Liu, B.: Uncertainty Theory, 2nd edn. Springer, Berlin (2007)
3. Liu, B., Liu, Y.K.: Expected value of fuzzy variable and fuzzy expected value models. IEEE Transaction on Fuzzy Systems 10, 445–450 (2002)
4. Liu, Y.K., Liu, B.: Fuzzy random variable: A scalar expected value operator. Fuzzy Optimization and Decision Making 2, 143–160 (2003)
5. Liu, Y.K., Liu, B.: On minimum-risk problems in fuzzy random decision systems. Computers & Operations Research 32, 257–283 (2005)
6. Nahmias, S.: Fuzzy variables. Fuzzy Sets and Systems 1(2), 97–111 (1978)
7. Wang, S., Watada, J.: T-independence condition for fuzzy random vector based on continuous triangular norms. Journal of Uncertain Systems 2, 155–160 (2008)
8. Wang, S., Watada, J.: Studying distribution functions of fuzzy random variables and its applications to critical value functions. International Journal of Innovative Computing, Information & Control 5, 279–292 (2009)
9. Watada, J., Wang, S.: Regression model based on fuzzy random variables. In: Rodulf, S. (ed.) Views on Fuzzy Sets and Systems from Different Perspectives, ch. 26. Springer, Berlin (2009)
10. Watada, J., Wang, S., Pedrycz, W.: Building confidence-interval-based fuzzy random regression models. In: IEEE Transactions on Fuzzy Systems (to be published)

A New Approach for Solving Fuzzy Maximal Flow Problems

Amit Kumar, Neha Bhatia, and Manjot Kaur

School of Mathematics and Computer Applications
Thapar University, Patiala, 147004, India
amitkumar@thapar.edu, neha26bhatia@gmail.com,
manjot.thaparian@gmail.com

Abstract. In conventional maximal flow problems, it is assumed that decision maker is certain about the flows between different nodes. But in real life situations, there always exist uncertainty about the flows between different nodes. In such cases, the flows may be represented by fuzzy numbers. In literature, there are several methods to solve such type of problems. Till now, no one has used ranking function to solve above type of problems. In this paper, a new algorithm is proposed to find fuzzy maximal flow between source and sink by using ranking function. To illustrate the algorithm a numerical example is solved and result is explained. If there is no uncertainty about the flow between source and sink then the proposed algorithm gives the same result as in crisp maximal flow problems.

Keywords: Fuzzy maximal flow problem, Ranking function, Triangular fuzzy number.

1 Introduction

A network consists of a set of nodes linked by arcs [9]. Network flow problems have many applications in the most diverse fields, such as: telecommunications, transportations, computation, manufacturing etc. The maximal flow problem is an important problem in the network flow problems, it consists of sending the biggest amount of flow between two nodes (source S and sink N), holding the capacity restrictions of each arc. In the literature, there are some efficient algorithms to solve the crisp problems [1]. But, there are problems that have uncertainties in their parameters (e.g.: costs, capacities and demands). This problem is called fuzzy maximal flow problem, where the graph has a crisp structure and fuzzy parameters.

In the literature, the number of papers dealing with fuzzy maximal flow problem is short [3,4,5,11]. The paper by Kim and Roush [11] is one of the first on this subject. The authors developed the fuzzy flow theory, presenting the conditions to obtain a optimal flow, by means of definitions on fuzzy matrices. But there were Chanas and

H. Sakai et al. (Eds.): RSFDGrC 2009, LNAI 5908, pp. 278–286, 2009.

Kolodziejczyk [3,4,5] who introduced the main works in the literature involving this subject. They approached this problem using the minimum cuts technique. Chanas and Kolodziejczyk [3] presented an algorithm for a graph with crisp structure and fuzzy capacities, i.e., the arcs have a membership function associated in their flow. This problem was studied by Chanas and Kolodzijczyk [4] again, in this paper the flow was a real number and the capacities have upper and lower bounds with a satisfaction function. Chanas and Kolodzijczyk [5] also studied the integer flow and proposed an algorithm. In this paper, a new algorithm is proposed to solve the fuzzy maximal flow problem occurring in real life. To illustrate the algorithm a numerical example is solved and result is explained. If there is no uncertainty about the flow between source and sink then the proposed algorithm gives the same result as in crisp maximal flow problem.

2 Preliminaries

In this section some basic definitions and arithmetic operations are reviewed.

2.1 Basic Definitions

Definition 2.1.1 [10]

The characteristic function μ_A of a crisp set $A \subseteq X$ assigns a value either 0 or 1 to each member in X. This function can be generalized to a function $\mu_{\tilde{A}}$ such that the value assigned to the element of the universal set X fall within a specified range $[0,1]$ i.e. $\mu_{\tilde{A}} : X \rightarrow [0,1]$. The assigned values indicate the membership grade of the element in the set A.

The function $\mu_{\tilde{A}}$ is called the membership function and the set $\tilde{A} = \{(x, \mu_{\tilde{A}}(x)) : x \in X\}$ defined by $\mu_{\tilde{A}}$ for each $x \in X$ is called a fuzzy set.

Definition 2.1.2 [10]

A fuzzy number $\tilde{A} = (a,b,c)$, shown in Fig. 1, is said to be a triangular fuzzy number if its membership function is given by

$$\mu_{\tilde{A}}(x) = \begin{cases} \dfrac{(x-a)}{(b-a)}, & a \leq x \leq b \\ 1 & x = b \\ \dfrac{(x-c)}{(b-c)}, & b \leq x \leq c \end{cases}$$

where $a, b, c \in R$

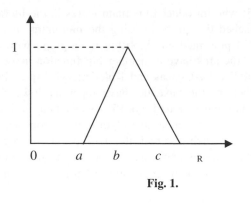

Fig. 1.

2.2 Arithmetic Operations on Triangular Fuzzy Numbers

Let $\tilde{A} = (a_1, b_1, c_1)$ and $\tilde{B} = (a_2, b_2, c_2)$ be two triangular fuzzy numbers then

(i) $\tilde{A} \oplus \tilde{B} = (a_1 + a_2,\ b_1 + b_2,\ c_1 + c_2)$

(ii) $\tilde{A} \ominus \tilde{B} = (a_1 - c_2,\ b_1 - b_2,\ c_1 - a_2)$

3 Ranking Function

A convenient method for comparing fuzzy numbers is by the use of ranking function [2,7,10,12]. A ranking function R: F(R) $\rightarrow$ R, where F(R) set of all fuzzy numbers defined on set of real numbers, maps each fuzzy number into a real number.

Let $\tilde{A}$ and $\tilde{B}$ be two triangular fuzzy numbers, then

$$\text{(i) } \tilde{A} \underset{R}{\geq} \tilde{B} \quad \text{if and only if } R(\tilde{A}) \geq R(\tilde{B})$$

$$\text{(ii) } \tilde{A} \underset{R}{\leq} \tilde{B} \quad \text{if and only if } R(\tilde{A}) \leq R(\tilde{B})$$

$$\text{(iii) } \tilde{A} \underset{R}{=} \tilde{B} \quad \text{if and only if } R(\tilde{A}) = R(\tilde{B})$$

For a triangular fuzzy number $\tilde{A} = (a, b, c)$, ranking function R is given by

$$R(\tilde{A}) = \frac{1}{4}(a + 2b + c).$$

4 Fuzzy Maximal Flow Algorithm

The proposed algorithm is a labeling technique. The idea of fuzzy maximal flow algorithm is to find a breakthrough path with positive net flow that connects the source and sink nodes. Take an arc (i, j) with initial approximate capacities ($\tilde{fc}_{ij}, \tilde{fc}_{ji}$). As the computations of the algorithm proceed, portions of these approximate capacities are committed to the flow in the arc, the residuals of the arc are updated. We use the notation ($\tilde{fd}_{ij}$, $\tilde{fd}_{ji}$). The fuzzy network with the updated excess approximate capacities will be referred to as the residual fuzzy network.

For a node j that receives flow from node i, attach a label $[\tilde{fa}_j, i]$, where $\tilde{fa}_j$ is the approximate flow from node i to j .

The source node is numbered 1 and the algorithm proceeds as follows:

Step 1

Let the index j refer to all nodes that can be reached directly from source node 1 by arc with positive excess capacities, i.e. $R(\tilde{fc}_{1j}) > 0$ and rank of $\tilde{fc}_{1j}$ is maximal for all j. On the diagram of the network, the node j is labeled with two numbers $[\tilde{fa}_j, 1]$, where $\tilde{fa}_j$ is the approximate positive excess capacity, and 1 means flow is coming from node 1. If in doing this the sink N is labeled, so that there is a branch of approximate positive excess capacity from source to the sink, then the approximate maximal flow along the path is given by $\tilde{f}_1 = \tilde{fc}_{1N}$, and the excess capacity due to this breakthrough path is decrease by $\tilde{f}_1$ in the direction of the flow and is increased by $\tilde{f}_1$ in the reverse direction. This means that for source nodes 1 and sink node N the excess flow is changed from the current

$$(\tilde{fc}_{1N}, \tilde{fd}_{N1}) \quad \text{to} \quad (\tilde{fc}_{1N} \ominus \tilde{f}_1, \ \tilde{fd}_{N1} \oplus \tilde{f}_1)$$

Step 2

In case in step 1, the sink is not labeled then again find all the nodes k that can be reached from node j and label these nodes as $[\tilde{fa}_k, j]$. Repeat this step until sink is labeled after that compute $\tilde{f}_1$.

$\tilde{f}_1$ = minimum of the excess approximate capacities on the path to the sink.

Subtract $\tilde{f}_1$ from excess approximate capacities on the arc in the direction of path and add $\tilde{f}_1$ from the excess approximate capacity in reverse direction. In this way the fresh excess approximate capacities are obtained.

Step 3

Step 1 or 2 gives first breakthrough. Compute freshly excess approximate capacities of all arcs which are changed due to first breakthrough. The process is repeated until, in a finite number of steps, we reach the state so that no additional nodes can be labeled to reach sink. The approximate maximal flow is computed by

$$\tilde{f} = \tilde{f}_1 \oplus \tilde{f}_2 \oplus \tilde{f}_3 \oplus ... \oplus \tilde{f}_p$$

where p is the number of iteration to get no breakthrough.

The approximate optimal flow in the arc (i, j) is computed as

$$(\tilde{\alpha}, \tilde{\beta}) = (\tilde{fc}_{ij} \ominus \tilde{fd}'_{ij}, \tilde{fc}_{ji} \ominus \tilde{fd}'_{ji})$$

where $\tilde{fc}_{ij}$ and $\tilde{fc}_{ji}$ are the initial approximate capacities, and $\tilde{fd}'_{ij}$, $\tilde{fd}'_{ji}$ are the final approximate excess capacities. If $R(\tilde{\alpha}) > 0$, the approximate optimal flow from i to j is $\tilde{\alpha}$. Otherwise, if $R(\tilde{\beta}) > 0$, the approximate optimal flow from j to i is $\tilde{\beta}$. Note that $R(\tilde{\alpha})$ and $R(\tilde{\beta})$ can not be positive together.

5 Illustrative Example

Consider the network shown in Fig. 2. The bidirectional approximate capacities are shown on the respective arcs. For example, for arc (3,4) the flow limit is approximately 10 say (5,10,15) units from 3 to 4 and approximately 5 units from 4 to 3 say (0,5,10) units. Determine the approximate maximal flow in this network between source 1 and sink 5.

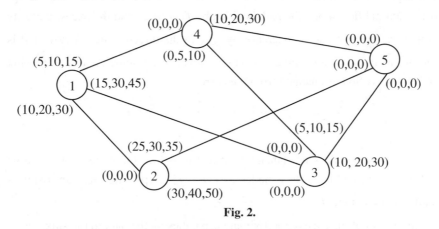

Fig. 2.

The algorithm is applied in the following manner.

Iteration 1

At the first step, find the nodes that can be reached directly from the source by arc of positive excess approximate capacity $R(\tilde{fc}_{ij}) > 0$. These nodes are 2, 3, 4. Label these nodes with the ordered pair of numbers $[\tilde{fa}_j, 1]$, where $\tilde{fa}_j = \tilde{fc}_{1j}$ and 1 means we have reached from node 1. Firstly we will choose the path having approximate maximal flow limit.

Since the rank of $(15, 30, 45)$ is maximum so we will choose the path form node 1 to 3. Node 3 is labeled as $[(15, 30, 45), 1]$. Still sink is not labeled. Again we will choose the path having maximal approximate flow limit i.e. $[(10, 20, 30), 3]$ from node 3 to node 5. Now, sink is reached and labeling process stops as we have got first breakthrough. The approximate flow in the network can be increased by

A New Approach for Solving Fuzzy Maximal Flow Problems 283

$$\tilde{f}_1 = \min\{(15,30,45),(10,20,30)\}$$

The value of $\tilde{f}_1$ indicates that increase of approximately 20 i.e. $(10,20,30)$ units can be made along the path traced out in a move from source to sink. We can easily work backward to find the path. The label on the sink shows that we came from node 3. From node 3 it is seen that we came from node 1. The path is $1\rightarrow 3\rightarrow 5$. After first iteration arc $(1,3)$ has fuzzy residual in direction of flow $(15,30,45)\ominus(10,20,30)=(-15,10,35)$ and $(0,0,0)\oplus(10,20,30)=(10,20,30)$ in opposite direction. Similarly, arc $(3,5)$ has $(-20,0,20)$ and $(10,20,30)$ in the direction of flow and in opposite direction.

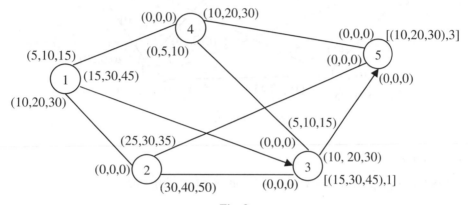

Fig. 3.

Iteration 2

We repeat the procedure described in the first iteration, at the starting node 1, the breakthrough path obtained is $1\rightarrow 2\rightarrow 3\rightarrow 4\rightarrow 5$ and $\tilde{f}_2=(5,10,15)$.

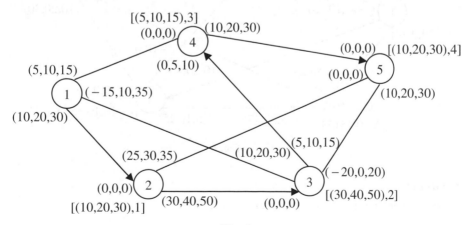

Fig. 4.

Iteration 3

Repeat the procedure. The breakthrough path obtained is $1 \to 2 \to 5$ and $\widetilde{f}_3 = (-5, 10, 25)$.

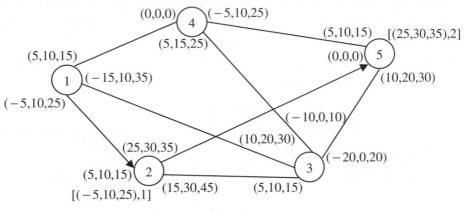

Fig. 5.

Iteration 4

A breakthrough path is $1 \to 3 \to 2 \to 5$ and $\widetilde{f}_4 = (-15, 10, 35)$.

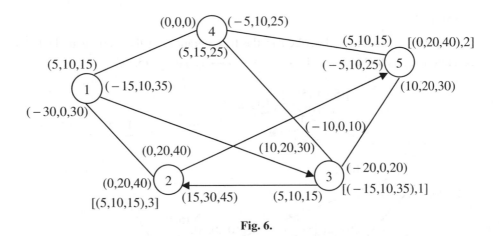

Fig. 6.

Iteration 5

A breakthrough path is $1 \to 4 \to 5$ and $\widetilde{f}_5 = (5, 10, 15)$.

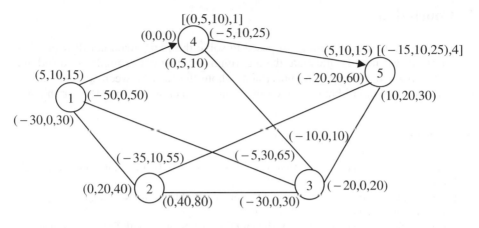

Fig. 7.

More iterations are not possible after 5th iteration a there is no way out to reach at sink from source. The approximate maximal flow is

$$\tilde{f} = \tilde{f}_1 \oplus \tilde{f}_2 \oplus \tilde{f}_3 \oplus \tilde{f}_4 \oplus \tilde{f}_5 = (10, 20, 30) \oplus (5, 10, 15) \oplus (-5, 10, 25) \oplus (-15, 10, 35) \oplus (5, 10, 15)$$
$$= (0, 60, 120)$$

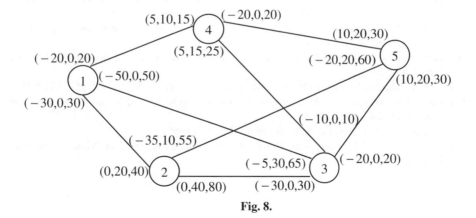

Fig. 8.

6   Result and Discussion

The obtained result can be explained as follow:

1) The flow between source and sink is greater than 0 and less than 120.
2) Maximum number of persons are in favour that flow will be 60.
3) The percentage of person increases when flow varies from 0 to 60 and decreases from 60 to 120.

7 Conclusion

In this paper, a new algorithm is proposed to solve the fuzzy maximal flow problem occurring in real life. To illustrate the algorithm a numerical example is solved and result is explained. If there is no uncertainty about the flow between source and sink then the proposed algorithm gives the same result as in crisp maximal flow problem.

References

[1] Ahuja, R.K., Magnanti, T.L., Orlin, J.B.: Network Flows, Theory, Algorithms and Applications. Prentice Hall, New Jersey (1993)

[2] Campos, L., Verdegay, J.L.: Linear Programming Problems and Ranking of Fuzzy Numbers. Fuzzy Sets and Systems 32, 1–11 (1989)

[3] Chanas, S., Kolodziejczyk, W.: Maximum Flow in a Network with Fuzzy Arc Capacities. Fuzzy Sets and Systems 8, 165–173 (1982)

[4] Chanas, S., Kolodziejczyk, W.: Real-valued Flows in a Network with Fuzzy Arc Capacities. Fuzzy Sets and Systems 13, 139–151 (1984)

[5] Chanas, S., Kolodziejczyk, W.: Integer Flows in Network with Fuzzy Capacity Constraints. Networks 16, 17–31 (1986)

[6] Dubois, D., Prade, H.: Fuzzy Sets and Systems, Theory and Applications. Academic Press, New York (1980)

[7] Fortemps, P., Roubens, M.: Ranking and Defuzzification Methods Based on Area Compensation. Fuzzy Sets and Systems 82, 319–330 (1996)

[8] Kalir, G.J., Yuhan, B.: Fuzzy Sets and Fuzzy Logic, Theory and Applications. Prentice Hall, Englewood Cliffs (1995)

[9] Kasana, H.S., Kumar, K.D.: Introductory Operations Research Theory and Application. Springer, Heidelberg (2008)

[10] Kaufmann, A., Gupta, M.M.: Introduction to Fuzzy Arithmetics. Theory and Application, New York (1991)

[11] Kim, K., Roush, F.: Fuzzy Flows on Network. Fuzzy Sets and Systems 8, 35–38 (1982)

[12] Liou, T.S., Wang, M.J.: Ranking Fuzzy Numbers with Integral Value. Fuzzy Sets and Systems 50, 247–255 (1992)

[13] Zadeh, L.: Fuzzy Sets. Inf. and Control 8, 338–353 (1965)

[14] Zimmermann, H.J.: Fuzzy Sets Theory and Its Application, 2nd edn. Kluwer Academic Publisher, Boston (1991)

Peptic Ulcer Detection Using DNA Nanorobot Based on Fuzzy Logic Rules

Sanchita Paul*, Abhimanyu K. Singh, and Gadhadhar Sahoo

Birla Institute of Technology, Mesra, Ranchi-835215, India
sanchita07@gmail.com, abhisingh.bit@gmail.com,
gsahoo@bitmesra.ac.in

Abstract. Ongoing development in nanotechnology and bioinformatics will enable the construction of nanorobot which will work at nano-scale. Nanorobot development has many challenges and limitations such as its control and behaviour in different environments. In this proposed work we present DNA nanorobot design, methodology for detection of *Helicobacter pylori* (*H. Pylori*) infection and DNA nanorobot control techniques for its movement in dynamic environment are described using Fuzzy Logic (FL) rules. Propose model will detect the CagA protein in blood and subsequently indicates towards *H. pylori* infection.

Keywords: ATP, DNA Nanorobot, Fuzzy Logic, Nanomedicine, Nanorobotics, Nanotechnology, *Helicobacter pylori*.

1 Introduction

According to A. Cavalcanti "One of the major factors for successfully developing nanorobot is to bring together professionals with interdisciplinary views of science and technologies". Today, nanorobots have a wide application in medical treatment. However conventional techniques of robotics can't help in the cell or molecular scale diseases like gastric ulcers and other duodenal diseases. This needs the development of nanoscale robots to perform the operation at nanoscale precision. Just as biotechnology extends the range and efficacy of treatment options available from nanomaterials, the advent of molecular nanotechnology will again expand enormously the effectiveness, precision and speed of future medical treatments while significantly reducing their risk, cost, and invasiveness at the same time. *H. pylori* is a Gram-negative, microaerophilic bacterium that inhabits various areas of the stomach and duodenum. It causes a chronic low-level inflammation of the stomach lining and is strongly linked to the development of duodenal and gastric ulcers and stomach cancer. More than 50% of the world's population harbour *H. pylori* in their upper gastrointestinal tract. The main objective of our paper is to propose a detection methodology for *H. pylori* infection at molecular level using DNA Nanorobot. Here, DNA nanorobot movement uses fuzzy approach. Fuzzy decision making deals with uncertainty and

---

* Corresponding author.

H. Sakai et al. (Eds.): RSFDGrC 2009, LNAI 5908, pp. 287–294, 2009.

vagueness in the dynamic environment. Fuzzy logic allows defining behaviour decision rules through linguistic terms that simplify expert knowledge encoding.

2   Literature Review

M. Hamdi et. al [2] in his paper, presents a molecular mechanics study of nanorobotics structures using molecular dynamics (MD) simulations coupled to virtual reality (VR) techniques. U. A. Dubbey et.al [3] in his paper describe, that Nanorobots would constitute any active structure (nano-scale) capable of actuation, sensing, signaling, information processing, intelligence, and swarm behavior at nano-scale. The idea of using DNA to build nanoscale objects has been pioneered by Nadrian Seeman at U.S. He reported the construction of a mechanical DNA-based device that might serve as the basis for a nanoscale robotic actuator [2] [3] [5]. The *H. pylori* infection is associated with the presence of a 145 KD immunodominant, cytotoxin-associated antigen known as CagA protein coded by CagA gene. *H. Pylori* injects CagA protein into the host gastric epithelial cells through its needle-like structure. Injected CagA hijacks physiological signal transduction and causes pathological cellular response such as increased cell proliferation, motility, apoptosis and morphological change through different mechanisms [6]. There are different conventional techniques to diagnose *H. pylori* infection like blood antibody test, carbon urea breath test, biopsy etc. but none of these tests are failsafe [7]. So detecting the *H. pylori* infection with the DNA nanorobot would be a fast and new method and since CagA protein is present in almost all cases of *H. pylori* infection, this method can be trusted.

3   DNA Nanorobot Design

A DNA nanorobot design is comprised of components such as DNA, ATP motor, CNT and Rhodopsin sensor. Nanorobot is a kind of molecular machine, which includes embedded and integrated device that can comprise the main sensing, actuation,

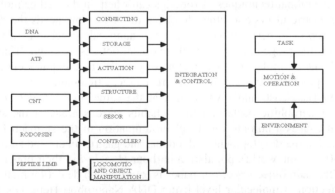

Fig. 1. Architecture of DNA Nanorobot design

data transmission, remote control uploading, and coupling power supply subsystem as given in Fig. 1. If all these different components were assembled together in the proper proportion and orientation, they would form DNA nanorobot with multiple degrees of freedom, able to apply forces and manipulate objects in the nano-scale world. Main goal of DNA nanorobot is ensuring sufficient biocompatibility for the nanorobot to avoid immune system attack. The control system must ensure a suitable performance. It can be demonstrated by giving some rules for the movement of DNA Nanorobot when it enters to the small vessel. In our work, we consider all DNA nanorobots will enter in small blood vessel for finding a small target area and will be controlled by means of Fuzzy control. The components DNA, ATP, CNT, Rhodopsin and Peptide limb are described below.

3.1 DNA

There has been a great interest and many reports in use of DNA specifically to actuate and assemble micro and nano sized system. Here, DNA could be act as motors, mechanical joints, transmission elements, sensors and information carrier.

3.2 ATP

ATP synthesis is the process within the mitochondria of a cell by which a rotary engine uses the potential difference across the lipid bilayer to power a chemical transformation of Adenosine Di-Phosphate (ADP) into Adenines Tri Phosphate (ATP). Therefore   it can be used as a motor in DNA nanorobotics. Here α and β units are opened they take in ADP, and a phosphate group. As the rotor continues to turn the section closes and the phosphate is chemically bound to the ADP to form ATP.

3.3 CNT

Carbon Nano Tubes (CNT) are cylindrical sheets of carbon. They can have tensile strength as high as sixty times larger than steel. They also show electronic stability. In DNA nanorobotics, CNT could be used as structural element while the passive/active joints are formed by appropriate designed DNA elements.

3.4 Rhodopsin

The DNA nanorobot uses sensors allowing it to detect nearby objects in the environment or its target region. Rhodopsin is a light sensor. Nanosensors include long distance sensor and short distance sensor, the former navigate the nanorobot to the target tissue, other nanorobots or obstacle. Biochemical sensors can perceive chemical grads, pH, temperature and radiation of the environment.

3.5 Peptide Limb

In above DNA nanorobot model, the arm of nanorobot is made of helical peptide which can be used for locomotion and object manipulation. Because our aim is to find the ulcer cell, so we discussed only those components which will applicable to this specific task.

4 Methodology for Detection of *H. Pylori* Infection

Main objective of our work is to detect *H. pylori* infection in body which gives a clue about occurrence of gastro duodenal diseases. Here we detect a protein produced by the bacteria called as cytotoxin associated geneA protein or CagA protein in blood. We will adopt SELEX or systematic evolution of ligands by exponential enrichment technique through which high affinity nucleotides will be synthesized [8] [1]. These oligonucleotides are called Aptemer (see fig. 2). In this technique first the pool of random sequences of DNA would be constructed as a molecular library corresponding to 'spontaneous mutation'. The library obtained would then be subjected to asymmetric PCR using only the forward primer for yielding a single-stranded DNA pool. Each single-stranded oligonucleotide would fold to form an independent three-dimensional structure according to the intra-molecular hydrogen bonds and hydrophobic interactions in each molecule. The specific oligonucleotide that could recognize the target molecule would then be selected, through the process of 'natural selection'. The next step which is 'proliferation' would be performed by PCR enrichment. Repeating the above three mentioned steps of spontaneous mutation, natural selection and proliferation, an oligonucleotide with a higher affinity for the target CagA protein would be obtained. This Aptemer is then attached with protein arm of Nanorobot and send to the Human body.

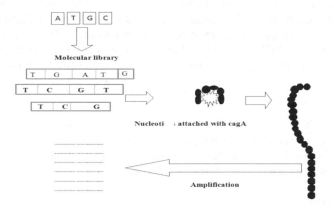

Fig. 2. SELEX technique for producing high affinity Aptamers

5 Nanorobot Movement

DNA nanorobot movement in unknown dynamic environment is one of the important factors for avoiding collision with obstacle or other DNA nanorobot. Here, we consider DNA nanorobots flowing in a blood vessel with small target area to avoid the collision with obstacle as in Fig. 3. The DNA nanorobots, designed with sensors must be capable of detecting other Nanorobots or obstacle. Basically, the DNA Nanorobot can go to these two paths. *(A) Global Path:* It is carried out based on the coverage of the particles. *(B) Local Path:* It is found for each nanorobot when obstacle or other DNA nanorobot is encountered on its currently selected path. Even though obstacle can be of different sizes, for simplicity, we have considered here both DNA nanorobot and the obstacle are circular type with radius r of same size [4]. Fig. 4 depicts two nanorobots S_i & S_j from a set of nanorobots S with radius r and coverage area πr^2. When a DNA nanorobot moves, if any obstacle or other DNA nanorobot comes in its coverage area, it detects the distance according to time t. Suppose other DNA nanorobot or obstacle has a certain position (x_i, y_i). We assume that both the DNA nanorobot and the obstacle having the same fluid velocity v_f $(v_f x_i, y_f y_i)$, now new position within time Δt can be calculated as follows:

$$x_j = x_i + v_f \, x_i * \Delta t, \qquad y_j = y_i + v_f \, y_i * \Delta t \tag{1}$$

The distance Δd between a DNA nanorobot and other DNA nanorobot or obstacle can be calculated within time Δt

$$\Delta d = \sqrt{((x_j + v_f \, x_j * \Delta t - x_i + v_f \, x_i * \Delta t)^2 + (y_j + v_f \, y_j * \Delta t - y_i + v_f \, y_i * \Delta t)^2)} \tag{2}$$

DNA nanorobot movement control Algorithm:

1. Inject all Nanorobots into the blood vessel.
2. Each Nanorobot will move to new location based on fluid velocity and the conditional statements checks if it has enough coverage based its neighbour.
3. For each neighbour S_i, $S_j \in S$, where i, j=1..n checks for the distance information.
4. If the coverage value < threshold value then
   4.1   It will find the new position based on *Fuzzy Rules*.
5. If DNA nanorobot finds the new position then
   5.1   Change the angle,
   5.2   Move to new position.
6. If nanorobot does not find any other nanorobot or obstacle then
   6.1 It will move to its path ahead until it does find the target protein.

Initially each DNA nanorobot will choose the *Global Path*, but when any obstacle is encountered at its path, it will change its direction and will move to *Local Path*. In DNA nanorobot movement control strategy, coverage area for each nanorobot has to be considered. Suppose there are two DNA nanorobots S_i, S_j with position S_i (x_i, y_i) and S_j (x_j, y_j) with radius r_i and r_j of sensor with coverage area πr^2. In the given algorithm we have applied Fuzzy Rules for the movement of DNA nanorobot when obstacle or other DNA nanorobot encountered at its path and come in DNA nanorobot

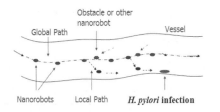

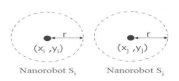

Fig. 3. DNA Nanorobots movement in small blood vessel

Fig. 4. Coverage Area of DNA nanorobot

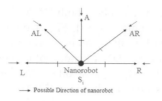

Fig. 5. DNA nanorobot movement angle

Table 1. Set of Fuzzy rule for Deviation

		Angle		
	L	AL	A	AR
Distance VN	A	AR	L	AL
N	AL	A	AL	A

coverage area. DNA nanorobot's sensor will detect the distance of the obstacle or other nanorobot when those come in the coverage area of DNA nanorobot. The obstacle or other DNA nanorobot could be at Near (N) or Very Near (VN) distance, and this distance will be given by Δd as in (2). DNA nanorobot makes decision for deviation on the basis of fuzzy rules only when obstacle or other DNA nanorobot is coming toward or at its path. Possible angle can be Left (L), Ahead Left (AL), Ahead (A), Ahead Right (AR), and Right (R) as given in Fig. 5.

6   Simulation and Result

A solution to the DNA nanorobot movement when obstacle is encountered at its path is based on Fuzzy Rules. Here each Fuzzy Rule has basically three conditions naming Distance, Angle and Deviation. *1. Distance-* The distance Δd is calculated when other nanorobot or obstacle is come in DNA nanorobot's coverage area. Distance will have two fuzzy values VN and N as given in Fig. 6. *2. Angle-* This is defined as the angle between the DNA nanorobot and other DNA nanorobot or obstacle. There are five fuzzy values considered here for the parameter angle L, AL, A, AR, and R as given in Fig. 7. *3. Deviation-* Deviation shows the movement in angle, when fuzzy rule is applied to the DNA nanorobot. Here each function is assumed as triangular membership function. When DNA nanorobot detects any obstacle in its coverage area, it calculates the distance in between DNA nanorobot and other nanorobot or obstacle, then DNA nanorobot deviate according to fuzzy rules. For deviation, there

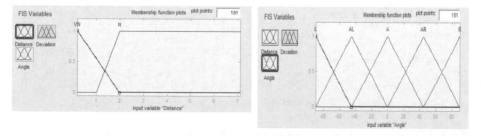

Fig. 6. Membership function for Distance   **Fig. 7.** Membership function for Angle

are five fuzzy values L, AL, A, AR, and R. In our experiment, there are two fuzzy values for Distance and five fuzzy values for Angle, so possible set of rules will be 2 x 5 or 10 combination as given in Table 1. For each combination we have given a rule. Based on these rules, DNA nanorobot will deviate its angle when obstacle will encounter in its coverage area. In the above Table 1, we can interpret, for example: If (distance is VN and angle is AL (-45°). In this condition according to fuzzy rules, deviation will be AR (45°) as given in Fig. 8. We have simulated our solution of DNA nanorobot movement with dynamic environment in MATLAB. For the deviation of DNA nanorobot, we can mention here that distance as well as angle membership functions play important roles for the deviation of DNA nanorobot.

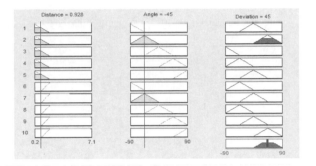

Fig. 8. Deviation using Fuzzy Rules

7   Conclusion

In this report we have proposed a method to detect *H. pylori* infection and subsequently any duodenal ulcer within human body using DNA nanorobot, and DNA nanorobot movement control algorithm for 2D dynamic environment. The simulations results have proved that the proposed scheme effectively construct an obstacle free self organized trajectory.

References

1. Turek, C., Gold, L.: Science 249, 505 (1990)
2. Hamdi, M., Ferreira, A.: DNA Nanorobotics, Laboratoire Vision et Robotique, ENSI Bourges – University d Orleans, 18000, Bourges, France, 14-15 December (2006)
3. Ummat, A., Dubey, A., Marvroidis, C.: Bio-Nanorobotics-A Field Inspired by Nature, Bar Cohen. Biommetics: Biologically Inspired Technologies DK3163 c007, 201–226 (June 22, 2005)
4. Hla, K.H.S., Choi, Y., Park, J.S.: Obstacle Avoidance Algorithm for Collective Movement in Nanorobots, Department of Computer Engineering, Korea Aerospace University, Korea. IJCSNS International Journal of Computer Science and Network security 8(11) (November 2008)

5. Sahu, S.: DNA-based self-Assembly and Nanorobotics: Theory and Experiments, Department of Computer Science Duke University
6. Asahi, M., et al.: Helicobacter pylori CagA Protein Can Be Tyrosine Phosphorylated in Gastric Epithelial Cells. The Journal of Experimental Medicine 191(4), 593–602 (2000)
7. Steustrom, B., Mendis, A., Marshall, B.: Helicobacter pylori-the latest in diagonosis and treatment. Aust Fam Physician 37(8), 608–612 (2008)
8. Ellington, A.D., Szostak, J.W.: Nature 346, 818 (1990)

Investigations on Benefits Generated by Using Fuzzy Numbers in a TOPSIS Model Developed for Automated Guided Vehicle Selection Problem

V.B. Sawant[1] and S.S. Mohite[2]

[1] Sr. Lecturer, Mechanical Engineering Department, Rajiv Gandhi Institute of Technology, Andheri (West) Mumbai, Maharashtra – India 400053
[2] Professor, Mechanical Engineering Department, Government College of Engineering, Vidyanagar, Karad, Maharashtra – India 415124
sawantvb@gmail.com, mohitess@yahoo.com

Abstract. Selection of the appropriate automated guided vehicle (AGV) for a manufacturing company is a very important but at the same time a complex problem because of the availability of wide-ranging alternatives and similarities among AGVs. Although, the available studies in the literature developed various fuzzy models, they do not propose any approaches to measure the benefits generated by incorporating fuzziness in their selection models. This paper aims to fill this gap by trying to quantify the level of benefit provided by employing the fuzzy numbers in the multi attribute decision making (MADM) models. Technique for Order Preference by Similarity to Ideal Solution (TOPSIS) is used as the MADM approach to rank the AGV in this paper. In the paper, by increasing the fuzziness level steadily in the fuzzy numbers, the obtained AGV rankings are compared with the ranking obtained with the crisp values. The statistical significance of the differences between the ranks is calculated using Spearman's rank-correlation coefficient. It can be observed from the results that as the vagueness and imprecision increases, fuzzy numbers instead of crisp numbers should be used. On the other hand, in situations where there is a low level of fuzziness or the average value of the fuzzy number can be guessed, using crisp numbers will be more than adequate.

Keywords: AGV selection, Multi Attribute Decision Making (MADM), Fuzzy numbers, Technique for Order Preference by Similarity to Ideal Solution (TOPSIS).

1 Introduction

AGVs are among the fastest growing classes of equipment in the material handling industry. They are battery-powered, unmanned vehicles with programming capabilities for path selection and positioning. They are capable of responding readily to frequently changing transport patterns and they can be integrated with fully automated intelligent control systems. These features make AGVs a viable alternative to other material handling methods, especially in flexible environments where the variety of products processed results in fluctuating transport requirements [1,2].

H. Sakai et al. (Eds.): RSFDGrC 2009, LNAI 5908, pp. 295–302, 2009.

The decision to invest in AGVs and other advanced manufacturing technology has been an issue in the practitioner and academic literature for over two decades. An effective justification process requires the consideration of many quantitative and qualitative attributes. AGV selection attribute is defined as a factor that influences the selection of an automated guided vehicle for a given application. These attributes include: costs involved, floor space requirements, maximum load capacity, maximum travel speed, maximum lift height, minimum turning radius, travel patterns, programming flexibility, labor requirements, expansion flexibility, ease of operation, maintenance aspects, payback period, reconfiguration time, company policy, etc.

Kim and Eom [3] introduced a material handling selection expert system. Fisher et al. [4] introduced MATHES, the 'material handling equipment selection expert systems', for the selection of material handling equipment from 16 possible choices. Chan et al. [5] described the development of an intelligent material handling equipment selection system called material handling equipment selection advisor (MHE-SA). Fonseca et al. [6] developed expert decision support systems for the selection of material handling equipments. Kulak [7] developed a decision support system called FUMAHES-fuzzy multi-attribute material handling equipment selection. Chakraborthy and Banik [8] focused on the application of the analytic hierarchy process (AHP) technique in selecting the optimal material handling equipment for a specific material handling equipment type.

However, the studies mentioned above mostly used crisp data except [7] and do not take into account the uncertainties and imprecision that may be associated with the decision-maker's judgments. Fuzzy MADM approaches are proposed for selection problems where vagueness and imprecision are involved in the literature. Although detailed descriptions are provided in the literature, various issues of the fuzzy MADM approaches are not explored yet. Such a key issue is the justification for the usage of fuzzy versions of the MADM approaches. A study that compares usage of the fuzzy and crisp numbers or one that provides recommendations about when one is preferred over the other is not available in the literature. This paper tries to fill this gap by determining a level of fuzziness (a threshold value) that warns the users to start using fuzzy MADM approach instead of its crisp version. It is expected that above the calculated threshold value, the level of uncertainties and imprecision is high enough to justify the usage of fuzzy MADM approaches. On the other hand, a fuzziness level below the threshold value indicates that the benefit of the usage of fuzzy numbers is minimal. The benefit of using fuzzy numbers instead of crisp ones can be measured with the statistical significance of the difference between the rankings obtained using fuzzy numbers and crisp values. When the statistical significance of the difference is above a pre-defined value, it is recommended to use the fuzzy numbers instead of crisp ones. In this study, the measure of fuzziness level is first defined and then various cases are developed by varying (steadily increasing) the level of fuzziness and compared with the ranking obtained using crisp numbers.

Although there are many different fuzzy MADM approaches to calculate the rating scores and rankings of the AGVs, the technique for order preference by similarity to ideal solution, developed by Hwang and Yoon [9], will be used as the ranking method in this paper. The advantage of this method is its simplicity and ability to yield an indisputable preference order Sen and Yang [10]. Steps and application details of the fuzzy TOPSIS approach are presented in [10, 11, 12, 13]. The rankings are obtained with the application of the TOPSIS approach for various cases. For each case, the statistical significance of the difference between the ranking obtained for the fuzzy criteria weights and the one obtained for the crisp weight values is determined using Spearman's rank-correlation test. Spearman's rank-correlation test, which is a special form of correlation test, is used when 'the actual values of paired data are substituted with the ranks which the values occupy in the respective samples [14]. In this study, Spearman's test evaluates the similarity of the outcomes (rankings of the AGVs for various cases) of the TOPSIS approach. In its application in the paper, to test the null hypothesis (H_0: there is no similarity between the two rankings), a test statistic, Z, is calculated using Eqs. (1) and (2) and compared with a pre-determined level of significant α value. For example, if 3.5, which corresponds to the critical Z-value at the level of significance of $\alpha = 0.0002$ is selected and the test statistic computed by Eqs. (2) exceeds 3.5, the null hypothesis is rejected and it is to be concluded that 'H_1: the two rankings are similar' is true. Z-value itself can also be used as a measure of similarity of rankings. A higher Z-value shows a higher similarity between any two rankings.

$$rs = 1 - [\, 6 \, \Sigma_{j=1}^{k} dj^2 \,/\, K\,(K^2 - 1)\,]$$ (1)

$$Z = r_s \sqrt{K - 1}$$ (2)

In Eqs. (1) and (2), d_j is the difference in alternative ranks for the jth AGV and K is the number of alternatives to be compared. r_s represent the Spearman's rank-correlation coefficient.

2 Application of the Fuzzy TOPSIS and Spearman's Rank-Correlation Approaches for Different Levels of Fuzziness

An example is developed to explain and illustrate the analysis of fuzziness. Nine AGVs and six criteria are selected for the example, and the AGVs' performance values at the selected criteria are provided in Table 1. TOPSIS approach requires the weights of criteria and AGV performance values at the criteria as inputs. It should be noted that, in the application of the TOPSIS approach, since the AGVs' performance values at criteria are crisp values (see Table 1), only criteria weights are required to be the fuzzy numbers.

2.1 Analysis of the Fuzziness in Terms of Spread Only While Keeping the Center of the Fuzzy Numbers Constant

In the first part of section 2, six cases (cases B–G) are developed by varying the lower and upper values of fuzzy numbers while keeping their centers (mean or average) constant at their crisp values given in case A (see Table 2). The level of fuzziness is determined with the 'spread' in the developed cases. In this study the spread is defined as the number of units that the lower and upper values apart in a trapezoidal or triangular fuzzy number and takes the values 0, 1, 2, . . ., 6. It is assumed that as the level of imprecision and uncertainty increases, this increase translates into an increase in the spread (difference) between the lower and upper values of the fuzzy number. For each case, the criteria weights are readjusted by equaling the number of units.

Table 1. AGV Selection Attribute data

Alternative AGV	MLC	MS	LH	PC	TR	PF
1	50	148	20	105	400	0.335
2	65	197	20	120	600	0.5
3	300	131	20	130	400	0.590
4	681	300	20	160	500	0.665
5	1815	300	0.5	140	1000	0.665
6	5443	525	20	180	1219	0.745
7	9072	350	20	190	1219	0.745
8	13,608	350	40	210	1219	0.865
9	22,680	350	30	235	1219	0.865

Max Load Capacity in Kg (**MLC**), Max Travel Speed in ft/min (**MS**), Maximum Lift Height in ft (**LH**), Purchasing Cost in 100x$ (**PC**), Minimum turning radius in mm (**TR**), Programming flexibility(**PF**).

For each case, the ranking scores and rankings are obtained for AGVs using the criteria weights (see Table 2) and AGV performance values at the selection criteria (see Table 1) and provided in Table 3. The obtained rankings for cases B–G are compared with the ranking obtained for case A (see Table 4). The comparison is performed by taking the difference of the ranks of the AGVs (the columns under the headings 'A–B', 'A–C', ..., 'A–G' in Table 4) and then calculating spearman's correlation coefficients (Z-values) for each difference (last row in Table 4).

The calculated Z-values are further illustrated in Fig. 1. It can be observed from the figure that as the level of fuzziness increases, the similarity of the rankings does not show any changes. However, even the lowest Z-value in Fig. 1, 2.781, corresponds to the level of significance = 0.0003 and indicates that using fuzzy numbers instead of crisp values does not provide any meaningful differences in the rankings of cases B–G compared to case A. This observation is supported by checking the differences of the ranks of the AGVs provided in Table 4. Cases D, E, F and G provided the exact ranking with case A. For case C the highest difference in rankings is 1 for AGV 7. To conclude the benefit resulted from changing the spread without changing the center of the fuzzy number is statistically insignificant; in such cases the user can use crisp values as the criteria weights confidently.

2.2 Analysis of the Fuzziness in Terms of Spread and Center (Average Value) Together in the Fuzzy Numbers

In the cases developed so far, it is assumed that the average values are known with certainty but the users are not sure about the spreads of the numbers. However, in different situations, the average values may also be unknown along with the spread of the fuzzy number which implies increased uncertainties and imprecision.

Table 2. Criteria Weights for Seven different Cases (A-G)

Criteria Weights	Crisp number, Case A spread value: 0	Triangular fuzzy number, Case B spread value: 1	Triangular fuzzy number, Case C spread value: 2	Trapezoidal fuzzy number, Case D spread value: 3	Trapezoidal fuzzy number, Case E spread value: 4	Trapezoidal fuzzy number, Case F spread value: 5	Trapezoidal fuzzy number, Case G spread value: 6
MLC in Kg	9	8,9,9	8,9,10	7,8,9,10	6,8,9,10	5,7,9,10	4,7,9,10
MS in ft/min	8	7,8,8	7,8,9	6,7,8,9	5,7,8,9	5,6,9,10	3,5,8,9
LH in ft	2	1,2,2	1,2,3	1,2,3,4	1,2,3,5	1,2,3,6	1,2,4,7
PC	3	2,3,3	2,3,4	2,3,4,5	1,3,4,5	1,2,3,6	1,2,5,7
TR in mm	4	3,4,4	3,4,5	3,4,5,6	2,4,5,6	2,3,4,7	2,4,7,8
PF	5	4,5,5	4,5,6	4,5,6,7	3,5,6,7	4,5,7,9	3,5,8,9

Table 3. Scores and ranks of AGVs for Cases (A-G)

Alternative AGV	Case A		Case B		Case C		Case D		Case E		Case F		Case G	
	S	R	S	R	S	R	S	R	S	R	S	R	S	R
1	0.933	1	0.941	1	0.913	1	0.915	1	0.905	1	0.894	1	0.876	1
2	0.933	2	0.941	2	0.913	2	0.915	2	0.905	2	0.894	2	0.876	2
3	0.903	3	0.909	3	0.875	3	0.883	3	0.877	3	0.859	3	0.842	3
4	0.804	4	0.808	4	0.745	4	0.794	4	0.791	4	0.770	4	0.771	4
5	0.782	5	0.784	5	0.724	5	0.773	5	0.776	5	0.761	5	0.757	5
6	0.572	6	0.574	6	0.486	7	0.565	6	0.567	6	0.547	6	0.555	6
7	0.547	7	0.549	7	0.510	6	0.538	7	0.539	7	0.530	7	0.521	7
8	0.385	8	0.388	8	0.374	8	0.375	8	0.373	8	0.367	8	0.350	8
9	0.164	9	0.163	9	0.214	9	0.162	9	0.163	9	0.174	9	0.157	9

S- Score, R-Rank

Table 4. The differences and correlation values of the Cases B–G compared with Case A

Alternative AGV	A-B	A-C	A-D	A-E	A-F	A-G
1	0	0	0	0	0	0
2	0	0	0	0	0	0
3	0	0	0	0	0	0
4	0	0	0	0	0	0
5	0	0	0	0	0	0
6	0	-1	0	0	0	0
7	0	1	0	0	0	0
8	0	0	0	0	0	0
9	0	0	0	0	0	0
$(d^k)^2$	0	2	0	0	0	0
r_s	1.0	0.983	1.0	1.0	1.0	1.0
Z	2.828	2.781	2.828	2.828	2.828	2.828

To incorporate the shift in the average values of the fuzzy numbers, new cases are developed. In the new cases, the fuzzy numbers are shifted to either left (Cases H–M) or right (Cases N–S) or both directions (Cases T–Y) while the spread is again increased steadily. New cases are again compared with Case A whose crisp criteria weights are not changed from the values provided in Table 2. The calculated Z-values along with the Z-values for the Cases B–G are provided in Fig. 1.

The comparisons of the Z-values obtained for 'shifted to left', 'shifted to right' and 'shifted to both directions' with the 'no-shift' criteria weights (Cases A–G) show that as the fuzziness, which is represented with the shift of the average and the spread together, increases, the similarity of the ranking with the one obtained with the crisp numbers has no effect.

It can be concluded that the statistical significance of the benefit, which is defined in terms of the difference between the ranking obtained with the fuzzy numbers and the one obtained with the crisp values, is minimal. Especially the user can use crisp criteria weights confidently at low fuzziness levels. On the other hand, the magnitude of benefit of using fuzzy numbers increases when the decision maker is not sure about the spread and the mean of the fuzzy number at the same time. In such a case, although the statistical significance values of the differences in AGV rankings are still very small, using fuzzy numbers are recommended to quantify the criteria weights. Even small changes in the rankings may lead to the elimination of the best AGV and selection of a less qualified AGV.

3 Results and Discussions

This paper aims to measure the benefits of using fuzzy numbers instead of crisp ones in a TOPSIS AGV selection model. In the model, the fuzziness level is presented in terms of the spread (the difference between the lower and upper values) and shift of the mean of the fuzzy numbers. Various cases are developed by increasing the spread and shifting the mean of fuzzy numbers for criteria weights. The benefit, which is defined in terms of the ranking differences, is measured as the statistical significance of the difference between the rankings obtained with the ranking obtained using crisp numbers. Fig. 1 clearly shows that as the fuzziness level increases, the similarity of the ranking has no effect in this problem. The calculated Z-values clearly show that the ranking differences are not statistically significant even when the level of fuzziness increased. It can be concluded that when the fuzziness level is low and especially when the mean value of the fuzzy number can be approximately guessed, the benefit of using fuzzy numbers is minimal.

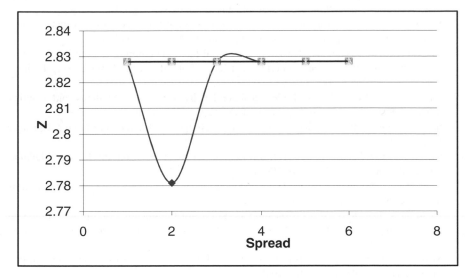

Fig. 1. Graphic illustration of Spearman rank-correlation test results for Cases A–Y

4   Conclusions

In this paper a method to determine benefits obtained by using fuzzy numbers in TOPSIS model for AGV selection problem has been presented. However, the problem considered for analysis does not reveal any major benefit of fuzzy number variation; different problems have to be analyzed so that study can be used in the real life decision making process.

Acknowledgements. The authors would like to thank three anonymous referees for their helpful comments.

References

1. Sule, D.R.: Manufacturing facilities: location, planning and design. PWS Publishing Company, Boston (1994)
2. Sujono, S., Lashkari, R.S.: A multi-objective model of operation allocation and material handling system selection in FMS design. International Journal of Production Economics 105, 116–133 (2007)

3. Kim, K.S., Eom, J.K.: Expert system for selection of material handling and storage systems. International Journal of Industrial Engineering 4, 81–89 (1997)
4. Fisher, E.L., Farber, J.B., Kay, M.G.: MATHES: an expert system for material handling equipment selection. Engineering Costs and Production Economics 14, 297–310 (1998)
5. Chan, F.T.S., Ip, R.W.L., Lau, H.: Integration of expert system with analytic hierarchy process for the design of material handling equipment selection system. Journal of Materials Processing Technology 116, 137–145 (2001)
6. Fonseca, D.J., Uppal, G., Greene, T.J.: A knowledge-based system for conveyor equipment selection. Expert Systems with Applications 26, 615–623 (2004)
7. Kulak, O.: A decision support system for fuzzy multi-attribute selection of material handling equipments. Expert Systems with Applications 29, 310–319 (2005)
8. Chakraborthy, S., Banik, D.: Design of a material handling equipment selection model using analytic hierarchy process. International Journal of Advanced Manufacturing Technology 28, 1237–1245 (2006)
9. Hwang, C.L., Yoon, K.: Multiple Attribute Decision Making—Methods and Applications—A State of Art Survey. Springer, Berlin (1982)
10. Sen, P., Yang, J.-B.: Multiple Criteria Decision Support in Engineering Design. Springer, London (1998)
11. Chu, T.-C., Lin, Y.-C.: A fuzzy TOPSIS method for robot selection. Int. J. Adv.. Manufacturing. Technology 21, 284–290 (2003)
12. Chen, C.-T.: Extensions of the TOPSIS for group decision-making under fuzzy environment. Fuzzy Sets System 114, 1–9 (2000)
13. Byun, H.S., Lee, K.H.: A decision support system for the selection of a rapid prototyping process using the modified TOPSIS method. Int. J. Adv. Manufacturing. Technology 26, 1338–1347 (2004)
14. Miller, I., Freund, J.E., Johnson, R.A.: Probability and Statistics for Engineers, 4th edn. Prentice Hall, Englewood Cliffs (1990)

A Multi-criteria Interval-Valued Intuitionistic Fuzzy Group Decision Making for Supplier Selection with TOPSIS Method

Kavita[1], Shiv Prasad Yadav[1], and Surendra Kumar[2]

[1] Department of Mathematics, Indian Institute of Technology Roorkee, Roorkee, India
[2] Department of Electrical Engineering, Indian Institute of Technology Roorkee, Roorkee, India
`kavita.iitr21@gmail.com, yadavfma@iitr.ernet.in,`
`kumarfee@iitr.ernet.in`

Abstract. The Technique for Order Preference by Similarity to Ideal Solution (TOPSIS) method, evaluation of alternatives based on weighted attributes play an important role in the best alternative selection. Practically it is difficult to precisely measure the exact values to the relative importance of the attributes and to the impacts of the alternatives on theses attributes. Therefore, the TOPSIS method has been extended for interval-valued intuitionistic fuzzy data in this paper, to tackle this problem. In addition, supplier selection problem a multi-criteria group decision making problem involving several conflicting criteria is solved with the proposed methodology.

Keywords: Interval-valued intuitionistic fuzzy set, TOPSIS, Decision making.

1 Introduction

Multi-criteria decision making (MCDM) is the most well-known branch of decision making. It is a branch of a general class of operations research models that deals with decision problems under the presence of a number of decision criteria. The MCDM approach requires that the selection be made among decision alternatives described by their attributes. MCDM problems are assumed to have a predetermined, limited number of decision alternatives. Solving a MCDM problem involves sorting and ranking. MCDM approaches can be viewed as alternative methods for combining the information in a problem's decision matrix together with additional information from the decision maker to determine a final ranking or selection from among the alternatives. Besides the information contained in the decision matrix, all but the simplest MCDM techniques require additional information from the decision maker to arrive at a final ranking or selection.

MCDM problems and their evaluation process usually involve subjective assessments, resulting with imprecise data in qualitative manner. Engineering or management decisions are generally made through available data and information that are mostly vague, imprecise, and uncertain by nature. The decision-making process in engineering schemes, developed in the concept-designing phase, is one of these typical occasions, which usually need some methods to deal with uncertain data and

H. Sakai et al. (Eds.): RSFDGrC 2009, LNAI 5908, pp. 303–312, 2009.

information that are hard to define. In designing phase, designers usually present many alternatives. However, the subjective characteristics of the alternatives are generally uncertain and need to be evaluated through decision maker's insufficient knowledge and judgments. The nature of this kind of vagueness and uncertainty is fuzzy rather than random, especially when subjective assessments are involved in the decision-making process. Fuzzy set theory [18] offers a possibility for handling these sorts of data and information involving the subjective characteristics of human nature in the decision-making process.

There, exist several methods to solve MADM problems, out of which the Technique for Order Preference by Similarity to Ideal Solution (TOPSIS) developed by Hwang and Yoon [7], is one of the well-known method. The basic principle of the TOPSIS method is that the chosen alternative should have the shortest distance from the positive ideal-solution and the farthest distance from the negative ideal- solution. There exist a large amount of literature involving TOPSIS theory and applications. In classical Multi Criteria Decision Making (MCDM) methods, the ratings and the weights of the criteria are known precisely. A survey of the methods has been presented in Hwang and Yoon [7]. In the process of TOPSIS the performance ratings and the weights of the criteria are given exact values. Jahanshahloo et al [8, 9, 10] extended the concept of TOPSIS to develop a methodology for solving MCDM problem with interval data. Triantaphyllou and Lin [11] develop a fuzzy version of the TOPSIS method based on fuzzy arithmetic operations. Wang and Elhag [12] extended TOPSIS to provide a fuzzy form of closeness co-efficient through α-cuts propagation. Guangtao Fu [6] proposed a fuzzy optimization method based on TOPSIS and demonstrated a case study of reservoir flood control operation. Most of the fuzzy versions of TOPSIS method are efficient in tackling the impreciseness and vagueness present in MCDM problems, but their results are not able to include the hesitation present in the information provided by the decision maker.

In real life, there exist many situations when available information is not sufficient for the exact definition of degree of membership for certain elements. There may be some hesitation degree between membership and non-membership. Atanassov [1] introduced intuitionistic fuzzy sets (IFSs) to tackle this hesitation degree, which is a generalization of the concept of fuzzy sets. IFSs has received a lot of attention. Gau and Buehrer [5] introduced the concept of vague sets (VSs), which is another generalization of fuzzy sets. But Bustince and Burillo [4] point out that the notion of VSs is the same as that of IFSs. Xu and Yager [16] developed some geometric aggregation operators, such as the intuitionistic fuzzy weighted geometric (IFWG) operator, the intuitionistic fuzzy ordered weighted geometric (IFOWG) operator and the intuitionistic fuzzy hybrid geometric (IFHG) operator, and gave an application of the IFHG operator to multicriteria decision-making problems with intuitionistic fuzzy information. Xu [14] developed some arithmetic aggregation operators, such as the intuitionistic fuzzy weighted averaging (IFWA) operator, the intuitionistic fuzzy ordered weighted averaging (IFOWA) operator and the intuitionistic fuzzy hybrid aggregation (IFHA) operator. Later, Atanassov and Gargov [2] introduced the concept of interval-valued intuitionistic fuzzy sets (IVIFSs) as a further generalization of that of IFSs. Atanassov [3] defined some operational laws of the IVIFSs. Recently, Xu and Chen [17]developed some arithmetic aggregation operators, such as the interval-valued intuitionistic fuzzy weighted averaging (IIFWA) operator, the interval-valued intuitionistic fuzzy ordered weighted averaging (IIFOWA) operator and the interval-valued intuitionistic fuzzy hybrid aggregation (IIFHA) operator, and gave an application of the IIFHA operator to multicriteria decision-making problems with interval-valued intuitionistic fuzzy information by using the score function and accuracy function of interval-valued intuitionistic fuzzy numbers.

This paper proposes an interval-valued intuitionistic fuzzy multi- criteria group decision making with TOPSIS method for supplier selection problem. The importance of the criteria and the impact of alternatives on criteria provided by decision makers are difficult to precisely express by crisp data in the selection of supplier problem. IVIFSs are efficient to deal this challenge and applied in many decision making problem under uncertain environment. In group decision making problems, aggregation of expert opinions is very important to appropriately perform evaluation process. Therefore, IIFWA is utilized to aggregate all individual decision makers opinions for rating the importance of criteria and the alternatives. TOPSIS method combined with IVIFSs has enormous chance of success for supplier selection process. In this paper we have developed TOPSIS with IVIFSs to solve MCDM problems in which the performance rating values as well as the weights of criteria are taken as IVIFSs. The remaining of this paper is organised as follows. In the next section, definition, notations and some arithmetical operations of IVIFSs sets are briefly introduced. TOPSIS method based on IVIFSs sets is then proposed in section3. A numerical example and a short conclusion are given in sections 4 and 5, respectively.

2 Interval Valued Intuitionistic Fuzzy Set

Let X be a non empty set of the universe, and $D[0,1]$ be the set of all closed subintervals of $[0,1]$, an interval-valued intuitionistic fuzzy set (IVIFS) A in X is defined by

$$A = \{< x, \mu_A(x), v_A(x) > | x \in X \} \tag{1}$$

where $\mu_A : X \to D[0,1]$, $v_A : X \to D[0,1]$, with the condition $0 \le \sup \mu_A(x) + \sup v_A(x) \le 1$ for any $x \in X$. The intervals $\mu_A(x)$ and $v_A(x)$ denote, respectively, the degree of membership and non-membership of the element x to the set A. Here for each $x \in X$, $\mu_A(x)$ and $v_A(x)$ are closed intervals and their lower and upper end points are denoted by $\mu_{AL}(x)$, $\mu_{AU}(x)$, $v_{AL}(x)$ and $v_{AU}(x)$, respectively. Therefore, now with these end points IVIFS A can be expressed as

$$A = \left\{ \left\langle x, [\mu_{AL}(x), \mu_{AU}(x)], [v_{AL}(x), v_{AU}(x)] \middle| x \in X \right\rangle \right\} \tag{2}$$

Also for each element x we can compute the unknown degree (hesitancy degree) of an IVIFS A defined as follows:

$$\pi_A(x) = 1 - \mu_A(x) - \nu_A(x)$$
$$= [1 - \mu_{AU}(x) - \nu_{AU}(x), 1 - \mu_{AL}(x) - \nu_{AL}(x)]. \tag{3}$$

Let IVIFS (X) is the set of all the IVIFSs in X. For any given x, the pair $(\mu_A(x), \nu_A(x))$ is called an interval valued intuitionistic fuzzy number (IVIFN) [15]. For convenience the pair is often denoted by $A = ([a,b],[c,d])$, where $[a,b] \in D[0,1]$, $[c,d] \in D[0,1]$ and $b + d \leq 1$. Let $A = ([a_1,b_1],[c_1,d_1])$ and $B = ([a_2,b_2],[c_2,d_2])$ be any two IVIFNs, then some of their arithmetic operations are as follows:

(1) $A + B = ([a_1 + a_2 - a_1 a_2, b_1 + b_2 - b_1 b_2], [c_1 c_2, d_1 d_2])$ \qquad (4)

(2) $A.B = ([a_1 a_2, b_1 b_2], [c_1 + c_2 - c_1 c_2, d_1 + d_2 - d_1 d_2])$ \qquad (5)

(3) $\lambda A = ([1-(1-a_1)^\lambda, 1-(1-b_1)^\lambda], [c_1^\lambda, d_1^\lambda]), \quad \lambda > 0$ \qquad (6)

3   Interval-Valued Intuitionistic Fuzzy TOPSIS

Suppose $A_1, A_2, \ldots, A_m$ are m possible alternatives among which decision makers have to choose, $C_1, C_2, \ldots, C_n$ are criteria with which alternative performance are measured. The decision problem is to select a most preferred alternative from given alternatives or obtain a ranking of all alternatives. More specifically, let $([a_{ij}, b_{ij}], [c_{ij}, d_{ij}])$ be the interval-valued intuitionistic fuzzy number, where $[a_{ij}, b_{ij}]$ indicates the degree that alternative A_i satisfies the criterion C_j given by the decision maker, $[c_{ij}, d_{ij}]$ indicates the degree that alternative A_i does not satisfies the criterion C_j, further $[a_{ij}, b_{ij}] \subset D[0,1]$, $[c_{ij}, d_{ij}] \subset D[0,1]$, and $b_{ij} + d_{ij} \leq 1$, $i = 1,2,\ldots,n, j = 1,2,\ldots,m$. Here $[a_{ij}, b_{ij}]$ is the lowest satisfaction degree of A_i with respect to C_j as given in the membership function and $[1 - c_{ij}, 1 - d_{ij}]$ is the highest satisfaction degree of A_i with respect to C_j, in the case that all hesitation is treated as membership or satisfaction.

Assume that decision group contains l decision makers. The importance/weights of the decision makers are considered as crisp terms. Now, aggregated interval-valued intuitionistic fuzzy decision matrix constructed based on the opinions of decision makers. Let $D^k = \left(r_{ij}^{(k)}\right)_{m \times n}$ is an interval-valued intuitionistic fuzzy decision matrix for k^{th} $(k = 1,2,\ldots,l)$ decision maker and $\lambda = \{\lambda_1,\lambda_2,\ldots,\lambda_l\}$ is the weight for

each decision matrix and $\sum_{k=1}^{l} \lambda_k = 1,\ \lambda_k \in [0,1]$. In group decision-making process,

all the individual decision opinions need to be fused into a group opinion to construct aggregated interval-valued intuitionistic fuzzy decision matrix. In order to do that, IIFWA operator proposed by Xu and Chen [17] is used. $D = \left(r_{ij}\right)_{n \times m}$, where

$$r_{ij} = IIFWA_\lambda \left(r_{ij}^{(1)}, r_{ij}^{(2)}, \ldots, r_{ij}^{(l)}\right)$$

$$= \left(\left[1 - \prod_{k=1}^{l}(1 - a_{ij}^{(k)})^{\lambda_k},\ 1 - \prod_{k=1}^{l}(1 - b_{ij}^{(k)})^{\lambda_k}\right], \left[\prod_{k=1}^{l}\left(c_{ij}^{(k)}\right)^{\lambda_k},\ \prod_{k=1}^{l}\left(d_{ij}^{(k)}\right)^{\lambda_k}\right]\right) \tag{7}$$

The aggregated intuitionistic fuzzy decision matrix can be defined as follows:

$$D = \begin{bmatrix} r_{11} & r_{12} & r_{13} & \cdots & r_{1n} \\ r_{21} & r_{22} & r_{23} & \cdots & r_{2n} \\ r_{31} & r_{32} & r_{33} & \cdots & r_{3n} \\ \vdots & \vdots & \vdots & \ddots & \vdots \\ r_{m1} & r_{m2} & r_{m3} & \cdots & r_{mn} \end{bmatrix}$$

All criteria may not be assumed to be of equal importance. Let W represents a set of grades of importance for given criteria's. In order to obtain W, all the individual decision maker opinions for the importance of each of criteria need to be combined. Let $w_j^{(k)} = \left(\left[a_j^{(k)}, b_j^{(k)}\right], \left[c_j^{(k)}, d_j^{(k)}\right]\right)$ be an IVIFN assigned to criterion C_j by the kth decision maker. Then the weights of the criteria are calculated by using IIFWA operator:

$$w_j = IIFWA_\lambda \left(w_j^{(1)}, w_j^{(2)}, \ldots, w_j^{(l)}\right)$$

$$= \left(\left[1 - \prod_{k=1}^{l}(1 - a_j^{(k)})^{\lambda_k},\ 1 - \prod_{k=1}^{l}(1 - b_j^{(k)})^{\lambda_k}\right], \left[\prod_{k=1}^{l}\left(c_j^{(k)}\right)^{\lambda_k},\ \prod_{k=1}^{l}\left(d_j^{(k)}\right)^{\lambda_k}\right]\right) \tag{8}$$

$W = [w_1, w_2, w_3, \ldots, w_l]$, here $w_j = \left(\left[a_j, b_j\right], \left[c_j, d_j\right]\right),\ j = 1,2,\ldots,n$. After the weights of criteria (W) and the aggregated interval valued intuitionistic fuzzy decision matrix are determined, the aggregated weighted interval-valued intuitionistic fuzzy decision matrix is constructed according to the definition [17]. The aggregated weighted interval-valued intuitionistic fuzzy decision matrix can be represented as follows:

$$D' = D \otimes W = \left(r_{ij}^{'} \right)_{m \times n} \tag{9}$$

$r_{ij}^{'} = \left([a_{ij}^{'}, b_{ij}^{'}], [c_{ij}^{'}, d_{ij}^{'}] \right)$ is an element of the aggregated weighted interval-valued intuitionistic fuzzy decision matrix. Let I be the collection of benefit attributes and O be the collection cost attributes. The interval-valued intuitionistic fuzzy positive-ideal solution, denoted as A^{+}, and the interval-valued intuitionistic fuzzy negative-ideal solution, denoted as A^{-}, are defined as follows:

$$A^{+} = \left(\left([a_1^{+}, b_1^{+}], [c_1^{+}, d_1^{+}] \right), \left([a_2^{+}, b_2^{+}], [c_2^{+}, d_2^{+}] \right), \ldots, \left([a_n^{+}, b_n^{+}], [c_n^{+}, d_n^{+}] \right) \right) \tag{10}$$

$$\left([a_j^{+}, b_j^{+}], [c_j^{+}, d_j^{+}] \right) = \left(\begin{array}{c} \left\langle \left[\max_i a_{ij}, \max_i b_{ij} \right], \left[\min_i c_{ij}, \min_i d_{ij} \right] \middle| j \in I \right\rangle, \\ \left\langle \left[\min_i a_{ij}, \min_i b_{ij} \right], \left[\max_i c_{ij}, \max_i d_{ij} \right] \middle| j \in O \right\rangle \end{array} \right), \quad j = 1, 2, \ldots, n. \tag{11}$$

$$A^{-} = \left(\left([a_1^{-}, b_1^{-}], [c_1^{-}, d_1^{-}] \right), \left([a_2^{-}, b_2^{-}], [c_2^{-}, d_2^{-}] \right), \ldots, \left([a_n^{-}, b_n^{-}], [c_n^{-}, d_n^{-}] \right) \right) \tag{12}$$

$$\left([a_j^{-}, b_j^{-}], [c_j^{-}, d_j^{-}] \right) = \left(\begin{array}{c} \left\langle \left[\min_i a_{ij}, \min_i b_{ij} \right], \left[\max_i c_{ij}, \max_i d_{ij} \right] \middle| j \in I \right\rangle, \\ \left\langle \left[\max_i a_{ij}, \max_i b_{ij} \right], \left[\min_i c_{ij}, \min_i d_{ij} \right] \middle| j \in O \right\rangle \end{array} \right), \quad j = 1, 2, \ldots, n. \tag{12}$$

The separation between alternatives will be found according to the Euclidean distance measure as follows:

$$S_i^{+}(A_i, A^{+}) = \left\{ \frac{1}{4} \sum_{j=1}^{n} \left[\left(a_{ij} - a_j^{+} \right)^2 + \left(b_{ij} - b_j^{+} \right)^2 + \left(c_{ij} - c_j^{+} \right)^2 + \left(d_{ij} - d_j^{+} \right)^2 \right] \right\}^{\frac{1}{2}} \tag{13}$$

$$S_i^{-}(A_i, A^{-}) = \left\{ \frac{1}{4} \sum_{j=1}^{n} \left[\left(a_{ij} - a_j^{-} \right)^2 + \left(b_{ij} - b_j^{-} \right)^2 + \left(c_{ij} - c^{-} \right)^2 + \left(d_{ij} - d_j^{-} \right)^2 \right] \right\}^{\frac{1}{2}} \tag{14}$$

The relative closeness of an alternative A_i with respect to the interval-valued intuitionistic fuzzy positive ideal solution A^{+} is defined by the following general formula:

$$RC_i^{+} = \frac{S_i^{-}}{S_i^{-} + S_i^{+}} \tag{15}$$

where $0 \leq RC_i^{+} \leq 1$ and $i = 1, 2, \ldots, m$. Then the preference order of alternatives can be ranked according to descending order of RC_i^{+}'s.

4 Numerical Example

A high-technology manufacturing company desires to select a suitable material supplier to purchase the key components of new products. After preliminary screening, four candidates (A_1, A_2, A_3, A_4) remain for further evaluation. A committee of three decision-makers D_1, D_2 and D_3,with weight vector $\lambda = (0.35, 0.35, 0.30)'$, has been formed to select the most suitable supplier. Four criteria are considered:

C_1: Product quality,

C_2 : Relationship closeness,

C_3 : Delivery performance,

C_4 : Price.

The proposed method is currently applied to solve this problem, the computational procedure is as follows: The decision makers $D_k (k = 1,2,3)$ compare each pair of the criteria's $C_i (i = 1,2,3,4)$, and construct, the following three interval-valued intuitionistic fuzzy judgment matrices, respectively:

$$D^1 = \begin{bmatrix} ([0.5,0.5],[0.5,0.5]) & ([0.6,0.7],[0.10.2]) & ([0.5,0.6],[0.2,0.3]) & ([0.3,0.5],[0.2,0.4]) \\ ([0.1,0.2],[0.6,0.7]) & ([0.5,0.5],[0.5,0.5]) & ([0.4,0.6],[0.1,0.2]) & ([0.6,0.7],[0.1,0.3]) \\ ([0.2,0.3],[0.5,0.6]) & ([0.1,0.2],[0.4,0.6]) & ([0.5,0.5],[0.5,0.5]) & ([0.3,0.4],[0.5,0.6]) \\ ([0.2,0.4],[0.3,0.5]) & ([0.1,0.3],[0.6,0.7]) & ([0.5,0.6],[0.3,0.4]) & ([0.5,0.5],[0.5,0.5]) \end{bmatrix}$$

$$D^2 = \begin{bmatrix} ([0.5,0.5],[0.5,0.5]) & ([0.2,0.3],[0.50.6]) & ([0.5,0.7],[0.1,0.2]) & ([0.2,0.4],[0.1,0.3]) \\ ([0.5,0.6],[0.2,0.3]) & ([0.5,0.5],[0.5,0.5]) & ([0.5,0.8],[0.1,0.2]) & ([0.3,0.6],[0.2,0.3]) \\ ([0.1,0.2],[0.5,0.7]) & ([0.1,0.2],[0.5,0.8]) & ([0.5,0.5],[0.5,0.5]) & ([0.4,0.6],[0.1,0.4]) \\ ([0.1,0.3],[0.2,0.4]) & ([0.2,0.3],[0.3,0.6]) & ([0.1,0.4],[0.4,0.6]) & ([0.5,0.5],[0.5,0.5]) \end{bmatrix}$$

$$D^3 = \begin{bmatrix} ([0.5,0.5],[0.5,0.5]) & ([0.4,0.5],[0.20.3]) & ([0.6,0.7],[0.1,0.2]) & ([0.5,0.7],[0.2,0.3]) \\ ([0.2,0.3],[0.4,0.5]) & ([0.5,0.5],[0.5,0.5]) & ([0.5,0.6],[0.2,0.4]) & ([0.7,0.8],[0.1,0.2]) \\ ([0.1,0.2],[0.6,0.7]) & ([0.2,0.4],[0.5,0.6]) & ([0.5,0.5],[0.5,0.5]) & ([0.6,0.7],[0.1,0.3]) \\ ([0.2,0.3],[0.5,0.7]) & ([0.1,0.2],[0.7,0.8]) & ([0.1,0.3],[0.6,0.7]) & ([0.5,0.5],[0.5,0.5]) \end{bmatrix}$$

Now the aggregated interval-valued intuitionistic fuzzy decision matrix based on the opinions of decision makers is constructed using IIFWA operator, as described in the section 3.

$$D = \begin{bmatrix} ([0.5,0.5],[0.5,0.5]) & ([0.42,0.5],[0.20.3]) & ([0.5,0.7],[0.1,0.23]) & ([0.3,0.5],[0.2,0.3]) \\ ([0.29,0.4],[0.4,0.47]) & ([0.5,0.5],[0.5,0.5]) & ([0.5,0.7],[0.1,0.25]) & ([0.6,0.7],[0.1,0.3]) \\ ([0.14,0.2],[0.5,0.66]) & ([0.13,0.3],[0.5,0.7]) & ([0.5,0.5],[0.5,0.5]) & ([0.4,0.6],[0.2,0.4]) \\ ([0.17,0.3],[0.3,0.51]) & ([0.14,0.3],[0.5,0.7]) & ([0.3,0.5],[0.4,0.55]) & ([0.5,0.5],[0.5,0.5]) \end{bmatrix}$$

The importance of the criteria provided by decision makers can be linguistic terms. These linguistic terms can be represented as interval-valued intuitionistic fuzzy numbers and can be aggregated by the operator IIFWA, as mention in section3. Let the interval-valued intuitionistic fuzzy weight of each criterion after aggregation of opinions of decision makers is:

$$W = \begin{pmatrix} ([0.25,0.45],[0.50,0.80]) \\ ([0.35,0.45],[0.65,0.75]) \\ ([0.30,0.55],[0.45,0.85]) \\ ([0.55,0.90],[0.60,0.95]) \end{pmatrix}$$

After the weights of the criteria and the rating of the alternatives has been determined, the aggregated weighted interval-valued intuitionistic fuzzy decision matrix is constructed as follows:

$$D = \begin{pmatrix} ([0.13,0.23],[0.75,0.90]) & ([0.15,0.24],0.73,0.83]) & ([0.16,0.37],[0.52,0.88]) & ([0.19,0.49],[0.66,0.97]) \\ ([0.07,0.18],[0.68,0.89]) & ([0.18,0.23],[0.83,0.88]) & ([0.14,0.38],[0.52,0.89]) & ([0.30,0.64],[0.65,0.96]) \\ ([0.03,0.11],[0.76,0.93]) & ([0.05,0.12],[0.81,0.92]) & ([0.15,0.28],[0.73,0.93]) & ([0.24,0.52],[0.67,0.97]) \\ ([0.04,0.15],[0.65,0.90]) & ([0.05,0.12],[0.82,0.92]) & ([0.08,0.25],[0.67,0.93]) & ([0.28,0.45],[0.80,0.98]) \end{pmatrix}$$

Product quality, relationship closeness, and delivery performance are benefit criteria $I = \{C_1, C_2, C_3\}$ and price is cost criteria $O = \{C_3\}$. Then interval-valued intuitionistic fuzzy positive-ideal solution and interval-valued intuitionistic fuzzy negative-ideal solution were obtained.

$A^+ = \{([0.13,0.23],[0.65,0.89]), ([0.18,0.24],[0.73,0.83]), ([0.16,0.38],[0.52,0.88]), ([0.19,0.45],[0.80,0.98])\}$
$A^- = \{([0.03,0.11],[0.76,0.93]), ([0.05,0.12],[0.82,0.92]), ([0.08,0.25],[0.73,0.93]), ([0.28,0.52],[0.67,0.97])\}$

Negative and positive separation measures based on Euclidean distance for each alternative were calculated in table1.

Table 1. The distances from ideal-solution and negative-ideal solution

Alternatives	S^+	S^-
A_1	0.0893	0.1864
A_2	0.1489	0.1766
A_3	0.2033	0.0413
A_4	0.1730	0.0998

Finally, using eq.(16), the value of relative closeness of each alternative for final ranking is:

$$RC_1^+ = 0.6762$$
$$RC_2^+ = 0.5426$$
$$RC_3^+ = 0.1688$$
$$RC_4^+ = 0.3658$$

Thus, the preference order of alternatives is A_1, A_2, A_4 and A_3 according to decreasing order of RC_i^+.

5 Conclusions

This study presents a multi-attribute group decision making for evaluation of supplier using interval-valued intuitionistic fuzzy TOPSIS. Interval-valued intuitionistic fuzzy sets are more suitable to deal with uncertainty than other generalised forms of fuzzy sets. In the evaluation process, the ratings of each alternative with respect to each criterion are taken as interval-valued intuitionistic fuzzy number. Also interval-valued intuitionistic fuzzy weighted averaging operator is utilised to aggregate the opinions of decision makers. After Interval-valued intuitionistic fuzzy positive-ideal solution and Interval-valued intuitionistic fuzzy negative-ideal solution are calculated based on the Euclidean distance measure, the relative closeness coefficients of alternatives are obtained and alternatives were ranked.

TOPSIS method combined with Interval-valued intuitionistic fuzzy set has enormous chance of success for multi-criteria decision-making problems due to containing vague perception of decision makers opinions. Therefore, in future, Interval-valued intuitionistic fuzzy set can be used for dealing with uncertainty in multi-criteria decision-making problems such as project selection, manufacturing systems, personnel selection, and many other areas of management decision problems.

References

[1] Atanassov, K.T.: Intuitionistic fuzzy sets. Fuzzy Sets and Systems 20, 87–97 (1986)
[2] Atanassov, K., Gargov, G.: Interval-valued intuitionistic fuzzy sets. Fuzzy Sets and Systems 31, 343–349 (1989)
[3] Atanassov, K.: Operators over interval-valued intuitionistic fuzzy sets. Fuzzy Sets and Systems 64, 159–174 (1994)
[4] Bustince, H., Burillo, P.: vague sets are intuitionistic fuzzy sets. Fuzzy Sets and Systems 79, 403–405 (1996)
[5] Gau, W.L., Buehrer, D.J.: Vague sets. IEEE Trans. on Systems, Man and Cybernetics 23, 610–614 (1993)
[6] Guangtao, F.: A fuzzy optimization method for multi criteria decision making: An application to reservoir flood control operation. Expert System with Application 34, 145–149 (2008)
[7] Hwang, C.L., Yoon, K.: Multiple Attribute Decision Making: Methods and Applications. Springer, Heidelberg (1981)
[8] Jahanshahlo, G.R., Hosseinzade, L.F., Izadikhah, M.: An algorithmic method to extend TOPSIS for decision making problems with interval data. Applied Mathematics and Computation 175, 1375–1384 (2006)
[9] Jahanshahloo, G.R., Hosseinzaadeh, F.L., Davoodi, A.R.: Extension of TOPSIS for decision-making problems with interval data: Interval efficiency. Mathematical and Computer Modelling, 49, 1137–1142 (2009)

312   Kavita, S.P. Yadav, and S. Kumar

[10] Jahanshahlo, G.R., Hosseinzade, L.F., Izadikhah, M.: Extension of the TOPSIS method for decision making problems with fuzzy data. Applied Mathematics and Computation 181, 1544–1551 (2006)

[11] Triantaphyllou, E., Lin, C.T.: Development and evaluation of five fuzzy multi attribute decision-making methods. International J. of Approximate Reasoning 14, 281–310 (1996)

[12] Wang, Y.M., Elhag, T.M.S.: Fuzzy TOPSIS method based on alpha level sets with an application to bridge risk assessment. Expert Systems with Applications 31, 309–319 (2006)

[13] Xu, Z.S.: Methods for aggregating interval-valued intuitionistic fuzzy information and their application to decision making. Control and Decision 22, 215–219 (2007)

[14] Xu, Z.S.: Intuitionistic fuzzy aggregation opterators. IEEE Transaction of Fuzzy Systems 15, 1179–1187 (2007)

[15] Xu, Z.S., Yager, R.R.: Dynamic intuitionistic fuzzy muti-attribute decision making. International journal of Approximate Reasoning 48, 246–262 (2008)

[16] Xu, Z.S., Yager, R.R.: Some geometric aggregation operators based on intuitionistic fuzzy sets. International J. of General System 35, 417–433 (2006)

[17] Xu, Z.S., Chen, J.: Approach to group decision making based on interval-valued intuitionistic judgment matrices. System Engineering – Theory & Practice 27, 126–133 (2007)

[18] Zadeh, L.A.: Fuzzy sets. Information and Control 8, 338–353 (1965)

The Lower System, the Upper System and Rules with Stability Factor in Non-deterministic Information Systems

Hiroshi Sakai[1], Kohei Hayashi[1], Michinori Nakata[2], and Dominik Ślęzak[3,4]

[1] Mathematical Sciences Section, Department of Basic Sciences,
Faculty of Engineering, Kyushu Institute of Technology
Tobata, Kitakyushu 804, Japan
sakai@mns.kyutech.ac.jp
[2] Faculty of Management and Information Science,
Josai International University
Gumyo, Togane, Chiba 283, Japan
nakatam@ieee.org
[3] Institute of Mathematics, University of Warsaw
Banacha 2, 02-097 Warsaw, Poland
[4] Infobright Inc., Poland
Krzywickiego 34 pok. 219, 02-078 Warsaw, Poland
slezak@infobright.com

Abstract. A rule in a *Deterministic Information System* (*DIS*) is often defined by an implication τ such that both $support(\tau) \geq \alpha$ and $accuracy(\tau) \geq \beta$ hold for the threshold values α and β. In a *Non-deterministic Information System* (*NIS*), there are *derived DISs* due to the information incompleteness. A rule in a *DIS* was extended to either a rule in the *lower system* or a rule in the *upper system* in a *NIS*. This paper newly introduces a criterion, i.e., *stability factor*, into rules in a *NIS*. Rules in the *upper system* are classified according to the stability factor.

Keywords: Rough sets, Non-deterministic information, Incomplete information, Rule generation, Apriori algorithm, Stability factor.

1 Introduction

We follow rule generation in *DISs* [8,12], and we describe rule generation in *NISs*. *NISs* were proposed by Pawlak [8], Orłowska [7] and Lipski [6] in order to handle information incompleteness in *DISs*, like null values, unknown values, missing values. Since the emergence of incomplete information research [3,5,6,7], *NISs* have been playing an important role. We have also focused on the semantic aspect of incomplete information, and proposed *Rough Non-deterministic Information Analysis* (*RNIA*) [9]. This paper continues the framework of rule generation in *NISs* [9,10,11], and we introduce *stability factor* into the upper system in *NISs*.

H. Sakai et al. (Eds.): RSFDGrC 2009, LNAI 5908, pp. 313–320, 2009.

2   Decision Rule Generation and Apriori Algorithm

A *Deterministic Information System* (*DIS*) is a quadruplet $(OB, AT, \{VAL_A| A \in AT\}, f)$ [8]. We usually identify a *DIS* with a standard table. A rule (more correctly, a candidate of a rule) is an appropriate implication in the form of $\tau : Condition\_part \Rightarrow Decision\_part$ generated from a table. We usually employ two criteria, *support*(τ) and *accuracy*(τ) for the appropriateness [1,8].

A definition of a rule generation in DISs
Find all implications τ satisfying *support*(τ) $\geq \alpha$ and *accuracy*(τ) $\geq \beta$ for the threshold values α and β $(0 < \alpha, \beta \leq 1)$.

Agrawal proposed *Apriori* algorithm [1] for such rule generation, and *Apriori* algorithm is now a representative algorithm for data mining [2].

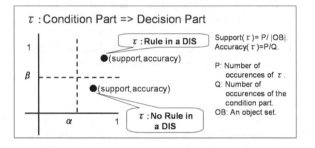

Fig. 1. A pair (*support,accuracy*) corresponding to the implication τ is plotted in a coordinate plane

3   Decision Rule Generation in NISs

A *Non-deterministic Information System* (*NIS*) is also a quadruplet $(OB, AT, \{VAL_A|A \in AT\}, g)$, where $g : OB \times AT \rightarrow P(\cup_{A \in AT} VAL_A)$ (a power set of $\cup_{A \in AT} VAL_A$). Every set $g(x, A)$ is interpreted as that there is an actual value in this set but this value is uncertain. For a $NIS=(OB, AT, \{VAL_A| A \in AT\}, g)$ and a set $ATR \subseteq AT$, we name a $DIS=(OB, ATR, \{VAL_A|A \in ATR\}, h)$ satisfying $h(x, A) \in g(x, A)$ a *derived DIS* (*for ATR*) *from a NIS*. In a *NIS*, there are derived *DISs* due to the information incompleteness.

   In Table 1, we can pick up $\tau_1 : [Temperature, high] \Rightarrow [Flu, yes]$ from objects 1, 2, 3, 4 and 8. We may use the notation τ^x from object x, for example, τ_1^1 (τ_1 from object 1) and τ_1^8 (τ_1 from object 8). Furthermore, we consider a set of derived *DISs* with τ^x, and let $DD(\tau^x)$ denote this set. For a set of attributes $\{Temperature, Flu\}$, there are 144 ($=2^4 \times 3^2$) derived *DISs*, $|DD(\tau_1^1)|=144$ and $|DD(\tau_1^8)|=48$ hold. If τ^x (for an object x) satisfies the condition of the criterion values, we see this τ is a rule.

Table 1. A Non-deterministic Information System (An artificial table data)

OB	Temperature	Headache	Nausea	Flu
1	{high}	{yes, no}	{no}	{yes}
2	{high, very_high}	{yes}	{yes}	{yes}
3	{normal, high, very_high}	{no}	{no}	{yes, no}
4	{high}	{yes}	{yes, no}	{yes, no}
5	{high}	{yes, no}	{yes}	{no}
6	{normal}	{yes}	{yes, no}	{yes, no}
7	{normal}	{no}	{yes}	{no}
8	{normal, high, very_high}	{yes}	{yes, no}	{yes}

A definition of a rule generation in NISs (A revised definition in [10])
Let us consider the threshold values α and β ($0 < \alpha, \beta \leq 1$).
(The lower system) Find all implications τ in the following: There exists an object x such that $support(\tau^x) \geq \alpha$ and $accuracy(\tau^x) \geq \beta$ hold in each $\psi \in DD(\tau^x)$.
(The upper system) Find all implications τ in the following: There exists an object x such that $support(\tau^x) \geq \alpha$ and $accuracy(\tau^x) \geq \beta$ hold in some $\psi \in DD(\tau^x)$.

In a DIS, $DD(\tau^x)$ means a singleton set, therefore the lower and the upper systems define the same implications in a DIS. Namely, the above definition is a natural extension from rule generation in $DISs$. Intuitively, the lower system defines *rules with certainty*, and the upper system defines *rules with possibility*. Especially, if $DD(\tau^x)$ is equal to the set of all derived $DISs$ and τ is a rule in the lower system, this τ is the most reliable.

4   The Minimum and the Maximum Criterion Values, and NIS-Apriori Algorithm

In Table 1, let us consider $\tau_1^1 : [Temperature, high] \Rightarrow [Flu, yes]$, again. In this case, $support(\tau_1^1)$ and $accuracy(\tau_1^1)$ take several values according to the derived $DISs$. We define $minsupp(\tau^x)$ (minimum support), $minacc(\tau^x)$ (minimum accuracy), $maxsupp(\tau^x)$ (maximum support) and $maxacc(\tau^x)$ (maximum accuracy) for each implication τ^x. For such criterion values, we have proved the next results.

Result 1 [10]. In a $NIS=(OB, AT, \{VAL_A | A \in AT\}, g)$, we can calculate these values by the next classes, i.e., $Descinf([A_i, val_{i,j}])$ and $Descsup([A_i, val_{i,j}])$.
(1) $Descinf([A_i, \zeta_{i,j}])=\{x \in OB | g(x, A)=\{\zeta_{i,j}\}\}$.
(2) $Descinf(\wedge_i [A_i, \zeta_{i,j}])=\cap_i Descinf([A_i, \zeta_{i,j}])$.
(3) $Descsup([A_i, \zeta_{i,j}])=\{x \in OB | \zeta_{i,j} \in g(x, A)\}$.
(4) $Descsup(\wedge_i [A_i, \zeta_{i,j}])=\cap_i Descsup([A_i, \zeta_{i,j}])$.
For example, if every attribute value is definite in $\tau^x : [CON, \zeta] \Rightarrow [DEC, \eta]$, $minsupp(\tau^x)=|Descinf([CON, \zeta]) \cap Descinf([DEC, \eta])|/|OB|$,

$$minacc(\tau^x)=\frac{|Descinf([CON,\zeta])\cap Descinf([DEC,\eta])|}{|Descinf([CON,\zeta])|+|OUTACC|},$$

$$(OUTACC=[Descsup([CON,\zeta])-Descinf([CON,\zeta])]-Descinf([DEC,\eta]),$$

$$maxsupp(\tau^x)=|Descsup([CON,\zeta])\cap Descsup([DEC,\eta])|/|OB|,$$

$$maxacc(\tau^x)=\frac{|Descinf([CON,\zeta])\cap Descsup([DEC,\eta])|+|INACC|}{|Descinf([CON,\zeta])|+|INACC|}.$$

$$(INACC=[Descsup([CON,\zeta])-Descinf([CON,\zeta])]\cap Descsup([DEC,\eta])).$$

Result 2 [10]. For each implication τ^x, there is a derived DIS_{worst}, where both $support(\tau^x)$ and $accuracy(\tau^x)$ are minimum in Table 2. Furthermore, there is a derived DIS_{best}, where both $support(\tau^x)$ and $accuracy(\tau^x)$ are maximum.

Table 2. Derived DIS_{worst} from Table 1, which causes the minimum support $1/8$ and the minimum accuracy $1/4$ of τ_1^1, and derived DIS_{best} which causes the maximum support $5/8$ and the maximum accuracy $5/6$ of τ_1^1 to the right

OB	Temperature	Flu		OB	Temperature	Flu
1	high	yes		1	high	yes
2	very_high	yes		2	high	yes
3	high	no		3	high	yes
4	high	no		4	high	yes
5	high	no		5	high	no
6	normal	no		6	normal	yes
7	normal	no		7	normal	no
8	normal	yes		8	high	yes

According to the above two results, we can handle the next definition.

An equivalent definition of a rule generation in NISs
Let us consider the threshold values α and β ($0 < \alpha, \beta \leq 1$).
(The lower system) Find all implications τ in the following: There exists an object x such that $minsupp(\tau^x) \geq \alpha$ and $minacc(\tau^x) \geq \beta$.
(The upper system) Find all implications τ in the following: There exists an object x such that $maxsupp(\tau^x) \geq \alpha$ and $maxacc(\tau^x) \geq \beta$.

In the first definition, we needed to examine *support* and *accuracy* in all derived $DD(\tau^x)$, however we can examine the same results by comparing ($minsupp$,

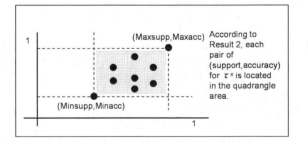

Fig. 2. A distribution of pairs (*support,accuracy*) for an implication τ^x

minacc) and (*maxsupp,maxacc*) with the threshold α and β. Like this, we extended rule generation in *DISs* to rule generation in *NISs*, and realized a software tool *NIS-Apriori* [10,11]. This is an adjusted *Apriori* algorithm to *NISs*, and it can handle not only deterministic information but also non-deterministic information. *NIS-Apriori* algorithm does not depend upon the number of derived *DISs*, and the complexity is almost the same except the following:

(1) Instead of the large item sets in *Apriori* algorithm, we need to manage *Descinf* and *Descsup*.
(2) *support*(τ^x) and *acuracy*(τ^x) are the same for each x in a *DIS*, however they are slightly different for x in a *NIS*. If there exists τ^x satisfying the criterion values, we see this τ is a rule. Therefore, we need to examine   criterion values for each τ^x

We are now coping with *SQL-NIS-Apriori* on Infobright ICE system [13], and we are discussing on Data Mining in Warehousing and Various Types of Inexact Data [4].

5   Introducing Stability Factor into the Upper System

Now, we consider the next two cases related to rules in the upper system.
(CASE 1) If there is just a *DIS* in $DD(\tau^x)$ satisfying the condition, this τ is picked up as a rule in the upper system.
(CASE 2) If most of the *DISs* in $DD(\tau^x)$ satisfy the condition, this τ is also picked up as a rule in the upper system.

Due to this example, the definition of the upper system seems weak. In order to distinguish two cases, we add another criterion, i.e., stability factor of τ.

Definition 1. Let us suppose $\tau : [CON, \zeta] \Rightarrow [DEC, \eta]$ is a rule in the upper system for α and β, and let $OBJ(\tau) = Descsup([CON, \zeta]) \cap Descsup([DEC, \eta])$. If *support*($\tau^x$) $\geq \alpha$ and *accuracy*(τ^x) $\geq \beta$ hold in $\psi \in DD(\tau^x)$, we say τ is *stable* in ψ. Let

$\quad DD(\tau) = \cup_{x \in OBJ(\tau)} DD(\tau^x)$,
$\quad ST(\tau, \alpha, \beta) = \{\psi \in DD(\tau) | \ \tau$ is stable in $\psi\}$.
Stability factor of τ is $STF(\tau, \alpha, \beta) = |ST(\tau, \alpha, \beta)| / |DD(\tau)|$.

Let us consider $\tau_2 : [Nausea, yes] \Rightarrow [Flu, no]$ in Table 1. In reality, for $\alpha = 0.3$ and $\beta = 0.7$, τ_2 is not picked up as a rule by the lower system, but τ_2 is picked up as a rule by the upper system. Here, $DD(\tau_2) = DD(\tau_2^4) \cup DD(\tau_2^5) \cup DD(\tau_2^6) \cup DD(\tau_2^7)$ holds. Since τ_2^5 is definite, $DD(\tau_2) = DD(\tau_2^5)$ and there are 64 derived *DISs*. In Fig.3, 64 pairs of (*support,accuracy*) are plotted in a coordinate plane, and "$\bullet$: *number*" implies the number of derived *DISs*. For example, 4 derived *DISs* cause a pair (3/8,0.5) in Fig.3. As for τ_2, $|ST(\tau_2, 0.3, 0.7)| = 8 + 2 = 10$ and $STF(\tau_2, 0.3, 0.7) = 10/64$ (about 16%) hold. According to this stability factor, we can assign the probability (of the reliability) to rules in the upper system.

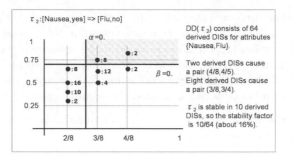

Fig. 3. A distribution (*support,accuracy*) of $\tau_2 : [Nausea, yes] \Rightarrow [Flu, no]$ in Table 1

This stability factor seems useful, but the number of elements in $DD(\tau)$ increases in exponential order. Therefore, the calculation of the stability factor will generally be hard. Even though, it will be possible to calculate this factor, if the next assupmtions hold.

(Assumption 1) $Descinf([CON, \zeta]) \cap Descinf([DEC, \eta]) \neq \{\}$.
(Assumption 2) The sizes of both $|DIFF_{[CON,\zeta]}|$ and $|DIFF_{[DEC,\eta]}|$ are small.
    Here, $DIFF_{[CON,\zeta]}$ denotes $Descsup([CON, \zeta]) - Descinf([CON, \zeta])$.

Due to (Assumption 1), there is a definite implication $[CON, \zeta] \Rightarrow [DEC, \eta]$, and $DD(\tau)$ is equal to the derived $DISs$ for attributes $CON \cup DEC$. According to (Assumption 2), the number of derived $DISs$ is restricted to small.

6   A Method for Calculating Stability Factor

In order to calculate $STF(\tau, \alpha, \beta)$, we propose a method using $Descinf$ and $Descsup$ in Result 1. In a $NIS=(OB, AT, \{VAL_A | A \in AT\}, g)$, let

$$OBJ_{[A,val],M}=Descinf([A, val]) \cup M \ (M \subset DIFF_{[A,val]}).$$

We define the next attribute values of x in $\{A\}$ related to $OBJ_{[A,val],M}$:

(Assignment 1) If $x \in OBJ_{[A,val],M}$, the value (of x for the attribute $\{A\}$) is val.
(Assignment 2) If $x \in Descsup([A, val]) - OBJ_{[A,val],M}$, the value is except val.
(Assignment 3) If $x \notin Descsup([A, val])$, the value is any value in $g(x, A)$.

Like this, it is possible to define derived $DISs$ for an attribute $\{A\}$. The (Assignment 3) occurs in each $OBJ_{[A,val],M}$, therefore it is enough to consider the cases defined by (Assignment 2).

Proposition 1. $OBJ_{[A,val],M}$ causes

$$NUM_{[A,val],M} : \prod_{y \in Descsup([A,val])-OBJ_{[A,val],M}} (|g(y, A)| - 1)$$

cases of derived $DISs$. For a conjunction of descriptors $\wedge_i[A_i, \zeta_{i,j}]$, it is possible to define the same result.

It is possible to calculate the following by $OBJ_{[CON,\varsigma],M}$ and $OBJ_{[DEC,\eta],M'}$.

$support(\tau)=|OBJ_{[CON,\varsigma],M} \cap OBJ_{[DEC,\eta],M'}|/|OB|$,
$accuracy(\tau)=|OBJ_{[CON,\varsigma],M} \cap OBJ_{[DEC,\eta],M'}|/|OBJ_{[CON,\varsigma],M}|$,

and the same values occur in $NUM_{[CON,\varsigma],M} \times NUM_{[DEC,\eta],M'}$ derived $DISs$. Finally, we have the next method to calculate $STF(\tau,\alpha,\beta)$.

An overview of the caluculation method for stability factor
According to the (Assumption 1), there is a definite $\tau^x \cdot [CON,\varsigma] \Rightarrow [DEC,\eta]$ for an object x. $DD(\tau)$ is a set of derived $DISs$ for attributes $CON \cup DEC$.

(1) Obtain $Descinf([CON,\varsigma])$, $Descsup([CON,\varsigma])$, $DIFF_{[CON,\varsigma]}$, $Descinf([DEC,\eta])$, $Descsup([DEC,\eta])$ and $DIFF_{[DEC,\eta]}$.
(2) For each pair $OBJ_{[CON,\varsigma],M}$ and $OBJ_{[DEC,\eta],M'}$, examine the condition by $support$ and $accuracy$. If a pair satisfies the condition, count the number of derived $DISs$ for $CON \cup DEC$.
(3) $STF(\tau,\alpha,\beta)$ is a ratio, (totally counted number of derived $DISs$)/$|DD(\tau)|$.

Now, we simulate the above calculation. Let us consider a rule $\tau_2^5 : [Nausea, yes] \Rightarrow [Flu,no]$ in the upper system ($\alpha=0.3$ and $\beta=0.7$). At first, we obtain $Descinf$ and $Descsup$, i.e.,

$Descinf([Nausea,yes])=\{2,5,7\}$, $Descsup([Nausea,yes])=\{2,5,7,4,6,8\}$,
$DIFF_{[Nausea,yes]}=\{4,6,8\}$, $Descinf([Flu,no])=\{5,7\}$,
$Descsup([Flu,no])=\{5,7,3,4,6\}$, $DIFF_{[Flu,no]}=\{3,4,6\}$.

Then, the conditions $support$ and $accuracy$ are examined for each pair

$OBJ_{[Nausea,yes],M}$ $(=\{2,5,7\} \cup M)$ $(M \subset \{4,6,8\})$,
$OBJ_{[Flu,no],M'}$ $(=\{5,7\} \cup M')$ $(M' \subset \{3,4,6\})$.

For example, $OBJ_{[Nausea,yes],\{4\}}$ and $OBJ_{[Flu,no],\{4\}}$ satisfy the conditions, and the following is calculated.

$NUM_{[Nausea,yes],\{4\}}=(|g(6,Nausea)|-1)| \times (|g(8,Nausea)|-1)=1$,
$NUM_{[Flu,no],\{4\}}=(|g(3,Flu)|-1)| \times (|g(6,Flu)|-1)=1$.

Like this, we examine the number of derived $DISs$ where τ_2 is stable. Finally, we obtain $|ST(\tau_2,0.3,0.7)|=10$ and $STF(\tau_2,0.3,0.7)=|ST(\tau_2,0.3,0.7)|/|DD(\tau_2)|= 10/64 \approx 16\%$.

Generally, the number of distinct M is 2 power $DIFF_{[CON,\varsigma]}$. In this example, we handled 8 cases for M. We are now considering another method without (Assumption 2), but it seems difficult. We have also realized a simple program for calculating the stability factor. In reality, we obtained the following two implications in the upper system for $\alpha=0.3$ and $\beta=0.7$.

$\tau_2 : [Nausea,yes] \Rightarrow [Flu,no]$, $STF(\tau_2,0.3,0.7)=10/64 \approx 16\%$.
$\tau_3 : [Headache,yes] \Rightarrow [Flu,yes]$, $STF(\tau_3,0.3,0.7)=16/32=50\%$.

From this result, we may conclude τ_3 is more reliable than τ_2.

7   Concluding Remarks

This paper proposed stability factor of a rule in the upper system. Due to this ratio, we obtained another criterion for rules in the upper system. We have already proposed two kinds of rule generation in $NISs$. The one is the consistency based rule generation, and the other is the criterion based rule generation [9,10]. This paper coped with the criterion based rule generation, and presented NIS-$Apriori$, the lower system, the upper system and the stability factor in the upper system.

Acknowledgment. The first author Hiroshi Sakai was partially supported by the Grant-in-Aid for Scientific Research (C) (No.16500176, No.18500214), Japan Society for the Promotion of Science. The fourth author Dominik Ślęzak was partially supported by the grants N N516 368334 and N N516 077837 from the Ministry of Science and Higher Education of the Republic of Poland.

References

1. Agrawal, R., Srikant, R.: Fast Algorithms for Mining Association Rules. In: Proceedings of the 20th Very Large Data Base, pp. 487–499 (1994)
2. Ceglar, A., Roddick, J.F.: Association mining. ACM Comput. Surv. 38(2) (2006)
3. Grzymala-Busse, J.: Data with Missing Attribute Values: Generalization of Indiscernibility Relation and Rule Induction. Transactions on Rough Sets 1, 78–95 (2004)
4. Infobright.org Forums: `http://www.infobright.org/Forums/viewthread/288/`, `http://www.infobright.org/Forums/viewthread/621/`
5. Kryszkiewicz, M.: Rules in Incomplete Information Systems. Information Sciences 113, 271–292 (1999)
6. Lipski, W.: On Semantic Issues Connected with Incomplete Information Data Base. ACM Trans. DBS. 4, 269–296 (1979)
7. Orłowska, E., Pawlak, Z.: Representation of Nondeterministic Information. Theoretical Computer Science 29, 27–39 (1984)
8. Pawlak, Z.: Rough Sets. Kluwer Academic Publisher, Dordrecht (1991)
9. Sakai, H., Okuma, A.: Basic Algorithms and Tools for Rough Non-deterministic Information Analysis. Transactions on Rough Sets 1, 209–231 (2004)
10. Sakai, H., Ishibashi, R., Nakata, M.: On Rules and Apriori Algorithm in Non-deterministic Information Systems. Transactions on Rough Sets 9, 328–350 (2008)
11. Sakai, H., Ishibashi, R., Nakata, M.: Lower and Upper Approximations of Rules in Non-deterministic Information Systems. In: Chan, C.C., Grzymala-Busse, J., Ziarko, W. (eds.) RSCTC 2008. LNCS (LNAI), vol. 5306, pp. 299–309. Springer, Heidelberg (2008)
12. Skowron, A., Rauszer, C.: The Discernibility Matrices and Functions in Information Systems. In: Intelligent Decision Support - Handbook of Advances and Applications of the Rough Set Theory, pp. 331–362. Kluwer Academic Publishers, Dordrecht (1992)
13. Ślęzak, D., Sakai, H.: Automatic extraction of decision rules from non-deterministic data systems: Theoretical foundations and SQL-based implementation. In: Proc. of DTA 2009, Jeju, Korea, December 10-12 (in print, 2009)

Learning Player Behaviors in Real Time Strategy Games from Real Data

P.H.F. Ng, S.C.K. Shiu, and H. Wang

Department of Computing
The Hong Kong Polytechnic University
Hum Hom, Kowloon, Hong Kong, P.R. China
{cshfng,csckshiu,cshbwang}@comp.polyu.edu.hk

Abstract. This paper illustrates our idea of learning and building player behavioral models in real time strategy (RTS) games from replay data by adopting a Case-Based Reasoning (CBR) approach. The proposed method analyzes and cleans the data in RTS games and converts the learned knowledge into a probabilistic model, i.e., a Dynamic Bayesian Network (DBN), for representation and predication of player behaviors. Each DBN is constructed as a case to represent a prototypical player's behavior in the game, thus if more cases are constructed the simulation of different types of players in a multi-players game is made possible. Sixty sets of replay data of a prototypical player is chosen to test our idea, fifty cases for learning and ten cases for testing, and the experimental result is very promising.

Keywords: Case-based Reasoning (CBR), Real Time Strategy (RTS) Games, Dynamic Bayesian Network (DBN), Junction Tree.

1 Introduction

DFC Intelligence which is a market research organization forecasts the online game market will reach of US\$57 billion in 2009. Players enjoy interacting with other real players in the virtual world. However, it is very difficult for game companies to maintain a huge number of players, with varied styles, online at the same time. Therefore, to improve attractiveness, many avatars and characters in the game or virtual community need to be controlled by AI techniques. Developing many different behavioral styles from scratch to simulate real players is difficult and time consuming. To help accomplish this goal, we develop a method to learn them from real data. We used the Blizzard Warcraft III Expansion: The Frozen Throne (W3X) which is a well known and best selling RTS game in recent years to test our idea. Experimental result is shown and discussed in this paper.

2 Related Work

Case-based reasoning has been studied for many years. Its application to computer games is becoming more popular recently. For example, Hsieh [1] used professional

H. Sakai et al. (Eds.): RSFDGrC 2009, LNAI 5908, pp. 321–327, 2009.

game players' data of up to 300 replays to train a case-based decision system in one single battlefield. Ontanon [2] introduced similar case-based reasoning framework for RTS game. Aha [3] and Ponsen [4] focused on strategies of building sequence in their case-based planning model. In general, the prediction accuracy in CBR systems depends largely on the number of learned cases and their quality, i.e., the more the quality cases, the better the accuracy. However, since in RTS games, response time is critical, thus there is always a tradeoff between accuracy and efficiency.

Bayesian network (BN) can be viewed as a mathematical model that describes the relationships between antecedents and consequences using conditional probabilities. Dynamic Bayesian network (DBN) is one form of BN that represents sequences of variables, and is usually time-invariant. Some related work of using DBN for user modeling is by Ranganathan [5] and Montaner [6]. They implemented BN into their SMART Agents and proved its possibility to analyze user behaviors. Kuenzer [7] and Schiaffino [8] also did similar user behavior modeling work on web applications using BN. Furthermore, Gillies [9] combined BN with finite-state machine to improve the use of motion capture data. Other uses of BN on games, include Albrecht's [10] BN structure to adventure games and Yeung's [11] BN technique to detect cheats in first person shooting games. In this research, we use DBN and junction tree algorithm as a case and similarity calculation respectively for predicting user behaviors. We believed that this is a promising direction for game developers, and publishers that require varied styles of avatar behaviors.

3   Proposed Approach

First, sets of replay data are collected over the internet. Information inside these replays will be filtered, cleaned and summarized. The information will be then used to build the DBN structure. Prediction of user behaviors will be carried out afterwards. The framework is shown as Fig. 1. We will use an example of a typical player called "Player A" to illustrate our approach.

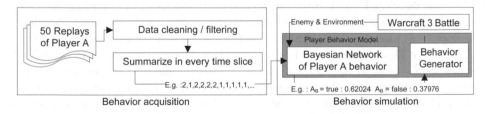

Fig. 1. Work flow of player behavior simulation model

3.1   Behavior Acquisition in Replay Data

Hundreds of thousands W3X replays can be collected on the Internet. All the player's actions are recorded in the replay data. In this research, Solo Ladder (1 versus 1) battle in Battlen.net is chosen. It is the official game site and provides a fair environment for the players to fight against each other. Fifty replays of a player called "Player A" (name is removed) are collected. They describe the behavior of Player A against different opponents in different maps. For the purpose of reuse the player behavior model in different maps and games, the data is analyzed with meanings as described in Table 1.

Table 1. Selected data for DBN structure

Player Action	
Set Name	**Description**
Attack (A)	All kinds of attack commands with target data. Data of player A (A) and opponent (A') will be both collected. For example: attack unit (A_U, A'_U), attack base (A_B, A'_B), etc. Each element of A ∈ [True, False].
Create Building (B)	All kinds of create building commands and their numbers in the same time slice. Data of player A (B) and opponent (B') will be both collected. For example: build base (B_B, B'_B), build research centre (B_R, B'_R), etc. Each element of B ∈ [0, 1, 2 ... Upper Limit] where upper limit is the maximum number of buildings that are created in the same time slice.
Create Unit (U)	All kinds of create unit command and their numbers in the same time slice. Data of player A (U) and opponent (U') will be both collected. For example: create piercing unit (U_P, U'_P), create siege unit (U_S, U'_S), etc. Each element of U ∈ [0, 1, 2 ... Upper Limit] where upper limit is the maximum number of units that are created in the same time slice.
Demographic Information	
Node Name	**Description**
Race (R)	There are 4 races that are provided by W3X, where R ∈ [1, 2, 3, 4]. Player A (R) and Opponent (R') will be both collected
Alive Unit Number (N)	Alive units in time slice t, rounding to the nearest 10, where N ∈ [10, 20 ... Upper Limit], where upper limit is the maximum number of units that are created in the replays. Player A (N) and Opponent (N') will be both collected
Map (M)	Official battle fields that are provided by Battle.net, where M ∈ [1, 2, 3, ..., 13]

As an observation, professional players in W3X usually focus on a few types of units in the battles. They seldom create many different kinds of unit because they wanted to save the resources for upgrading. Therefore, to reduce the complexity of DBN, all unused commands of player A are filtered, i.e., if player A does not create any siege unit in 50 replays, the field of siege unit (U_S) will be neglected and will not become a component of the DBN. Moreover, different from the research done by Aha [3] and Ponsen [4], resource (gold and lumber) is neglected in our model as it does not have a significant impact on prediction accuracy. The reason is because professional players in W3X usually used up all resources to create units or buildings. So, we suggest using alive unit number (N) to estimate the player current situation.

Then, actions of Player A are summarized in every fixed time slice (Δt). In this study, Δt is set to 15 seconds which is the minimum time to create a unit in W3X.

Selected fields for each Δt will become components in DBN and will be represented as a set of numeric data (e.g.: 2,1,2,2,2,2,1,1,1,1,1,5,2,2,2,2,2,2,1,1,1,1,10,6...). The average time of Player A's replay is 20 minutes. Therefore, there are around 80 instances in each replay. Upper limit for field (B), (U) and (N) will also be set according to the maximum number of productions in Δt. It can reduce the parameters of the tabular nodes in the DBN.

3.2   Dynamic Bayesian Network Structure and Parameters Learning

A DBN consists of a structure and a number of parameters. The analyzed fields in the previous process will become the tabular nodes of the DBN structure. Intra-slice topology (within a slice) and inter-slice topology (between two slices) are defined according to the game play of W3X. For example, if the player wants to create certain units, he needs to build certain buildings first. The relationship is shown in Fig. 2.

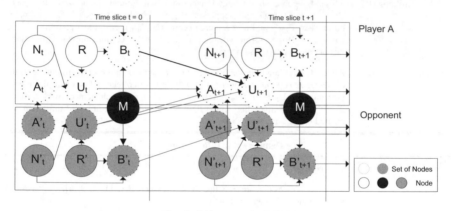

Fig. 2. Structure of a DBN

Parameters of DBN are represented as conditional probability distribution (CPD). It defines the probability distribution of a node given its parents, i.e., $P(A_{t+1}, U_{t+1}, ...)$ = $P(A_{t+1} \mid$ Parents of $(A_{t+1}))\, P(U_{t+1} \mid$ Parents of $(U_{t+1}))....$ All instances from 50 replays will be used to perform parameters learning. As the data from the replays are fully observed and the structure is known, maximum likelihood estimation algorithm is used to compute a full DBN. A DBN that contains multidimensional CPD in each node is considered as a "case" to represent player A's behavior.

3.3   Prediction in Dynamic Bayesian Network

Having created the DBN of Player A, it can be used for prediction. In this research, junction tree algorithm is used. Its purpose is to find out the probability of attack ($P(A_{t+1})$), create building ($P(B_{t+1})$) and create unit command ($P(U_{t+1})$) of Player A based on his previous behavior. The calculation of probabilities is based on all the previous time slices of their parent. In every Δt, enemy and environment situation information

(M, R, R', N, N', A' B' U') will be summarized and sent to the DBN as a fact to compute the marginal distribution for each node (A, B, U). Marginal distribution contains the probabilities of all parameters in each node. For example the attack base command (A_B) only contains 2 parameters (A $\in$ [True, False]). Therefore, the marginal distribution of A_B will be represented as A_B = true : 0.62024 and A_B = false : 0.37976. The parameters with the highest probability will be chosen as the Player A's behavior and passed to the behavior generator, i.e.: attack base command.

4   Experimental Result

To calculate the predication accuracy, ten new cases of Player A are prepared for testing. The simulation was run using Matlab version 2008b with the BN toolbox that was written by Kevin Murphy. The machine used is a Core 2 Duo 2.13GHz with 4 GB Ram PC. In this simulation, 18 nodes (8 in (B), 5 in (U) and 5 in (A)) of Player A is required to be predicted in every Δt. The running times for the learning and the average prediction for each Δt are shown in Table 2. The prediction time is nearly unchanged as it only depends on the complexity of the DBN structure (Dimensions of the CPD) and is independent of the number of learning instances. The constant performance of this prediction time fits the game implementation requirement.

Table 2. Running time of Learning and Prediction in DBN

Number of cases	10	20	30	40	50
Learning Time (S)	245.15	516.35	649.90	770.38	968.72
Prediction Time (S)	0.1143	0.1373	0.1025	0.1033	0.1038

The highest prediction probability of each node will be taken as the predicted command. Predicted command was compared with the actual command of Player A. The accuracy is calculated in every Δt using equation (1) which is similar to Albrecht's [10] approach.

$$\frac{1}{n}\frac{1}{N}\sum_{i=1}^{n}\sum_{j=1}^{N}\begin{cases}1 & \text{Predicted commad} = \text{actual command of Player A} \\ 0 & \text{Otherwise}\end{cases} \qquad (1)$$

Where n is number of testing replays, i.e., n = 10 and N is the number of nodes that is required to be predicated in every Δt, i.e., N = 18. The accuracy against time slices with different numbers of learning instances are plotted in Fig. 3. We observed that if the number of learning instances are insufficient, the probabilities of computed marginal distribution would be closed to 1/(number of choices in the node). (For example: AB = true : 0.51, AB = false : 0.49 where AB $\in$ [True, False]). As a result, the accuracy decreases. We suggested do not take this command, and only executes the predicted command if the probability reaches a certain threshold level. For example: 20

% increment (P(A$_B$) > 0.6) or 40 % increment (P(A$_B$) > 0.7). However, in this case, the behavior generator may not perform fast enough responses according to the situation. Accuracy could be improved if the number of learning instances increases. As shown in Fig 3b, the curves are similar in shape and closer to higher threshold.

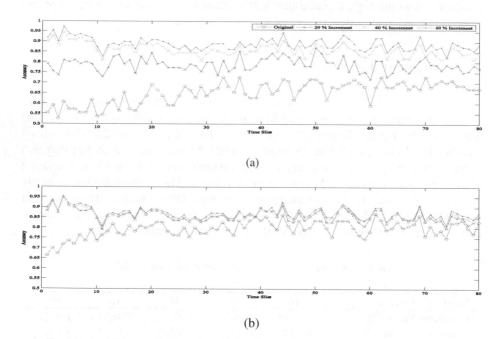

(a)

(b)

Fig. 3. a) Accuracy of 10 learning instances b) Accuracy of 50 learning instances

5   Conclusion

In this paper, we presented a CBR framework to learn players' behavior in RTS. The advantage of applying DBN and junction tree technique to RTS games is shown. Many players in multiplayer online games are not looking for challenging AIs but varied AIs. Our approach shows a possibility to develop varied AIs by converting different players' replays. It is efficient and useful. Future work of our research will focus on a larger scale of learning of many different types of replays and construction of a useful case library for simulating various players of different styles. We plan to combine DBN with other soft computing techniques, such as artificial neural networks and genetic algorithms for faster learning, indexing and similarity calculations in player models.

Acknowledgement

This project is supported by the HK Polytechnic University Grant A-PA6N.

References

1. Hsieh, J.L., Sun, C.T.: Building a player strategy model by analyzing replays of real-time strategy games. In: IJCNN, WCCI, Hong Kong, China, pp. 3106–3111 (2008)
2. Ontanon, S., Mishra, K.: Case-Based Planning and Execution for Real-Time Strategy Games. In: Weber, R.O., Richter, M.M. (eds.) ICCBR 2007. LNCS (LNAI), vol. 4626, pp. 164–178. Springer, Heidelberg (2007)
3. Aha, D.W., Molineaux, M.: Learning to Win: Case-Based Plan Selection in a Real-Time Strategy Game. In: Muñoz-Ávila, H., Ricci, F. (eds.) ICCBR 2005. LNCS (LNAI), vol. 3620, pp. 5–20. Springer, Heidelberg (2005)
4. Ponsen, M.: Improving adaptive game AI with evolutionary learning. MSc Thesis, Delft University of Technology (2004)
5. Ranganathan, A., Campbell, R.: A middleware for context-aware agents in ubiquitous computing environments. In: International Middleware Conference, Rio de Janeiro, Brazil (2003)
6. Montaner, M.: A taxonomy of recommender agents on the internet. Artificial intelligence review 19(4), 285–330 (2003)
7. Kuenzer, A., Schlick, C.: An empirical study of dynamic bayesian networks for user modeling. In: Proc. of the UM 2001 Workshop on Machine Learning for User Modeling (2001)
8. Schiaffino, S., Amandi, A.: User profiling with case-based reasoning and bayesian networks. In: International Joint Conference, 7th Ibero-American Conference, 15th Brazilian Symposium on AI, IBERAMIA-SBIA, Open Discussion Track Proceedings on AI, pp. 12–21 (2000)
9. Gillies, M.: Learning Finite-State Machine Controllers From Motion Capture Data. IEEE Transactions on Computational Intelligence and AI in Games 1(1), 63–72 (2009)
10. Albrecht, D., Zukerman, I.: Bayesian models for keyhole plan recognition in an adventure game. User modeling and user-adapted interaction 8(1), 5–47 (1998)
11. Yeung, S.F., Lui, J.C.S.: Detecting cheaters for multiplayer games: theory, design and implementation. In: Consumer Communications and Networking Conference, pp. 1178–1182 (2006)

An Efficient Prediction Model for Diabetic Database Using Soft Computing Techniques

Veena H. Bhat[1], Prasanth G. Rao[1], P. Deepa Shenoy[1],
K.R. Venugopal[1], and L.M. Patnaik[2]

[1] University Visvesvaraya College of Engineering, Bangalore University,
Bangalore, India
[2] Vice Chancellor, Defence Institute of Advanced Technology, Deemed University,
Pune, India
veena.h.bhat@gmail.com, prasanthgrao@gmail.com, shenoypd@yahoo.com

Abstract. Organizations aim at harnessing predictive insights, using the vast real-time data stores that they have accumulated through the years, using data mining techniques. Health sector, has an extremely large source of digital data - patient-health related data-store, which can be effectively used for predictive analytics. This data, may consists of missing, incorrect and sometimes incomplete values sets that can have a detrimental effect on the decisions that are outcomes of data analytics. Using the PIMA Indians Diabetes dataset, we have proposed an efficient imputation method using a hybrid combination of CART and Genetic Algorithm, as a preprocessing step. The classical neural network model is used for prediction, on the preprocessed dataset. The accuracy achieved by the proposed model far exceeds the existing models, mainly because of the soft computing preprocessing adopted. This approach is simple, easy to understand and implement and practical in its approach.

1 Introduction

Real time information, based on transactional data store, differentiated the competitive business organizations from others in their field, enabling them to make timely and relevant business decisions. This no longer holds good. Today, the ability to forecast the direction of the business trends, to predict the effect of the various variables involved in the complex situations, allowing business organizations to make proactive, knowledge driven decisions, is what differentiates the leaders in the business organizations. Gartner, Inc. has revealed that by 2012, business units will devote at least 40 % of their total budget for business intelligence. This is so, as they foresee the impact of the current economic turndown, to result in under-investment of information infrastructure and business tools, required to make informed and responsive decisions. Predictive modeling is a part of business intelligence.

Data Mining refers to knowledge discovery in large and complex volumes of data. Soft computing involves information processing, with methods aimed to exploit the data tolerance for imprecision, uncertainty, approximate reasoning

H. Sakai et al. (Eds.): RSFDGrC 2009, LNAI 5908, pp. 328–335, 2009.

and partial truth, in order to achieve tractability, robustness and low cost solutions. Neural networks, rough sets, fuzzy logic, genetic algorithm are a few of the methods that are termed as soft computing methods. Combining the two technologies, synergetically, the flexibility in processing information, from large volumes of data, which have all the limitations of the data collected in real-life ambiguous situations, can be harvested. This paper discusses a prediction model which works on real-life data, specifically the data from the medical field, which contains both numeric and textual data, often redundant attributes, where the domain of values contain erroneous, misspelled, incomplete and sometimes even missing values.

Predictive models have been used even before the advent of data mining, but with large data stores available, the demand for an accurate prediction model which works on the extremely large data store has led to predictive data mining, a goal based, supervised approach. Sample cases stored over time with known results, having unseen characteristic patterns for achieving or not achieving the outcome, form the representative data set for the model. The objective of prediction mining is to find patterns in the data that gives correct and accurate outcomes on the newer, unseen cases.

The model, based on the sample data will work well, when ideally, the data the model builds itself on, is complete, accurate and valid in the domain of values. Of all the factors that affect the accuracy of the prediction model, research has identified that the handling of missing and outlier values does play a crucial role. Prediction models have been built, after eliminating or ignoring the tuples with missing values (termed case deletion method). But when the data used to build the model, does contain null or zero, instead of just missing value(s) for an attribute, the proposed model will have a detrimental effect on the prediction of the outcome(s). This is because, the null or zero value, may be representing any value in the data domain for that attribute.

This paper is focused on building a prediction model, which uses a combination of classification-regression-genetic-neural network, to ensure a higher degree of accuracy as compared with the earlier works in this field. As the first step, it handles the missing and outlier values in the dataset, efficiently, by replacing the missing values with appropriate values from the data domain of the corresponding attribute, evaluating the accuracy of the replaced value, by minimizing the error function, using the genetic algorithm. The dataset thus preprocessed, is input to build the neural network model, for prediction.

2   Motivation

Healthcare information systems today, amass a large amount of digital data in various forms, but are hardly used for effective preventive diagnosis, disease transmission and outbreaks, cost effective patient care or even for tracking the patients critical parameters and medicine intake for long term effects. An intelligent and robust prediction model can also help in the identification of long-term probable health issues. This can help the attending doctors or nurse care coordinators to work with the patient to give quality treatment and also help to identify the lifestyle changes required in the patients routine, to prevent short and long tem ill-health complexities.

3   Related Work

Predictive modeling in health sector is a valuable research area as it focuses on real-time application and solutions. The industry is keen to model, develop and adopt prediction models for its strategic decision making, enabling it to be highly effective. Prediction models have been developed for diagnosis of fileria, heart disease, thyroid tracking, melanoma, asthma etc. Paper [6] has discussed an intelligent heart disease prediction model, using three data mining techniques Decision Trees, Nave Bayes and Neural Network, from a historical heart disease database.

Any dataset in real-time data store contains missing values as well as outliers, which affect the prediction accuracy of the model, developed based on the dataset. The three categories of missing values, identified by statisticians [2] are MCAR: Missing Completely At Random where the probability of missing a value is the same for all the variables; MAR: Missing At Random where the probability of missing a value is only dependent on other variables and NMAR: Not Missing At Random where the probability of missing a value is dependent on the value of the missing variable. Depending on the categorization of the missing value and the application/context, imputation methods have been researched [9]. The three broad categories of handling missing data are case deletion, parameter estimation (maximum likelihood method, mean/mode Imputation, all possible values imputation, regression methods, cold/hot deck imputation) and imputation techniques (k-nearest neighbor imputation, multiple imputation, internal treatment method for C4.5, Bayesian iteration imputation) [7]. Imputation techniques use the present information in the dataset, recognize the relationship between the values in the dataset and to estimate and replace the missing values.

Imputation methods are further categorized into three groups: global imputation based on missing attribute-where known value(s) are used to impute the missing data of the attribute, global imputation based on non-missing attribute-where the relationship between the missing and the non-missing attributes are used to predict the missing data and the third category local imputation where the sample data is divided into clusters and the missing data is estimated. The presence of missing values in rates of less than 1% is generally considered as trivial, 1-5% manageable. However, 5-15% require sophisticated methods to handle, and more than 15% may severely impact any kind of interpretation. The selected dataset for our work has a missing/incorrect value range of upto 48% [11]. Our model, with a view to give a cost effective imputation method, uses pre-placing method of imputation, assesses the accuracy of the imputed values and then builds the prediction model, using the now-corrected dataset, with no missing values and incorrect zero values. The accuracy of the model is also established.

4   Implementation

We have adapting the SEMMA methodology from SAS Enterprise Miner for our model, mapping our process as Sample, Explore, Modify, Model and Access which refers to the core process of conducting data mining, to build and evaluate the predictive data mining model. Fig. 1 illustrates the features of the proposed model.

4.1   Sample

The dataset selected for building the predictive model is PIMA Indians Diabetes data set from the Machine Learning Database Repository at the University of California, Irvine. As per previous studies, the Pima Indians may be genetically predisposed to diabetes and it was noted that their diabetic rate was 19 times that of any typical town. The National Institute of Diabetes and Digestive and Kidney Diseases of the NIH originally owned the Pima Indian Diabetes Database (PIDD). In 1990 it was received by the UC-Irvine Machine Learning Repository and can be downloaded at www.ics.uci.edu/ mlearn/MLRepository. The dataset contains 768 patient data records and each record is described by 8 attributes, while the 9th attribute associates the class attribute - the onset of diabetes within 5 years, with a zero indicating the onset and one indicating the non-occurrence. All the patients considered here are females and at least 21 years old of Pima Indian heritage. The population lives near Phoenix, Arizona, USA.

4.2   Explore

Data in a database or data warehouse is usually explored either graphically or analytically. The quality of data being explored is an important requirement for data mining. The presence of missing values, outliers obstruct the pattern of the data mined. Various methods are used to handle the noisy and missing data like replacing the incorrect values with the means, medians or even user defined values. Adopted the visualization methodology for data exploration, the boxplot as indicated in Fig. 2, obtained from the data indicates the quartiles, median as well as the outlier values for each of the attributes.

The dataset, when examined manually, is seen to contain some incorrect values for specific attributes, as listed in Table 1. There are 500 non-diabetic patients (class = negative) and 268 diabetic ones (class = positive) for an incidence rate of 34.9%. Thus if you simply guess that all are non-diabetic, your accuracy rate is 65.1% (or error rate of 34.9%). Using the simple strategy of selecting tuples that contain complete and correct data values, we statistically represented the dataset selected, using the boxplot to study the dataset. The selected data tuples, is the base on which the regression model is built to predict the incorrect values. The tuples containing the erroneous/incorrect values are then fed to the regression model and the values are predicted, as a part of our approach for the preprocessing of the data.

Table 1. Attribute-value study outcome, indicating attributes with incorrectly present 'zero' value

Sl.No.	Attribute Name	No. of Zero value instances for attributes listed, present out of 768 tuples.
1.	Plasma glucose concentration, in 2 hours, in an oral glucose tolerance test	5
2.	Diastolic blood pressure (mm Hg)	35
3.	Triceps skin fold thickness (mm)	227
4.	2-hour serum insulin (mu U/ml)	374
5.	Body mass index (weight in kg/(height in m)^2)	11

4.3   Model

An Approach to Handle Missing Values

A regression model, using Classification and Regression Trees (CART), is developed, using tuples, which have complete correct values for all the attributes. These tuples form the training set for the model. The records that need values to be imputed, are then input to the model. The outlier set of the original dataset is compared with the generated values outlier set. A second order quadratic error function is generated.

Genetic Algorithms is used to optimize the error function in any given domain. Using a Genetic Algorithm function call,

$$x = ga(fitnessfcn, nvars) \qquad (1)$$

we find a local unconstrained minimum, x, to the objective function, $fitnessfcn$. $nvars$ is the dimension (number of design variables) of $fitnessfcn$. The objective function, $fitnessfcn$, accepts a vector x of size 1-by-$nvars$, and returns a scalar

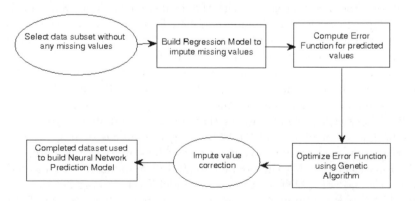

Fig. 1. Architecture of the proposed prediction model

evaluated at x. The predicted value for the incorrect/missing value is corrected using this optimization through genetic algorithms. A box-plot is once again used to inspect the mean and the outliers in the dataset, for any improvement shown.

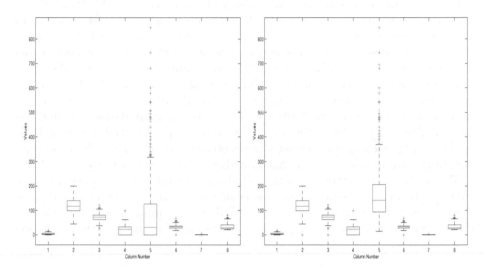

Fig. 2. Boxplots of the attribute values before and after soft computing preprocessing

Prediction Model
The model selected for the prediction study is the neural network supervised model. The completely preprocessed data is then input to the model. Our research shows that the neural network model works best with this data and the output statistics from the proposed model, based on our preprocessing approach is discussed in the performance section.

4.4   Assess

The model is then assessed using the test dataset, which is the randomly selected 20% part of the preprocessed, completed dataset. The proposed model gave an average of 80.22% as against a prediction model built on the un-preprocessed data. The random split used in the selection of the training dataset and the test dataset is the cause for the variance in the accuracy of the prediction model proposed in this work. The accuracy of the model is discussed in detail in the performance section.

5   Performance Analysis

The PIMA database, was used as it is available at the Machine Learning Database Repository at the University of California, Irvine, with no preprocessing step at all, to model the neural network prediction model. This means that the dataset contains a missing/incorrect value range comprising of 4% (taking only the records containing zero values to the attribute -diastolic blood pressure) to 48% (taking only the records containing zero values to the attribute - 2 hour serum insulin presence). The prediction accuracy of this neural network model is 75.816%. On implementing our architecture comprising of extensive data pre-processing and prediction, the prediction - neural network model is built on the completed, error optimized dataset, by splitting the dataset, using the random split, of 80-20 split where 80% (615 records) is the training dataset volume and 20% (153 records) is the testing dataset volume, thus validating the model. To avoid bias, the records for each set was selected randomly. This splitting of data facilitates easy validation of the model with the focus on reliability.

The performance for the 4 runs are as indicated in Table 2. The sensitivity (the true positive rate), specificity (the true negative rate) and correctness calculated from the confusion matrix is shown in Table 3., which emphasizes that the higher the sensitivity, specificity and correctness is, the more accurate is the prediction model.

Table 2. Prediction Accuracy details

Sl.No.	Accuracy	Sensitivity	Specificity	Correctness
1.	79.084%	69.64%	85%	79.49%
2.	81.046%	70%	86.41%	79.49%
3.	78.431%	84.31%	75.49%	76.92%
4.	82.353%	73.77%	88.04%	80.77%

6   Conclusions

This work proposes a fresh approach for preprocessing real-time data, used to build prediction models, with practical application, specifically taking a dataset from the health sector. The preprocessing, which involves imputation using CART along with optimization of the error function using a Genetic Algorithm, takes care to map imputed values in the valid domain of values for each attribute. The accuracy of proposed model is higher, by a maximum of 82.33%, as against the accuracy of a neural network prediction model, build on the unprocessed data which is 75.816%.

This model is to be tested on other similar health sector data, which tracks ailments and the factors that influence the occurrence or recurrence of the disease. We can test out this model, in the Indian diabetes patients context and predict the possibility of the patient to be affected by diabetes, in which case, changes required in the routine or life style can be adopted, to postpone the immediate chance of the disease.

The prepossessing steps that have been discussed here can be used on any prediction model development framework, to impute missing, incorrect and even incomplete values. As the approach is simple to implement, and the results easy to comprehend, real-time decision making or any other categories of applications that require imputation, can adopt these steps effectively.

References

1. Mitra, S., Acharya, T.: Data Mining, Multimedia, Soft-computing and Bioinformatics. Wiley Interscience, Hoboken (2004)
2. Little, R.J.A., Rubin, D.B.: Statistical Analysis with Missing Data. John Wiley, New York (1987)
3. Zhang, S., Qin, Z., Ling, C.X., Sheng, S.: Missing is Useful: Missing Values in Cost Sensitive Decision Trees. IEEE Transactions on Knowledge and Data Engineering 17(12) (2005)
4. Rady, E.A., Abd, El-Monsef, M.M.E., Abd, El-Latif, W.A.: A Modified Rough Set Approach to Incomplete Information Systems. Journal of Applied Mathematics and Decision Sciences 2007, article ID 58248 (2007)
5. Satya Kumar, D.V.R., Sriram, K., Rao, K.M., Murty, U.S.: Management of Filariasis Using Prediction Rules Derived from Data Mining. In: Bioinformation by Biomedical Informatics Publishing Group (2005)
6. Palaniappan, S., Awang, R.: Intelligent Heart Disease Prediction System using Data Mining Techniques. International Journal of Computer Science and Network Security 8(8) (2008)
7. Liu, P., Lei, L.: A Review of Missing Data Treatment Methods. Intelligent Information Management Systems and Technologies 1(3), 412–419 (2005)
8. Mehala, B., Ranjit Jeba Thangaiah, P., Vivekanandan, K.: Selecting Scalable Algorithms to Deal with Missing Values. International Journal of Recent Trends in Engineering 1(2) (2009)
9. Adbdella, M., Marwala, T.: Treatment of Missing Data Using Neural Networks and Genetic Algorithms. In: International Joint Conference on Neural Networks, Canada (2005)
10. Magnani, M.: Techniques for Dealing with Missing Data in Knowledge Discovery Tasks, Department of Computer Science, University of Bologna (2004)
11. Acuna, E., Rodriguez, C.: The Treatment of Missing Values and its Effect in the Classifier Accuracy. In: Multiscale Methods in Science and Engineering, pp. 639–647. Springer, Heidelberg (2004)

A Decision Making Model for Vendor-Buyer Inventory Systems

Hui-Ming Wee[1,2], Jie Lu[1], Guangquan Zhang[1], Huai-En Chiao[2], and Ya Gao[1]

[1] Decision Systems & e-Service Intelligence (DeSI) lab
Centre for Quantum Computation & Intelligent Systems (QCIS)
Faculty of Engineering & Information Technology
University of Technology Sydney, PO Box 123, NSW 2007, Australia
{jielu,zhangg,yagao}@it.uts.edu.au
[2] Department of Industrial and Systems Engineering
Chung Yuan Christian University, Chungli 32023, Taiwan, ROC
weehm@cycu.edu.tw

Abstract. In a vendor-buyer supply chain, the buyer's economic order quantity and the vendor's optimal number of deliveries are derived either independently or collaboratively. In this paper, we establish a two-stage vendor-buyer inventory system decision model by using bi-level decision making approach. The experimental result shows that the proposed bi-level decision model can effectively handle two-stage vendor-buyer inventory problems and obtain better results than the existing methods.

Keywords: Bi-level decision-making, two-stage supply chain, vendor–buyer inventory system, optimization.

1 Introduction

Supply chain activities transform natural resources, raw materials and components into a finished product that is delivered to the end customer. A two-stage vendor-buyer inventory system is a typical supply chain system of organizations, people, technology, activities, information and resources involved in moving a product or service from vendor to buyer. In a two-stage supply chain problem, both vendor and buyer aim to minimize their individual costs in the inventory system but their decisions are constrained with each other. Researchers have developed various models to describe this problem and tried to find optimized solutions. For example, Yang et al. [17] presented a global optimal policy using classical optimization technique for vendor–buyer integrated inventory systems with a just-in-time environment. Banerjee [2] derived a joint economic lot size model for a single vendor -- single buyer system where the vendor has a finite production rate. Goyal [5] extended Banerjee's model by relaxing the lot-for-lot production assumption. There are some other methods to solve the inventory system models such as the algebraic optimization approach popularized by Grubbström and Erdem [6], Yang and Wee

H. Sakai et al. (Eds.): RSFDGrC 2009, LNAI 5908, pp. 336–343, 2009.
© Springer-Verlag Berlin Heidelberg 2009

[16], and Wee and Chung [14]. Although these methods can model the two-stage supply chain problem, they didn't consider the hierarchical relationship between the decisions of the vendor and the buyer when they aim to minimize their individual costs in an inventory system. That is, in most previous researches of two-stage supply chain, the buyer's economic order quantity and the vendor's optimal number of deliveries are derived either independently or collaboratively. We have found that in fact, the two-stage supply chain problem reflects the features of two level decision making: non-cooperative, two-player. In the problem, both buyer and vendor considered other party's decision variable but cannot control it.

Bi-level programming (BLP) aims at solving decentralized decision problems within a bi-level hierarchical organization [1, 7]. The decision entity at the upper level is termed as leader, and the lower level, follower. Each decision entity tries to optimize their own objective functions but any of their decisions will affect the objective value of the other level [9, 15]. Since the two-stage supply chain problem in the vendor–buyer inventory systems involves two hierarchical optimizations (buyer and vendor), obviously, bi-level programming is naturally suitable for modeling the features of the problem, in which the buyer is as the leader and the vendor as the follower, or vice versa.

We therefore established a two-stage vendor-buyer inventory system decision model using bi-level programming method. As both vendor and buyer's objectives are nonlinear functions, the two-stage vendor–buyer inventory system decision model becomes a nonlinear bi-level programming problem. This study has two main contributions. First, we apply bi-level programming technique to establish a vendor–buyer inventory system decision model. The second is the results obtained using the model is much better than using existing methods.

The rest of our paper is organized as follows: Section 2 gives a general bi-level decision making model. Section 3 proposes a two-stage vendor–buyer inventory system decision model by using bi-level programming. Experiment of solving the two-stage supply chain problems using the proposed model is presented in Section 4. Experiment results are very positive to support the proposed bi-level vendor–buyer inventory system decision model. Finally, conclusions and further studies are highlighted in Section 5.

2 Bi-level Decision Making Model

A bi-level decision problem can be viewed as a static version of the non-cooperative, two-player (decision entity) game. The decision entity at the upper level is termed as leader, and the lower level, follower. The control for the decision variables is partitioned amongst the decision entities who seek to optimize their individual objective function [1, 3, 4, 15].

Bi-level programming typically models bi-level decision problems, in which the objectives and the constraints of both the upper and the lower level decision entities (leader and follower) are expressed by linear or nonlinear functions, as follows [1, 3, 4, 8]:

For $x \in X \subset R^n$, $y \in Y \subset R^m$, $F : X \times Y \to R^1$, and $f : X \times Y \to R^1$,

$$\min_{x \in X} F(x, y)$$

subject to $G(x, y) \leq 0$

$$\min_{y \in Y} f(x, y)$$

subject to $g(x, y) \leq 0$,

where the variables x, y are called the leader and the follower variables respectively, and $F(x, y)$ and $f(x, y)$ the leader's and the follower's objective functions.

This model aims to find a solution to the upper level problem $\min_{x \in X} F(x, y)$ subject to its constraints $G(x, y) \leq 0$ where, for each value of leader's variable x, y is the solution of the lower level problem $\min_{y \in Y} f(x, y)$ under its constraints $g(x, y) \leq 0$.

Under the general bi-level programming model, there are some special situations. For example, in the leader's objective function $F(x, y)$ and/or constraint function $G(x, y)$, y can be absent.

There have been many approaches and algorithms proposed for solving BLP problems since the field caught the attention of researchers in the mid-1970s, including the well-known Kuhn-Tucker approach [3], the Branch-and-bound algorithm [7], penalty function approach [15], and the Kth-best approach [1, 3, 4],. Furthermore, some fuzzy BLP approaches [10, 12, 18, 19], multi-follower BLP approaches [8,9], and multi-objective BLP approaches [11, 13, 19] have been recently developed to deal with more complex cases of bi-level decision problems in a real-world.

However, most of them mainly deal with linear bi-level programming, but there are less research results reported for non-linear bi-level programming and less applications in two-stage vendor-buyer inventory system.

3   A Bi-level Vendor-Buyer Inventory System Decision Model

In a two-stage supply chain problem of a vendor–buyer inventory system, both vendor and buyer aim to minimize their costs in the inventory system but their decisions are related with each other in a hierarchical way. We first assume to give the priority to the buyer's decision. Therefore, the buyer is assigned as leader and the vendor as follower in a bi-level model. The buyer attempts to optimize his/her objective, i.e.: minimize cost, the vendor, under the constraint of the leader, tries to find his/her minimized cost, based on the decisions made by the buyer. We therefore have two decision variables to consider in the problem:

(1) the number of deliveries from vendor to buyer;
(2) the number of buyer's lot size per delivery.

In Order to Compare the Results with Yang et al. [17], We Use the Same Nomenclature as Follows.

Q = Buyer's lot size per delivery

n = Number of deliveries from the vendor to the buyer per vendor's replenishment interval

S = Vendor's setup cost per set up

A = Buyer's ordering cost per order

C_v = Vendor's unit production cost

C_b = Unit purchase cost paid by the buyer

r = Annual inventory carrying cost per dollar invested in stocks

P = Annual production rate

D = Annual demand rate

$TC(Q,n)$ = integrated total cost of the vendor and the buyer when they collaborate

Yang et al.[17] proposed the following total cost of the integrated two-stage inventory system for the vendor-buyer as:

$$TC(Q,n) = \frac{DA}{Q} + \frac{rQC_b}{2} + \frac{DS}{nQ} + \frac{rQC_v}{2}\left[(n-1)(1-D/P)+D/P\right] \tag{1}$$

The buyer's total cost is:

$$TC_b(Q) = \frac{DA}{Q} + \frac{rQC_b}{2} \tag{2}$$

The vendor's total cost is:

$$TCv(Q,n) = \frac{DS}{nQ} + \frac{rQC_v}{2}\left[(n-1)(1-D/P)+D/P\right] \tag{3}$$

In Equation (2), the buyer controls Q, the number of order quantity. In Equation (3), the vendor controls n, the number of delivery. All other parameters are known for a specific problem.

As we assume that the buyer is the leader and the vendor the follower, by combining Equations (2) and (3), we establish a bi-level decision model where Equations (2) and (3) can be re-written as:

$$\operatorname*{Min}_{Q} TC_b(Q) = \frac{DA}{Q} + \frac{rQC_b}{2}$$

subject to $Q > 0$ (4)

$$\operatorname*{Min}_{n} TCv(Q,n) = \frac{DS}{nQ} + \frac{rQC_v}{2}\left[(n-1)(1-D/P)+D/P\right]$$

subject to $Q > 0, n > 0,$

In this model, both the buyer and the vendor try to adjust their controlling variable, wishing to minimize their respective cost under specific constraints. The buyer is the leader who tries to minimize his/her total cost by taking the derivative of Equation (2) with respect to Q and setting it to zero; one can obtain the expression for the economic order quantity Q as:

$$Q* = \sqrt{\frac{2DA}{rC_b}} \tag{5}$$

Once the buyer has made his/her decision, the vendor will optimize his/her strategy base on the buyer's decision. Substitute the information from Equation (5) into Equation (3), and then take the derivative with respect to n and setting it to zero, one can obtain the expression to the global minimum of the decision variable n as:

$$n* = \sqrt{\frac{2DS}{rQ*^2\,C_v\left(1-D/P\right)}} \tag{6}$$

Substitute the optimal values of n and Q into Equation (3), the following optimal condition of Equation (3) occurs:

When $TC(n*-1) \geq TC(n*) \leq TC(n*+1)$, the optimal delivery number of n is

$$n* = \left\lfloor \left| \sqrt{\frac{2DS}{rQ*^2C_v\left(1-D/P\right)}} \right| \right\rfloor, \tag{7}$$

where $\lfloor\ \rfloor$ is the integer operation where the integer value is equal or less than its argument.

For each $n \geq 1$, there is a feasible and optimal solution to the lot size (Q).

All feasible values of Q and n are derived by the vendor. Optimal value of Q and n are chosen after considering the benefit to both the buyer and the vendor.

Now we assume that the vendor is the leader and the buyer the follower. By combining Equations (2) and (3), we establish another bi-level decision model where Equations (2) and (3) can be re-written as:

$$\underset{n}{\text{Min}}\,TCv(Q,n) = \frac{DS}{nQ} + \frac{rQC_v}{2}\left[(n-1)(1-D/P)+D/P\right]$$

subject to $n > 0$, $\tag{8}$

$$\underset{Q}{\text{Min}}\,TC_b(Q) = \frac{DA}{Q} + \frac{rQC_b}{2}$$

subject to $Q > 0$

4 Experiment Analysis

This section presents a numerical example to validate the effectiveness and working process of the proposed bi-level decision models.

We consider a two-stage vendor-buyer inventory problem with the following parameters:

Annual demand rate, $D = 1000$.
Annual production rate, $P = 3200$.
Buyer's ordering cost, $A = \$25$ per order.
Vendor's setup cost, $S = \$400$ per setup.

Buyer's unit purchase cost, $C_b = \$25$.
Vendor's unit production cost, $C_v = \$20$.
Annual inventory carrying cost per dollar, $r = 0.2$.
Using (4), the optimal value of $Q = 100$ units.
Using (5), the value of $n = 5.39$.

We first consider that the buyer is the leader and establish a two-stage vendor-buyer inventory system decision model.

Since $TC(n-5)$-$TC(n-6) - 1412.5\ 1416.7 < 0$; the optimal number of deliveries is 5. Since condition (8) is satisfied, there is a solution for the problem. Using conditions (2) and (3), one can obtain the optimal values for Q^* and n^*, as 100 units and 5 deliveries, respectively. Substituting the optimal values of $Q^* = 100$ units and $n^* = 5$ deliveries into equation (2) and (3), the optimal value for the total cost for the buyer and vendor is \$500 and \$1412.5 per year respectively. The lot size for the vendor is $(n^*) \times (Q^*) = 500$ units, and the total vendor-buyer cost is \$1912.5 per year. Table 1 shows these values.

We can obtain similar results when the roles of the buyer/vendor are interchanged as shown in Table 1.

We also compare the results of the analysis with those of Yang et al. [17], Banerjee [2] and Goyal [5].

From Table 1 we can see that we propose two alternatives in business dealings:

(1) The traditional supply chain arrangement where the buyer is fully responsible for the lot size decision and the vendor is fully responsible for the number of deliveries.

Table 1. Summary of results

Models	Banerjee (lot for lot assumption)	Goyal (relaxed lot for lot assumption)	Yang & Wee (integrated model)	This Study (buyer as leader)	This Study (vendor as leader)
n	1	2	5	5	5
Buyer's order size	369	198	110	100	106
Vendor's lot size	369	396	550	500	530
Buyer's annual cost	\$ 990.3	\$621.3	\$ 502.4	\$ 500	\$ 500.8
Vendor's annual cost	\$1314.6	\$1653.6	\$1400.9	\$1412.5	\$1406.2
Total vendor - buyer cost	\$2304.9	\$2274.9	\$1903.3	\$1912.5	\$1907

(2) When the vendor is responsible to the lot size and the number of deliveries which will result in optimal inventory and delivery costs. The alternative is usually termed vendor-management-inventory (VMI).

In our two-stage vendor-buyer inventory system decision model, we also try to consider the other player's benefit. After deriving all possible solutions using VMI, we finalize the vendor's decisions, in term of lot size and the number of delivery that will be beneficial to the buyer as well.

From our experimental data, we see that the vendor as leader outperforms the buyer as leader. This is because vendor as leader improves the actual consumption rates; the lot size decision by the vendor ensures production matches demand more closely, reduces inventory and improves business performance. This is why the VMI has become very popular in recent years. The total costs for the vendor-buyer system from our study are more than the optimal integrated model by Yang and Wee. This is understandable because the model by Yang and Wee is a global solution that minimizes the total costs for the vendor-buyer system, and in practice this is difficult for implementation. Obviously, our results have lower costs than those of Banjeree and Goyal's studies, and it is easily acceptable by both buyer and vendor in practice.

5   Conclusion and Further Study

This paper presents the development of an inventory system for a supply chain with bi-level decision making model. The results obtained using the model is much better than using existing methods. Due to the relative simple experiment data in this study, we can derive optimal global solution quite easily. For more complex problems, evolutionary methods such as genetic algorithms and swam optimization methods are suggested. These meta-heuristic methods do not guarantee a global solution, and few algorithms have been published in literature. Moreover, there is no commercial solver to do the job. Our results provide insights into how firms can choose the best supply chain arrangement to optimize supply chain efficiency.

Future research includes (1) to handle more complex vendor-buyer inventory system decision problems where multiple followers (multi-buyers) appear and they may have different objectives and variables; (2) to extend the proposed bi-level vendor-buyer inventory system decision model to a tri-level model to deal with three level supply chain; (3) to improve the developed bi-level decision support software to implement the multi-follower tri-level programming algorithms and apply it in more real-world supply chain management applications.

Acknowledgment. The work presented in this paper was supported by Australian Research Council (ARC) under discovery project DP0557154.

References

1. Anandalingam, G., Friesz, T.: Hierarchical optimization: An introduction. Annals of Operations Research 34, 1–11 (1992)
2. Banerjee, A.: A joint economic lot size model for purchaser and vendor. Decision Sciences 17, 292–311 (1986)

3. Bard, J., Falk, J.: An explicit solution to the multi-level programming problem. Computer and Operations Research 9, 77–100 (1982)
4. Bard, J., Moore, J.T.: A branch and bound algorithm for the bilevel programming problem. SIAM Journal on Scientific and Statistical Computing 11, 281–292 (1990)
5. Goyal, S.K.: A joint economic lot size model for purchaser and vendor: A comment. Decision Sciences 19, 236–241 (1988)
6. Grubbström, R.W., Erdem, A.: The EOQ with backlogging derived without derivatives. International Journal of Production Economics 59, 529–530 (1999)
7. Hansen, P., Jaumard, B., Savard, G.: New branch-and-bound rules for linear bilevel programming. SIAM Journal on Scientific and Statistical Computing 13, 1194–1217 (1992)
8. Lu, J., Shi, C., Zhang, G., Dillon, T.: Model and extended Kuhn-tucker approach for bilevel multi-follower decision making in a referential-uncooperative situation International Journal of Global Optimization 38, 597–608 (2007)
9. Lu, J., Shi, C., Zhang, G.: On bilevel multi-follower decision-making: general framework and solutions. Information Sciences 176, 1607–1627 (2006)
10. Sakawa, M., Nishizaki, I., Uemura, Y.: Interactive fuzzy programming for multilevel linear programming problems with fuzzy parameters. Fuzzy Sets and Systems 109, 3–19 (2000)
11. Shi, X., Xia, H.: Interactive bilevel multi-objective decision making. Journal of the Operational Research Society 48, 943–949 (1997)
12. Shih, H.S., Lai, Y.J., Lee, E.S.: Fuzzy approach for multi-level programming problems. Computers and Operations Research 23, 73–91 (1996)
13. Sinha, S., Sinha, S.B.: KKT transformation approach for multi-objective multi-level linear programming problems. European Journal of Operational Research 143, 19–31 (2002)
14. Wee, H.M., Chung, C.J.: A note on the economic lot size of the integrated vendor–buyer inventory system derived without derivatives. European Journal of Operational Research 177, 1289–1293 (2007)
15. White, D., Anandalingam, G.: A penalty function approach for solving bi-level linear programs. Journal of Global Optimization 3, 397–419 (1993)
16. Yang, P.C., Wee, H.M.: The economic lot size of the integrated vendor–buyer inventory system derived without derivatives. Optimal Control Applications and Methods 23, 163–169 (2002)
17. Yang, P.C., Wee, H.M., Yang, H.J.: Global optimal policy for vendor–buyer integrated inventory system within just in time environment. Journal of Global Optimization 37, 505–511 (2007)

Software Reliability Prediction Using Group Method of Data Handling

Ramakanta Mohanty[1], V. Ravi[2,*], and Manas Ranjan Patra[1]

[1] Computer Science Department, Berhampur University,
Berhampur-760 007, Orissa, India
[2] Institute for Development and Research in Banking Technology,
Castle Hills Road #1, Masab Tank, Hyderabad – 500 057 (AP) India
Tel.: +91-40-2353 4981; Fax: +91-40-2353 5157
ramakanta5a@gmail.com, rav_padma@yahoo.com,
mrpatra12@gmail.com

Abstract. The main purpose of this paper is to propose the use of Group Method of Data Handling (GMDH) to predict software reliability. The GMDH algorithm presented in this paper is a heuristic self-organization method. It establishes the input-output relationship of a complex system using multilayered perception type structure that is similar to a feed forward multilayer neural network. The effectiveness of GMDH is demonstrated on a dataset taken from literature. Its performance is compared with that of multiple linear regression (MLR), back propagation trained neural networks (BPNN), threshold accepting trained neural network (TANN), general regression neural network (GRNN), pi-sigma network (PSN), dynamic evolving neuro-fuzzy inference system (DENFIS), TreeNet, multivariate adaptive regression splines (MARS) and wavelet neural network (WNN) in terms of normalized root mean square error (NRMSE). Based on experiments conducted, it is found that GMDH predicted reliability with least error compared to other techniques. Hence, GMDH can be used a sound alternative to the existing techniques for software reliability prediction.

Keywords: Software reliability forecasting, Group method of Data handling, Intelligent Techniques.

1 Introduction

Reliability is the important factor in accessing the software quality. It is related with defects and faults. If more faults are encountered, the reliability of software decreases. Therefore, Reliability is defined as the probability of a system or component to work properly for a particular period of time under certain conditions. According to Lyu [1], Software reliability consists of three main activities viz. error prevention, fault detection and recovery and measurement criteria. Software reliability is an important factor which affects system reliability. It differs from hardware reliability in that it reflects the design perfection, rather than manufacturing perfection. Software

---

* Corresponding author.

H. Sakai et al. (Eds.): RSFDGrC 2009, LNAI 5908, pp. 344–351, 2009.

modeling techniques can be divided into two categories viz. prediction modeling and estimation modeling. But both modeling techniques are based on observing and accumulating failure data and analyzing it with statistical inference. In the past few years much research work has been carried out in software reliability and forecasting. But none of the models could give accurate prediction about software reliability. All those models are data driven. In recent years, neural network (NN) approaches have proven to be universal approximators for any non-linear continuous function with arbitrary accuracy [1, 2, 3]. Many papers published in the literature observe that neural network could offer promising approaches to software reliability estimation and modeling [2-18]. For example, Karunanithi et al. [20, 21] first applied neural network architecture to estimate the software reliability and used the execution time as input and cumulative number of detected faults as desired output, and encoded the input and output into the binary bit string. Furthermore, they also illustrated the usefulness of connectionist models for software reliability growth prediction and showed that the connectionist approach is capable of developing models of varying complexity. Sherer [19] employed neural networks to predict software faults in several NASA projects. She found that identification of fault-prone modules through initial testing could guide subsequent testing efforts provided software faults tend to cluster. Besides, Sitte [16] compared predictive performance of 'neural networks' and 'recalibration for parametric models' for software reliability. She found that neural networks are much simpler to use than the recalibration method.

Khoshgoftaar et al. [11], Khoshgoftaar and Szabo [11] used the neural network for predicting the number of faults in programs. They introduced an approach for static reliability modeling and concluded that the neural networks produce models with better quality of fit and predictive quality. The neural networks used for software reliability modeling can be classified into two classes. One used cumulative execution time as inputs and the corresponding accumulated failures as desired outputs. This class focuses on modeling software reliability modeling by varying different kind of neural network such as recurrent neural network [20, 21], Tian and Noore, [17, 22] and Elman network [16]. The other class models the software reliability based on multiple-delayed input single-output neural network. Cai et al. [2, 4] employed neural network to predict the next failure time by using 50 inter-failure times as the multiple-delayed-inputs. Tian and Noore [17] proposed software cumulative failure time prediction on multiple-delayed-input single-output architecture by using an evolutionary neural network. However, there is a problem in this kind of approaches. We have to predetermine the network hidden architecture in terms of the number of neurons in each layer and the numbers of hidden layers. Cai's [2, 4] experimented the effect of the number of input neurons, the number of neurons in hidden layer and the number of hidden layers by independently varying the network architecture. Another problem is that since several fast training algorithms are investigated for reducing the training time, these advanced algorithms focus on the model fitting and this will cause the over-fitting. Pai and Hong [23] have applied Support Vector Machines (SVMs) for forecasting software reliability where Simulated Annealing (SA) algorithm was used to select the parameters of the SVM model. Su and Huang [24] applied neural networks to predict software reliability and built a dynamic weighted combinational model (DWCM). Recently, Rajkiran and Ravi [25] proposed an ensemble model to forecast software reliability. They used MLR, MARS, dynamic evolving neuro-fuzzy inference system (DENFIS) and TreeNet to develop the ensemble. They found that non-linear ensemble outperformed all other ensembles and intelligent techniques. Later, Rajkiran and Ravi [26] employed Wavelet Neural Networks (WNN) to predict software reliability. They used two kinds of wavelets – More wavelet and Gaussian wavelet as transfer function resulting in two variants of WNN. It was found that Wavelets Neural Networks based on normalized root mean square error (NRMSE) obtained on test data outperformed all other techniques. Most recently, Ravi et al. [27] proposed the use of Threshold accepting trained wavelet neural network (TAWNN) to predict operational risk in banks and firms by predicting software reliability. They employed two types of TAWNN as transfer function viz. Morlet wavelet and Gaussian wavelet. They compared the performance of TAWNN variants with that of multiple linear regression (MLR), Multivariate adaptive regression splines (MARS), Back propagation trained neural network (BPNN), Threshold accepting trained neural network (TANN), Pi-sigma network (PSN), General regression neural network (GRNN), Dynamic evolving neuro-fuzzy inference system (DENFIS) and TreeNet in terms of normalized root mean square. From the experiments, they found that WNN based models outperformed all the individual techniques over all the lags.

In this paper, we explore the usefulness of GMDH in predicting software reliability and compare its performance with that of several intelligent methods.

The rest of the paper is organized in the following manner. A brief discussion about overview of GMDH is presented in section 2. Section 3 presents the experimental methodology. Section 4 presents a detailed discussion of the results. Finally, Section 5 concludes the paper.

2   Overview of GMDH

GMDH [28] is a heuristic self organizing method that models the input-output relationship of a complex system using a multilayered Rosenblatt's perception-type network structure. Each element in the network implements a non-linear equation of two inputs and its coefficients are determined by a regression analysis. Self selection thresholds are given at each layer in the network to delete those useless elements which can not estimate the correct output. Only those elements whose performance indices exceed the threshold are allowed to pass to succeeding layers, where more complex combination is formed. These steps are repeated until the convergence criterion is satisfied or a predetermined number of layers are reached. GMDH approach can be useful because:

- A small training set of data is required.
- The computational burden is reduced.
- The procedure automatically filters out input properties that provide little information about the location and shape of hyper surface.
- A multilayer structure is a computationally feasible way to implement multinomials of high degree.

The concept of GMDH algorithm [28] used in this paper is described as follows:
GMDH algorithm can be represented as a set of neurons in which different pairs of them in each layer are connected through a quadratic polynomial and thus produce

new neuron in the next layer. Such representation can be used to map inputs to output. The formal definition of the identification is to find a function f in order to predict output Y for a given input vector $x = (x_1 x_2 x_3,\ldots\ldots\ldots, x_n)$ as close as possible to its actual output Y. Therefore, assume the output variable Y is a function of the input variable $(x_1 x_2 x_3,\ldots\ldots\ldots x_n)$, as in the following equation

$$Y = f(x_1, x_2, \ldots\ldots, x_n) \tag{1}$$

At each layer, GMDH will build a polynomial like the following:

$$\hat{Y} = f(x_j, x_k) = a_{0i} + a_{1i}x_j + a_{2i}x_k + a_{3i}x^2_{\ j} + a_{4i}x^2_{\ k} + a_{5i}x_j x_k$$

$$i = 1,\ldots q; \ j = 1,\ldots n; k = 1\ldots\ldots n - 1; q = \frac{n(n-1)}{2} \tag{2}$$

Where a_0 is a constant

$a_1, a_2, a_3 \ldots\ldots\ldots\ldots\ldots$ are coefficients,

and x_i, x_j, x_k ....are input variables.

In the first layer, the input neurons are the problem input and the second layer neuron consists of polynomials like in Eqn. (3). When third layer is built, the input to the third layer polynomials can either be the original problem inputs, the polynomial from the second layer, or both. If inputs are polynomial, then a much more complicated polynomial will be built. Successive layers take inputs from either the original inputs or the polynomials from immediately from proceeding layer.

In general, The Kolmogorov-Gabor polynomial [28] can simulate

$$\hat{Y} = a_0 + \sum_{i=1}^{n} a_i x_i + \sum_{i=1}^{n}\sum_{j=1}^{n} a_{ij} x_i x_j + \sum_{i=1}^{n}\sum_{j=1}^{n}\sum_{k=1}^{n} a_{ijk} x_i x_j x_k + \ldots \tag{3}$$

The input-output relationship perfectly and has been widely used as a complete description of the system model. By combining the so called partial polynomial of two variables in the multilayer, the GMDH can easily solve these problems. In fact, the GMDH network is not like regular feed forward network and was not originally represented as a network. In Neuroshell2 [21], the GMDH network is implemented with polynomial terms in the links and a genetic component to decide how many layers are built. The result of training at the output layer can be represented as a polynomial function of all or some of inputs Ivakhnenko [29]. The main idea behind GMDH is that it is trying to build a function (called a polynomial model) that would behave in such a way that the predicted value of the output would be as close as possible to the actual value of output [31].

3 Experimental Design

In our experiment, we followed general time series forecasting model because software reliability forecasting has only one dependent variable and no explanatory variables in strict sense. The general time series can be presented as

$$X_t = f(X')   \tag{4}$$

Where X' is vector of lagged variables { $x_{t-1}, x_{t-2}, ..., x_{t-p}$ }. Hence the key to finding the solution to the forecasting problems is to approximate the function ' f '. This can be done by iteratively adjusting the weights in the modeling process.

Below is an illustration of how training patterns can be designed in the neural network modeling process [3].

X				Y
x_1	x_2	$\cdots$	x_p	x_{p+1}
x_2	x_3	$\cdots$	x_{p+1}	x_{p+2}
x_3	x_4	$\cdots$	x_{p+2}	x_{p+3}
.	.	$\cdots$	.	.
.	.	$\cdots$	.	.
.	.	$\cdots$	.	.
x_{t-p}	x_{t-p+1}	$\cdots$	x_{t-1}	x_t

Where p denotes the number of lagged variables and $(t - p)$ denotes the total number of training samples. In this representation, X is a set of $(t - p)$ vectors of dimension p and Y is a vector of dimension $(t - p)$. X and Y represent the vector of explanatory variables and dependent variable in the transformed datasets respectively.

In this paper, the software failure data obtained from Musa [30]. It is used to demonstrate the forecasting performance of GMDH. The data contains 101 observations of the pair (t, Y_t) pertaining to software failure. Here Y_t represents the time to failure of the software after the tth modification has been made. We created five datasets viz. lag # 1,2,3,4 and 5 in view of the foregoing discussion on generating lagged data sets out of a time series. The experiments are performed by dividing the data into training and test set in the ratio of 80:20. The value of Normalized Root Mean Square Error (NRMSE) is used as the measurement criteria.

$$NRMSE = \sqrt{\frac{\sum_{i=1}^{n}(d_i - \hat{d}_i)^2}{\sum_{i=1}^{n} d_i^2}}   \tag{5}$$

4 Results and Discussions

We employed GMDH available in Neroshell2 tool [31] to predict software reliability on datasets taken from literature [30]. NeuroShell2 includes a very powerful architecture, called Group Method of Data Handling (GMDH) or polynomial nets. The parameters used in GMDH are: Scale Function-[0, 1], GMDH type - *Advanced,* Optimization - *Full,* Maximum variable - X_1, X_2, X_3, Selection Criteria-*Regularly,* Missing values to be - *Error Condition* for obtaining the results.

We compared the performance of GMDH with that of BPNN, TANN, PSN, MARS, GRNN, MLR, TreeNet, DENFIS, Morlet based WNN and Gaussian based WNN [26, 27].

Table 1. NRMSE values on test data for different techniques

Techniques	Lag1	Lag2	Lag3	Lag4	Lag5
BPNN	0.171375	0.166086	0.151429	0.144949	0.145541
TANN	0.179309	0.183735	0.158407	0.152008	0.150355
PSN	0.186867	0.176708	0.165935	0.164855	0.157922
MARS	0.170584	0.17091	0.161343	0.154821	0.15267
GRNN	0.210247	0.211408	0.176769	0.179869	0.166883
MLR	0.171448	0.167776	0.156537	0.151152	0.147881
TreeNet	0.168286	0.167865	0.168105	0.156998	0.161121
DENFIS	0.170907	0.167306	0.15425	0.148379	0.147641
Morlet based WNN	0.119402	0.118788	0.122810	0.115742	0.116238
Gaussian based WNN	0.124162	0.120630	0.118937	0.118610	0.111520
GMDH	**0.150875**	**0.152475**	**0.107226**	**0.103601**	**0.098616**

Table 1 illustrates the NRMSE values of different lags of data obtained over different techniques. The parameters of all the algorithms are tweaked until the least NRMSE values could be obtained and the same are presented in Table 1. We observed from Table 1 that the NRMSE value of different techniques is gradually decreasing with increase in lag number and also we can clearly deduce from Table 1 that GMDH yielded better NRMSE values compared to BPNN, TANN, PSN, MARS, GRNN, MLR, TreeNet, and DENFIS. But GMDH could not predict well in lag1 and lag2 as compared to Morlet based WNN and Gaussian based WNN. However, GMDH outperformed all techniques in case of Lag3, Lag4 and Lag5. In fact, GMDH yielded the best result of NRMSE value of 0.098616 in Lag5. We further observe here that the Morlet/Gaussian based WNN yielded the best result in literature so far. However, this study shows that GMDH outperformed even them in lag3,   lag4, and lag5 by a good margin. Further, we observe that Pai and Hong [23] used the same data set to test the efficacy of their support vector machine simulated annealing (SVMSA) method. However, our results cannot be compared with theirs since they did not use the lagged data in their experimentation. They divided the data set of 101 observations into training (33 observations), validation (8 observations) test (60 observations, which is not standard method of splitting the data set for experimentation. So we did not compare our results with theirs.

5 Conclusions

In this paper, Group Method of Data Handling (GMDH) is proposed first time to predict software reliability. We compared GMDH with that of MLR, PNN, TANN, GRNN, PSN, DENFIS, TreeNet, MARS, Morlet and Gaussian wavelet Neural network (WNN) in terms of normalized root mean square error (NRMSE). Based on the normalized root mean square error (NRMSE) values, we conclude that GMDH yielded better results compared to MLR, PNN, TANN, GRNN, PSN, DENFIS, TreeNet, MARS but GMDH could not yield better results compared to Morlet and Gaussian wavelet neural network (WNN) in lags 1 and 2. However, in lags 3, 4 and 5, GMDH yielded the best results so far in literature.

References

1. Lyu, M.R.: Handbook of Software Reliability Engineering. McGraw-Hill, New York (1996)
2. Cai, K., Yuan, C., Zhang, M.L.: A critical review on software reliability modeling. Reliability engineering and Systems Safety 32, 357–371 (1991)
3. Xu, K., Xie, M., Tang, L.C., Ho, S.L.: Application of neural networks in forecasting engine systems reliability. Applied Soft Computing 2, 255–268 (2003)
4. Cai, K.Y., Cai, L., Wang, W.D., Yu, Z.Y., Zhang, D.: On the neural network approach in software reliability modeling. The Journal of Systems and Software 58, 47–62 (2001)
5. Dohi, T., Nishio, Y., Osaki, S.: Optional software release scheduling based on artificial neural networks. Annals of Software engineering 8, 167–185 (1999)
6. Karunanithi, N., Whitley, D., Malaiya, Y.K.: Prediction of software reliability using neural networks. In: International Symposium on Software Reliability, pp. 124–130 (1991)
7. Khoshgoftaar, T.M., Szabo, R.M.: Predicting software quality, during testing using neural network models: A comparative study. International Journal of Reliability, Quality and Safety Engineering 1, 303–319 (1994)
8. Khoshgoftaar, T.M., Szabo, R.M.: Using neural networks to predict software faults during testing. IEEE Transactions on Reliability 45(3), 456–462 (1996)
9. Khoshgoftaar, T.M., Lanning, D.L., Pandya, A.S.: A neural network modeling for detection of high-risk program. In: Proceedings of the Fourth IEEE International Symposium on Software reliability Engineering, Denver, Colorado, pp. 302–309 (1993)
10. Khoshgoftaar, T.M., Rebours, P.: Noise elimination with partitioning filter for software quality estimation. International Journal of Computer Application in Technology 27, 246–258 (2003)
11. Khoshgoftaar, T.M., Pandya, A.S., More, H.B.: A neural network approach for predicting software development faults. In: Proceedings of the third IEEE International Symposium on Software Reliability Engineering, Los Alamitos, CA, pp. 83–89 (1992)

12. Khoshgoftaar, T.M., Allen, E.B., Hudepohl, J.P., Aud, S.J.: Application of neural networks to software quality modeling of a very large telecommunications system. IEEE Transactions on Neural Networks 8(4), 902–909 (1997)
13. Khoshgoftaar, T.M., Allen, E.B., Jones, W.D., Hudepohl, J.P.: Classification –Tree models of software quality over multiple releases. IEEE Transactions on Reliability 49(1), 4–11 (2000)
14. Lyu, M.R., Nikora, A.: Using software reliability models more effectively. IEEE Software, 43–53 (1992)
15. Musa, J.D., Iannino, A., Okumoto, K.: Software Reliability, Measurement, Prediction and Application. McGraw-Hill, New York (1997)
16. Sitte, R.: Comparison of Software-Reliability-Growth Predictions: Neural Networks vs. Parametric-Recalibration. IEEE Transactions on Reliability 48(3), 285–291 (1999)
17. Tian, L., Noore, A.: Evolutionary neural network modeling for software cumulative failure time prediction. Reliability Engineering and System Safety 87, 45–51 (2005b)
18. Thwin, M.M.T., Quah, T.S.: Application of neural networks for software quality prediction using object-oriented metrics. Journal of Systems and Software 76, 147–156 (2005)
19. Sherer, S.A.: Software fault prediction. Journal of Systems and Software 29(2), 97–105 (1995)
20. Karunanithi, N., Malaiya, Y.K., Whitley, D.: The scaling problem in neural networks for software reliability prediction. In: Proceedings of the Third International IEEE Symposium of Software Reliability Engineering, Los Alamitos, CA, pp. 76–82 (1992a)
21. Karunanithi, N., Whitley, D., Malaiya, Y.K.: Prediction of software reliability using connectionist models. IEEETransactions on Software Engineering 18, 563–574 (1992b)
22. Tian, L., Noore, A.: On-line prediction of software reliability using an evolutionary connectionist model. The Journal of Systems and Software 77, 173–180 (2005a)
23. Pai, P.F., Hong, W.C.: Software reliability forecasting by support vector machine with simulated annealing algorithms. Journal of System and Software 79(6), 747–755 (2006)
24. Su, Y.S., Huang, C.Y.: Neural-network-based approaches for software reliability estimation using dynamic weighted combinational models. The Journal of Systems and Software (2006), doi:10.1016/j.jss.2006.06.017
25. Rajkiran, N., Ravi, V.: Software Reliability prediction by soft computing technique. The Journal of Systems and Software 81(4), 576–583 (2007)
26. Rajkiran, N., Ravi, V.: Software Reliability prediction using wavelet Neural Networks. In: International Conference on Computational Intelligence and Multimedia Application (ICCIMA 2007), vol. 1, pp. 195–197 (2007)
27. Ravi, V., Chauhan, N.J., Raj Kiran, N.: Software reliability prediction using intelligent techniques: Application to operational risk prediction in Firms. International Journal of Computational Intelligence and Applications 8(2), 181–194 (2009)
28. Farlow, S.J.: Self-Organizing Methods in Modeling: GMDH type Algorithm. Marcel Dekker Inc., New York (1984)
29. Ivakhnenko, A.G.: The GMDH.: A rival of stochastic approximation. Sov. Autom. Control 3, 43 (1968)
30. Musa, J.D.: Software reliability data. IEEE Computer Society- Repository (1979)
31. Neuroshell2 tool: http://www.inf.kiew.ua/gmdh-home

Mining Temporal Patterns for Humanoid Robot Using Pattern Growth Method

Upasna Singh, Kevindra Pal Singh, and G.C. Nandi

Indian Institute of Information Technology, Allahabad
{upasnasingh,gcnandi}@iiita.ac.in, kevindra.singh@gmail.com

Abstract. In this paper, we have projected an efficient mining method for a temporal dataset of humanoid robot HOAP-2 (Humanoid Open Architecture Platform). This method is adequate to discover knowledge of intermediate patterns which are hidden inside different existing patterns of motion of HOAP-2 joints. Pattern-growth method such as FP (Frequent Pattern) growth, unfolds many unpredictable associations among different joint trajectories of HOAP-2 that can depict various kinds of motion. In addition, we have cross-checked our methodology over Webots, a simulation platform for HOAP-2, and found that our investigation is adjuvant to predict new patterns of motion in terms of temporal association rules for HOAP-2.

Keywords: Temporal Association Rules, HOAP-2, Pattern growth method, FP-Growth, Webots.

1 Introduction

Since late 90's, the field of data mining is emerging very fast and till today it has exploded many research areas by merging itself in various areas[1]. As day-by-day size of the databases increases, data analysis takes higher domain of complexity. If the data is real-time and containing many unknown hidden patterns inside it, which are almost impossible to extract manually due to its increased volume, the major need arises to develop some technique which can automatically figure out the pattern in the data and also comprehend some meaning from it. For such purpose data mining provides various statistical and intelligent paradigms which can handle large and bulky databases. Some of its major capabilities are: associations, classifications, clustering etc [2].

Through past researches in [3, 4, 5], association rule mining has proved itself to be most prominent technique to disclose effective hidden patterns from bulky databases. One of the complex category of dataset is temporal dataset which contains time unit as one of its attribute and some associations w.r.t. time. Thus, it would be interesting if those associations can be captured and then further be used purposefully.

Temporal database is a dynamic dataset which is generated on the basis of time [6]. According to [7], the time can be in terms of any granularity like month, day, hour, minute or second or it could be millisecond also. In such datasets, each transaction is

H. Sakai et al. (Eds.): RSFDGrC 2009, LNAI 5908, pp. 352–360, 2009.
© Springer-Verlag Berlin Heidelberg 2009

basically an event at a particular instant of time. Since the dataset is temporal, the associations should be in terms of several kinds of rules such as: calendar association rules[8], sequential association rules[9], episode or periodic association rules[10], cyclic association rules [11] etc. and thus the ruleset is called temporal association rules or simply temporal patterns. It generally depends on the nature of dataset that which kind of rule is to be generated.

Till today, various researches like [12, 13] follow apriori-based method for generating temporal association rules. In our method we have proposed pattern-growth-based association rules which have used FP(Frequent Pattern) Growth algorithm, an esteemed pattern-growth method, to overcome the limitations of apriori-based method by avoiding candidate generations for impenetrable datasets [4]. Section 2 demonstrates various steps necessitated for applying the proposed method and then Section 3 explained the experimental application of the method over the real time dataset. Section 4 concludes the objective of the investigation followed by future perspectives.

2   Temporal Association Rule Mining

For mining temporal association rule in temporal datasets we have divided our method into following parts which can be shown in fig. 1. Step-by-step functioning of each part is explained as:

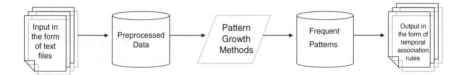

Fig. 1. Temporal Association Rule Mining method

Step 1. Take the input dataset in .txt format.

Step 2. Preprocess the data to convert it into readable form. It means that if some attributes are not contributing in associations then those attributes are eliminated and also if the attribute is numeric then it should be converted into string so as to generate some meaningful information in terms of association rule.

Step 3. Apply any pattern growth method to it in order to obtain frequent patterns. This is the vital part of the method as it contains one of the fastest algorithm of data mining such as FP-growth or H-Mine for mining frequent pattern. According to [4, 5] these algorithms are found to be quickest w.r.t. time and space efficient for generating frequent patterns.

Step 4. This step is the initial phase of temporal association rule mining, which is to develop frequent patterns w.r.t. minimum support threshold. These patterns will be in the form of maximum occurring associations of the attributes within the dataset.

Step 5. This step holds the final phase of temporal association rule mining which is to mine that association rule from frequent patterns which is based on certain measures: minimum temporal support and minimum temporal confidence threshold.

3   Experimental Results and Discussion

3.1   Data Description

We have tested aforesaid methodology over the temporal dataset of Humanoid Robot HOAP-2. In this dataset we have several files for several motions of HOAP-2. These files contain 25 attributes. These attributes store the value of 25 joints of HOAP-2. Each value has certain maximum and minimum range. Their values and other informations are taken from [14]. In the dataset, one transaction (or record) represents the movement of 25 joints in 1 millisecond. We are considering the dataset of Walk pattern which contains 26000 records. It means, it is representing a walk pattern for 26 seconds. All this information is gathered from [14] and Webots [15], a simulated platform for HOAP-2. This platform provides various user-friendly frameworks of robotics. The walk pattern in Webots for HOAP-2 is shown as in fig. 2. In this figure three windows are cascaded. The leftmost window shows the simulated HOAP-2 which is programmed in C taking .csv (comma separated value) file as an input. That csv file is shown in the rightmost window. It contains decimal value of 25 joints. At the bottom window, the position of sensors per movement of HOAP-2 is displayed.

Through association rule mining, we aimed to find out some intermediate pattern w.r.t. time that helps us to generate some new pattern. Intermediate patterns could be movement of hands, movement of legs etc. for a particular instant of time.

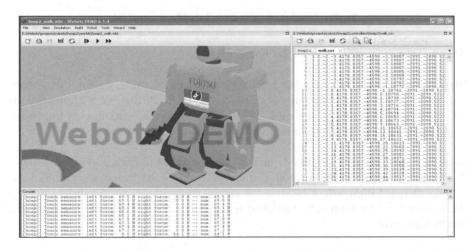

Fig. 2. Simulated Model of HOAP-2 in Webots

3.2 Implementation

The method given in section 3 can be implemented for HOAP-2 dataset as:

3.2.1 Input
Initially the dataset was in terms of .csv format. But since for pattern-growth methods, text formats are most feasible, so we have converted that data into .txt formats.

3.2.2 Preprocessing
The input file contains 27 attribute from which we have analysed that 6 attributes are of no use as those values remained constant, thus we have eliminated those attributes from the dataset as they would not be contributing in associations. The remaining attribute set is

{rleg _joint_1, rleg _joint_2, rleg _joint_3, rleg _joint_4, rleg _joint_5, rleg _joint_6, rarm_joint_1, rarm_joint_2, rarm_joint_3, rarm_joint_4, lleg _joint_1, lleg _joint_2, lleg _joint_3, lleg _joint_4, lleg _joint_5, lleg _joint_6, larm_joint_1, larm_joint_2, larm_joint_3, larm_joint_4, body_joint_1}

Also the data is in the form of numeric values. Seeing the raw data one couldn't find any meaning in its associations and thus we have converted it into meaningful form by appending attribute name with its value. The resulting input file for finding patterns is given in fig. 3. In this file we have 21 attributes, but for applying FP-growth algorithm we have appended transaction ID and time attribute in the transactional dataset.

The finalized input file will be taken as given in table 1. It has 23 attributes which includes 21 attributes (mentioned above), transaction ID and time respectively.

In this dataset, the time factor is taken on the basis of timestamps. We have considered two files for capturing different patterns, Walk-pattern and Sumo-pattern. For walk-pattern, one interval is of 50 timestamps i.e. 50 transactions. The time attribute

0	-30	-30
1	-60	-60
2	-90	-90

⟹

rleg_joint_1_0	rleg_joint_2_-30	rleg_joint_3_-30
rleg_joint_1_1	rleg_joint_2_-60	rleg_joint_3_-60
rleg_joint_1_2	rleg_joint_2_-90	rleg_joint_3_-90

Fig. 3. Data Transformation

Table 1. Input Transactional dataset

T1	time1	rleg_joint_1_0	rleg_joint_2_0	rarm_joint_1_0	....
T2	time1	rleg_joint_1_1	rleg_joint_2_0	rarm_joint_1_0	....
T3	time1	rleg_joint_1_2	rleg_joint_2_0	rarm_joint_1_0	....
T4	time1	rleg_joint_1_3	rleg_joint_2_0	rarm_joint_1_0	....
T5	time1	rleg_joint_1_4	rleg_joint_2_0	rarm_joint_1_0	....
.	.	.	.	.	....

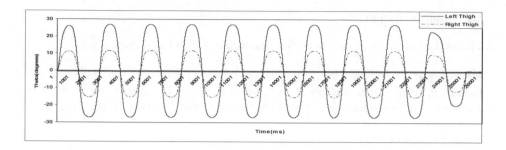

Fig. 4. Thigh-joint (Leg Joint 4) trajectory for walk-pattern

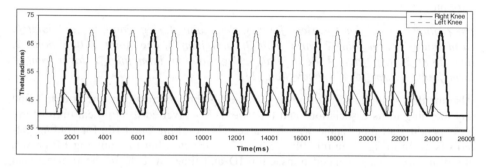

Fig. 5. Knee-joint (Leg Joint 2) trajectory for walk-pattern

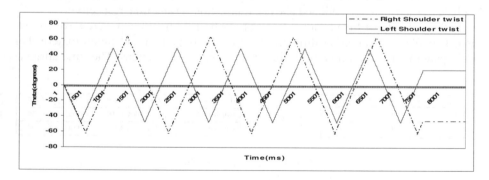

Fig. 6. Shoulder-twist-joint (Arm Joint 3) trajectory for Sumo-pattern

has the values *time1, time2, time3*............so on. In original dataset one record represents 1 millisecond and likewise the simulation model of Webots defined 'control-step' unit which is of 50 milliseconds, so on that basis we have defined our time interval. Similarly for sumo-pattern, one interval is of 64 timestamps since in Webots model one control-step is of 60 milliseconds. Also we assume the trajectory, formed from some joints which create associations in joint space for walk-pattern and sumo-pattern respectively. Each trajectory is obtained with repeated sequences of joint values for associated joints. In fig.4, Left_Thigh_joint and Right_Thigh_joint are associated with each other with some repeated sequences of their joint values. By seeing figure only one cannot judge that what values of joints are responsible to create this pattern and up to what time a particular association is being repeated so as to follow this pattern. Similar queries are there for fig.5 and fig.6 respectively. Thus, in order to see the repeated sequences for each association of joints for particular instant of time, there is a need to generate some frequent pattern which tells most frequently occurred sequence w.r.t. time in terms of temporal association rule.

3.2.3 Pattern Growth Methods

One of the most important aspect of the overall method is to apply pattern-growth method. In order to dredge frequently occurred association of sequences we have applied FP growth method over the dataset given in Table 2. This method finds frequent patterns on the basis of minimum support threshold which usually represents the minimum possible frequency of any sequence w.r.t. time.

Table 2. Input Transactional dataset for leg-Joint 4

T1	Time1	rleg_joint_4_8357	lleg_joint_4_-8360
T2	Time1	rleg_joint_4_8357	lleg_joint_4_-8360
T3	Time1	rleg_joint_4_8357	lleg_joint_4_-8360
T4	Time1	rleg_joint_4_8357	lleg_joint_4_-8360
T5	Time1	rleg_joint_4_8357	lleg_joint_4_-8360
.	.	.	.

Thus, for the dataset given in Table 2, we have defined any temporal sequence as

Definition 1: A sequence is said to be a temporal sequence iff:

For Sequence S1:(A,B,T)
If ((Support(A)>= Minimum Support Threshold)>(Support(B)>= Minimum Support Threshold)>(Support(T)>= Minimum Support Threshold))
Then Support(S1)=P(A∪B∪T)

where Support(A)=n(A), Support(B)=n(B), Support(T)=n(T) since for every attribute value the support of that value is its frequency in that dataset; Minimum Support threshold is provided by the user.

Following above definition we have temporal sequences as frequent patterns which are generated after applying FP-growth algorithm. In the resultant format the frequent pattern is represented as:

*lleg_joint_2_-3 : 886 rleg_joint_2_-3 : 762 time521 : 50 **

It represents that left leg joint 2 having value -3 with support 886 is associated with right leg joint 2 having value -3 with support 762 in the time interval 521 with support 50 when the minimum support threshold value is 2. Also '*' denotes the end of the pattern. In this way we have several frequent patterns for walk-pattern trajectory are created and shown in fig. 7 (a) and (b) respectively.

(a)
```
time488 : 50  lleg_joint_2_-1696 : 4  rleg_joint_2_-2503 : 4  *
time488 : 50  lleg_joint_2_-1695 : 2  rleg_joint_2_-2502 : 2  *
time488 : 50  rleg_joint_2_-2502 : 2  lleg_joint_2_-1693 : 2  *
rleg_joint_2_-3210 : 48  time437 : 4  lleg_joint_2_-2401 : 2  *
rleg_joint_2_-3210 : 48  time387 : 4  lleg_joint_2_-2402 : 2  *
rleg_joint_2_-3210 : 48  time340 : 2  lleg_joint_2_-2402 : 2  *
rleg_joint_2_-3210 : 48  time291 : 2  lleg_joint_2_-2402 : 2  *
rleg_joint_2_-3210 : 48  time238 : 6  lleg_joint_2_-2403 : 4  *
lleg_joint_2_-3 : 886  rleg_joint_2_-3 : 762  time521 : 50  *
lleg_joint_2_-3 : 886  rleg_joint_2_-3 : 762  time520 : 50  *
lleg_joint_2_-3 : 886  rleg_joint_2_-3 : 762  time519 : 50  *
```

(b)
```
time472 : 50  rleg_joint_4_8455 : 2  lleg_joint_4_-10133 : 2  *
time472 : 50  rleg_joint_4_8458 : 2  lleg_joint_4_-10168 : 2  *
time468 : 50  rleg_joint_4_8808 : 2  lleg_joint_4_-13386 : 2  *
time468 : 50  rleg_joint_4_8811 : 2  lleg_joint_4_-13412 : 2  *
rleg_joint_4_8357 : 4594  time329 : 37  lleg_joint_4_-10575 : 2  *
rleg_joint_4_8357 : 4594  time329 : 37  lleg_joint_4_-10534 : 2  *
rleg_joint_4_8357 : 4594  time329 : 37  lleg_joint_4_-10510 : 2  *
rleg_joint_4_8357 : 4594  lleg_joint_4_-8360 : 479  time8 : 50  *
rleg_joint_4_8357 : 4594  lleg_joint_4_-8360 : 479  time7 : 50  *
rleg_joint_4_8357 : 4594  lleg_joint_4_-8360 : 479  time6 : 50  *
rleg_joint_4_8357 : 4594  lleg_joint_4_-8360 : 479  time5 : 50  *
```

Fig. 7. (a) Frequent Temporal sequences for Thigh-Joint trajectory, (b) Frequent Temporal sequences for Knee-Joint trajectory

(a)
```
Minimum Temporal Support = .001% and Minimum Temporal Confidence = 50%
time506 : 35 -> rleg_joint_2_-3 : 886 rleg_joint_2_-2 : 104 Supp: 0.134347 % Conf: 70 %
time505 : 41 -> lleg_joint_2_-3 : 886 rleg_joint_2_-2 : 104 Supp: 0.157378 % Conf: 82 %
time521 : 50 -> lleg_joint_2_-3 : 886 rleg_joint_2_-3 : 762 Supp: 0.191924 % Conf: 100 %
time520 : 50 -> lleg_joint_2_-3 : 886 rleg_joint_2_-3 : 762 Supp: 0.191924 % Conf: 100 %
time519 : 50 -> lleg_joint_2_-3 : 886 rleg_joint_2_-3 : 762 Supp: 0.191924 % Conf: 100 %
time518 : 50 -> lleg_joint_2_-3 : 886 rleg_joint_2_-3 : 762 Supp: 0.191924 % Conf: 100 %
time517 : 50 -> lleg_joint_2_-3 : 886 rleg_joint_2_-3 : 762 Supp: 0.191924 % Conf: 100 %
time516 : 50 -> lleg_joint_2_-3 : 886 rleg_joint_2_-3 : 762 Supp: 0.191924 % Conf: 100 %
time515 : 50 -> lleg_joint_2_-3 : 886 rleg_joint_2_-3 : 762 Supp: 0.191924 % Conf: 100 %
time514 : 50 -> lleg_joint_2_-3 : 886 rleg_joint_2_-3 : 762 Supp: 0.191924 % Conf: 100 %
time513 : 50 -> lleg_joint_2_-3 : 886 rleg_joint_2_-3 : 762 Supp: 0.191924 % Conf: 100 %
time512 : 50 -> lleg_joint_2_-3 : 886 rleg_joint_2_-3 : 762 Supp: 0.191924 % Conf: 100 %
time511 : 50 -> lleg_joint_2_-3 : 886 rleg_joint_2_-3 : 762 Supp: 0.191924 % Conf: 100 %
time510 : 50 -> lleg_joint_2_-3 : 886 rleg_joint_2_-3 : 762 Supp: 0.191924 % Conf: 100 %
time509 : 48 -> lleg_joint_2_-3 : 886 rleg_joint_2_-3 : 762 Supp: 0.184247 % Conf: 96 %
time508 : 47 -> lleg_joint_2_-3 : 886 rleg_joint_2_-3 : 762 Supp: 0.180408 % Conf: 94 %
time507 : 27 -> lleg_joint_2_-3 : 886 rleg_joint_2_-3 : 762 Supp: 0.103639 % Conf: 54 %
```

(b)
```
Minimum Temporal Support = .001% and Minimum Temporal Confidence = 50%
time506 : 35 -> lleg_joint_2_-3 : 886 rleg_joint_2_-2 : 104 Supp: 0.134347 % Conf: 70 %
time505 : 41 -> lleg_joint_2_-3 : 886 rleg_joint_2_-2 : 104 Supp: 0.157378 % Conf: 82 %
time521 : 50 -> lleg_joint_2_-3 : 886 rleg_joint_2_-3 : 762 Supp: 0.191924 % Conf: 100 %
time520 : 50 -> lleg_joint_2_-3 : 886 rleg_joint_2_-3 : 762 Supp: 0.191924 % Conf: 100 %
time519 : 50 -> lleg_joint_2_-3 : 886 rleg_joint_2_-3 : 762 Supp: 0.191924 % Conf: 100 %
time518 : 50 -> lleg_joint_2_-3 : 886 rleg_joint_2_-3 : 762 Supp: 0.191924 % Conf: 100 %
time517 : 50 -> lleg_joint_2_-3 : 886 rleg_joint_2_-3 : 762 Supp: 0.191924 % Conf: 100 %
time516 : 50 -> lleg_joint_2_-3 : 886 rleg_joint_2_-3 : 762 Supp: 0.191924 % Conf: 100 %
time515 : 50 -> lleg_joint_2_-3 : 886 rleg_joint_2_-3 : 762 Supp: 0.191924 % Conf: 100 %
time514 : 50 -> lleg_joint_2_-3 : 886 rleg_joint_2_-3 : 762 Supp: 0.191924 % Conf: 100 %
time513 : 50 -> lleg_joint_2_-3 : 886 rleg_joint_2_-3 : 762 Supp: 0.191924 % Conf: 100 %
time512 : 50 -> lleg_joint_2_-3 : 886 rleg_joint_2_-3 : 762 Supp: 0.191924 % Conf: 100 %
time511 : 50 -> lleg_joint_2_-3 : 886 rleg_joint_2_-3 : 762 Supp: 0.191924 % Conf: 100 %
time510 : 50 -> lleg_joint_2_-3 : 886 rleg_joint_2_-3 : 762 Supp: 0.191924 % Conf: 100 %
time509 : 48 -> lleg_joint_2_-3 : 886 rleg_joint_2_-3 : 762 Supp: 0.184247 % Conf: 96 %
time508 : 47 -> llez_joint_2_-3 : 886 rlez_joint_2_-3 : 762 Supp: 0.180408 % Conf: 94 %
```

Fig. 8. (a) Temporal Association Rules for Thigh-Joint trajectory, (b) Temporal Association Rules for Knee-Joint trajectory

3.2.4 Temporal Association Rule Generation

Finalized output of Hoap-2 dataset is in terms of temporal association rule which is generated by taking frequent temporal sequences as an input. For our dataset, we can define temporal rules as:

Definition 2: A rule R is said to be temporal association rule iff:

For any frequent temporal sequence S_F:(A, B, T) when
  $Support(S_F) = P(TUAUB)$
  $Conf(S_F) = P((TUAUB)/T)*100$
If ($Support(S_F) \geq$ Minimum temporal support & $Conf(S_F) \geq$ Minimum temporal confidence)
Then Rule R : $T \rightarrow (A,B)$ is a Temporal association Rule

where T is a time interval in which the sequence (A,B) is occurred. Minimum temporal support and Minimum temporal confidence is provided by the user.

Thus following aforesaid, the temporal association rule is formatted as:

*time520 : 50 → lleg_joint_2_-3 : 886   rleg_joint_2_-3 : 762   Supp: 0.191924 %
Conf: 100 %*

It means that for time interval 520 which has support 50, the sequence (lleg_joint_2_-3, rleg_joint_2_-3) is frequent temporal sequence in that time interval with support 0.191924% and confidence 100%. Here the minimum temporal support is .001% and minimum temporal confidence is 50%. Similarly we can generate similar kind of rules for all the frequent temporal sequences of every trajectory. Some of the rules for walk-pattern trajectories are shown in Fig. 8 (a) and (b) respectively. Due to space construct we have not shown sumo-pattern results here, but the criteria of dredging frequent temporal sequences and then generating temporal association rules remains same for every joint-trajectories.

4 Conclusion and Future Work

The proposed demonstration highlights key features of pattern growth method for real-time temporal datasets. Observational analysis show that the contribution of FP-growth algorithm is very useful and productive for Hoap-2 dataset as it generates temporal sequence associations of joints in terms of temporal association rules. In furtherance these rules can be used to classify variety of patterns in terms of association based temporal classifications.

Acknowledgments. We thank our summer trainee Anuj Singh, pursuing B.Tech from Manipal Institute of Technology, Manipal, for assisting us in various experimental analysis throughout the research.

References

1. Han, J., Kamber, M.: Book: Data Mining Concept & Technique (2001)
2. Dunham, M.: Data Mining: Introductory and Advanced Topics. Prentice Hall, Englewood Cliffs (2003)
3. Agrawal, R., Imielienski, T., Swami, A.: Mining Association Rules between Sets of Items in Large Databases. In: Proc. Conf. on Management of Data, pp. 207–216. ACM Press, New York (1993)
4. Han, J., Pei, J., Yin, Y.: Mining Frequent Patterns without Candidate Generation. In: Intl. Conference on Management of Data, ACM SIGMOD (2000)
5. Pei, J., Han, J., Lu, H., Nishio, S., Tang, S., Yang, D.: H-mine: hyper-structure mining of frequent patterns in large database. In: Proceedings of the IEEE International Conference on Data Mining, San Jose, CA (November 2001)
6. Tansel, A.U., Ayan, N.F.: Discovery of Association Rules in Temporal Databases. In: Fourth Int'l Conference on KDD Workshop on Distributed Data Mining (August 1998)
7. Bettini, C., Sean Wang, X., Jajodia, S., Lin, J.-L.: Discovering Frequent Event Patterns with Multiple Granularities in Time Sequences. IEEE TOKDE 10, 222–237 (1998)
8. Vyas, O.P., Verma, K.: Efficient Calendar Based Temporal Association Rule. ACM SIGMOD 34(3), 63–71 (2005)
9. Agrawal, Srikant: Mining sequential patterns. In: Proc.11th Int. Conf. Data Engineering, Taipei, Taiwan, R.O.C., September 1995, pp. 3–14 (1995)
10. Mannila, H., Toivonen, H., Verkamo, A.I.: Discovering Frequent Episodes in Sequences. In: KDD 1995, August 1995, pp. 210–215. AAAI, Menlo Park (1995)

11. Ozden, B., Ramaswamy, S., Silberschatz, A.: Cyclic association rules. In: Proc. 15th Int. Conf. Data Engineering, Orlando, February 1998, pp. 412–421 (1998)
12. Alie, J.M., Rossi, G.H.: An approach to discovering temporal association rules. In: Proc. of the 2000 ACM symposium on Applied computing, Come, Italy, vol. 1, pp. 294–300 (2000)
13. Lee, C.-H., Lin, C.-R., Chen, M.-S.: On Mining General Temporal Association Rules in a Publication Database. In: Proc. IEEE International Conference on Data Mining, pp. 337–344 (2001)
14. Fujitsu Automation, Humanoid Robot HOAP-2 Specification, Fujitsu Corporation, http://jp.fujitsu.com/group/automation/downloads/en/services/humanoidrobot/hoap2/spec.pdf
15. http://www.cyberbotics.com

Multiscale Comparison of Three-Dimensional Trajectories: A Preliminary Step

Shoji Hirano and Shusaku Tsumoto

Department of Medical Informatics, Shimane University, School of Medicine
89-1 Enya-cho, Izumo, Shimane 693-8501, Japan
hirano@ieee.org, tsumoto@computer.org

Abstract. In this paper, we propose a multiscale comparison method for three-dimensional trajectories based on the maxima on the curvature scale space. We define a segment as a partial trajectory between two adjacent maxima where curvature becomes locally maximal. Then we trace the place of maxima across the scales in order to obtain the hierarchy of segments. By applying segment-based matching technique, we obtain the best correspondences between partial trajectories. We demonstrate in a preliminary experiment that our method could successfully capture the structural similarity of three-dimensional trajectories.

1 Introduction

Recent advances in hospital information systems enable us to collect huge amount of time-series medical data used for diagnosis and treatment of diseases. For example, our university hospital stores more than 150,000 records of laboratory tests per year, usually including dozens items for each test. These multivariate time-series data can be viewed as multidimensional trajectories containing temporal information about health status of patients; therefore, through the cross-patient analysis, i.e. clustering of the trajectories, it may be able to obtain interesting knowledge such as temporal relationships between examination items, common course of disease progress, and characteristics of exceptional cases. However, it is still difficult to perform such a large-scale analysis due to the following problems: (1) both acute and chronic changes may coexist in time series; therefore multiscale observation scheme is required. (2) in order to recognize implicit correlation among variables that may reflect some related phenomena in human body, comparison of multidimensional trajectory is needed.

Multiscale comparison of time series or planar curves has been widely studied since 80's mainly in the area of pattern recognition. Based on the Witkin's framework of scale space filtering [1], many methods have been proposed [2]. Ueda et al. [3] enabled the use of discrete scales and comparison of largely distorted curves by introducing a segment-based matching scheme, where a segment corresponds to a subsequence between two adjacent inflection points. Based on these methods, we have developed multiscale comparison and clustering methods for one-dimensional time-series and two-dimensional trajectories of medical data [4,5]. However, comparison of trajectories grater than three-dimension is still a challenging problem because the zero-crossing of curvature (inflection point), that plays an important role in recognizing segment hierarchy, is difficult to be determined for space curves as curvature is always positive.

H. Sakai et al. (Eds.): RSFDGrC 2009, LNAI 5908, pp. 361–368, 2009.

In this paper, we propose a multiscale comparison method for three-dimensional trajectories based on the maxima on a curvature scale space. We define a segment as a partial trajectory between two adjacent maxima where curvature becomes locally maximal. Then we trace the place of maxima across the scales in order to obtain the hierarchy of segments. By applying segment-based matching technique, we obtain the best correspondences between partial trajectories. We demonstrate in a preliminary experiment that our method could successfully capture structural similarity in simple three-dimensional trajectories.

The remainder of this paper is organized as follows. Section 2 briefly introduces the conventional segment-based matching method for two-dimensional trajectories. Section 3 describes the method for comparing three-dimensional trajectories. Section 4 shows the results of matching experiments, and Section 5 is a conclusion of this paper.

2 Multiscale Comparison of Two-Dimensional Trajectories

2.1 Multiscale Representation

Let us denote by $c(t) = \{x(t), y(t)\}$ two-dimensional trajectories composed of two time series $x(t)$ and $y(t)$. Also let us denote by σ an observation scale of the trajectory. Time series $x(t)$ at scale σ, $X(t, \sigma)$, is then derived by the discrete convolution of $x(t)$ and smoothing kernel $I_n(\sigma)$ as follows [6].

$$X(t, \sigma) = \sum_{n=-\infty}^{\infty} e^{-\sigma} I_n(\sigma) x(t - n)$$

where $I_n(\sigma)$ denotes the modified Bessel function of order n, which has better properties for dealing with discrete scales than a sampled Gaussian kernel [6]. By applying this convolution independently to $x(t)$ and $y(t)$, we obtain the trajectory at scale σ as $C(t, \sigma) = \{X(t, \sigma), Y(t, \sigma)\}$. By changing σ, we can represent the trajectory at various observation scales. Figure 1 shows an example of multiscale representation of two-dimensional trajectories. An increase of σ causes an increase of weights for temporally distant points, together with the decrease of weights around the neighbors. Therefore it produces a more smoothed trajectories with less inflection points.

2.2 Hierarchy of Inflection Point

For each trajectory we locate the curvature zero-crossings (inflection points) and represent the trajectory as a set of convex/concave segments. A segment is defined as a partial trajectory between adjacent inflection points. Next, we chase the cross-scale correspondence of inflection points successively from top scale to bottom scale. It defines the hierarchy of segments and guarantees the connectivity of segments across scales. Details of the algorithm for checking segment hierarchy is available in ref. [3]. In order to apply the algorithm to an open trajectory, we modified it to allow the replacement of odd number of segments at start and end, since cyclic property of a set of inflection points can be lost. In Figure 1, segments are represented by $\{a_i^{(k)} \mid i = 1, 2, \cdots, n\}$, where k and n denotes the scale and the number of segments at k, respectively.

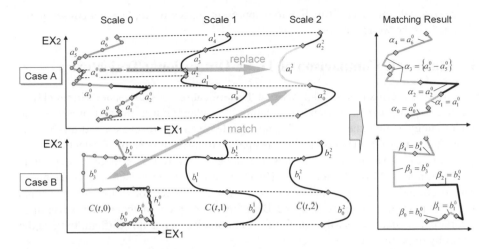

Fig. 1. An illustrative example of multiscale comparison for 2D trajectories

2.3   Matching

The main procedure of multiscale matching is to search the best set of segment pairs that satisfies both of the following conditions: (1) Complete match: By concatenating all segments, the original trajectory must be completely formed without any gaps or overlaps. (2) Minimal difference: The sum of segment dissimilarities over all segment pairs should be minimized.

The search is performed throughout all scales. For example, in Figure 1, three contiguous segments $a_3^{(0)} - a_5^{(0)}$ at the lowest scale of case A can be integrated into one segment $a_1^{(2)}$ at upper scale 2, and the replaced segment well matches to one segment $b_3^{(0)}$ of case B at the lowest scale. Thus the set of the three segments $a_3^{(0)} - a_5^{(0)}$ and one segment $b_3^{(0)}$ will be considered as a candidate for corresponding segments. On the other hand, segments such as $a_6^{(0)}$ and $b_4^{(0)}$ are similar even at the bottom scale without any replacement. Therefore they will be also a candidate for corresponding segments. In this way, if segments exhibit short-term similarity, they are matched at a lower scale, and if they present long-term similarity, they are matched at a higher scale.

2.4   Segment Dissimilarity

The dissimilarity between segments can be defined alternatively. In [5], we used three shape parameters: (1) Gradient at starting point $g(a_m^{(k)})$, (2) Rotation angle $\theta(a_m^{(k)})$, and (3) Velocity $v(a_m^{(k)})$, and defined the local dissimilarity $d(a_m^{(k)}, b_n^{(h)})$ between two segments $a_m^{(k)}$ at scale k and $b_n^{(h)}$ at scale h as

$$d(a_m^{(k)}, b_n^{(h)}) = \sqrt{\left(g(a_m^{(k)}) - g(b_n^{(h)})\right)^2 + \left(\theta(a_m^{(k)}) - \theta(b_n^{(h)})\right)^2}$$
$$+ \left|v(a_m^{(k)}) - v(b_n^{(h)})\right| + \gamma\left\{cost(a_m^{(k)}) + cost(b_n^{(h)})\right\}$$

where $cost()$ denotes a cost function for suppressing excessive replacement of segments [3], and γ is the weight of costs.

3 Multiscale Comparison of Three-Dimensional Trajectories

Curvature zero-crossings is widely used in the applications of scale-space filtering [1,7], because it constitutes a fundamental feature of a planar curve [8] and preserves the monotonicity against the change of scale [9]. However, approaches based on the curvature zero-crossings may not be directly applied to three-dimensional trajectories. For space curves, curvature takes only positive value; therefore it is difficult to determine inflection points. Mokhatarian et al. [10] focused on torsion, which is another major property of the space curve, and proposed multiscale comparison of three-dimensional object shapes using torsion scale space. But it involves a problem that the zero-crossings of torsion may not necessarily satisfy the monotonicity; hence it is difficult to trace the hierarchy of partial trajectories across scales.

In this work, we focus on the maxima of curvature that also satisfies the monotonicity against the change of scale [6], and propose a multiscale comparison method that utilizes *maxima on curvature scale space* for splitting partial trajectories (segments) and recognizing their hierarchy. Its matching procedure is basically similar to the two-dimensional case, but different in following points:

1. A segment is defined as a partial trajectory not between adjacent inflection points but between adjacent maxima.
2. Polarity of a segment (the sign of curvature) is no longer taken into account when matching two segments because every segment has positive sign.
3. Not only odd number of segments, but also even number of segments can be replaced into one segment when scale increases.

In the followings we describe the way of constructing multiscale representation of three-dimensional trajectories, making segments and tracing segment hierarchy based on the maxima of curvature, and defining dissimilarity between segments. By incorporating these procedure with the matching algorithm described in Section 2.3, we finally obtain the best correspondence between segments.

3.1 Multiscale Representation of Three-Dimensional Trajectories

Let us denote by $c(t) = \{x(t), y(t), z(t)\}$ a three-dimensional trajectory constituted of three time series $x(t)$, $y(t)$ and $z(t)$. Similarly to the two-dimensional case, the trajectory $C(t, \sigma)$ at scale σ is derived by the discrete convolution of each time series and the modified Bessel smoothing kernel $I_n(\sigma)$ as $C(t, \sigma) = \{X(t, \sigma), Y(t, \sigma), Z(t, \sigma)\}$.

3.2 Derivation of Curvature Maxima

Next, for each trajectory we compute the curvature of each point and locate their local maxima. Curvature $\kappa(t, \sigma)$ of $C(t, \sigma)$ is defined by

$$\kappa(t, \sigma) = \frac{\sqrt{(Z''Y' - Y''Z')^2 + (X''Z' - Z''X')^2 + (Y''X' - X''Y')^2}}{(X'^2 + Y'^2 + Z'^2)^{3/2}}$$

where X' and X'' respectively denote the first- and second-order derivatives of $X(t, \sigma)$ about t that are defined as follows. Similar treatment applies to $Y(t, \sigma)$ and $Z(t, \sigma)$.

$$X'(t, \sigma) = \sum_{n=-\infty}^{\infty} \frac{-n}{\sigma} e^{-\sigma} I_n(\sigma) \cdot x(t - n)$$

$$X''(t, \sigma) = \sum_{n=-\infty}^{\infty} e^{-\sigma} (I_{n+1}(\sigma) - 2I_n(\sigma) + I_{n-1}(\sigma)) \cdot x(t - n)$$

3.3 Construction of the Maxima Scale Space and Trace of Segment Hierarchy

Curvature maxima are then plotted on a two-dimensional plane of time and scale. We call this plane *curvature maxima scale space*. Figure 2 shows an example of the maxima scale space for three-dimensional trajectory $T2n_1$ used in our experiments. The horizontal axis denotes time t, the vertical axis denotes scale σ, and '+' denotes a point of maximal curvature. For each scale, a segment is defined as a set of data points separated by two adjacent curvature maxima. The number of maxima will decrease when scale increases because of the smoothing, and segments at a fine scale will be merged into one segment at a coarse scale. Maxima are successively linked from the top scale to the bottom scale based on the minimal distance criterion for forming segment hierarchy.

3.4 Matching

After constructing segments and recognizing their hierarchy across scales, segment matching is performed in the same way with the two-dimensional case. Since structural feature points are changed from zero-crossings (inflection points) to curvature maxima, the shape of a segment also changes from convex/concave shape to 's' shape. We model this shape as a straight vector connecting both ends of the segment and define the dissimilarity between segments as follows. First, let us denote by $a_m^{(k)}$ and $b_n^{(h)}$ two segments to be compared. Next, let us denote by $\mathbf{v}(a_m^{(k)})$ a three-dimensional vector connecting both ends of $a_m^{(k)}$, and similarly denote $\mathbf{v}(b_n^{(h)})$ for $b_n^{(h)}$ as shown in

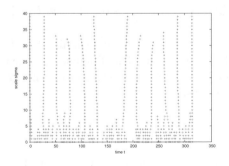

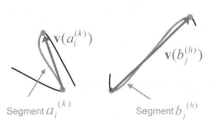

Fig. 2. Maxima scale space for $T2n_1$ **Fig. 3.** Vector representation of segments

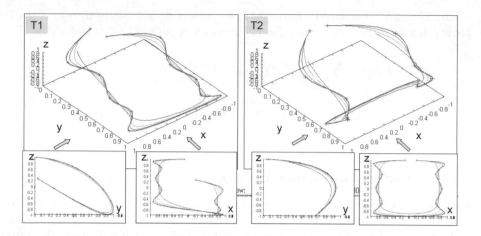

Fig. 4. Multiscale representations of test trajectories T1 (left) and T2 (right). The red curves represent the original shapes. Maxima points are denoted by '+'. Two sub-figures located at the bottom show the views from $y - z$ and $x - z$ planes respectively.

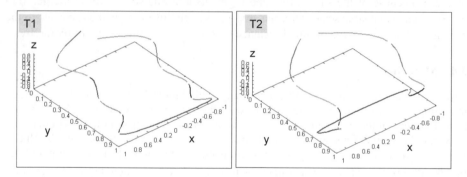

Fig. 5. Matching results. Matched segments are represented in the same color.

Figure 3. Then we define the dissimilarity between $a_m^{(k)}$ and $b_n^{(h)}$ by

$$d(a_m^{(k)}, b_n^{(h)}) = |\mathbf{v}(a_m^{(k)})||\mathbf{v}(b_n^{(h)})| \times cosdist(\mathbf{v}(a_m^{(k)}), \mathbf{v}(b_n^{(h)})) + \gamma \left\{ cost(a_m^{(k)}) + cost(b_n^{(h)}) \right\}$$

where $cosdist()$ denotes the cosine distance between two vectors. It quantifies the difference of directions of segments weighted by their lengths. The difference between large segments is more emphasized than that of small segments. The cost of segment replace is added similarly to the two dimensional case.

4   Experimental Results

We have conducted a preliminary matching experiment for checking the basic functionality of the proposed method. We generated two simple three-dimensional trajectories T1 and T2 by using triangular functions as follows.

T1	T2
$x = \sin(t) + \frac{1}{3}\sin(3t) + \frac{1}{5}\sin(5t)$	$x = \sin(t) + \frac{1}{3}\sin(3t) + \frac{1}{5}\sin(5t)$
$y = \sin(\frac{t}{2})$	$y = \sin(\frac{t}{2}) + \frac{1}{3}\sin(\frac{3}{2}t)$
$z = \cos(\frac{4}{5}t)$	$z = \cos(t)$

The parameters used for matching were: starting scale = 1.0, scale interval = 1.0, number of scales = 40, weight for replacement cost = 3.0. Figure 4 shows the shapes of trajectories T1 and T2 and their multiscale representation. They were divided into segments by the curvature maxima denoted by '+'. Figure 5 shows the matching result. Matched segments are represented in the same color. The original T1 and T2 were different around $z = 0$ in terms of their y-direction changes, but they were successfully matched according to the structural similarity of the entire shapes.

Next, we conducted a matching experiment for noisy trajectories. We generated two noisy trajectories denoted by T2n1 and T2n2 by adding Gaussian noise to T2. Figure 6 shows multiscale representations of their shapes and Figure 7 shows their

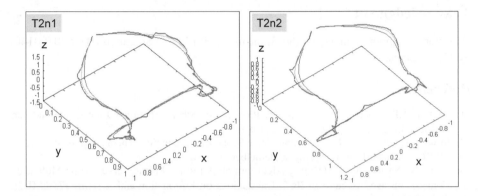

Fig. 6. Multiscale representations of noisy test trajectories T2n1 (left) and T2n2 (right). The red curve represents their original shapes.

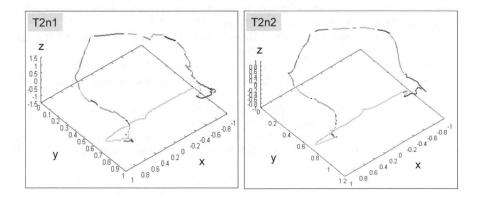

Fig. 7. Matching results. Matched segments are represented in the same color.

matching results. We could confirm that their structural similarity was successfully captured in a global scale if there existed local differences at fine scales caused by noise.

5 Conclusions

In this paper we have presented a multiscale comparison method for three-dimensional trajectories. In order to deal with the problem that zero-crossings of curvature cannot be determined for space curve, we focused on the maxima of curvature. The hierarchy of partial trajectories was recognized by tracing the positions of maxima across scales. Then we performed segment-by-segment matching across the scales, and obtained the best correspondence of segments. In the experiments we could observe that reasonable correspondences were obtained on the simple but noisy trajectories.

This work is still at an early stage and there are lots of work to be done. We will continue to tackle the following issues: (1) investigation of the characteristics of maxima on the curvature scale space, (2) refinement of the segment dissimilarity, (3) quantitative evaluation of the performance, (4) application to the real medical data.

Acknowledgment

This work is supported in part by the Grant-in-Aid for Young Scientists (B) (#20700140) by MEXT, Japan.

References

1. Witkin, A.P.: Scale-space filtering. In: Proc. of the 8th International Joint Conference on Artificial Intelligence, pp. 1019–1022 (1983)
2. Mokhtarian, F., Bober, M.: Curvature Scale Space Representation: Theory, Applications, and MPEG-7 Standardization. Springer, Heidelberg (2003)
3. Ueda, N., Suzuki, S.: A Matching Algorithm of Deformed Planar Curves Using Multiscale Convex/Concave Structures. IEICE Transactions on Information and Systems J73-D-II(7), 992–1000 (1990)
4. Hirano, S., Tsumoto, S.: Clustering Time-series Medical Databases based on the Improved Multiscale Matching. In: Hacid, M.-S., Murray, N.V., Raś, Z.W., Tsumoto, S. (eds.) ISMIS 2005. LNCS (LNAI), vol. 3488, pp. 612–621. Springer, Heidelberg (2005)
5. Hirano, S., Tsumoto, S.: Cluster analysis of trajectory data on hospital laboratory examinations. In: AMIA Annual Symp Proc., vol. 11, pp. 324–328 (2007)
6. Lindeberg, T.: Scale-Space for Discrete Signals. IEEE Transactions on Pattern Analysis and Machine Intelligence 12(3), 234–254 (1990)
7. Mokhtarian, F., Mackworth, A.K.: Scale-based Description and Recognition of planar Curves and Two Dimensional Shapes. IEEE Transactions on Pattern Analysis and Machine Intelligence PAMI-8(1), 24–43 (1986)
8. Dudek, G., Tostsos, J.K.: Shape Representation and Recognition from Multiscale Curvature. Comp. Vis. Img. Understanding 68(2), 170–189 (1997)
9. Babaud, J., Witkin, A.P., Baudin, M., Duda, O.: Uniqueness of the Gaussian kernel for scale-space filtering. IEEE Transactions on Pattern Analysis and Machine Intelligence 8(1), 26–33 (1986)
10. Mokhtarian, F.: Multi-scale description of space curves and three-dimensional objects. In: Proceedings of the IEEE Computer Society Conference on Computer Vision and Pattern Recognition, pp. 298–303 (1988)

Time Series Forecasting Using Hybrid Neuro-Fuzzy Regression Model

Arindam Chaudhuri[1] and Kajal De[2]

[1] Research Scholar (Computer Science), Netaji Subhas Open University, Kolkata, India
and
Lecturer (Mathematics & Computer Science), Meghnad Saha Institute of Technology, Kolkata
[2] Professor in Mathematics, School of Science, Netaji Subhas Open University, Kolkata, India

Abstract. During the past few decades various time-series forecasting methods have been developed for financial market forecasting leading to improved decisions and investments. But accuracy remains a matter of concern in these forecasts. The quest is thus on improving the effectiveness of time-series models. Artificial neural networks (ANN) are flexible computing paradigms and universal approximations that have been applied to a wide range of forecasting problems with high degree of accuracy. However, they need large amount of historical data to yield accurate results. The real world situation experiences uncertain and quick changes, as a result of which future situations should be forecasted using small amount of data from a short span of time. Therefore, forecasting in these situations requires techniques that work efficiently with incomplete data for which Fuzzy sets are ideally suitable. In this work, a hybrid Neuro-Fuzzy model combining the advantages of ANN and Fuzzy regression is developed to forecast the exchange rate of US Dollar to Indian Rupee. The model yields more accurate results with fewer observations and incomplete data sets for both point and interval forecasts. The empirical results indicate that performance of the model is comparatively better than other models which make it an ideal candidate for forecasting and decision making.

Keywords: Artificial neural networks, Fuzzy regression, Fuzzy time-series, Neuro-Fuzzy model, Exchange rate forecasting.

1 Introduction

Time series forecasting is a key element of financial and managerial decision making. It is highly utilized in predicting economic and business trends for improved decisions and investments. Financial data presents challenging and complex problem to forecast. Many forecasting methods have been developed to study financial data in last few decades. Artificial neural networks (ANN) serve as powerful computational framework have gained much popularity in business applications. ANN have been successfully applied to loan evaluation, signature recognition, time series forecasting, and many other pattern recognition problems [TF]. The major advantage of ANN is their flexible non-linear modeling capability and no need to specify particular model form. Rather, the model is adaptively formed based on features presented in data. This

H. Sakai et al. (Eds.): RSFDGrC 2009, LNAI 5908, pp. 369–381, 2009.

data-driven approach is suitable for many empirical data sets where no theoretical guidance is available to suggest an appropriate data generating process. However, ANN require large amount of data in order to yield accurate results. No definite rule exists for sample size requirement. The amount of data for ANN training depends on network structure, training method and problem complexity or amount of noise in data. With large enough sample, ANN can model any complex structure in data. ANN can thus benefit more from large samples than linear statistical models. Forecasting using Fuzzy sets are suitable under incomplete data conditions and require fewer observations than other forecasting models, but their performance is not always satisfactory. Fuzzy theory was originally developed to deal with problems involving linguistic terms [Zad1, Zad2, Zad3] and have been successfully applied to various applications such as university enrollment forecasting [CH1, CH2, Ch], financial forecasting [HTW, HY1, HY2, Yu], temperature forecasting [CH1] etc. Tanaka et al. [Tan, TI, TUA] have suggested Fuzzy regression to solve fuzzy environment and to avoid modeling error. The model is an interval prediction model with the disadvantage that prediction interval can be very wide if some extreme values are present. Watada [Wata] gave an application of Fuzzy regression to fuzzy time-series analysis.

Combining strengths of ANN and Fuzzy sets leads to the development of hybrid Neuro-Fuzzy model which improves forecasting accuracy. The basic idea of the model in forecasting is to use each model's unique feature to capture different patterns in data. Theoretical and empirical findings suggest that combining different techniques can be an effective and efficient way to improve forecasts [AMM, LRS, KHB]. The notable works on time series forecasting include Fuzzy auto regressive integrated moving average (FARIMA) method by Tseng et al. [TTYY] hybrid Genetic algorithm and high-order Fuzzy time-series approach for enrollment forecasting by Chen et al. [CC] and Huarng et al. [HY3] combined methodology using ANN to forecast fuzzy time-series. In this work, a hybrid Neuro-Fuzzy model is developed based on the concepts of ANN and Fuzzy regression models to time-series forecasting under incomplete data conditions. The time-series data considered here is exchange rate of US dollar to Indian rupee. ANN is used to preprocess raw data and provide necessary background to apply Fuzzy regression model. The Fuzzy regression model eliminates the disadvantage of large requirement of historical data. The effectiveness of method is demonstrated by applying it to an important problem in financial markets viz. exchange rate forecasting. The performance of method is empirically compared with other forecasting models such as auto regressive integrated moving average, Chen's fuzzy time-series (first and higher order) [Ch, CC], Yu's fuzzy time-series [Yu1], FARIMA [TTYY] and ANN which gives improved forecasting results for the present method. This paper is organized as follows. In section 2, ANN approach to time series is given. This is followed by brief discussion on Fuzzy regression in section 3. In the next section, Neuro-Fuzzy forecasting model is presented. Section 5 illustrates the simulation results. Finally, in section 6 conclusions are given.

2   Artificial Neural Networks Approach to Time Series

ANN flexibly models wide range of non-linear problems [WVJ]. They are universal approximators approximating large class of functions with accuracy. Their power lies in parallel processing of information. No prior assumption of model form is required;

instead the network model is largely determined by characteristics of data. Single hidden layer feed forward network is most widely used form for time-series modeling and forecasting [ZM], which is characterized by network of three layers of simple processing units connected by acyclic links (figure 1). The output y_t and inputs $(y_{t-1},..........,y_{t-p})$ are represented by relationship:

$$y_t = w_0 + \sum_{j=1}^{q} w_j \cdot g(w_{0,j} + \sum_{i=1}^{p} w_{i,j} \cdot y_{t-i}) + \varepsilon_t \tag{1}$$

where, $(w_{i,j}; i = 0,1,2,........, p, j = 1,2,........,q)$ and $(w_j; j = 0,1,2,........,q)$ are model parameters called connection weights and biases, p is number of input nodes and q is number of hidden nodes. The logistic function used as hidden layer transfer function is

$$Sig(x) = \frac{1}{1+e^{-x}} \tag{2}$$

Thus, Equation (1) performs non-linear functional mapping from past observations to future values y_t, i.e.

$$y_t = f(y_{t-1},..........,y_{t-p},w) + \varepsilon_t \tag{3}$$

where, w is vector of all parameters and $f(\cdot)$ is a function determined by network structure and connection weights. Hence, ANN is equivalent to a non-linear autoregressive model. The Equation (1) implies that one output node in output layer is used for one step ahead forecasting and is able to approximate the arbitrary function as number of hidden nodes when q is sufficiently large [SJ]. The simple network structure that has small number of hidden nodes performs well in out of sample forecasting which is primarily due to over-fitting effect. The choice of q is data dependent and

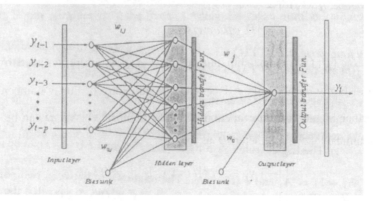

Fig. 1. Neural Network Structure $[N^{(p-q-1)}]$

there is no systematic rule in determining this parameter. Alongwith choosing an appropriate number of hidden nodes, an important task of ANN modeling of time-series is selection of number of lagged observations p and dimension of input vector [TE].

3 Fuzzy Regression Model

The Fuzzy sets introduced by Zadeh [Zad1, Zad2, Zad3] provides a powerful framework to cope with vague or ambiguous situations and expresses linguistic values and human subjective judgments of natural language. Fuzzy sets were applied to time-series leading to Fuzzy time-series by Song and Chissom [SC1, SC2]. Among the notable Fuzzy time-series model where calculations are easy and forecasting performance is good is by Chen [Ch]. Statistical models use the concept of measurement error to deal with difference between estimators and observations. These data are precise values and is identical to Fuzzy regression model suggested by Tanaka et al. [TI]. In Fuzzy regression residuals between estimators and observations are not produced by measurement errors, rather by parameter uncertainty in the model. The generalized Fuzzy linear regression is

$$\tilde{Y} = \tilde{\beta}_0 + \tilde{\beta}_1 x_1 + \ldots\ldots\ldots + \tilde{\beta}_n x_n = \sum_{i=0}^{n} \tilde{\beta}_i x_i = x' \tilde{\beta} \qquad (4)$$

where, x' denotes transpose of vector of independent variables, n is number of variables and $\tilde{\beta}_i$ the fuzzy sets representing i^{th} parameter of the model which is of form of L-type Fuzzy numbers [DP], possibility distribution for which is given by

$$\mu_{\tilde{\beta}_i}(\beta_i) = L\{(\alpha_i - \beta_i)/c\} \qquad (5)$$

Fuzzy parameters in form of triangular Fuzzy numbers are used,

$$\mu_{\tilde{\beta}_i}(\beta_i) = \begin{cases} 1 - \dfrac{|\alpha_i - \beta_i|}{c_i}, \alpha_i - c_i \leq \beta \leq \alpha_i + c_i \\ 0, otherwise \end{cases} \qquad (6)$$

where, $\mu_{\tilde{\beta}_i}(\beta_i)$ is membership function of Fuzzy set is represented by β_i, α_i is centre of Fuzzy number and c_i is spread around centre of Fuzzy number. By Extension Principle, membership function of Fuzzy number $\tilde{y}_t = x_t'\tilde{\beta}$ can be defined using pyramidal Fuzzy parameter β as follows:

$$\mu_{\tilde{y}}(y_t) = \begin{cases} 1 - |y_t - x_t \alpha|/c' |x_t|, & x_t \neq 0 \\ 1, & x_t = 0, y_t = 0 \\ 0, & x_t = 0, y_t \neq 0 \end{cases} \qquad (7)$$

where, α and c denote vectors of model values and spreads for all model parameters respectively, $t = 1,2,\ldots\ldots,k$. Finally, the method uses criterion of minimizing total vagueness, S defined as sum of individual spreads of Fuzzy parameters of model, such that *Minimize*

$$S = \sum_{t=1}^{k} c^{'} \mid x_t \mid \tag{8}$$

Here the membership value of each observation y_t is greater than an imposed threshold level $h \in [0,1]$. The h level value influences widths c of Fuzzy parameters

$$\mu_{\tilde{y}}(y_t) \geq h \text{ for } t = 1,2,\ldots\ldots,k \tag{9}$$

Index t refers to number of non-fuzzy data. The problem of finding Fuzzy regression parameters was formulated by Tanaka et al. [TI] as linear programming problem as follows: *Minimize* $S = \sum_{t=1}^{k} c^{'} \mid x_t \mid$ subject to

$$\begin{cases} x_t^{'} \alpha + (1-h)c^{'} \mid x_t \mid \geq y_t, \, t = 1,2,\ldots\ldots,k \\ x_t^{'} \alpha - (1-h)c^{'} \mid x_t \mid \leq y_t, \, t = 1,2,\ldots\ldots,k \\ c \geq 0 \qquad\qquad\qquad otherwise \end{cases} \tag{10}$$

where $\alpha^{'} = (\alpha_1, \alpha_2, \ldots\ldots, \alpha_n)$ and $c^{'} = (c_1, c_2, \ldots\ldots, c_n)$ are a vector of unknown variables and S is total vagueness.

4 Neuro-Fuzzy Forecasting Model

Due to uncertainty from environment, forecasting is performed using little data over short span of time. Thus, efficient forecasting methods under incomplete data conditions are required. Fuzzy regression model is suitable for interval forecasting where historical data is inadequate. The parameters of ANN model viz. weights $w_{i,j}$ and biases w_j are crisp in nature. Now, Fuzzy parameters in form of π_2 Fuzzy numbers for weights $(\tilde{w}_{i,j} ; i = 0,1,2,\ldots\ldots, p, j = 1,2,\ldots\ldots, q)$ and biases $(\tilde{w}_j ; j = 0,1,2,\ldots\ldots, q)$ are used for related parameters of layers. Considering the methodology formulated by Isibuchi and Tanaka [IT] which includes wide spread of forecasted model, proposed model uses Fuzzy function with Fuzzy parameters:

$$\tilde{y}_t = f(\tilde{w}_0 + \sum_{j=1}^{q} \tilde{w}_j \cdot g(\tilde{w}_{0,j} + \sum_{i=1}^{p} \tilde{w}_{i,j} \cdot y_{t-i})) \tag{11}$$

where y_t are observations. Equation (1) is rewritten as:

$$\tilde{y}_t = f(\tilde{w}_0 + \sum_{j=1}^{q} \tilde{w}_j \cdot \tilde{X}_{t,j}) = f(\sum_{j=0}^{q} \tilde{w}_j \cdot \tilde{X}_{t,j}) \tag{12}$$

where $\tilde{X}_{t,j} = g(\tilde{w}_{0,j} + \sum_{i=1}^{p} \tilde{w}_{i,j} \cdot y_{t-i})$. π_2 Fuzzy numbers

$\tilde{w}_{i,j} = (a_{i,j}, b_{i,j}, c_{i,j}, d_{i,j})$ are: $\mu_{\tilde{w}_{i,j}}(w_{i,j}) = \begin{cases} \dfrac{a_{i,j}}{b_{i,j} + a_{i,j} - w_{i,j}}, w_{i,j} < b_{i,j} \\ 1, b_{i,j} \le w_{i,j} \le c_{i,j} \\ \dfrac{d_{i,j}}{w_{i,j} - c_{i,j} + d_{i,j}}, w_{i,j} > c_{i,j} \end{cases} \tag{13}$

where, $\mu_{\tilde{w}}(w_{i,j})$ is membership function of Fuzzy set. By using Extension Principle [MD, MK, MVG], membership of $\tilde{X}_{t,j} = g(\tilde{w}_{0,j} + \sum_{i=1}^{p} \tilde{w}_{i,j} \cdot y_{t-i})$ in Equation (2) is given by Equation (14).

$$\mu_{\tilde{X}_{t,j}}(x_{t,j}) = \begin{cases} \dfrac{g(\sum_{i=0}^{p} a_{i,j} \cdot y_{t,i})}{g(\sum_{i=0}^{p} b_{i,j} \cdot y_{t,i}) + g(\sum_{i=0}^{p} a_{i,j} \cdot y_{t,i}) - X_{t,j}}, X_{t,j} < g(\sum_{i=0}^{p} b_{i,j} \cdot y_{t,i}) \\ 1, g(\sum_{i=0}^{p} b_{i,j} \cdot y_{t,i}) \le X_{t,j} \le g(\sum_{i=0}^{p} c_{i,j} \cdot y_{t,i}) \\ \dfrac{g(\sum_{i=0}^{p} d_{i,j} \cdot y_{t,i})}{X_{t,j} - g(\sum_{i=0}^{p} c_{i,j} \cdot y_{t,i}) + g(\sum_{i=0}^{p} d_{i,j} \cdot y_{t,i})}, X_{t,j} > g(\sum_{i=0}^{p} c_{i,j} \cdot y_{t,i}) \end{cases} \tag{14}$$

where,

$y_{t,i} = 1(t = 1,2,\ldots\ldots,k, i = 0)$, $y_{t,i} = y_{t-i}(t = 1,2,\ldots\ldots,k, i = 1,2,\ldots\ldots,p)$.

Considering π_2 fuzzy numbers $\tilde{X}_{t,j}$ with membership function given by Equation (14) π_2 Fuzzy parameters $\tilde{w}_j$ are:

$$\mu_{\tilde{w}_j}(w_j) = \begin{cases} \dfrac{d_j}{e_j + d_j - w_j}, w_j < e_j \\ 1, e_j \le w_j \le f_j \\ \dfrac{g_j}{w_j - f_j + g_j}, w_j > f_j \end{cases} \tag{15}$$

The membership function of $\tilde{y}_t = f(\tilde{w}_0 + \sum\limits_{j=1}^{q} \tilde{w}_j \cdot \tilde{X}_{t,j}) = f(\sum\limits_{j=0}^{q} \tilde{w}_j \cdot \tilde{X}_{t,j})$ is

given as,

$$\mu_{\tilde{Y}}(y_t) \cong \begin{cases} \dfrac{-B_1}{2A_1} + \left[\left(\dfrac{B_1}{2A_1}\right)^2 - \dfrac{C_1 - f^{-1}(y_t)}{A_1}\right]^{1/2}, f^{-1}(y_t) < C_1 \\ 1, C_1 \le f^{-1}(y_t) \le C_2 \\ \dfrac{B_2}{2A_2} + \left[\left(\dfrac{B_2}{2A_2}\right)^2 - \dfrac{C_2 - f^{-1}(y_t)}{A_2}\right]^{1/2}, f^{-1}(y_t) > C_2 \end{cases} \tag{16}$$

where, A_1, B_1, A_2, B_2, C_1, C_2 are functions of variables given in Equations (14) and (15). Considering threshold level h for all membership function values of observations, the non-linear programming problem is given as:

$$Min \sum\limits_{t=1}^{k} \sum\limits_{j=0}^{q} (f_j \cdot g(\sum\limits_{i=0}^{p} c_{i,j} \cdot y_{t,i})) - (d_j \cdot g(\sum\limits_{i=0}^{p} a_{i,j} \cdot y_{t,i})) \quad \text{subject to}$$

$$\begin{cases} \dfrac{-B_1}{2A_1} + \left[\left(\dfrac{B_1}{2A_1}\right)^2 - \dfrac{C_1 - f^{-1}(y_t)}{A_1}\right]^{1/2} \le h, f^{-1}(y_t) < C_1, t = 1,2,\dots\dots,k \\ 1, C_1 \le f^{-1}(y_t) \le C_2 \\ \dfrac{B_2}{2A_2} + \left[\left(\dfrac{B_2}{2A_2}\right)^2 - \dfrac{C_2 - f^{-1}(y_t)}{A_2}\right]^{1/2} \le h, f^{-1}(y_t) > C_2, t = 1,2,\dots\dots,k \end{cases} \tag{17}$$

The nature of π_2 Fuzzy numbers are symmetric, output neuron transfer function is taken to be linear and connected weights between input and hidden layers are of crisp form. The membership function of y_t is transformed as follows:

$$\mu_{\underset{y}{\cdot}}(y_t) = \begin{cases} 1 - \dfrac{\left| y_t - \sum\limits_{j=0}^{q} \alpha_j \cdot X_{t,j} \right|}{\sum\limits_{j=0}^{q} c_j \mid X_{t,j} \mid}, X_{t,j} \neq 0 \\ 0, otherwise \end{cases} \tag{18}$$

Parameter y_t represents the degree to which the model should be satisfied by all data points $y_1, \ldots\ldots\ldots, y_k$. The h value influences widths of fuzzy parameters:

$$\mu_{\underset{y}{\cdot}}(y_t) \geq h \quad \forall t = 1, \ldots\ldots\ldots, k \tag{19}$$ Fuzziness S included in model is:

$$S = \sum_{j=0}^{q} \sum_{t=1}^{k} c_j \mid w_j \parallel X_{t,j} \mid \tag{20}$$ Now, the linear programming problem is reformulated

as: $Minimize \ S = \sum\limits_{j=0}^{q} \sum\limits_{t=1}^{k} c_j \mid w_j \parallel X_{t,j} \mid$

subject to
$$\begin{cases} \sum\limits_{j=0}^{q} \alpha_j X_{t,j} + (1-h)(\sum\limits_{j=0}^{q} c_j \mid X_{t,j} \mid) \geq y_t, t = 1, \ldots\ldots\ldots, k \\ \sum\limits_{j=0}^{q} \alpha_j X_{t,j} - (1-h)(\sum\limits_{j=0}^{q} c_j \mid X_{t,j} \mid) \leq y_t, t = 1, \ldots\ldots\ldots, k \\ c_j \geq 0, j = 0,1, \ldots\ldots\ldots, q \end{cases} \tag{21}$$

The procedure of the model comprises of three phases.

Phase I: Train network using available information. The parameters $w^* = (w_j^*; j = 0,1,2, \ldots\ldots, q$, $w_{i,j}^*; i = 0,1,2, \ldots\ldots, p, j = 0,1,2, \ldots\ldots, q)$ and output value of hidden neuron serves as input data sets for next phase.

Phase II: Determine minimal fuzziness using Equation (7) and $w^* = (w_j^*; j = 0,1,2, \ldots\ldots, q, w_{i,j}^*; i = 0,1,2, \ldots\ldots, p, j = 0,1,2, \ldots\ldots, q)$.

Phase III: The data around model's upper bound and lower bound when proposed model has outliers with wide spread are deleted in accordance with Isibuchi's recommendations. To make the model include all possible conditions c_j has wide spread when data set includes significant difference. Ishibuchi and Tanaka [IT] suggested that data around model's upper and lower boundaries are deleted so that the model can be reformulated.

5 Simulation Results

This section demonstrates efficiency of the proposed method considering exchange rate data of US dollar (US $) against Indian national rupee (INR) and comparing it with other forecasting models. The information used here consists of 50 daily observations of exchange rates of US $ versus INR from 8th May to 15th October, 2008. Here, 40 observations are first used to formulate the model and last 10 observations to evaluate performance of method. Considering the three phases as illustrated in previous section, Phase I *trains the neural network model* using [RBE]. The best fitted network architecture which presented best forecasting accuracy with test data composed of two inputs, three hidden and one output neurons, whose weights and biases are given in table 1. Phase II *determines the minimal fuzziness* considering weights

$$(w_0, w_1, w_2, w_3) = (-985.2, 0.5947, 985.33, 0.2787)$$ .Fuzzy parameters are

obtained from Equation (17) with $(h = 0)$. From the results plotted in figure2, actual values are located in Fuzzy intervals but string of Fuzzy intervals is wide enough for smooth macro-economic environment. This problem is resolved using method of Isibuchi and Tanaka [IT] to provide narrower interval for the decision maker. Phase III *deletes the outliers around the model's upper bound and lower bounds.* From the results it is evident that observation of 22nd August, 2008 is located at lower bound, so linear programming constrained equation that is produced by this observation is deleted to renew phase II (figure 3). The future values of exchange rates for next ten transaction days are also forecasted (table 2). The results of forecast are satisfactory and Fuzzy intervals are narrower. The performance can be improved with larger data sets.

Now we give a comparison of performance of the discussed model using single time-series viz., exchange rate (US $/INR) with other forecasting models such as ARIMA, Chen's fuzzy time-series (first-order) [Ch], Chen's fuzzy time-series (high-order) [CC], Yu's fuzzy time-series [Yu1], FARIMA [TTYY] and neural networks. To measure forecasting performance, MAE (mean absolute error) and MSE (mean squared error) are employed as performance indicators, computed as:

$$MAE = \frac{1}{N} \sum_{i=1}^{N} |e_i|$$ (22)

$$MSE = \frac{1}{N} \sum_{i=1}^{N} (e_i)^2$$ (23)

where, e_i is individual forecast error and N is number of error terms. Based on results in Table 3, it is seen that predictive power of discussed model is rather encouraging and that possible interval by the model is narrower than 95 % of confidence interval of ANN. The width of forecasted interval by Neuro-Fuzzy model is 1.9 Rupees which indicates an 86.9 % improvement upon the 95 % of confidence interval of ANN. However, model requires fewer observations than ANN and is an interval forecaster that yields more information. Evidence shows that performance of Neuro-Fuzzy model is better than that of other models and interval obtained is narrower than that obtained by FARIMA. Its performance is superior to FARIMA by the 42.6 %.

Neuro-Fuzzy model is also better than other models such as [Ch], [CC] and [Yu1] as given in table 4. Thus, Neuro-Fuzzy model obtains accurate results under incomplete data conditions, forecasts where little historical data are available and provides best and worst possible situations for decision makers.

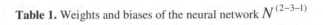

Table 1. Weights and biases of the neural network $N^{(2-3-1)}$

Input weights			Hidden weights	Biases	
$w_{i,1}$	$w_{i,2}$	$w_{i,3}$	w_j	$w_{0,j}$	w_0
3.786	2.3752	4.5530	0.59466	-6.5937	-
					985.1925
42.1044	-11.4969	-26.0886	985.3296	11.4486	
-155.2669	172.2537	158.1926	0.27868	-27.1696	

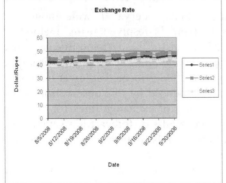

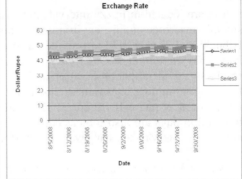

Fig. 2. Results obtained from Neuro-Fuzzy model (Series1 denote Upper bound of Exchange Rate; Series2 denote Actual value of Exchange Rate; Series3 denote Lower bound of Exchange Rate)

Fig. 3. Results of Neuro-Fuzzy Model after deleting the 22nd August, 2008 Lower bound (Series1 denote Upper bound of Exchange Rate; Series2 denote Actual value of Exchange Rate; Series3 denote Lower bound of Exchange Rate)

Table 2. Results of Neuro-Fuzzy model for the test data

Date	Actual	Lower Bound	Upper Bound
1st October, 2008	46.47	44.30	48.50
2nd October, 2008	46.63	44.50	48.66
3rd October, 2008	47.01	45.00	49.09
6th October, 2008	47.77	45.28	49.79
7th October, 2008	47.78	45.65	49.80
8th October, 2008	48.00	45.96	50.10
9th October, 2008	48.00	45.99	50.40
10th October, 2008	48.41	46.36	50.46
14th October, 2008	47.93	45.80	49.97
15th October, 2008	48.53	46.36	50.57

Table 3. Comparison of forecasted interval widths by the Neuro-Fuzzy model with other forecasting models

Model	Forecasted interval width	Related performance		
		ANN	Fuzzy ARIMA	Neuro-Fuzzy Model
ANN (95% confidence interval)	15.4	0	-	-
Fuzzy ARIMA	3.7	75.7 %	0	-
Neuro-Fuzzy Model	1.9	86.9 %	42.6 %	0

Table 4. Comparison of the performance of Neuro-Fuzzy model with other forecasting models

Model	Exchange Rate	
	MAE	MSE
Auto Regressive Integrated Moving Average	0.925	1.242
Chen's Fuzzy time-series (first order)	0.750	0.778
Chen's Fuzzy time-series (second order)	0.750	0.778
Yu's Fuzzy time-series	0.750	0.778
Artificial Neural Networks (ANN)	0.690	0.684
Neuro-Fuzzy Model	0.580	0.521

6 Conclusion

In today's competitive scenario, quantitative techniques have become important tool for financial market forecasting and for improving decisions and investments. One of the most important factors in choosing forecasting technique is its accuracy. The thrust is on improving the effectiveness of time series models. The real world environment experiences uncertain and quick changes, such that future situations should be forecasted using small amount of data from short span of time. This work proposes a Neuro-Fuzzy model combining advantages of ANN and Fuzzy regression to forecast exchange rate of US dollar to Indian national rupee. The disadvantage of large volume of historical data is removed through investing on advantages of fuzzy regression models. The Neuro-Fuzzy model requires fewer observations to obtain accurate results and also obtains narrower possible intervals than other interval forecasting models by exploiting advantage of ANN. The empirical results indicate that model is suitable for use in incomplete data conditions. Results indicate that the model performs better than other models. It is also suitable for both point and interval forecasts with incomplete data. Thus, Neuro-Fuzzy model makes good forecasts under best and worst situations which make it more suitable for decision making over other techniques.

References

[AMM] Armano, G., Marchesi, M., Murru, A.: A hybrid genetic-neural architecture for stock indexes forecasting. Inform. Sci. 170, 3–33 (2005)

[CC] Chen, S.M., Chung, N.Y.: Forecasting enrollments using high-order fuzzy time series and genetic algorithms. Internat. J. Intell. Syst. 21, 485–501 (2006)

[CH1] Chen, S.M., Hwang, J.R.: Temperature prediction using fuzzy time series. IEEE Trans. Systems Man Cybernet., B 30(2), 263–275 (2002)

[CH2] Chen, S.M., Hsu, C.C.: A new method to forecast enrollments using fuzzy time series. Appl. Sci. Engg. 2, 234–244 (2004)

[Ch] Chen, S.M.: Forecasting enrollments based on fuzzy time series. Fuzzy Sets and Systems 81(3), 311–319 (1996)

[CLH] Chen, A., Leung, M.T., Hazem, D.: Application of neural networks to an emerging financial market: forecasting and trading the Taiwan Stock Index. Comp. Oper. Res. 30, 901–923 (2003)

[DP] Dubois, D., Prade, H.: Theory and Applications. Fuzzy Sets and Systems. Academic Press, New York (1980)

[HTW] Hsu, Y.Y., Tse, S.M., Wu, B.: A new approach of bivariate fuzzy time series analysis to the forecasting of a stock index. Internat. J. Uncert. Fuzzi. Know. Bas. Syst. 11(6), 671–690 (2003)

[HY1] Huarng, K., Yu, H.K.: An nth order heuristic fuzzy time series model for TAIEX forecasting. Internat. J. Fuzzy Systems 5(4) (2003)

[HY2] Huarng, K., Yu, H.K.: An type 2 fuzzy time series model for stock index forecasting. Phys. A 353, 445–462 (2005)

[HY3] Huarng, K., Yu, T.H.K.: The application of neural networks to forecast fuzzy time series. Phys. A 336, 481–491 (2006)

[IT] Ishibuchi, H., Tanaka, H.: Interval regression analysis based on mixed 0-1 integer programming problem. J. Japan Soc. Ind. Engg. 40(5), 312–319 (1988)

[KHB] Khashei, M., Hejazi, S.R., Bijari, M.: A new hybrid artificial neural networks and fuzzy regression model for time series forecasting. Fuzzy Sets and Systems 159(7), 769–786 (2008)

[LRS] Luxhoj, J.T., Riis, J.O., Stensballe, B.: A hybrid econometric neural network modeling approach for sales forecasting. Internat. J. Prod. Econom. 43, 175–192 (1996)

[MD] Maier, H.R., Dandy, G.C.: The effect of internal parameters and geometry on the performance of back-propagation neural networks: an empirical study. Environ. Model. & Soft. 13, 193–209 (1998)

[MK] Ma, L., Khorasani, K.: A new strategy for adaptively constructing multilayer feedforward neural networks. Neurocomputing 51, 361–385 (2003)

[MVG] Martin, D., Varo, A., Guerrero, J.E.: Non-linear regression methods in NIRS quantitative analysis. Talanta 72, 28–42 (2007)

[RBE] Rafiq, M.Y., Bugmann, G., Easterbrook, D.J.: Neural network design for engineering applications. Comput. Struct. 79, 1541–1552 (2001)

[SC1] Song, Q., Chissom, B.S.: Forecasting enrollments with fuzzy time series – part I. Fuzzy Sets and Systems 54(1), 1–9 (1993)

[SC2] Song, Q., Chissom, B.S.: Forecasting enrollments with fuzzy time series – part II. Fuzzy Sets and Systems 62(1), 1–8 (1994)

[SJ] Smith, K., Jatinder, N.D.: Neural networks in business: techniques and applications for the operations researcher. Comp. Oper. Res. 27, 1045–1076 (2000)

[Tan] Tanaka, H.: Fuzzy data analysis by possibility linear models. Fuzzy Sets and Systems 24(3), 363–375 (1987)

[TE] Thawornwong, S., Enke, D.: The adaptive selection of financial and economic variables for use with artificial neural networks. Neurocomputing 31, 1–13 (2000)

[TF] Tang, Z., Fishwick, P.A.: Feed forward neural networks as models for time series forecasting. ORSA J. Comp. 5, 374–385 (1993)

[TI] Tanaka, H., Ishibuchi, H.: Possibility regression analysis based on linear programming. In: Kacprzyk, J., Fedrizzi, M. (eds.) Fuzzy regression analysis, pp. 47–60. Omnitech Press, Warsaw (1992)

[TTYY] Tseng, F., Tzeng, G., Yu, H.C., Yuana, B.J.C.: Fuzzy ARIMA model for forecasting the foreign exchange market. Fuzzy Sets and Systems 118, 9–19 (2001)

[TUA] Tanaka, H., Uejima, S., Asai, K.: Linear regression analysis with fuzzy model. IEEE Trans. Systems Man Cybernet. 12(6), 903–907 (1982)

[Wata] Watada, J.: Fuzzy time series analysis and forecasting of sales volume. In: Kacprzyk, J., Fedrizzi, M. (eds.) Fuzzy regression analysis, pp. 211–227. Omnitech Press/Physica-Verlag, Warsaw/Heidelberg (1992)

[WVJ] Wong, B.K., Vincent, S., Jolie, L.: A bibliography of neural network business applications research: 1994 – 1998. Comp. Oper. Res. 27, 1023–1044 (2000)

[Yu1] Yu, H.K.: Weighted fuzzy time-series models for TAIEX forecasting. Phys. A 349, 609–624 (2004)

[Yu2] Yu, H.K.: A refined fuzzy time-series model for forecasting. Phys. A 349, 609–624 (2005)

[Zad1] Zadeh, L.A.: The concept of a linguistic variable and its application to approximate reasoning I. Inform. Sci. 8, 199–249 (1975)

[Zad2] Zadeh, L.A.: The concept of a linguistic variable and its application to approximate reasoning II. Inform. Sci. 8, 301–357 (1975)

[Zad3] Zadeh, L.A.: The concept of a linguistic variable and its application to approximate reasoning III. Inform. Sci. 8, 43–80 (1976)

[ZM] Zhang, P., Min Qi, G.: Neural network forecasting for seasonal and trend time series. Europe. J. Oper. Res. 160, 501–514 (2005)

A Fast Support Vector Machine Classification Algorithm Based on Karush-Kuhn-Tucker Conditions

Ying Zhang, Xizhao Wang, and Junhai Zhai

Key Lab. for Machine Learning and Computational Intelligence,
College of Mathematics and Computer Science,
Hebei University, Baoding 071002, China
zhangying03071984@163.com

Abstract. Although SVM have shown potential and promising performance in classification, they have been limited by speed particularly when the training data set is large. In this paper, we propose an algorithm called the fast SVM classification algorithm based on Karush-Kuhn-Tucker (KKT) conditions. In this algorithm, we remove points that are independent of support vectors firstly in the training process, and then decompose the remaining points into blocks to accelerate the next training. From the theoretical analysis, this algorithm can remarkably reduce the computation complexity and accelerate SVM training. And experiments on both artificial and real datasets demonstrate the efficiency of this algorithm.

Keywords: Support Vector Machine, Karush-Kuhn-Tucker (KKT) conditions, Agglomerative hierarchical clustering algorithm, Remove samples.

1 Introduction

Support vector machine (SVM) is developed by Vapnik and his co-workers in Russia. It is based on the theory of the VC dimension and the structural risk minimization principle [1]. Recently SVM have shown promising performance in many applications. However, they require the use of an iterative process such as quadratic programming (QP) to identify the support vectors from the labeled training set. When the number of samples in the training set is huge, sometimes it is impossible to use all of them for training, so some heuristic methods have to be used to speed up the process.

A typical solution to accelerate SVM training is to decompose the QP into a number of sub-problems so that the overall SVM training complexity can be reduced from $O(N^3)$ to $O(N^2)$ [3,4]. However, when the number of data points N is very large, the time complexity is still unsatisfactory and needs further improvement.

In the literature [9], a new incremental learning algorithm based on the Karush-Kuhn-Tucker (KKT) conditions was proposed. The whole learning process is divided into the initial learning process and the incremental learning process in this method. It is according to that the optimal solution of the QP allows each sample to satisfy the KKT conditions. The incremental sample set and the initial sample set are equal in status in this algorithm.

H. Sakai et al. (Eds.): RSFDGrC 2009, LNAI 5908, pp. 382–389, 2009.

Based on the above ideas, we proposed the block-learning algorithm based on the KKT conditions (B-KKT), which is similar to the incremental learning algorithm described in [9], and the difference is the training process is carried out in the training data set in one time but not an incremental learning process. And mainly, a new algorithm called the fast support vector machine classification algorithm based on the KKT conditions (FC-KKT) is proposed in this paper. Experimental results show that it remarkably reduce the computation complexity and accelerate SVM training.

The rest of the paper is organized as follows. Section 2 briefly introduces the optimization problem involved in training SVM, followed by the theory of the B-KKT algorithm. In section 3, the FC-KKT algorithm is introduced in detail. In section 4   the experimental results were reported on both the artificial data and the real data. Finally, Section 5 gives the conclusions of the paper.

2   Support Vector Machine and the B-KKT Algorithm

2.1   Support Vector Machine

Given l training points $(x_1, y_1), \cdots, (x_l, y_l)$, where $x_i \in R^n$ is an input vector labeled by $y_i \in \{+1, -1\}$ for $i = 1, \cdots, l$, support vector machine searches for a separating hyperplane with largest margin, which is called the optimal hyperplane $w^T x + b = 0$. This hyperplane can classify the input pattern x according to the following function

$$f(x) = \text{sgn}(w^T x + b) \tag{1}$$

In order to maximize the margin, we find the solution for the following QP problem

$$\text{Min} \quad \frac{1}{2}\|w\|^2 + C\sum_{i=1}^{l} \xi_i$$
$$\text{Subject to } y_i \cdot (w^T x_i + b) \geq 1 - \xi_i, \forall i = 1, \cdots, l \tag{2}$$
$$\xi_i \geq 0$$

Where $\xi_i \geq 0, i = 1, \cdots, l$ are slacking variables. The non-zero $\xi_i \geq 0, i = 1, \cdots, l$ are the training patterns that do not satisfy the constraints in (2).

The problem is usually posed in its Wolfe dual form involved the Lagrange multipliers $\alpha_i \in [0, C], i = 1, \cdots, l$, which can be solved by standard quadratic optimization packages. The bias b can easily be calculated by the vectors x_i satisfying $0 < \alpha_i < C$. The decision function is therefore given by

$$f(x) = \text{sgn}(w^T x + b) = \text{sgn}(\sum_i \alpha_i y_i x_i^T x + b) \tag{3}$$

In a typical classification task, only a small number of the Lagrange multipliers α_i tend to be greater than zero, and the corresponding training vectors are called support vectors. The hyperplane $f(x)$ depends on these points.

The basic idea of nonlinear SVM is to map points from the input space to a high-dimensional feature space using a nonlinear mapping Φ, and then to proceed pattern classification using linear SVM. The mapping Φ is performed by employing Kernel functions $K(x_i, x)$, which obeys Mercer conditions [6], to compute the inner products between support vectors $\Phi(x_i)$ and the pattern vector $\Phi(x)$ in the feature space. For an unknown input pattern x, we have the following decision function [6],

$$f(x) = \text{sgn}(\sum_i^l \alpha_i y_i x_i^T x + b) \tag{4}$$

2.2   The B-KKT Algorithm

Suppose that $\alpha = [\alpha_1, \alpha_2, \cdots, \alpha_l]$ is the optimal solution of the dual problem. The KKT conditions which make each sample satisfy the QP problem is

$$\alpha_i = 0 \Rightarrow y_i f(x_i) \geq 1 \ ; \ 0 < \alpha_i < C \Rightarrow y_i f(x_i) = 1 \ ;$$
$$\alpha_i = C \Rightarrow y_i f(x_i) \leq 1, i = 1, \cdots, l \tag{5}$$

When α_i tend to be greater than zero, the corresponding x_i are called support vectors.

Theorem 1: Q is the Hession Matrix in the linear constrained QP, if Q is semi-definite, the QP is convex. Suppose α is a feasible solution of the QP, if and only if every point x satisfies the KKT conditions, α is the optimal solution of the QP [6].

From Theorem 1, we know that, if the additional sample x violates the KKT conditions obtained in the initial learning process, the corresponding α is not the overall optimal solution after the addition of new samples. According to the relationship of α and the points x, we have the following Theorem 2.

Theorem 2: Suppose $f(x)$ is the decision function of the SVM classifier, and $\{x_i, y_i\}$ is an additional sample. The additional samples that violate the KKT condition will change the support vectors set training in the initial data set.

The samples that violate the KKT condition are divided into three categories:

(1) The samples that is between the two boundary lines and could be correctly classified by the initial SVM classifier are the samples which satisfy $0 \leq y_i f(x_i) < 1$;
(2) The samples that is between the two boundary lines but are wrongly classified by the original SVM classifier are the samples which satisfy $-1 \leq y_i f(x_i) \leq 0$;
(3) The samples that are outside the two boundary lines and are wrongly classified by the original SVM classifier are the samples which satisfy $y_i f(x_i) < -1$.

In summary the samples that violate the KKT conditions satisfy $y_i f(x_i) < 1$.

Based on the above theories, in the literature [9], a new incremental learning algorithm based on the Karush-Kuhn-Tucker (KKT) conditions is proposed. The B-KKT algorithm is similar to the algorithm proposed in literature [9], it is described as follow:

(1) The entire training data is divided into two subsets X_0, X_1, $X_0 \cap X_1 = \Phi$; T_0 is trained in subset X_0, and the corresponding support vector set is X_0^{SV};

(2) Test whether exists samples violating the KKT conditions of T_0 in subset X_1, if not; the algorithm stops, and T_0 is the classifier in the whole training data. Otherwise, according to the test results, X_1 is divided into two subsets X_1^S and X_1^{NS}, where X_1^S is the set of samples satisfying the KKT conditions obtained in subset X_0 and X_1^{NS} is set of samples violating the KKT conditions obtained in subset X_0;

(3) T_1 is obtained by training in subset X_1, and the corresponding support vector set is X_1^{SV};

(4) Test whether exists samples violating the KKT conditions of T_1 in subset X_0, if not; the algorithm stops, and T_1 is the classifier in the whole training data. Otherwise, according to the test results, X_0 is divided into two subsets X_0^S and X_0^{NS}, where X_0^S is the set of samples satisfying the KKT conditions and X_0^{NS} is the set of samples violating the KKT conditions;

(5) $X = X_0^{SV} \cup X_1^{SV} \cup X_0^{NS} \cup X_1^{NS}$, the set X is trained, the final training classifier T is obtained, which is the classifier in the overall training data set.

The B-KKT algorithm decomposes the QP problem into a number of sub-problems so that the overall SVM training is accelerated. However, when the number of training dataset is very large, the time complexity is still unsatisfactory and needs further improvement, as follows, we propose the new fast algorithm to further accelerate the SVM training under the premise of no loss of the classification accuracy.

3 The FC-KKT Algorithm

The theoretical basis of this algorithm is the hyperplane constructed by SVM is dependent on support vectors that lie closed to the decision boundary (hyperplane). Thus, removing any training samples that are not relevant to support vectors may have no effect on building the hyperplane. In this method, we remove non-relevant samples from the training set and then decompose the remaining samples into blocks to train.

The algorithm is described as follows:

1) Using the agglomerative hierarchical clustering [10] on training data set to extract the underlying data structure. The distance metric we use is called the Average Linkage distance. The cluster number is $k^+ = round(\sqrt{n^+})$ in positive class and the cluster number is $k^- = round(\sqrt{n^-})$ in negative class, where n^+ is the number of samples in positive class of the training data and n^- is the number of samples in negative class in two-class classification problem;

2) Find the nearest point to the center point in each cluster respectively, and then use these points to represent points of the whole cluster.

$$x^p(D_i) = \arg\min_{x \in D_i} \left\| x - \frac{1}{n_i} \sum_{k=1}^{n_i} x_k \right\|_2 \tag{6}$$

where $\|\cdot\|_2$ means 2-norm and n_i is the number of points in the cluster D_i. An initial SVM classifier T_0 is obtained by training in the points $x^p(D_i)$;

3) The points in positive class are divided into two subsets X_{NSV}^+ and X_{SV}^+ by the classifier T_0, where the points satisfy $y_i f(x_i) > 1$ belonging to the set X_{NSV}^+ and the points satisfy $y_i f(x_i) \leq 1$ belonging to the set X_{SV}^+, likewise, the points in negative class is divided into two parts X_{NSV}^- and X_{SV}^-, where the points satisfy $y_i f(x_i) < -1$ belonging to the set X_{NSV}^- and the points satisfy $y_i f(x_i) \geq -1$ belonging to the set X_{SV}^-. The points in the sets X_{NSV}^+ and X_{NSV}^- are removed, respectively;

4) The remaining points are divided into two subsets X_1 and X_2, $X_1 \cap X_2 = \Phi$, and then to train in the two subsets X_1 and X_2, respectively;

5) T_1 is obtained by training in subset X_1, and the corresponding support vector set is X_1^{SV};

6) Test whether exists samples violating the KKT conditions of T_1 in subset X_2, if not; the algorithm stops, and T_1 is the classifier in the whole training data set. Otherwise, according to the test results, X_2 is divided into two subsets X_2^S and X_2^{NS}, where X_2^S is the samples satisfying the KKT conditions and X_2^{NS} is the samples violating the KKT conditions;

7) T_2 is obtained by training in subset X_2, and the corresponding support vector set is X_2^{SV};

8) Test whether exists samples violating the KKT conditions of T_1 in subset X_2, if not; the algorithm stops, and T_2 is the classifier in the whole training data set. Otherwise, according to the test results, X_1 is divided into two subsets X_1^S and X_1^{NS}, where X_1^S is the set of samples satisfying the KKT conditions and X_1^{NS} is the set of samples violating the KKT conditions;

9) $X = X_1^{SV} \cup X_2^{SV} \cup X_1^{NS} \cup X_2^{NS}$, the set X is trained, the final training classifier T is obtained, which is the classifier in the overall training data set.

4   Experiments and Results

Experiments on both artificial and real datasets demonstrate the effectiveness of this algorithm. The experiment is carried out under the hardware of PC (Pentium (R) 4 CPU 2.93GHz DDR512MB RAM) and the software environment is Windows XP/Matlab 7.1.

4.1 The Experiment Results in Artificial Datasets

In artificial data experiments, we proved that the method to remove points referred in this paper is feasible and effective. A part of points are removed and the training accuracy is improved in this method, which is shown in Fig.1 (by comparing the classification of points near the hyperplane after removing points with before). In particular, most of the non-support vectors are removed in the data set that is clear, which is shown in Fig.2.

Fig. 1. (a). Before removing. (b). After removing.

Fig. 2. (a). Before removing. (b). After removing.

Fig. 1. A part of points are removed and the training accuracy is improved in this method of removing points.

Fig. 2. Most of the non-support vectors can be removed in the data set that is clear under the premise of no loss of the classification accuracy in this method of removing points.

4.2 The Experiment Results in Real Datasets

The standard SVM algorithm (SVM), the B-KKT algorithm and the FC-KKT algorithm are compared through the experiments in two real datasets: bupa and pima. From the number of training samples (TRS), the number of support vectors (SV), the training time (TT), and the testing accuracy (TA), the efficiency of FC-KKT is verified in these experiments. In experiments of B-KKT, the number of training samples is the number of points in each block after decomposing the whole training data set into blocks, and in experiments of FC-KKT, it is the number of points in each block of the remaining training data set after removing points. From Table 1 and Table 2, we know that the

Table 1. The experiment results on the bupa data set

Algorithm		TRS	SV	TES	TT(s)	TA
SVM		276	172	69	0.85938	62.32%
B-KKT	X_1	138	93	69	0.71876	63.22%
	X_2	138	96			
	X	218	181			
FC-KKT	X_1	82	51	69	0.53117	63.88%
	X_2	83	49			
	X	128	106			

Table 2. The experiment results on the pima data set

Algorithm		TRS	SV	TES	TT(s)	TA
SVM		614	309	154	5.9375	72.73%
B-KKT	X_1	307	142	154	4.0469	74.03%
	X_2	307	151			
	X	388	301			
FC-KKT	X_1	178	107	154	1.5993	75.43%
	X_2	177	102			
	X	261	212			

training time is reduced remarkably especially in the larger data set by using this method FC-KKT. The test accuracy is also improved, because we use the agglomerative hierarchical clustering algorithm to extract the underlying data structure of the training data in the training process.

5 Conclusions

Support Vector Machine (SVM) have gained wide acceptance because of the high generalization ability for a wide range of classification applications. Although SVM have shown potential and promising performance in classification, they have been limited by speed particularly when the training data set is large. The hyperplane constructed by SVM is dependent on only a portion of the training samples called support vectors that lie close to the decision boundary (hyperplane). Thus, removing any training samples that are irrelevant to support vectors might have no effect on building the proper decision function. Based on the above theories, we propose an algorithm called the fast SVM classification algorithm based on the KKT conditions. This algorithm can remarkably reduce the computational complexity and accelerate SVM training. Experiments on both artificial and real datasets demonstrate the efficiency of this algorithm.

Acknowledgements. This research is supported by the Natural Science Foundation of Hebei Province (F2008000635), by the key project foundation of applied fundamental research of Hebei Province (08963522D), by the plan of 100 excellent innovative scientists of the first group in Education Department of Hebei Province.

References

1. Vapnik, V.N.: The Nature of Statistical Learning Theory. Springer, Heidelberg (1995)
2. Boley, D., Cao, D.: Training support vector machine using adaptive clustering. In: Proceedings of the SIAM International Conference on Data Mining, Lake Buena Vista, FL, USA, pp. 126–137 (2004)
3. Joachims, T.: Making large-scale support vector machine learning practical. In: Smola, A.J., Scholkopf, B., Burges, C. (eds.) Advances in Kernel Methods: Support Vector Machines. MIT Press, Cambridge (1998)

4. Platt, J.: Fast training of support vector machines using sequential minimal optimization. In: Smola, A.J., et al. (eds.) Advances in Kernel Methods: Support Vector Machines. MIT Press, Cambridge (1998)
5. Wang, D., Yeung, D.S., Eric, C.C.T.: Sample Reduction for SVM via Data Structure Analysis. In: IEEE International Conference on In System of Systems Engineering, pp. 1–6 (2007)
6. Naiyang, D., Yingjie, T.: The new data mining method–Support Vector Machine. Science Press, Beijing (2004)
7. Lin, K. M., Lin, C. J.: A study on reduced support vector machines. IEEE Transactions on Neural Networks (2003) (to appear)
8. Yeung, D., Wang, D.-F., Ng, W., Tsang, E., Wang, X.-Z.: Structured large margin machines: sensitive to data distribution. Machine Learning 68, 171–200 (2007)
9. Wen-hua, Z., Jian, M.: An Incremental Learning Algorithm for Support Vector Machine and its Application. Computer Integrated Manufacturing Systems, Special Magazine 9, 144–148 (2001)
10. Salvador, S., Chan, P.: Determining the number of clusters/segments in hierarchical clustering/segmentation algorithms. In: Proceedings of the 16th IEEE international conference on tools with AI, pp. 576–584 (2004)

Data Mining Using Rules Extracted from SVM: An Application to Churn Prediction in Bank Credit Cards

M.A.H. Farquad[1,2], V. Ravi[1,*] , and S. Bapi Raju[2]

[1] Institute for Development and Research in Banking Technology, Castle Hills Road #1,
Masab Tank, Hyderabad – 500 057 (AP) India
[2] Department of Computer & Information Sciences, University of Hyderabad,
Hyderabad – 500 046 (AP) India
Tel.: +91-40-2353 4981
farquadonline@gmail.com, rav_padma@yahoo.com,
bapics@uohyd.ernet.in

Abstract. In this work, an eclectic procedure for rule extraction from Support Vector Machine is proposed, where Tree is generated using Naïve Bayes Tree (NBTree) resulting in the SVM+NBTree hybrid. The data set analyzed in this paper is about churn prediction in bank credit cards and is obtained from Business Intelligence Cup 2004. The data set under consideration is highly unbalanced with 93.11% loyal and 6.89% churned customers. Since identifying churner is of paramount importance from business perspective, sensitivity of classification model is more critical. Using the available, original unbalanced data only, we observed that the proposed hybrid SVM+NBTree yielded the best sensitivity compared to other classifiers.

Keywords: Churn prediction in credit cards, Support Vector Machine, Rule Extraction, Naive Bayes Tree.

1 Introduction

Data mining (also called as Knowledge Discovery in Database) is a process that consists of applying data analysis and discovery algorithms that produce particular enumeration of pattern (or model) over the data [1]. Data mining has been efficiently used in wide range of profiling practices, such as manufacturing [2], fraud detection [3].

Increasing number of customers has made the banks conscious of the quality of the services they offer. The problem of customers shifting loyalties from one bank to another has become common. This phenomenon, called '*churn*' occurs due to reasons such as availability of latest technology, customer-friendly staff and proximity of geographical location, etc. Hence, there is a pressing need to develop models that can predict which existing '*loyal*' customer is going to churn out in near future [4].

Research shows that customers with longer relationships with the firm have higher prior cumulative satisfaction ratings [5] than online bank customers [6]. It is more profitable to segment and target customers on the basis of their (changing) purchase

---

* Corresponding author.

H. Sakai et al. (Eds.): RSFDGrC 2009, LNAI 5908, pp. 390–397, 2009.
© Springer-Verlag Berlin Heidelberg 2009

behavior and service experiences rather than on the basis of their (stable) demographics [7]. Churn management consists of developing techniques that enable firms to keep their profitable customers and aims at increasing customer loyalty [8]. Churn prediction and management is one of the important activities of Customer Relationship Management.

Credit Card Database and PBX Database with Data Mining by Evolutionary Learning (DMEL) [9], emphasize on the natural differences between Savings and Investment (SI) products in banks to cross-sell in terms of both maximizing the customers' retention proneness and their preferences [10]. Chu et al., [11] reported that management should prepare an anti-churn strategy that is usually far less expensive than acquiring new customers.

In this paper we address an important issue of rule extraction from SVM and investigate its usefulness in credit card churn prediction in banks. The proposed approach is carried out in two major steps. (1) Extraction of support vectors and obtaining predictions for training instances and support vectros. (2) Rule generation. Incidentally, by using the predictions of SVM for training set and support vectors, we ensure that the rules generated are basically mimicking the behavior of SVM.

The rest of the paper is organized as follows. Section 2 presents the literature review of rule extraction from SVM approaches. Section 3 describes the data set used. Section 4 presents the proposed hybrid rule extraction approach. Section 5 presents results and discussion. Finally section 6 concludes the paper.

2 Rule Extraction from SVM

SVM [13] recently became one of the most popular classification methods. They have been used in wide variety of applications such as Text classification [14], Facial expression recognition [15], and Gene analysis [16] so on. Despite superior performance of SVM, they are often regarded as black box models. Converting this black box, high accurate models to transparent model is *"Rule Extraction"* [17].

Recently attempts have been made to extract rules from SVMs [18]. Intensive work has been done towards developing rule extraction techniques for neural networks but less work has been done for extracting rules from SVM. Some of the approaches proposed towards rule extraction from SVM are; SVM+Prototype [19], RulExtSVM [20], Extracting rules from trained support vector machines [21], Hyper rectangle Rules Extraction (HRE) [22], Fuzzy Rule Extraction (FREx) [23], Multiple Kernel-Support Vector Machine (MK-SVM) [24], SQRex-SVM [25], sequential covering approach [26] and Recently a new Active Learning-Based Approach (ALBA) [27] are some of the approaches proposed towards rule extraction from SVM.

Incidentally Farquad et al. [28, 29] also proposed a hybrid rule extraction approach using SVM and the extracted rules are tested for bankruptcy prediction in banks. They first extracted the support vectors then they used these support vectors to train Fuzzy Rule Based System (FRBS), Decision Tree and Radial Basis Function Network. They concluded that the hybrid SVM+FRBS outperformed the stand-alone classifiers.

3 Data Set Description

The dataset is from a Latin American bank that suffered from an increasing number of churns with respect to their credit card customers and decided to improve its retention system. Two groups of variables are available for each customer: sociodemographic and behavioural data, which are described in Table 1. The dataset comprises 22 variables, with 21 predictor variables and 1 class variable. It consists of 14814 records, of which 13812 are nonchurners and 1002 are churners, which means there are 93.24% nonchurners and 6.76% churners. Hence, the dataset is highly unbalanced in terms of the proportion of churners versus non-churners [30].

Table 1. Feature description of churn prediction data set

Feature	Description	Value
Target	Target Variable	0-NonChurner 1-Churner
CRED_T	Credit in month T	Positive real number
CRED_T-1	Credit in month T-1	Positive real number
CRED_T-2	Credit in month T-2	Positive real number
NCC_T	Number of credit cards in months T	Positive integer value
NCC_T-1	Number of credit cards in months T-1	Positive integer value
NCC_T-2	Number of credit cards in months T-2	Positive integer value
INCOME	Customer's Income	Positive real number
N_EDUC	Customer's educational level	1 - University student
		2 - Medium degree
		3 - Technical degree
		4 - University degree
AGE	Customer's age	Positive integer
SX	Customers sex	1 - male
		0 - Female
E_CIV	Civilian status	1-Single 2-Married
		3-Widow 4-Divorced
T_WEB_T	Number of web transaction in months T	Positive integer
T_WEB_T-1	Number of web transaction in months T-1	Positive integer
T_WEB_T-2	Number of web transaction in months T-2	Positive integer
MAR_T	Customer's margin for the company in months T	Real Number
MAR_T-1	Customer's margin for the company in months T-1	Real Number
MAR_T-2	Customer's margin for the company in months T-2	Real Number
MAR_T-3	Customer's margin for the company in months T-3	Real Number
MAR_T-4	Customer's margin for the company in months T-4	Real Number
MAR_T-5	Customer's margin for the company in months T-5	Real Number
MAR_T-6	Customer's margin for the company in months T-6	Real Number

4 Proposed Rule Extraction Approach

In this research work we propose a hybrid rule extraction procedure for solving large scale classification problem in the framework of data mining using rules extracted from support vector machine. In the churn prediction problem, sensitivity alone is the important criteria, higher the sensitivity better is the model. The proposed hybrid is composed of two major steps (i) support vector extraction and obtaining the predictions of the training instances and support vectors extracted (ii) rule generation using

NBTree [12]. The data set used in this study is highly unbalanced. However, we did not employ any balancing technique to balance the data.

The current approach in this paper is distinct from earlier studies where this data set was analyzed [4, 31] in the following ways:

- Extraction of the support vectors makes the sample size very much small.
- Application of the approach is extended to a data mining problem.

Also, the hybrid approach presented here is different from [28, 29] in the following ways:

- Dealing with unbalanced large scale data set.
- Using the predictions of support vectors using SVM model i.e. *Case-SP* to generate rules with NBTree.

4.1 Support Vectors Extraction and Predictions of SVM

Figure 1 depicts the extraction of support vectors and the resulting 3 variants of the hybrid. The predictions for training set and support vectors are obtained from the developed SVM model. Here *Case-A* and *Case-SA* are two sets viz. Training and SVs set with their corresponding actual target values respectively. Predictions of SVM are obtained for *Case-A* and *Case-SA* and the actual target values are then replaced by the predicted target values to get *Case-P* and *Case-SP* respectively. By using the newly generated *Case-P* and *Case-SP* we ensure that the rules extracted are actually from SVM.

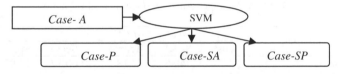

Fig. 1. Support vectors extraction and predictions of SVM

4.2 Rule Generation

During rule generation phase, depicted in Fig. 2, we analyzed 4 different data sets, (i) *Case-A*, (ii) *Case-P*, (iii) *Case-SA* and (iv) *Case-SP* separately. Rules are generated using NBtree hybrid [12]. NBTree attempts to utilize the advantages of both decision trees (i.e. segmentation) and naïve bayes (evidence accumulation from multiple attributes). A decision tree is built with univariate splits at each node, but with Navie-Bayes classifiers at the leaves. Rules are generated under 10-fold cross validation method and the average sensitivity is presented.

Fig. 2. Rule generation phase

4.3   Experimental Setup

The available large scale unbalanced dataset is first divided into two parts of 70:30 ratios. 70% of the data is then used for 10-Fold Cross Validation (10-FCV) and 30% of the data is named as validation set and stored for evaluating the efficiency of the rules generated using 10-FCV at later point of time. The class distribution in the training and validation data sets is as same as that in the original data i.e. 93.11% for loyal customers and 6.89% for churned customers. The accuracy and validity of the rules generated during 10-FCV are then tested against the validation set.

5   Results and Discussions

We used the SVM library viz., LibSVM [32] for building SVM model and support vector extraction. LibSVM is integrated software for support vector classification and is developed in MATLAB. RapidMiner4.5 community edition [33] is used for generating NBTree. Many business decision makers, dealing with churn prediction problem, place high emphasis on sensitivity alone because higher sensitivity leads to greater success in identifying potential churners correctly and thereby contributing to the bottom-line of the CRM viz., retaining extant loyal customers. Consequently in this paper, sensitivity is accorded top priority ahead of specificity. The results of hybrid in various cases (*Case-A, Case-P, Case-SA and Case-SP*) are presented in Table. 2.

Table 2. Average results of 10-fold cross validation

Classifier	Test under 10-FCV			Validation		
	Sens*	Spec*	Acc*	Sens*	Spec*	Acc*
SVM (*Case-A*)	64.2	74.86	74.13	60.17	74.92	73.92
NBTree (*Case-A*)	55.5	98.99	96.06	61.21	99.02	96.46
SVM + NBTree (*Case-P*)	68.62	78.45	77.78	68.52	78	77.07
SVM + NBTree (*Case-SA*)	0	100	93.11	0	100	93.11
SVM + NBTree (*Case-SP*)	68.33	74.38	75.18	68.04	75.34	75.11
Kumar and Ravi(2008) [9]	62.07	98.51	96.05	NA	NA	NA
Naveen et al. (2009) [10]	41.62	79.6	77.03	NA	NA	NA

Note: * Sens = sensitivity; Spec = specificity; Acc = accuracy

It is observed from the results that the hybrid SVM+NBTree using *Case-P* yielded the average sensitivity under 10-FCV and against validation set is 68.62% and 68.52% respectively. The hybrid SVM+NBTree using *Case-SP* obtained the average sensitivity under 10-FCV and against validation set is 68.33% and 68.04% respectively. Stand alone SVM and NBTree using *Case-A* yielded the average sensitivity of 64.2% and 55.5% respectively.

Working on the same data set, Kumar and Ravi [5] reported 62.07% average sensitivity achieved using decision tree classifier, whereas, Naveen et al. [31] reported 41.62% average sensitivity obtained using FuzzyARTMAP. Our results are not strictly comparable to their results as they performed 10-FCV without partitioning the

Table 3. Rule set extracted by SVM+NBTree hybrid using *Case-SP*

#	Rule Antecedents	Consequent
1	CRED_T<=598.1 and MAR_T-2<=14.045	Churner
2	CRED_T<=598.1 and MAR_T-2<=14.045 and INCOME<=1035 and MAR_T-4<=15.135 and T_WEB-T<=7.5	Churner
3	CRED_T>598.1 and T_WEB_T-2<=2.5	Non-Churner
4	CRED_T<=598.1 and MAR_T-2<=14.045 and E_CIV>1.5 and NCC_T-2>0.5 and T_WEB-T>7.5	Non-Churner
5	CRED_T<=598.1 and MAR_T-2<=14.045 and E_CIV>1.5 and NCC_T-2>0.5	Non-Churner
6	CRED_T<=598.1 and MAR_T-2<=14.045 and MAR_T-5<=9.73	Churner
7	CRED_T<=598.1 and MAR_T-2<=14.045 and E_CIV>1.5 and NCC_T-2>0.5 and INCOME>1035 and N_EDUC<=3.5	Non-Churner

original data set into training and validation set. From the above discussions, it is observed that the proposed hybrid SVM+NBTree using *Case-P* and *Case-SP* stand as the best performers compared to other classifiers evaluated in this study. The example rule set obtained by the hybrid SVM+NBTree using *Case-SP* is presented in Table 3.

The tree obtained using NBTree has naïve bayes classifiers at leaf nodes that indicates the probability of each class available in the data set used, instead of prediction of any single class. For better understanding of the tree we modified the rules and the class with higher probability assigned by the naïve bayes classifier at leaf node is considered the consequent of the rule. The number of rules extracted using our approach i.e. SVM+NBTree is very much less and rule length is smaller when compared to those of Kumar and Ravi [4].

It is observed that the number of SVs extracted is 67.8% less than the actual number of training instances. Still, we got a decent sensitivity of 68.52% in *Case-SP* which is a significant outcome of the present study. Hence it is recommended to use support vectors instead of using all the training instances to generate rules.

6 Conclusions

In this paper, we present a rule extraction approach from SVM using NBTree to solve customer churn prediction problem concerning bank credit cards. The data set is highly unbalanced data with 93.11% loyal customers and 6.89% churned customers. We did not employ any balancing technique for balancing the data. Instead we analyzed the original data. We infer that the proposed approach SVM+NBTree using *Case-P* and *Case-SP* outperformed all other classifiers tested and achieved best average sensitivity of 68.52% and 68.03% respectively. The following recommendations are offered from this work. (i) it is better to extract support vectors and use *Case-SP* to generate rules because the number of instances is drastically reduced in the form of support vectors, (ii) the resultant sensitivity yielded by *Case-SP* is almost similar to that of the sensitivity yielded by *Case-P*, (iii) time taken for generation of rules is cut short by more than 60% and (iv) the number of rules extracted and the antecedents per rule are small thereby improving the comprehensibility of the rules.

References

1. Usama, F., Piatetsky-Shapiro, G., Smyth, P.: From Data Mining to Knowledge Discovery in Databases. American Association for Artificial Intelligence (1996)
2. Ravi, V., Arul Shalom, S.A., Manickavel, A.: Sputter Process Variables Prediction via Data Mining. In: Proceedings of the 2004 IEEE Conference on Cybernetics and Intelligent Systems, Singapore (2004)
3. Senator, T., Goldberg, H.G., Wooton, J., Cottini, M.A., Umarkhan, A.F., Klinger, C.D., Llamas, W.M., Marrone, M.P., Wong, R.W.H.: The Financial Crimes Enforcement Network AI System (FAIS): Identifying Potential Money Laundering from Reports of Large Cash Transactions. AI Magazine 16(4), 21–39 (1995)
4. Kumar, D.A., Ravi, V.: Predicting credit card customer churn in banks using data mining. International Journal for Data Analysis, Techniques and Strategies 1(1), 4–28 (2008)
5. Bolton, R.N.: A Dynamic model of the Duration of the customer's relationship with a continuous service provider: The Role of Satisfaction. Marketing Science 17(1), 45–65 (1998)
6. Mols, N.P.: The Behavioral consequences of PC banking. International Journal of Bank Marketing 16(5), 195–201 (1998)
7. Bolton, R.N., Kannan, P.K., Bramlett, M.D.: Implications of Loyalty Program Membership and Service Experiences for Customer Retention and Value. Journal of the Academy of Marketing Science 28(1), 95–108 (2000)
8. Lejeune, M.A.P.M.: Measuring the impact of data mining on churn management. Electronic Networking Applications and Policy 11(5), 375–387 (2001)
9. Au, W.-H., Chan, K.C.C., Yao, S.: A Novel Evolutionary Data Mining Algorithm with Applications to Churn Prediction. IEEE Transactions on Evolutionary Computation 7(6), 532–545 (2003)
10. Larivie're, B., Van den Poel, D.: Predicting customer retention and profitability by using random forests and regression forests techniques. Expert Systems with Applications 29(2), 472–484 (2005)
11. Chu, B.H., Tsai, M.-S., Ho, C.-S.: Toward a hybrid data mining model for customer retention. Knowledge-Based Systems 20(8), 703–718 (2007)
12. Ron, K.: Scaling Up the Accuracy of Naïve-Bayes Classifiers: a Decisoin-Tree Hybrid. In: Proceedings of KDD 1996, Portland, USA (1996)
13. Vapnik, V.N.: The Nature of Statistical Learning Theory. Springer, New York (1995)
14. Joachims, T.: Text categorization with support vector machines: Learning with many relevant features. In: Proceedings of the European Conference on Machine Learning. Springer, Heidelberg (1998)
15. Michel, P., Kaliouby, R.E.: Real time facial expression recognition in video using support vector machines. In: Proceedings of ICMI 2003, Vancouver, British Columbia, Canada, November 5-7, pp. 258–264 (2003)
16. Guyon, I., Weston, J., Barnhill, S., Vapnik, V.N.: Gene selection for cancer classification using support vector machines. Machine Learning 46(1-3), 389–422 (2002)
17. Gallant, S.: Connectionist expert systems. Communications of the ACM 31(2), 152–169 (1988)
18. Barakat, N.H., Diederich, J.: Eclectic rule-extraction from support vector machines. International journal of Computer Intelligence 2(1), 59–62 (2005)
19. Nunez, H., Angulo, C., Catata, A.: Rule extraction from support vector machines. In: European Symposium on Artificial Neural Networks Proceedings, pp. 107–112 (2002)

20. Fung, G., Sandilya, S., Bharat, R.R.: Rule extraction from linear support vector machines. In: Proceeding of the Eleventh ACM SIGKDD International Conference on Knowledge Discovery in Data Mining, pp. 32–40. ACM Press, New York (2005)
21. Barakat, N.H., Diederich, J.: Learning-based Rule-Extraction from Support Vector Machines. In: Proceedings of the 14th International Conference on Computer Theory and applications ICCTA 2004, Alexandria, Egypt (2004)
22. Zhang, Y., Su, H., Jia, T., Chu, J.: Rule Extraction from Trained Support Vector Machines. In: Ho, T.-B., Cheung, D., Liu, H. (eds.) PAKDD 2005. LNCS (LNAI), vol. 3518, pp. 61–70. Springer, Heidelberg (2005)
23. Barakat, N.H., Bradley, A.P.: Rule Extraction from Support Vector Machines: Measuring the Explanation Capability Using the Area under the ROC Curve. In: The 18th International Conference on Pattern Recognition (ICPR 2006), Hong Kong (2006)
24. Chen, Z., Li, J., Wei, L.: A multiple kernel support vector machine scheme for feature selection and rule extraction from gene expression data of cancer tissue. Artificial Intelligence in Medicine 41, 161–175 (2007)
25. Chaves, Ad.C.F., Vellasco, M.M.B.R., Tanscheit, R.: Fuzzy rule extraction from support vector machines. In: Fifth International Conference on Hybrid Intelligent Systems, Rio de Janeiro, Brazil, November 06-09 (2005)
26. Barakat, N.H., Bradley, A.P.: Rule Extraction from Support Vector Machines: A Sequential Covering Approach. IEEE Transactions on Knowledge and Data Engineering 19(6), 729–741 (2007)
27. Martens, D., Baesens, B., Gestel, T.V.: Decompositional Rule Extraction from Support Vector Machines by Active Learning. IEEE Transactions on Knowledge and Data Engineering 21(2), 178–191 (2009)
28. Farquad, M.A.H., Ravi, V., Bapi, R.S.: Support Vector Machine based Hybrid Classifiers and Rule Extraction Thereof: Application to Bankruptcy Prediction in Banks. In: Soria, E., Martín, J.D., Magdalena, R., Martínez, M., Serrano, A.J. (eds.) Handbook of Research on Machine Learning Applications and Trends: Algorithms, Methods and Techniques. IGI Global (2008)
29. Farquad, M.A.H., Ravi, V., Bapi, R.S.: Rule Extraction using Support Vector Machine Based Hybrid Classifier. In: Presented in TENCON-2008, IEEE region 10 Conference, Hyderabad, India, November 19-21 (2008)
30. Business Intelligence Cup-2004: Organized by the Univeristy of Chile, http://www.tis.cl/bicup_04/text-bicup/BICUP/202004/20public/20data.zip
31. Naveen, N., Ravi, V., Kumar, D.A.: Application of fuzzyARTMAP for churn prediction in bank credit cards. International Journal of Information and Decision Sciences 1(4), 428–444 (2009)
32. Chang, C.C., Lin, C.J.: LIBSVM: a library for support vector machines (2001) Software, http://www.csie.ntu.edu.tw/~cjlin/libsvm
33. Mierswa, I., Wurst, M., Klinkenberg, R., Scholz, M., Euler, T.: YALE: Rapid Prototyping for Complex Data Mining Tasks. In: Proceedings of the 12th ACM SIGKDD International Conference on Knowledge Discovery and Data Mining, KDD 2006 (2006)

Interval Set Cluster Analysis: A Re-formulation

Yiyu Yao[1], Pawan Lingras[2], Ruizhi Wang[3], and Duoqian Miao[3]

[1] Department of Computer Science, University of Regina
Regina, Saskatchewan, Canada S4S 0A2
yyao@cs.uregina.ca
[2] Department of Mathematics and Computing Science, Saint Mary's University
Halifax, Nova Scotia, Canada B3H 3C3
pawan@cs.smu.ca
[3] Department of Computer Science and Technology
The Key Laboratory of Embedded System and Service Computing
Tongji University, Shanghai, 201804, P.R. China

Abstract. A new clustering strategy is proposed based on interval sets, which is an alternative formulation different from the ones used in the existing studies. Instead of using a single set as the representation of a cluster, each cluster is represented by an interval set that is defined by a pair of sets called the lower and upper bounds. Elements in the lower bound are typical elements of the cluster and elements between the upper and lower bounds are fringe elements of the cluster. A cluster is therefore more realistically characterized by a set of core elements and a set of boundary elements. Two types of interval set clusterings are proposed, one is non-overlapping lower bound interval set clustering and the other is overlapping lower bound interval set clusterings, corresponding to the standard partition based and covering based clusterings.

1 Introduction

Cluster analysis focuses on grouping objects of similar kind into categories and organizing data into meaningful structures [1]. Objects are sorted into groups so that objects in the same group show a high degree of association and objects in different group show a low degree of association. A common assumption underlying many cluster analysis methods is that a cluster can be represented by a set with crisp boundary. The requirement of a sharp boundary leads to easy analytical results, but may be too restrictive for some practical applications. Several proposals have been made to remove such a stringent assumption.

In fuzzy cluster analysis, it is assumed that a cluster is represented by a fuzzy set that models a gradually changing boundary [3]. However, a fuzzy clustering provides a quantitative characterization of the unsharp cluster boundary at the expense of losing the qualitative characterization that better shows the structures provided by a clustering. To resolve this problem, Lingras and his associates [6,7,8,9] propose and systematically study rough clustering and interval set clustering. The basic idea is to derive and describe a cluster by a pair of

H. Sakai et al. (Eds.): RSFDGrC 2009, LNAI 5908, pp. 398–405, 2009.

lower and upper approximations. By describing a cluster in terms of a pair of
crisp sets, one recovers the qualitative characterization of a cluster.

There exists a semantic gap in the studies by Lingras and his associates. On
the one hand, rough clustering algorithms are explained in rough set terminology.
On the other hand, an equivalence relation that is needed for defining approximations is not explicitly referred to. The main objective of this paper is to fill
in such a semantic gap by representing a cluster as an interval set defined by
a pair of bounds. This leads to the introduction of interval set cluster analysis. Elements in the lower bound of an interval set are typical elements of the
cluster and elements between the upper and lower bounds are fringe elements of
the cluster. That is, a cluster is more realistically characterized by a set of core
elements and a set of fringe elements.

The strategy of interval set cluster analysis does not require an equivalence relation. A set of properties of an interval set clustering is proposed and examined.
Based on these properties, two types of interval set clusterings are proposed, one
is non-overlapping lower bound interval set clustering and the other is overlapping lower bound interval set clusterings. They correspond to the standard
partition based and covering based clusterings.

2   Overview of Interval Sets

In cluster analysis, a cluster may be interpreted as the extension of a concept,
that is, the set of objects that are instances of the concept. In some situations,
an object may actually be either an instance or not an instance of a concept.
On the other hand, due to a lack of information and knowledge, one can only
express the state of instance and non-instance for some objects, instead of all
objects. That is, one has a partially known concept defined by a lower bound
and upper bound of its extension. This leads to the interval set representation
of a partially known set [16].

Interval sets are defined and interpreted in a similar way that interval numbers are introduced in interval analysis [10]. The notion of interval sets represents a new kind of sets, defined by a pair of sets, namely, its lower and upper
bounds [13,16]. Mathematically, interval sets are defined as follows. Let U be a
finite set, called the universe or the reference set, and 2^U be its power set. A
subset of 2^U of the form,

$$\mathcal{A} = [A_l, A_u] = \{A \in 2^U \mid A_l \subseteq A \subseteq A_u\}, \tag{1}$$

is called a closed interval set, where it is assumed that $A_l \subseteq A_u$. Being an
interval of the power set lattice 2^U, an interval set $\mathcal{A}$ is also a lattice, with the
minimum element A_l, the maximum element A_u, and the standard set-theoretic
operations. The set of all closed interval sets is denoted by:

$$I(2^U) = \{[A_l, A_u] \mid A_l, A_u \subseteq U, A_l \subseteq A_u\}. \tag{2}$$

A degenerate interval set of the form $[A, A]$ is equivalent to the ordinary set A.

Semantically, an interval set, when interpreted as a family of sets of objects, provides an appropriate means to represent a partially known concept [5,12,13,16,19]. Although the extension of a concept is actually a subset of U, a lack of knowledge makes us unable to specify this subset. We can only provide a lower bound A_l and an upper bound A_u. Any subset A that lies between A_l and A_u, namely, $A_l \subseteq A \subseteq A_u$, can be the actual extension of the concept. The set,

$$\mathrm{BND}([A_l, A_u]) = A_u - A_l, \tag{3}$$

is called the boundary of the interval set $[A_l, A_u]$. For those elements, we are unable to tell if they are instances or non-instances of the concept.

Interval sets are subsets of the universe U. The symbols $\in, \subseteq, =, \cap, \cup$ may be used, in their usual set-theoretic sense, to represent relationships between elements of 2^U and an interval set, and between different interval sets. Thus, $A \in [A_l, A_u]$ means that A is a subset of U such that $A_l \subseteq A \subseteq A_u$. We write $[A_l, A_u] \subseteq [B_l, B_u]$ if the interval set $[A_l, A_u]$ as an ordinary set is contained in $[B_l, B_u]$ as an ordinary set. In other words, by $[A_l, A_u] \subseteq [B_l, B_u]$ we mean that $B_l \subseteq A_l \subseteq A_u \subseteq B_u$. Similarly, two interval sets are equal, written $\mathcal{A} = \mathcal{B}$, if they are equal in set-theoretic sense, that is $\mathcal{A} = \mathcal{B}$ if and only if $A_l = B_l$ and $A_u = B_u$.

Let $\cap, \cup$ and $-$ be the usual set intersection, union and difference defined on 2^U, respectively. Following the results of power algebras [2] and interval analysis [10], we can lift set operations into interval set operations. Specifically, for two interval sets $\mathcal{A} = [A_l, A_u]$ and $\mathcal{B} = [B_l, B_u]$ we have:

$$\mathcal{A} \sqcap \mathcal{B} = \{A \cap B \mid A \in \mathcal{A}, B \in \mathcal{B}\},$$
$$\mathcal{A} \sqcup \mathcal{B} = \{A \cup B \mid A \in \mathcal{A}, B \in \mathcal{B}\},$$
$$\mathcal{A} \setminus \mathcal{B} = \{A - B \mid A \in \mathcal{A}, B \in \mathcal{B}\}. \tag{4}$$

These operations are referred to as interval set intersection, union and difference. They are closed on $I(2^U)$, namely, $\mathcal{A} \sqcap \mathcal{B}$, $\mathcal{A} \sqcup \mathcal{B}$ and $\mathcal{A} \setminus \mathcal{B}$ are interval sets. They can be explicitly computed by using the following formulas [13,16]:

$$\mathcal{A} \sqcap \mathcal{B} = [A_l \cap B_l, A_u \cap B_u],$$
$$\mathcal{A} \sqcup \mathcal{B} = [A_l \cup B_l, A_u \cup B_u],$$
$$\mathcal{A} \setminus \mathcal{B} = [A_l - B_u, A_u - B_l]. \tag{5}$$

Interval set complement $\neg[A_l, A_u]$ of $[A_l, A_u]$ is defined as $[U, U] \setminus [A_l, A_u]$. It is equivalent to $[U - A_u, U - A_l] = [A_u^c, A_l^c]$, where $A^c = U - A$ denote the usual set complement operation. Clearly, we have $\neg[\emptyset, \emptyset] = [U, U]$ and $\neg[U, U] = [\emptyset, \emptyset]$.

3   Interval Sets, Fuzzy Sets and Rough Sets

Interval sets model concepts that are partially known; they are related to, but different from, fuzzy sets [18] and rough sets [11]. A brief comparison of the three notions will provide an argument supporting the proposed framework of interval set cluster analysis.

Fuzzy sets model concepts with gradual memberships [18]. Suppose $\mu_A : U \longrightarrow [0,1]$ is a fuzzy membership function. Given a number $\alpha \in [0,1]$, an α-cut of μ_A is defined by:

$$\mu_A^\alpha = \{x \in U \mid \mu_A(x) \geq \alpha\}. \tag{6}$$

For a pair of numbers $0 \leq \beta \leq \alpha \leq 1$, the pair of (α, β) cuts of μ_A gives rise to an interval set $[\mu_A^\alpha, \mu_A^\beta]$ with $\mu_A^\alpha \subseteq \mu_A^\beta$. Thus, an interval set may be used as a qualitative approximation of a fuzzy set [13].

Rough sets model the approximations of concepts under indiscernibility [11]. Suppose an equivalence relation on U is used to formally represent the indis cernibility of elements in U. The pair $apr = (U, E)$ is called an approximation space [11]. The equivalence relation E induces a partition of U, denoted by U/E. The equivalence class containing x is given by $[x] = \{y \in U \mid xEy\}$. The equivalence classes of E are the basic building blocks to construct rough set approximations. For a subset $A \subseteq U$, its lower and upper approximations are defined by [11]:

$$\begin{aligned} \underline{apr}(A) &= \{x \in U \mid [x] \subseteq A\}; \\ \overline{apr}(A) &= \{x \in U \mid [x] \cap A \neq \emptyset\}. \end{aligned} \tag{7}$$

The pair $(\underline{apr}(A), \overline{apr}(A))$ is referred to as a rough set generated by A. For a subset $A \subseteq U$, we have $\underline{apr}(A) \subseteq A \subseteq \overline{apr}(A)$. It follows that A induces an interval set $[\underline{apr}(A), \overline{apr}(A)]$. By applying the ideas of (α, β)-cuts of a fuzzy set, one can define probabilistic rough set approximations in a decision-theoretic rough set model [15,17].

Consider now the reverse process of constructing a fuzzy set or a rough set from an interval set. Given an interval set $[A_l, A_u]$, we can define a fuzzy set as follows:

$$\mu_A(x) = \begin{cases} 0, & x \in U - A_u, \\ 0.5, & x \in A_u - A_l, \\ 1, & x \in A_l. \end{cases} \tag{8}$$

If $\min, \max$ and $1 - ()$ are used to define fuzzy set intersection, union, and complement, respectively, we express interval set operations in terms of such three-valued fuzzy sets.

In the case of rough sets, given an interval set $[A_l, A_u]$, in general we may not be able to find a set A so that $A_l = \underline{apr}(A)$ and $A_u = \overline{apr}(A)$. Iwiński [4] suggests another formulation of rough sets, which is closely related to interval sets [14]. Let $\mathrm{Def}(U)$ denote the family of all definable subsets of U given by:

$$\mathrm{Def}(U) = \{A \subseteq U \mid A = \underline{apr}(A) = \overline{apr}(A)\}. \tag{9}$$

For a pair of sets $\underline{A}, \overline{A} \in \mathrm{Def}(U)$ with $\underline{A} \subseteq \overline{A}$, Iwiński refers to the pair $\langle \underline{A}, \overline{A} \rangle$ as a rough set. By definition, it corresponds to the interval set $[\underline{A}, \overline{A}]$. Conversely, for an interval set $[A_l, A_u]$ with $A_l, A_u \in \mathrm{Def}(U)$, we have an Iwiński rough set $\langle A_l, A_u \rangle$. Thus, the family of all Iwiński rough sets corresponds to a sub-family of all interval sets. Furthermore, their set-theoretic operations are the same [4,14].

Although an interval set may be induced from either a fuzzy set or a rough set, and the reverse is also true under certain conditions, it does have to be interpreted in this way. The interpretation of an interval set as the bounds of a partially known set makes it different from fuzzy sets and rough sets. This interpretation seems to be appropriate for the task of clustering. A cluster may be considered to be a partially known set; we know that certain elements must be in the cluster (e.g., elements in a small neighborhood), and certain elements may be in the cluster (e.g., elements in a large neighborhood).

4   Strategies of Interval Set Clustering

A main task of cluster analysis is to group objects in a universe so that objects in the same cluster are more similar to each other and objects in different clusters are dissimilar. There are two basic strategies of clustering that produce flat non-overlapping and overlapping clusters, respectively.

Suppose

$$\mathbf{C} = (C^1, C^2, \ldots, C^m) \tag{10}$$

is a family of clusters of U, that is, $\mathbf{C}$ is a clustering of the universe. Formally, a non-overlapping clustering is defined by the properties:

(i) $C^i \neq \emptyset, 0 \leq i \leq m,$

(ii) $\bigcup_{C^i \in \mathbf{C}} C^i = U,$

(iii) $C^i \cap C^j = \emptyset, i \neq j.$

Property (i) requires that each cluster cannot be empty. Property (ii) states that every $x \in U$ belongs to at least one cluster, and property (iii) states that x belongs to at most one cluster. Together they require that every $x \in U$ belongs to exactly one cluster. In this case, $\mathbf{C}$ is a partition of the universe. On the other hand, an overlapping clustering only requires properties (i) and (ii). For overlapping clustering, it is possible that an element belongs to more than one cluster. The family $\mathbf{C}$ is only a covering of the universe.

An underlying assumption of such a clustering is that one can precisely form a family of clusters with well defined boundary. A questioning of this assumption has led to other clustering strategies. For example, fuzzy clustering produces a family of fuzzy sets, where each cluster is a fuzzy set with gradually changing boundary. Given a pair of numbers $0 \leq \beta \leq \alpha \leq 1$, the (α, β)-cuts of a fuzzy set can be viewed as an interval set [16]. This immediately motivates the introduction of interval set clustering, although in general an interval set clustering can be interpreted without direct reference to a fuzzy clustering.

We assume that each cluster C^i is partially known based on the available information. One may use an interval set to represent such a partially known cluster, namely, C^i is represented by an interval set $[C^i_l, C^i_u]$ satisfying the constraint:

$$C^i_l \subseteq C^i \subseteq C^i_u. \tag{11}$$

The constraint reflects the fact that we do not know the exact cluster C^i but a pair of lower and upper bounds within which C^i lies. Any set in the family $[C^i_l, C^i_u] = \{X \mid C^i_l \subseteq X \subseteq C^i_u\}$ may be the actual cluster C^i. The elements in C^i_l may be interpreted as typical elements of the cluster C^i and elements in $C^i_u - C^i_l$ as fringe elements. With respect to the family of clusters $\mathbf{C} = (C^1, C^2, \ldots, C^m)$, we have the following family of interval set clusters:

$$\mathbf{IC} = ([C^1_l, C^1_u], [C^2_l, C^2_u], \ldots, [C^m_l, C^m_u])$$
$$= \{(C^1, C^2, \ldots, C^m) \mid C^i_l \subseteq C^i \subseteq C^i_u, 1 \le i \le m\}.$$

That is, an interval set cluster is interpreted as a pair of bounds of a family of possible crisp clusters and an interval set clustering is interpreted as bounds of a family of crisp set clusterings.

Based on interval set operations, corresponding to properties (i)-(iii), we adopt the following properties for an interval set clustering:

(I) $C^i_l \ne \emptyset, 0 \le i \le m$,

(II) $\bigcup_{[C_l, C_u] \in \mathbf{IC}} C_u = U$,

(III) $C^i_l \cap C^j_l = \emptyset, i \ne j$.

Property (I) requires that the lower bound must not be empty. It implies that the upper bound is not empty, namely, $C^i_u \ne \emptyset$. Thus, $C^i_l \ne \emptyset$ may be viewed as a strong version and $C^i_u \ne \emptyset$ as a weak version. It is reasonable to assume that each cluster must contain at least one typical element and hence its lower bound is not empty. We therefore adopt the strong version, instead of the weak version, in order to make sure that an interval set clustering is physically meaningful. Property (II) states that any element of U belongs to the upper bound of a cluster, which ensures that every element is properly clustered. Property (III) demands that the lower bounds of clusters are pairwise disjoint; a typical element of one cluster cannot, as the same time, be a typical element of another cluster.

Additional support for adopting properties (I), (II), and (III) is given by the following theorem that shows the connection of a standard clustering and an interval set clustering.

Theorem 1. *Suppose* $\mathbf{IC} = ([C^1_l, C^1_u], [C^2_l, C^2_u], \ldots, [C^m_l, C^m_u])$ *is an interval set clustering. If* $\mathbf{IC}$ *satisfies properties* (I), (II), *and* (III), *then there exists a family of clusters* $\mathbf{C} = (C^1, C^2, \ldots, C^m)$ *that satisfies the constraint* $C^i_l \subseteq C^i \subseteq C^i_u$ *and properties* (i), (ii), *and* (iii). *If* $\mathbf{IC}$ *satisfies properties* (I) *and* (II), *there exists a family of clusters* $\mathbf{C} = (C^1, C^2, \ldots, C^m)$ *that satisfies the constraint* $C^i_l \subseteq C^i \subseteq C^i_u$ *and properties* (i) *and* (ii).

Proof. The theorem can be proved constructively by building a family of clusters **C** from **IC**. Assume that **IC** satisfies properties (I), (II), and (III), one can construct a **C** as follows. We first construct a family of clusters $\{C^i = C^i_l \mid 1 \leq i \leq m\}$ based on typical elements of clusters. For each element x in the set of the fringe elements, $F = \bigcup\{C^i_u - C^i_l \mid 1 \leq i \leq m\}$, we assign it to only one of the clusters C^i's that satisfies the condition $x \in C^i_u - C^i_l$. By the property (I), it follows that **C** satisfies property (i); by the properties (II) and (III) and the construction procedure, it follows that **C** satisfies properties (ii) and (iii). To prove the second part of the theorem, we follow the same procedure except that we may assign each fringe element to a set of clusters instead of one. It can be easily seen that the resulting **C** satisfies properties (i) and (iii).

Based on the results from the theorem, an interval set clustering **IC** is called a lower bounds non-overlapping interval set clustering if it satisfies properties (I), (II), and (III); it is called a lower bounds overlapping interval set clustering if it only satisfies properties (I) and (II). They suggest different interval set clustering algorithms.

There are several differences between rough set clustering and interval set clusterings. Rough set clustering requires an underlying equivalence and hence is only applicable to non-overlapping clustering. In general, one may use a non-equivalence relation to obtain an overlapping clustering. In this case, it is necessary to refer to this underlying relation in order to properly interpret the rough set lower and upper approximations. In contrast, interval set clustering does not require such an underlying relation. In some earlier studies of rough set cluster analysis, it is assumed that a fringe element must belong to the upper bounds of at least two clusters [6,7,8,9], which is motivated by properties of the upper approximations in the rough set theory. With interval set clustering, we no longer need to impose such a constraint. It is possible that a fringe element belongs to the upper bound of only one cluster.

5   Conclusion

There is a growing interest in rough set cluster analysis. An important issue that has not received enough attention is a semantic interpretation of the derived clusters. Since rough set approximations must satisfy certain properties, their directly application to cluster analysis may be unnecessarily restrictive. In this paper, we outline a framework of interval set cluster analysis, which is motivated by, and different from, rough set cluster analysis.

The clarification of rough set cluster analysis and interval set cluster analysis have both theoretical and practical values. Although the results from both clustering methods are intervals in the power set of a set, they have different semantic interpretations. Rough set approximations are approximation of known sets in an approximation space defined by an underlying equivalence or non-equivalence relation. In order to explain rough set cluster analysis, we need to refer to the relation. In contrast, interval sets are approximations of partially known sets; interval set cluster analysis does not require such a relation. With interval set clustering, an object can belong to the upper bound of one cluster, which is different from rough set clustering where an object, if in the upper approximation of one cluster, must be in the upper approximation of at least one more cluster.

References

1. Anderberg, M.R.: Cluster Analysis for Applications. Academic Press, New York (1973)
2. Brink, C.: Power structures. Algebra Universalis 30, 177–216 (1993)
3. Höppner, F., Klawonn, F., Kruse, R., Runkler, T.: Fuzzy Cluster Analysis: Methods for Classification, Data Analysis and Image Recognition. Wiley, Chichester (1999)
4. Iwiński, T.B.: Algebraic approach to rough sets. Bulletin of the Polish Academy of Sciences, Mathematics 35, 673–683 (1987)
5. Marek, V.W., Truszczyński, M.: Contributions to the theory of rough sets. Fundamenta Informaticae 39, 389–409 (1999)
6. Lingras, P.: Rough K-Medoids clustering using GAs. In: Proceedings of the 8th IEEE International Conference on Cognitive Informatics, pp. 315–319 (2009)
7. Lingras, P., Hogo, M., Snorek, M.: Interval set clustering of web users using modified Kohonen self-organizing maps based on the properties of rough sets. Web Intelligence and Agent Systems: An International Journal 2, 217–230 (2004)
8. Lingras, P., Hogo, M., Snorek, M., West, C.: Temporal analysis of clusters of supermarket customers: conventional versus interval set approach. Information Sciences 172, 215–240 (2005)
9. Lingras, P., West, C.: Interval set clustering of web users with rough K-Means. Journal of Intelligent Information Systems 23, 5–16 (2004)
10. Moore, R.E.: Interval Analysis. Prentice-Hall, Englewood Cliffs (1966)
11. Pawlak, Z.: Rough sets. International Journal of Computer and Information Sciences 11, 341–356 (1982)
12. Wang, Y.Q., Zhang, X.H.: Some implication operators on interval sets and rough sets. In: Proceedings of 2009 IEEE International Conference on Cognitive Informatics, pp. 328–332 (2009)
13. Yao, Y.Y.: Interval-set algebra for qualitative knowledge representation. In: Proceedings of the Fifth International Conference on Computing and Information, pp. 370–374 (1993)
14. Yao, Y.Y.: Two views of the theory of rough sets in finite universes. International Journal of Approximation Reasoning 15, 291–317 (1996)
15. Yao, Y.Y.: Probabilistic rough set approximations. International Journal of Approximation Reasoning 49, 255–271 (2008)
16. Yao, Y.Y.: Interval sets and interval-set algebras. In: Proceedings of the 8th IEEE International Conference on Cognitive Informatics, pp. 307–314 (2009)
17. Yao, Y.Y., Wong, S.K.M.: A decision theoretic framework for approximating concepts. International Journal of Man-machine Studies 37, 793–809 (1992)
18. Zadeh, L.A.: Fuzzy sets. Information and Control 8, 338–353 (1965)
19. Zhang, X.H., Jia, X.Y.: Lattice-valued interval sets and t-representable interval set t-norms. In: Proceedings of 2009 IEEE International Conference on Cognitive Informatics, pp. 333–337 (2009)

Rough Entropy Based k-Means Clustering

Dariusz Małyszko and Jarosław Stepaniuk

Department of Computer Science
Bialystok University of Technology
Wiejska 45A, 15-351 Bialystok, Poland
{malyszko,jstepan}@wi.pb.edu.pl

Abstract. Data clustering algorithmic schemes receive much careful research insight due to the prominent role that clustering plays during data analysis. Proper data clustering reveals data structure and makes possible further data processing and analysis. In the application area, k-means clustering algorithms are most often exploited in almost all important branches of data processing and data exploration. During last decades, a great deal of new algorithmic techniques have been invented and implemented that extend basic k-means clustering methods. In this context, fuzzy and rough k-means clustering presents robust modifications of basic k-means clustering that are aimed at better apprehension of data structure that advantageously incorporate notions from fuzzy and rough set theories. In the paper, an extension of rough k-means clustering into rough entropy domain has been introduced. Experimental results suggest that proposed algorithm outperforms standard k-means clustering methods applied in the area of image segmentation.

Keywords: Data clustering, image clustering, k-means clustering, rough k-means clustering, rough entropy k-means clustering.

1 Introduction

In image analysis the problem of robust, proper and optimal data grouping in the form of image data segmentation, presents the most important stage upon that quality of image analysis systems is primarily dependent.

In image segmentation routines, data clustering with k-means algorithm has been widely used in almost great majority of applications. Image segmentation presents process of partitioning image data into disjoint regions that exhibit within group similarity according to some predefined criteria and exhibit between group dissimilarity. Image segmentation presents difficult combinatorial problem and exact optimal segmentations are of high computational cost that is not practically attainable in real applications. In this context, image segmentation is based on some predefined criteria that make them feasible. High demand on robust image segmentation routines springs from the development of new emerging technologies, their accessibility, their higher complexity and dimensionality.

H. Sakai et al. (Eds.): RSFDGrC 2009, LNAI 5908, pp. 406–413, 2009.

Fuzzy set theory assumes that data objects do not belong to one group, concept or other notion but that they may participate in certain number of groups or concepts. Rough set theory represents a new paradigm in dealing with uncertainty, vagueness, and incompleteness. The spectrum of rough set based practical applications during last decade has extended into rough–fuzzy rule extraction, reasoning with uncertainty, rough–fuzzy modeling, image analysis. In [2] authors have introduced a new clustering method, called rough k-means, which prototypes each cluster by center and a pair of lower and upper approximations. The lower and upper approximations are weighted different parameters to compute the new centers. In [1] authors propose extended version of rough k-means algorithm not requiring prior specification of the number of clusters.

In the paper, extension of the rough k-means algorithm has been proposed that incorporates the notion of rough entropy of the cluster lower and upper approximations. Rough entropy framework in image segmentation has been primarily introduced in [6] in the domain of image thresholding routines. Authors proposed rough entropy measure for image thresholding into two objects: foreground and background object. This type of thresholding has been extended into multilevel thresholding for one-dimensional and two-dimensional domain in [4]. Further, rough entropy thresholding has been employed in image data clustering setting in [5].

The paper material consists of outlined in Section 2 rough k-means clustering algorithmic approaches. In Section 3 rough entropy k-means clustering is investigated. Experimental setup and results have been presented in Section 4. Algorithm summarization and concluding remarks are given in Section 5.

2   Rough k-Means Clustering Framework

2.1   Rough Set Theory

An information system is a pair (U, A) where U represents a non-empty finite set called the universe and A a non-empty finite set of attributes. Let $B \subseteq A$ and $X \subseteq U$ and AS_B is an approximation space (see e.g [7], [9]). Taking into account these two sets, it is possible to approximate the set X making only the use of the information contained in B by the process of construction of the lower and upper approximations of X and further to express numerically the roughness $R(AS_B, X)$ of a set X with respect to B by assignment

$$R(AS_B, X) = 1 - \frac{Card(LOW(AS_B, X))}{Card(UPP(AS_B, X))}. \tag{1}$$

In this way, the value of the roughness of the set X equal 0 means that X is crisp with respect to B, and conversely if $R(AS_B, X) > 0$ then X is rough (i.e., X is vague with respect to B). Detailed information on rough set theory is provided in [7], [8], [9].

Shannon entropy notion describes uncertainty of the system and is defined as follows $E(p_1, \ldots, p_k) = \sum_{l=1}^{k} -p_l \cdot \log(p_l)$ where p_l represents probability of the state l and $l = 1, \ldots, k$. In this context, combined rough entropy formula is given as

$$RE(AS_B, \{X_1, \ldots, X_k\}) = \sum_{l=1}^{k} -\frac{e}{2} \cdot R(AS_B, X_l) \cdot \log(R(AS_B, X_l)) \quad (2)$$

where $R(AS_B, X_l)$ represents roughness of the cluster X_l, $l \in \{1, \ldots, k\}$ indexes the set of all clusters ($\bigcup_{l=1}^{k} X_l = U$ and for any $p \neq l$ and $p, l \in \{1, \ldots, k\}$ $X_p \cap X_l = \emptyset$). Fuzzy membership value $\mu_{C_l}(x_i) \in [0, 1]$ for the data point $x_i \in U$ in cluster X_l (equivalently C_l) is given as

$$\mu_{X_l}(x_i) = \frac{d(x_i, X_l)^{-2/(\mu-1)}}{\sum_{j=1}^{k} d(x_i, X_j)^{-2/(\mu-1)}} \quad (3)$$

where a real number μ represents fuzzifier value that should be greater than 1.0 and $d(x_i, X_l)$ denotes distance between data object x_i and cluster (center) X_l.

2.2   Rough k-Means Clustering

In rough k-means algorithm proposed in [2], the data objects assigned to the given cluster are divided into two sets, lower and upper approximation. Lower and upper approximation have assigned weights that determine the importance of the approximations. If the upper bound of each cluster is equal to its lower bound then the cluster is standard conventional crisp cluster. After cluster approximations have been determined for all clusters, approximation weights are assigned and new cluster centers are calculated. The procedure is iteratively repeated until predefined criteria are met.

2.3   Adaptive Rough k-Means Clustering

In [10] authors proposed the method of adaptive parameter selection during algorithm run. Adaptive parameter selection is based on the assumption that during iterative data partitioning and subsequent assigning to clusters, initially upper approximations contribute more to cluster formation as data objects are not strongly uniform. During subsequent cluster center recalculation, the weight of the lower approximation increases with simultaneous decreasing the importance - weight for upper approximations. The author proposed the formulae for the lower and upper approximation weights.

2.4   Rough Fuzzy k-Means Clustering

In [3] authors proposed extension of the rough k-means algorithm into rough fuzzy k-means domain that considers lower and upper approximations as fuzzy sets as opposed to rough approximations in rough k-means clustering. Authors have incorporated both fuzzy and rough sets in k-means clustering algorithmic setting referred to as rough - fuzzy k-means (or rough - fuzzy k-means RFCM).

In this solution the concept of fuzzy membership of fuzzy sets, and lower and upper approximations of rough sets has been extended into k-means algorithm. The notion of the membership of fuzzy sets present the tool in efficient handling of overlapping partitions, and the rough sets are aimed at interpretation of uncertainty, vagueness, and incompleteness in class definition. The proposed rough-fuzzy k-means clustering method partitions a set of n objects into k clusters by minimizing the objective function.

3   Rough Entropy k-Means Clustering

3.1   Rough Measures

Crisp measure, crisp threshold, difference metric. Standard rough entropy calculation as proposed in [5] incorporates determination of lower and upper approximations for the given cluster centers and considering these two set cardinalities during calculation of roughness and further rough entropy clustering measure. Rough measure general calculation routine has been given in Algorithm 1. In all presented algorithms, before calculations, cardinalities of the lower and upper cluster approximations should be set to zero. For each data point x_i, distance to the closest cluster C_l is denoted as $d_{dist}^{min} = d(x_i, C_l)$ and approximations are increased by value 1 of clusters C_m that satisfy the condition:

$$|d(x_i, C_m) - d(x_i, C_l)| \leq \epsilon_{dist}.$$

3.2   Rough Entropy k-Means Clustering

Proposed clustering algorithm takes as input data objects and number of clusters and k. After creation of initial data clustering, predefined number of weight pairs is created, for each weigh pair a new offspring clustering C_i is determined. For each clustering C_i, rough entropy is calculated. From offspring clusterings a partition is selected with the highest rough entropy measure for further algorithm iterations. The procedure are repeated predefined number of iterations or until other termination criteria are met. Proposed solution does take into account lower and upper approximations during recalculation of cluster centers as opposed to standard rough k-means clustering methods that consider lower approximations and boundary regions. The formulae for calculation of new cluster centers based on lower and upper cluster approximations is given as

$$v_l = W_l \times \mathcal{L}_l + W_u \times \mathcal{U}_l$$

where v_l denotes new center for cluster C_l, W_l, W_u - weights for lower and upper approximations, lower approximation and upper approximation centers are denoted as

$$\mathcal{L}_l = \frac{1}{Card(LOW(C_l))} \sum_{x_j \in LOW(C_l)} x_j, \quad \mathcal{U}_l = \frac{1}{Card(UPP(C_l))} \sum_{x_j \in UPP(C_l)} x_j$$

Algorithm 1. Crisp - Crisp Difference Rough Entropy

---

foreach *Data object* x_i **do**
    Determine the closest cluster C_l **for** x_i
    Increment LOW(C_l) and UPP(C_l) by 1.0
    foreach *Cluster* $C_m \neq C_l$ *with* $|d(x_i, C_m) - d(x_i, C_l)| \leq \epsilon_{dist}$ **do**
        | **Increment UPP(C_m) by 1.0**
    end

for $l = 1$ **to** k *(number of data clusters)* **do**
    | $roughness(C_l) = 1 - LOW(C_l)/UPP(C_l)$
Fuzzy_RE = 0
for $l = 1$ **to** *number of data clusters* **do**
    | $Fuzzy\_RE = Fuzzy\_RE - \frac{e}{2} \cdot roughness(C_l) \cdot log(roughness(C_l))$

Algorithm 2. Rough Entropy k-means Clustering Algorithm

Assign initial cluster centers $v_i, i = 1, 2, \ldots, k$.
for $l = 1$ **to** I *(number of ierations)* **do**
    For each cluster determine cluster lower and upper
    approximations.
    Create weights W_i **for lower and upper approximations e.g.**
    $(0, 1), (0.1, 0.9), (0.2, 0.8), \ldots, (1, 0)$.
    for $l = 1$ **to** *number of weight pairs* **do**
        | **Recalculate cluster centers Determine lower and upper**
        | **approximations Calculate and remember rough entropy**
        | **measure**
    end
    Select clustering with the highest rough entropy measure by
    taking its cluster centers as current solution $v_i, i = 1, 2, \ldots, k$.
end

Time complexity of rough entropy k-means clustering is comparable to other rough k-means clustering algorithms.

4   Experimental Setup and Results

4.1   Image Datasets

In this Section experimental setup and experimental results have been presented. In the experiments, three color images from Berkeley image dataset have been chosen. The images identifiers are 27059, 86000 and 78004. In the paper images have been segmented in $R - B$, $R - G$, $G - B$ and $R - G - B$ bands. Additionally, ground truth images are appended in the Berkeley database that have been considered during computation of maximized Vinet index in order to assess segmentation quality. The three examined images are presented in Figure 1.

Fig. 1. Berkeley dataset images: (a) 27059 image, (b) 78004 image, (c) 86000 image

4.2   Segmentation Validity Indices

Quantitative Measure: β-index. Measure in the form of β-index denotes the ratio of the total variation and within-class variation. Define n_i as the number of pixels in the i-th ($i = 1, 2, \ldots, k$) region from segmented image. Define X_{ij} as the gray value of j-th pixel ($j = 1, \ldots, n_i$) in the region i and $\overline{X_i}$ the mean of n_i values of the i-th region. The β-index is defined in the following way

$$\beta = \frac{\sum_{i=1}^{k} \sum_{j=1}^{n_i} \left(X_{ij} - \overline{X}\right)^2}{\sum_{i=1}^{k} \sum_{j=1}^{n_i} \left(X_{ij} - \overline{X_i}\right)^2} \tag{4}$$

where n is the size of the image and $\overline{X}$ represents the mean value of the image pixel attributes. This index defines the ratio of the total variation and the within-class variation. In this context, important notice is the fact that index-b value increases as the increase of k number. The value of β-index should be maximized.

Vinet index. The Vinet index compares two different segmentations on the basis of the Hamming distance between non-maximally intersecting regions. Given S_1 and S_2 considered to be two segmentations of the same image, and $S_1 = C_1^1, \ldots, C_m^1$ and $S_2 = C_1^2, \ldots, C_n^2$ where C_i^j corresponds to the set of pixels in region i from segmentation S_j - $j = 1, 2$. Each region C_i^1 generates a region C_k^2 such that the area $|C_i^1 \mathbin{/} C_k^2|$ is maximal. The Vinet index between two segmentations has been defined as

$$V_s(S_1 => S_2) = \sum_{C_i^1 \in S_1} \sum_{C_j^2 \neq C_k^2} |C_i^1 / C_j^2|$$

which denotes the sum of areas of intersection for all non-maximally intersecting regions. The measure is in the range $[0, 1]$, where N denotes the number of data objects. In this context, index values that are closer to one denote a better segmentation compared to segmentations with lower index values.

4.3   Experimental Results

In the first experiment rough entropy k-means algorithm - RKM has been compared to three standard k-means algorithms: hard k-means clustering - KM, fuzzy k-means clustering - FKM and possibilistic k-means clustering - PKM. All algorithms have been performed independently for 100 runs. Average from all runs have been presented in Table 1 for three selected datasets in two and three bands.

Table 1. Rough entropy clusterings against standard k-means clusterings for three selected datasets image measured by β-index values - averaged values over 100 independent runs

Bands	27059				78004				86000			
	KM	FKM	PKM	RKM	KM	FKM	PKM	RKM	KM	FKM	PKM	RKM
RB	13.22	13.37	13.37	13.73	36.04	35.95	35.94	35.73	9.23	9.14	9.22	9.27
BG	13.50	13.35	13.38	13.71	45.65	45.61	45.61	43.48	16.15	15.87	15.97	15.91
RG	25.82	25.77	25.78	24.20	48.83	48.76	48.74	48.42	9.37	9.51	9.55	9.78
RGB	13.47	13.77	13.79	13.87	40.57	40.53	40.52	39.44	8.76	8.46	8.45	8.78

In the second experiment rough entropy k-means algorithm - RKM has been examined and compared to three standard k-means algorithms: hard k-means clustering - KM, fuzzy k-means clustering - FKM and possibilistic k-means clustering - PKM. All algorithms have been performed independently for 100 runs. Average Vinet index values calculated on the basis of Berkeley database ground truth segmentations for all runs have been presented in Table 2 for three selected datasets in two and three bands.

Table 2. Rough entropy clusterings against standard k-means clusterings for three selected images compared on the basis of ground truth Vinet index - averaged values over 100 independent runs

Bands	27059				78004				86000			
	KM	FKM	PKM	RKM	KM	FKM	PKM	RKM	KM	FKM	PKM	RKM
RB	0.874	0.906	0.909	0.893	0.970	0.967	0.967	0.971	0.980	0.979	0.980	0.982
BG	0.889	0.922	0.923	0.914	0.971	0.971	0.971	0.974	0.984	0.984	0.984	0.895
RG	0.921	0.919	0.919	0.925	0.969	0.967	0.966	0.970	0.982	0.982	0.982	0.983
RGB	0.892	0.920	0.924	0.898	0.969	0.967	0.967	0.972	0.986	0.986	0.896	0.986

5   Future Research

In the paper, a new algorithmic clustering method based on rough entropy k-means algorithm has been presented. The algorithm has been precisely defined and compared relative to standard k-means clustering algorithms on the basis of

β-index values as segmentation quality and independently on the basis of ground truth segmentations from Berkeley image database. Experimental results suggest that proposed algorithm yields high quality segmentations and outperforms existing k-means methods giving possible area for future applications in real image segmentation systems.

Acknowledgments

The research is supported by the grants N N516 0692 35 and N N516 3774 36 from the Ministry of Science and Higher Education of the Republic of Poland and the Rector's grant of Bialystok University of Technology - W\WI\3\09. Computational experiments were performed on a cluster built by the Faculty of Computer Science, Bialystok University of Technology.

References

1. Asharaf, S., Murty, M.N.: A Rough Fuzzy Approach to Web Usage Categorization. Fuzzy Sets and Systems 148, 119–129 (2004)
2. Lingras, P., West, C.: Interval Set Clustering of Web Users with Rough k-Means. Journal of Intelligent Information Systems 23(1), 5–16 (2004)
3. Maji, P., Pal, S.K.: RFCM: A Hybrid Clustering Algorithm Using Rough and Fuzzy Sets. Fundamenta Informaticae 80(4), 477–498 (2007)
4. Malyszko, D., Stepaniuk, J.: Granular Multilevel Rough Entropy Thresholding in 2D Domain. In: IIS 2008, 16th International Conference Intelligent Information Systems, Zakopane, Poland, June 16-18, pp. 151–160 (2008)
5. Malyszko, D., Stepaniuk, J.: Standard and Fuzzy Rough Entropy Clustering Algorithms in Image Segmentation. In: Chan, C.-C., Grzymala-Busse, J.W., Ziarko, W.P. (eds.) RSCTC 2008. LNCS (LNAI), vol. 5306, pp. 409–418. Springer, Heidelberg (2008)
6. Pal, S.K., Shankar, B.U., Mitra, P.: Granular computing, rough entropy and object extraction. Pattern Recognition Letters 26(16), 2509–2517 (2005)
7. Pawlak, Z., Skowron, A.: Rudiments of rough sets. Information Sciences 177(1), 3–27 (2007); Rough sets: Some extensions. Information Sciences 177(1), 28–40 (2007); Rough sets and Boolean reasoning. Information Sciences 177(1), 41–73 (2007)
8. Pedrycz, W., Skowron, A., Kreinovich, V. (eds.): Handbook of Granular Computing. John Wiley & Sons, New York (2008)
9. Stepaniuk, J.: Rough–Granular Computing in Knowledge Discovery and Data Mining. Springer, Heidelberg (2008)
10. Zhou, T., Zhang, Y.N., Lu, H.L.: Rough k-means Cluster with Adaptive Parameters. In: Proceedings of the Sixth International Conference on Machine Learning and Cybernetics, Hong Kong, August 19-22, pp. 3063–3068 (2007)

Fast Single-Link Clustering Method Based on Tolerance Rough Set Model

Bidyut Kr. Patra and Sukumar Nandi

Department of Computer Science & Engineering,
Indian Institute of Technology Guwahati, Assam 781039, India
{bidyut,sukumar}@iitg.ernet.in

Abstract. The single-link (SL) clustering method is not scalable with the size of the dataset and needs many database scans. This is potentially a severe problem for large datasets. One way to speed up the SL method is to summarize the data efficiently and subsequently apply the SL method to the summary of the data. In this paper, we propose a summarization scheme based on a tolerance rough set theory called *data-sphere (DS)*. The SL method is modified to work with *data spheres*. The proposed clustering method takes considerably less time compared to the classical single-link method which is applied to the dataset directly. The clustering results produced by the proposed method is very close to that of the SL method. We also show that proposed summarization scheme outperforms recently introduced *data bubbles (DB)* as a summarization scheme when single-link is applied to it at clustering quality. Experiments are conducted with two synthetic and two real world datasets to show effectiveness of the proposed method.

1 Introduction

Clustering problem appears in many different fields like Data Mining, Pattern Recognition, Bio-informatics, *etc*. The Clustering problem can be defined as follows. Let $\mathcal{D} = \{x_1, x_2, x_3, \ldots, x_n\}$ be the set of n patterns, where each x_i is a N-dimensional vector in the given feature space. The clustering activity is to find groups of patterns, called clusters of data in such a way that patterns in a cluster are similar to each other than patterns in distinct clusters. The clustering methods are mainly divided into two categories *viz.*, partitional and hierarchical method [1].

Partitional clustering methods create a single partition of the dataset optimizing a criterion function. Hierarchical clustering methods create a sequence of nested partitions of the dataset. The hierarchical clustering methods (*eg:* OPTICS [2], Single-link (SL) [3], Complete-link) do not scale well with the size of the dataset and scan the dataset several times. One remedy to these problems is to create a summary of the data first which subsequently is used to find the clusters present in the dataset.

Rough set theory [4] has been extensively used in many applications in recent years [5]. Tolerance rough set model is a generalization of the rough set theory. In this paper, we propose a summarization scheme which is based on the tolerance rough set model (TRSM). The proposed summarization scheme is called *data-sphere (DS)* . This scheme uses the *leaders clustering* method to collect the statistics of each data sphere. Subsequently, these data spheres are used with the single-link clustering method to derive the

H. Sakai et al. (Eds.): RSFDGrC 2009, LNAI 5908, pp. 414–422, 2009.

clusters of data. The proposed clustering method is called *the data sphere single-link (DS-SL)* method. Various empirical evidences are shown to establish the effectiveness of the DS-SL method over the classical single-link method. We also show that our proposed clustering method outperforms in clustering results compared to single-link which uses a recently proposed *data bubble (DB)* [6] as a summarization scheme. The proposed method produces consistent clustering unlike DB based single-link method. The proposed method is especially a suitable one to work with large datasets.

The rest of the paper is organized as follows. Section 2 discusses some background of the proposed clustering method. Section 3 describes a summary of related research works. Section 4 describes the proposed summarization scheme (called the DS scheme) and the proposed clustering method (called the DS-SL method). Experimental evaluations and conclusion are discussed in Section 5 and Section 6, respectively.

2 Background of the Proposed Method

In this section we discuss briefly the rough set theory and tolerance rough set model (TRSM) for clustering methods. Leaders clustering and single-link clustering method are discussed in this section. We propose our clustering method exploiting these two clustering methods and TRSM model.

The fundamental idea of rough set theory is based on an approximation space $A = (U, R)$, where U is a nonempty set of objects and R is an equivalence relation called indiscernibility relation on U [4]. R creates a partition U/R of U, *i.e.* $U/R = \{X_1, \ldots, X_i, \ldots, X_p\}$ where each X_i is an equivalence class of R. These equivalence classes and the empty set are considered as the elementary sets in A. These elementary sets form the basic granules of knowledge. Any arbitrary set $X \subseteq U$ can be defined by two crisp sets called lower and upper approximation of X. More formally, one can define the lower and upper approximation as follow. $\underline{R}(X) = \bigcup_{X_i \subseteq X} X_i$; $\overline{R}(X) = \bigcup_{X_i \cap X \neq \emptyset} X_i$. The set $BND(X) = \overline{R}(X) - \underline{R}(X)$ is called *boundary* of X in A. Sets $\underline{Edg}(X) = X - \underline{R}(X)$ and $\overline{Edg}(X) = \overline{R}(X) - X$ are called *internal* and *external* edge of X in A, respectively.

It is found that the transitive property does not hold in certain application domains (*e.g.* document clustering). In that case, a tolerance relation (reflexive, symmetric) is used [7,8]. In tolerance relation based rough set model (TRSM), the basic granules of knowledge are the tolerance classes, which are intermingled. Therefore, the tolerance relation T does not create a partition of U. A set $X \subseteq U$, can be characterized by the lower and upper approximations as follow.

$$\underline{T}(X) = \{x \in X : T(x) \subseteq X\}; \quad \overline{T}(X) = \{x \in U : T(x) \cap X \neq \emptyset\}$$

where $T(x)$ is a tolerance class. In accordance with the classical rough set theory, we can define the set $T_{BND} = \overline{T}(X) - \underline{T}(X)$ as *tolerant boundary*. The sets $T_{Edg} = X - \underline{T}(X)$ and $T_{\overline{Edg}} = \overline{T}(X) - X$ are termed as *tolerant internal* and *tolerant external* edge of X, respectively.

The leaders clustering [9] is a single data-scan partitional clustering method. For a given threshold distance τ, leaders method produces a set of leaders $\mathcal{L} = \{l_1, l_2, \ldots, l_m\}$, where $l_i \in \mathcal{D}$, incrementally. Each leader can be seen as a representative for the cluster of patterns which are grouped with it. The time and space complexity of this method are $O(mn), O(m)$, respectively, where $m = |\mathcal{L}|, n = |\mathcal{D}|$. However, it can find only convex type clusters. The single-link (SL) [3] is an agglomerative hierarchical clustering method. In single link, distance between two clusters C_1 and C_2 is the minimum of distances between all pairs in $C_1 \times C_2$. The time and space complexity of the SL method are $O(n^2)$. The SL method scans the dataset many times. These are the serious drawback while working with the large dataset.

3   Related Work

Different data summarization schemes have been evolved to compress the large data and apply the existing clustering methods only to the compressed data. One of the widely used compression schemes is to use the CF tree constructed by the BIRCH [10] clustering method. However, it is suitable only for the k-means type of clustering methods. For hierarchical clustering methods there exist a very few data compression schemes which can speed them up. Breunig et al [11] proposed a data summarization scheme called Data Bubble (DB) to speed up the hierarchical clustering (OPTICS) method. Subsequently, Zhou and Sander [6] introduced "directional" notion to the data bubble in order to measure distance between data bubbles more accurately and to handle the 'gaps' in a data bubble. In this approach, patterns of a data bubble are divided in the directions of all other data bubbles and statistics of patterns are stored w.r.t. all other data bubbles. They showed that this approach of DB outperformed other approaches of DB [11] when the OPTICS method is applied to it.

  T. Ho et al. [12] introduced a document clustering method which is based on TRSM of a tolerance class of the index terms of all documents. P. Kumar et al. [13] introduced a fast hierarchical agglomerative clustering method based on the TRSM for sequential data. They showed that their method outperformed the complete-link method. However, all these clustering methods assume that entire dataset remains in main memory of the machine. This assumption might not be feasible for large datasets.

3.1   Application of Single-Link Method to the DB

As a summarization scheme DB works fine when OPTICS is applied to it. However, it is not clear whether it can produce consistent clusterings when SL is applied to it. We tested the DB with a dataset (Spiral) of size 3330. The clustering results of DB based single-link are compared with the classical single-link using the Rand Index [14]. We found that clustering results are very much inconsistent (Table 1). This is due to the following facts.

 – DB selects patterns randomly. Selected patterns do not cover all clusters.
 – DB may detect gaps present in a data bubble. However, reassignment of some patterns (restructing the data bubble) are not suggested. Therefore, a cluster loses some patterns to other clusters.

Table 1. Results produced by DB based SL method

Dataset	#DB	Rand Index
Spiral	266	0.499-1.000
	208	0.503-0.767

4   The Proposed Summarization and Clustering Method

To overcome the deficiencies of DB when SL is applied to it, we propose a summarization scheme, which is based on the TRSM. This new summarization scheme is called *data sphere (DS)*. We have used *DS* to speed up the single link clustering method.

4.1   The Summarization Scheme

The new summarization scheme utilizes the leaders clustering method and tolerance rough set theory. Let the leaders threshold distance be τ and $\mathcal{L} = \{l_1, l_2, \ldots, l_m\}$ be the set of all leaders of the dataset $\mathcal{D}$. For a pattern x and leader l_1, l_2 ($||l_1 - l_2|| <= 2\tau$), we may observe a scenario such that $||l_1 - x|| <= \tau$, $||l_2 - x|| <= \tau$. Then x is eligible to be the follower of both leaders. However, leaders clustering method assigns it to a leader which one is observed first (say l_1). The through study shows that clapping of pattern x with leader l_1 may restrict to have proper cluster-structure in data. If we allow x to be the member of l_2 also, we may recover the cluster-structure in further analysis. Therefore, we virtually allow l_1, l_2 to share pattern x. Let $\mathcal{L}^r = \{l_1^r, l_2^r, \ldots, l_m^r\}$ be the set of leaders where a leader can share patterns with others. We call $l_i^r \in \mathcal{L}^r$ as a *rough leader* and the $\mathcal{L}^r$ as a set of rough leaders. Note that there is a one to one correspondence between the set $\mathcal{L}$ and $\mathcal{L}^r$.

Now we can apply the TRSM to the dataset $\mathcal{D}$. We consider dataset $\mathcal{D} = U$ for our TRSM. Let $T \subseteq \mathcal{D} \times \mathcal{D}$ be a tolerance relation. One can define a tolerance class based on the proximity of the patterns in the dataset.

Definition 1 (Tolerance class). *If $x \in \mathcal{D}$, then the tolerance class of x is*
$$T(x) = \{x_j \in \mathcal{D} \mid \quad ||x - x_j|| <= \delta\}_{\delta \in \mathbb{R}^+} \qquad \square$$

Therefore, tolerance class of a pattern x is a set of patterns whose distance from x is less than a given threshold δ. We assume the value of $\delta = \tau/2$. One can define lower approximation and tolerant internal edge of a leader $\{l^r\} \subseteq \mathcal{D}$ are as follow.

Definition 2 (Lower approximation). *Let T be a tolerance relation on $\mathcal{D}$ and l^r be a leader obtained by leader threshold distance τ. The lower approximation of $\{l^r\}$ is*
$$\underline{T}(\{l^r\}) = \{x_j \in \mathcal{D} \mid \quad ||l^r - x_j|| <= \delta\} \qquad \square$$

Definition 3 (Tolerant internal edge). *Let T be a tolerance relation on $\mathcal{D}$ and l^r be a leader obtained using threshold distance τ. The tolerant internal edge of $\{l^r\}$ is defined as $\underline{T_{Edg}}(\{l^r\}) = \{x_j \in \mathcal{D} \mid \quad \tau/2 < ||l^r - x_j|| <= \tau\}$* \qquad $\square$

Definition 4. *Let l_1 and l_2 be the two leaders. The set of followers of l_1 in the direction of l_2 and set of followers of l_2 in the direction of l_1 are $l_1^{(l_2)}$ and $l_2^{(l_1)}$, respectively.*

$$l_1^{(l_2)} = \{x \in \{l_1\} : ||l_2 - x|| \leq ||l_1 - l_2||\}; l_2^{(l_1)} = \{x \in \{l_2\} : ||l_1 - x|| \leq ||l_1 - l_2||\}\ \square$$

We modified the leaders clustering in such a way that each leader stores the statistics of its followers in the direction of (w.r.t.) all other leaders. The leader l_1 stores statistics w. r. t. l_2 in the form

$$(k^{(l_2)}, l_1, ld^{(l_2)}, \quad sd^{(l_2)}, \underline{T}^{(l_2)}(\{l_1^r\}), \quad |T_{\underline{Edg}}^{(l_2)}(\{l_1^r\})|, \quad |Com_{l_1}^{(l_2)}|, ld_{Edg}^{(l_2)}, sd_{Edg}^{(l_2)})$$

where $k^{(l_2)} = |l_1^{(l_2)}|$,

$$ld^{(l_2)} = \sum_{j=1}^{k^{(l_2)}} d_j, \quad sd^{(l_2)} = \sum_{j=1}^{k^{(l_2)}} d_j^2, \quad d_j = ||l_1 - x_j||, \text{where} \quad x_j \in l_1^{(l_2)}$$

$$\underline{T}^{(l_2)}(\{l_1^r\}) = \{x_j \in \underline{T}(\{l_1^r\}) : ||l_2^r - x_j|| \leq ||l_1 - l_2||\}.$$

$$T_{\underline{Edg}}^{(l_2)}(\{l_1^r\}) = \{x_j \in T_{\underline{Edg}}(\{l_1^r\}) : ||l_2^r - x_j|| \leq ||l_1 - l_2||\},$$

$$Com_{l_1}^{(l_2)} = \{x \in \mathcal{D} : \tau/2 < ||x - l_1||, ||x - l_2|| \leq \tau\}; \quad ld_{Edg}^{(l_2)} = \sum_{i=1} d_i;$$

$$sd_{Edg}^{(l_2)} = \sum_{i=1} d_i^2, d_i = ||l_1 - x_i||, \text{where } x_i \in l_1^{(l_2)}, ||l_1 - x_i|| > \tau/2.$$

The set of shared patterns $Com_{l_1}^{(l_2)}$ between (l_1^r, l_2^r) are actually grouped with leader $l_1 \in \mathcal{L}$ as l_1 is observed prior to l_2. A leader l with statistics of its followers w.r.t. all other leaders is termed as *data sphere (DS)*, l being the representative pattern. The set $\mathcal{S}$ is the summary of the whole dataset. It is noted that followers of leader $l \in \mathcal{L}$ are also the members of a data sphere $S \in \mathcal{S}$, whose representative is l. In the next section, we discuss how this summary of the data can be clustered using the single-link method.

4.2   The Proposed Clustering Method

To apply the single-link clustering method to the set $\mathcal{S}$, we need to define a distance function $dist : \mathcal{S} \times \mathcal{S} \rightarrow \mathbb{R}_{\geq}$. Let the average and standard deviation of distances from l_1 to its followers w.r.t. l_2 be $\mu_{l_1}^{(l_2)}$ and $\sigma_{l_1}^{(l_2)}$, respectively. Similarly, $\mu_{l_2}^{(l_1)}$ and $\sigma_{l_2}^{(l_1)}$ be the average and standard deviation of distances from l_2 to its followers w.r.t. l_1, respectively. The $\mu_{l_1}^{(l_2)}$ and $\sigma_{l_1}^{(l_2)}$ are calculated as follow.

$$\mu_{l_1}^{(l_2)} = \frac{ld^{(l_2)}}{k^{(l_2)}}; \quad \sigma_{l_1}^{(l_2)} = \sqrt{\frac{sd^{(l_2)}}{k^{(l_2)}} - (\mu_{l_1}^{(l_2)})^2}; \text{ Similarly, one can calculate } \mu_{l_2}^{(l_1)} \text{ and } \sigma_{l_2}^{(l_1)} \text{ for the data sphere } S_2.$$

Definition 5 (Distance between two data spheres). *The distance between a pair of data spheres S_1 and S_2 is defined as*
$$dist(S_1, S_2) = \max(||l_1 - l_2|| - (\mu_{l_1}^{(l_2)} + 2\sigma_{l_1}^{(l_2)}) - (\mu_{l_2}^{(l_1)} + 2\sigma_{l_2}^{(l_1)}), 0).$$

We have few cases while finding the effective distance between a pair of data spheres S_1, S_2 (S_1 appears prior to S_2) as follow.

1. $dist(S_1, S_2) = 0$, if $||l_1 - l_2|| < 2\tau$ and the following conditions are satisfied simultaneously,

   (a) If lower approximation of l_1^r intersects with l_2^r and vice-versa.

   (b) If there is any pattern in tolerant internal edge of l_2^r except the patterns in lower approximation of l_1^r.

2. If $||l_1 - l_2|| < 2\tau$, then one data sphere (say S_1) may contain patterns from two different clusters. As a result, S_1 may have gap (a region having no patterns) in the direction of $l_2 \in S_2$ or vice versa (Fig. 1). We handle these gaps reassigning correct position of patterns from one DS to other DS. We have few scenarios as follow.

   (a) S_1 contains gap in the direction of S_2 if the following conditions are satisfied simultaneously, (i) If lower approximation of l_1^r does not intersect with l_2^r. (ii) If lower approximation of l_2^r intersect with l_1^r. (iii) If there is no pattern in the tolerant internal edge of l_1^r except the shared pattern pattern $Com_{l_1}^{l_2}$. In this case, the region of l_2^r is having more patterns towards l_1^r than l_2^r w.r.t. l_2^r. Therefore, the shared patterns $(Com_{l_1}^{l_2})$ should be with S_2. The method removes the shared patterns from S_1 and adds to S_2. Update the statistics.
   $$ld^{l_2} = ld^{l_2} - |Com_{l_1}^{(l_2)}| * (\tfrac{ld^{l_2}}{k^{l_2}}); \quad sd^{l_2} = sd^{l_2} - |Com_{l_1}^{(l_2)}| * (\tfrac{ld^{l_2}}{k^{l_2}})^2; ld^{l_1} = ld^{l_1} + |Com_{l_1}^{(l_2)}| * (||l_1 - l_2|| - \tfrac{ld^{l_2}}{k^{l_2}}); sd^{l_1} = sd^{l_1} + |Com_{l_2}^{(l_1)}| * (||l_1 - l_2|| - \tfrac{ld^{l_2}}{k^{l_2}})^2.$$

   (b) S_2 contains gap in the direction of S_1, if the converse of the above conditions $((a)(i), (ii), (iii))$ hold simultaneously (Fig 1b). However, re-assignment is not required as shared pattern $(Com_{l_1}^{l_2})$ are already grouped with S_1.

   (c) S_1 contains gap in the direction of S_2 if the following conditions hold (Fig. 1c) (i) If only tolerant edges of l_1^r and l_2^r intersect. (ii) Let number of patterns of internal edge of l_1^r and l_2^r excluding the shared patterns $(Com_{l_1}^{l_2})$ be c_1 and c_2, respectively. If the ratio of c_1 and c_2 is not more than a given threshold $h(0.5)$. In this case re-assignment is necessary. The proposed method removes shared patterns from S_1 and adds to S_2. It updates the corresponding statistics (Same as $2(a)$).

Therefore, $dist(S_1, S_2) = \max(||l_1 - l_2|| - (\mu_{l_1}^{(l_2)} + 2\sigma_{l_1}^{(l_2)}) - (\mu_{l_2}^{(l_1)} + 2\sigma_{l_2}^{(l_1)}), 0)$.

Having calculated the all-pair distances of data sphere and subsequently re-structed the data spheres, we apply the single-link clustering method to the data spheres. This gives a set of clusterings of the data spheres. Next we replace each data sphere by its

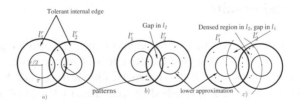

Fig. 1. (a) Lower approx. of l_1^r and tolerant internal edge of l_2^r intersect each other.; (b) Lower approx. of l_1^r intersects with tolerant internal edge of l_2^r, $\underline{T}^{(l_1)}(l_2^r) \cap T_{Edg}^{(l_2)}(\{l_1^r\}) = \emptyset$; (c) Intersection of two tolerant internal edges.

Algorithm 1. DS-single-link($\mathcal{D}, \tau$)

Apply leaders clustering method to $\mathcal{D}$. Let $\mathcal{L} = \{l_1, l_2, \ldots, l_m\}$. be the set of leaders.
Compute the statistics of each leader l_1 w. r. t. to every other leader l_2
for each leader $l_1^r \in \mathcal{S}$ **do**
  for each leader $l_2^r \in \mathcal{S} : l_1 \neq l_2$ **do**
    if ($\|l_1^r - l_2^r\| \leq 2\tau$) **then**
      if (lower approx. of $l_1^r (l_2^r)$ intersects with $l_2^r (l_1^r)$ and intersection of internal edges $\neq \emptyset$)
      then
        $dist(S_1, S_2) = 0$;   {Two data spheres are part of a cluster.}
      else if (lower approx. of l_1^r intersects with l_2^r and ($l_1^r \cap$ common edges) $\neq \emptyset$) **then**
        $dist(S_1, S_2) = 0$;   {No reassignment of patterns}
      else if (lower approx. of l_1^r not intersect with l_2^r but lower of l_2^r does and ($l_2^r \cap$ common
      edges) $\neq \emptyset$) **then**
        Add $Com_{l_1}^{(l_2)}$ and lower approx. of l_2^r to S_2; Remove these from S_1.
        Update $ld^{l_2}, sd^{l_2}, ld^{l_1}, sd^{l_1}, k^{(l_1)}, k^{(l_2)}$; {There is a gap in S_1 w.r.t. l_2}
      end if
      if ($Com_{l_1}^{(l_2)} \neq \emptyset$ and ($\frac{|T_{Edg}^{(l_2)}(\{l_1^r\}) \backslash Com_{l_1}^{(l_2)}|}{|T_{Edg}^{(l_1)}(\{l_2^r\}) \backslash Com_{l_1}^{(l_2)}|}$) $<= h$) **then**
        Add $Com_{l_1}^{(l_2)}$ to S_2 and remove from S_1; Updates $ld^{l_2}, sd^{l_2}, ld^{l_1}, sd^{l_1}, k^{(l_1)}, k^{(l_2)}$
      end if
    end if
    Calculate $\mu_{l_1}^{(l_2)}, \sigma_{l_1}^{(l_2)}, \mu_{l_2}^{(l_1)}, \sigma_{l_2}^{(l_1)}$
    $dist(S_1, S_2) = \max(\|l_1 - l_2\| - (\mu_{l_1}^{(l_2)} + 2\sigma_{l_1}^{(l_2)}) - (\mu_{l_2}^{(l_1)} + 2\sigma_{l_2}^{(l_1)}), 0)$
  end for
end for
Apply SL to $\mathcal{S}$. Let the result be $\pi_{\mathcal{S}} = \{\pi_1, \pi_2, \ldots, \pi_{|\mathcal{S}|}\}$, set of clusterings of $\mathcal{S}$.
Expand each $S_i \in \pi_i \in \pi_{\mathcal{S}}$. Therefore, $\quad \pi_{\mathcal{D} = \{\pi_{D_1}, \pi_{D_2}, \ldots, \pi_{D_{|\mathcal{S}|}}\}}$. Output $\pi_{\mathcal{D}}$

---

members. Finally, we get a set of clustering of the dataset $\mathcal{D}$. The whole scheme of summarization and subsequently the clustering method are noted as in Algorithm 1. The overall time and space complexity of *DS-SL* are $O(mn)$ and $O(m^2)$, respectively, $m = |\mathcal{L}|$. The proposed method scans the dataset twice.

5   Experimental Results

Experiment are conducted with two synthetic and two real world datasets (UCI) after removing the class labels. To show the effectiveness of our proposed summarization, we implemented three clustering method namely, classical SL, DB based single-link (DB-SL) and our proposed *DS-SL* method in Intel Xeon CPU ($3.6GHz$) with $8GB$ RAM Server. Table 2 shows the detailed performance of the three clustering methods for different datasets. The Compression Ratio (CR) is defined as $CR = n/m$ where, m is the cardinality of the representative set of the data. We used Rand Index (RI) [14] to compare the clusterings produced by the three methods. For this purpose, we considered the clustering of each method in which number of clusters are same as number of classes

Table 2. Results for different datasets

Table 3. Results for *Circle* dataset

Dataset	Size (n)	#Feature	#Class	CR $\frac{n}{m}$	Method	Time (in sec.)	Rand Index (RI)
Spiral (Synthetic)	3330	2	2	12.5	DS-SL	0.1	1.000
					DB-SL	0.1	0.499-1.000
					SL	40.1	–
Circle (Synthetic)	28000	2	4	50.9	DS-SL	1.5	1.000
					DB-SL	0.9	1.000-1.000
					SL	36,415.3	–
Pendigits	7494	16	10	12.2	DS-SL	1.6	0.985
					DB-SL	1.5	0.742-0.985
					SL	692.90	–
a8a	32561	123	2	12.4	DS-SL	145.5	0.998
					DB-SL	141.6	0.849-0.866
					SL	74,108	–

#cluster	Method	Rand Index (RI)
2	DS-SL	1.000
	DB-SL	1.000
3	DS-SL	0.749
	DB-SL	0.745
4	DS-SL	1.000
	DB-SL	1.000
5	DS-SL	0.973
	DB-SL	0.968
6	DS-SL	0.958
	DB-SL	0.946
7	DS-SL	0.928
	DB-SL	0.913

of the datasets. RI (Table 2) is computed between the partitions produced by SL, *DS-SL* and SL, *DB-SL* methods. The results produced by the proposed method are same or very close to that of the SL method. The clustering quality (RI) at different levels of hierarchy produced by DS-SL is also superior than that of the DB-SL method(Table 3).

6   Conclusion

The single-link is not scalable method and scans the dataset many times. DB could not produce a consistence results if SL is applied to it. In this paper we studied thoroughly the shortcoming of the DB as a summarization scheme. We proposed a TRSM based summarization scheme to overcome the shortcoming of the DB. Our proposed (*DS-SL*) method produces a good clusterings consistently. It takes considerably less time compared to that of the classical SL method. The clustering results produced by the *DS-SL* method is very close to that of the SL method. The *DS-SL* method outperforms DB based SL method at clustering quality.

Acknowledgments. Bidyut Kr. Patra is supported by CSIR, New Delhi, India.

References

1. Jain, A.K., Murty, M.N., Flynn, P.J.: Data Clustering: A Review. ACM Computing Surveys 31, 264–323 (1999)
2. Ankerst, M., Breunig, M.M., Kriegel, H.P., Sander, J.: Optics: Ordering points to identify the clustering structure. SIGMOD Rec. 28, 49–60 (1999)
3. Sneath, A., Sokal, P.H.: Numerical Taxonomy. Freeman, London (1973)
4. Pawlak, Z.: Rough sets. Int.J. of Computer and Information Sc. 11, 341–356 (1982)
5. Lin, T.Y., Cercone, N. (eds.): Rough Sets and Data Mining: Analysis of Imprecise Data. Kluwer Academic Publishers, Norwell (1996)
6. Zhou, J., Sander, J.: Data bubbles for non-vector data: speeding-up hierarchical clustering in arbitrary metric spaces. In: VLDB 2003, pp. 452–463 (2003)
7. Skowron, A., Stepaniuk, J.: Tolerance approximation spaces. Fundamenta Informaticae 27, 245–253 (1996)

8. Slowinski, R., Vanderpooten, D.: A generalized definition of rough approximations based on similarity. IEEE Trans. on Knowl. and Data Eng. 12, 331–336 (2000)
9. Hartigan, J.A.: Clustering Algorithms. John Wiley & Sons, Inc., New York (1975)
10. Zhang, T., Ramakrishnan, R., Livny, M.: Birch: An efficient data clustering method for very large databases. In: ACM SIGMOD Conference, pp. 103–114 (1996)
11. Breunig, M.M., Kriegel, H.P., Sander, J.: Fast hierarchical clustering based on compressed data and optics. In: Zighed, D.A., Komorowski, J., Żytkow, J.M. (eds.) PKDD 2000. LNCS (LNAI), vol. 1910, pp. 232–242. Springer, Heidelberg (2000)
12. Ho, T.B., Nguyen, N.B.: Nonhierarchical document clustering based on a tolerance rough set model. Int. J. Intell. Syst. 17, 199–212 (2002)
13. Kumar, P., Krishna, P.R., Bapi, R.S., De, S.K.: Rough clustering of sequential data. Data Knowl. Eng. 63, 183–199 (2007)
14. Rand, W.M.: Objective Criteria for Evaluation of Clustering Methods. J. of American Statistical Association 66, 846–850 (1971)

A Novel Possibilistic Fuzzy Leader Clustering Algorithm

Hong Yu and Hu Luo

Institute of Computer Science and Technology, Chongqing University of Posts and
Telecommunications, Chongqing, 400065, P.R. China
yuhongcq@yahoo.com.cn, luohui2120163.com

Abstract. The clusters tend to have vague or imprecise boundaries in
some fields such as web mining, since clustering has been widely used.
Fuzzy clustering is sensitive to noises and possibilistic clustering is sensitive to the initialization of cluster centers and generates coincident
clusters. Based on combination of fuzzy clustering and possibilistic clustering, a novel possibilistic fuzzy leader (PFL) clustering algorithm is
proposed in this paper to overcome these shortcomings. Considering the
advantages of the leader algorithm in time efficiency and the initialization of cluster, the framework of the leader algorithm is used. In addition,
a λ-cut set is defined to process the overlapping clusters adaptively. The
comparison of experimental results shows that our proposed algorithm
is valid, efficient, and has better accuracy.

Keywords: Fuzzy clustering, possibilistic clustering, leader clustering,
possibilistic fuzzy leader clustering.

1 Introduction

Cluster analysis has been widely applied in many areas such as data mining,
web mining, geographical data processing, medicine, classification of statistical
findings in social studies and so on. When talking about web mining, clustering
faces several additional challenges, compared to traditional applications [5]. The
clusters tend to have vague or imprecise boundaries. There is a likelihood that
an object may be a candidate for more than one cluster. In addition, due to noise
in the recording of data and incomplete logs, the possibility of the presence of
outliers in the data set is quite high.

Some researchers have focused on solving the uncertain clustering with fuzzy
sets theory, and one of the most popular and efficient clustering algorithms in
conventional applications is Fuzzy C-Means clustering (FCM) algorithm proposed by Bezdek [3], and most fuzzy clustering approaches are derived from the
FCM algorithm. The FCM uses the probabilistic constraint that the membership of a data point across classes sum to 1. The constraint is used to generate
the membership update equations for an iterative algorithm. The memberships
resulting from the FCM and its derivatives, however, do not always correspond
to the intuitive concept of degree of belonging or compatibility. Moreover, the
algorithms have considerable trouble in noisy environments.

H. Sakai et al. (Eds.): RSFDGrC 2009, LNAI 5908, pp. 423–430, 2009.

To overcome the shortcoming, Krishnapuram and Keller propose a Possibilistic C-Means clustering (PCM) algorithm [6] based on possibilistic theory, where they construct an appropriate objective function instead of the inherently probabilistic constraint used in the FCM. Noise points will have low degrees of compatibility in all clusters in the PCM, which makes their effect on the clustering negligible. However, the PCM is very sensitive to the initialization of cluster centers and generates coincident clusters [2].

To combat the shortcomings of the FCM sensitive to noises and the PCM sensitive to the initialization of cluster centers, a Possibilistic Fuzzy c-Means Clustering (PFCM) Algorithm is proposed [8], where the number of the cluster centers are decided by the fields experts. However, it is difficult to know there are how many clusters in many complicated and uncertain fields in advance. In this paper, we will focus on the uncertain clustering problem where clusters have vague or imprecise boundaries.

Some algorithms [1,4,7,11] have been proposed to cluster uncertain data sets. Lingras [7] proposes a k-means cluster algorithm based on the rough sets theory. Asharaf and Murty [1] proposes an adaptive rough fuzzy leader algorithm based on the rough sets theory, where the upper and lower approximates are expressed by the threshold values which are decided by the experiential experts. Wu and Zhou [11] proposes a possibilistic fuzzy algorithm based on c-means clustering, and Chen [4] proposes a possibilistic fuzzy algorithm based on the uncertainty membership. However, the efficiency of the methods are not satisfied because there are more than one scan the data set and there are too many iterations.

Considering that the leader clustering algorithm [9] makes only a single pass through the data set and finds a set of leaders as the cluster representatives, we will use the framework of leader cluster in our work to improve the time efficiency. In order to process the vague or imprecise boundaries, a λ-cut set is defined to partition the recorders "soft". In short, based on combination of fuzzy clustering and possibilistic clustering, a novel Possibilistic Fuzzy Leader (PFL) cluster method will be studied in this paper. The PFL solves the noise sensitivity defect of the FCM, and overcomes the coincident clusters problem of the PCM.

The rest of this paper is structured as follows. First, we introduce some basic concepts about clustering. A novel possibilistic fuzzy leader (PFL) clustering algorithm is proposed in Section 3. The experiment results in Section 4 show that the PFL algorithm is valid, efficient, and better accuracy. Some conclusions will be given in Section 5.

2   Basic Concepts

Firstly, let us review the concept of clustering. Clustering labeled data set $\mathbf{X} = \{CP_1, \ldots, CP_i, \ldots, CP_n\}$ is the partitioning of $\mathbf{X}$ into $c \in [1, n]$ subgroups such that each subgroup represents "natural" substructure in $\mathbf{X}$. This is done by assigning *labels* to the vectors in $\mathbf{X}$. A *c-partition* of $\mathbf{X}$ is a set of (cn) values u_{ik} that can be conveniently arrayed as a matrix $\mathbf{U} = [u_{ik}]_{c \times n}$. Generally, we use $\mathbb{L} = \{L_1, \ldots, L_c\}$ to denote a vector of (unknown) cluster centers (weights or prototypes).

The popularity of the leader clustering method [9] is due to the fact that the algorithm makes only a single pass through the data set and finds a set of leaders as the cluster representatives. Here, a cluster is also called a leader.

The leader clustering algorithm uses a user specified threshold τ and the framework of the algorithm can be stated as follows:

1. Start with any of the patterns as the initial leader
2. For each pattern in the data set do
   (1) Find the nearest leader L_j for the current pattern CP_i from the set of all currently available Leaders
   (2) If the distance $d(CP_i, L_j) < \tau$ assign CP_i to the cluster represented by L_j, else add CP_i as a new leader

At any step, the algorithm assigns the current pattern(feature points) CP_i to the most similar cluster (leader) or the pattern itself may get added as a leader if its similarity with the current set of leaders does not qualify it to get added to any of the clusters based on a user specified threshold. The found set of leaders acts as the prototype set representing the clusters and is used for classifying the test data.

3   Possibilistic Fuzzy Leader Clustering Algorithm

This section explains some ideas and notations used in the possibilistic fuzzy leader (PFL) clustering algorithm, which divides the data set into a set of over-lapping clusters.

3.1   Membership Function

Generally speaking, the similarity between the current pattern CP_i and the cluster L_k is more likelihood to measure the membership (in sense of "belonging" or "typicality") of CP_i to cluster $\{L_k\}$. Hence we extend the equation used in [3] to define the membership function.

Definition 1. Let $sim(CP_i, L_k)$ is the similarity of the current pattern CP_i to the cluster $\{L_k\}$. The similarity can be user-defined according to the fields. N_l is the number of the current clusters(leaders), and $1 \leq k \leq N_l \leq c$. $m \in [1, \infty)$ is a weighting exponent called the fuzzifier. Then the *membership function* is:

$$u_{ik} = \left(\sum_{j=1}^{N_l} (sim(CP_i, L_k)/sim(CP_i, L_j))^{2/(m-1)} \right)^{-1} \tag{1}$$

The membership function can help the patterns group to the cluster center, which is a good clustering.

3.2   Possibility Distribution Function

As we have discussed, the possibility theory is combined here. To extend the equation used in [6], we have the following definition.

Definition 2. The *possibility distribution function* is

$$t_{ik} = \left[1 + \left(\frac{sim^2(CP_i, L_k)}{\eta_i} \right)^{\frac{1}{p-1}} \right]^{-1} \tag{2}$$

Here, the value of p determines the fuzziness of the final possibilistic c-partition and the shape of the possibility distribution. A value of 2 for p yields a very simple equation for the membership updates. Fortunately, $p = 2$ seems to give good results in practice.

The value of η_i needs to be chosen depending on the desired bandwidth of the possibility distribution for each cluster. According to the theorem in [6], we define the η_i as followed.

$$\eta_i = K \frac{\sum_{k=1}^{N_l} u_{ik}^m sim^2(CP_i, L_k)}{\sum_{k=1}^{N_l} u_{ik}^m} \tag{3}$$

Eq.(3) makes η_i proportional to the average fuzzy intra-cluster distance of cluster $\{L_i\}$. Here $K > 0$, typically K is chosen to be 1.

3.3 λ-Cut Set

In order to process the overlapping feature points(pattern), the concept of a λ-cut set is used here. Let $\mathcal{A}$ be a fuzzy set on universe $\mathbf{X}$, given a number λ in $[0, 1]$, an λ-cut, or λ-level set, of a fuzzy set is defined by[12]: $\mathcal{A}_\lambda = \{x \in \mathbf{X} \mid u_{\mathcal{A}} \geq \lambda\}$, which is a subset of $\mathbf{X}$.

Let $u_{ic_1} = max\{u_{ik}\}$, $t_{ic_2} = max\{t_{ik}\}$. Obviously, the CP_i is most likely belonging to the leader L_{c_1} or L_{c_2}. Hence, in the PFL algorithm, we assign the current pattern CP_i to the L_{c_1} or L_{c_2} determinately. Then, we meet the following questions. Does the CP_i belong to others cluster? What clusters are they?

According to the quality of the λ-cut set of a fuzzy set, $u_{ik} \geq \lambda$ means that the current pattern CP_i is likelihood belonging to the cluster $\{L_k\}$, so the CP_i can be assigned to the cluster $\{L_k\}$.

To reduce the number of thresholds, we define the adaptive λ formula as followed.

Definition 3. The *adaptive threshold efficiency* λ is defined as:

$$\lambda = \left(1 - \frac{Sim(CP_i, L_{c^*})}{AU} \right) \times min(u_{ic_1}, u_{ic_2}) \tag{4}$$

Here, $u_{ic^*} = min(u_{ic_1}, u_{ic_2})$, $AU = \sum_{s=1}^{N_u} Sim(CP_i, L_s)$, and L_{c^*} means the cluster which have the smaller u_{ik}. N_u means the number of unvisited leaders(clusters) after the CP_i is assigned to L_{c_1} or L_{c_2} determinately. That is:

$$N_u = \begin{cases} N_l - 1 & \text{if } c_1 = c_2, CP_i \text{ is assigned to cluster } \{L_{c_1}\} \\ N_l - 2 & \text{if } c_1 \neq c_2, CP_i \text{ is assigned to cluster } \{L_{c_1}\} \text{ and } \{L_{c_2}\} \end{cases}$$

3.4   Description of the PFL Algorithm

The framework of the leader clustering is used here to improve the time efficiency, since there are only one single scanning the data set.

In order to process the vague or imprecise boundaries, namely, to divide the data set into a set of overlapping clusters, there are more thresholds used in the literatures mentioned above, which makes the results more depending on experience. In contrast, there is only one threshold τ used in the framework of the PFL just because it is used in the framework of the leader clustering. An adaptive method is proposed to reduce the number of thresholds.

Firstly, according to the values of membership function and possibility distribution function, we decide the current pattern CP_i whether assign to one or two exist clusters determinately. Then, to decide the CP_i go to the clusters represented by the λ-cut sets, where the threshold λ is adaptive decided. The description of the PFL algorithm is as following.

PFL Algorithm : Possibilistic Fuzzy Leader Clustering Algorithm
Input : $\mathbf{X} = (CP_1, CP_2, \ldots, CP_i, \ldots, CP_n)$.
Output: the result of clustering $\mathbb{L} = \{L_1, \ldots, L_k, \ldots, L_c\}$.
begin
Produce a random number $r \in [1, n]$, exchange CP_1 and CP_r;
Initially, the leader $\{L_1\} = \{CP_1\}$;
$N_l = 1$; //the number of leaders
for $(i = 2; i \leq n; i++)$ **do**
    for $(k = 1; k \leq N_l; k++)$ **do** Compute $Sim(CP_i, L_k)$;
    for $(k = 1; k \leq N_l; k++)$ **do** Update the u_{ik} according to Eq.(1);
    Update the value of η_i according to Eq.(3);
    for $(k = 1; k \leq N_l; k++)$ **do** Update the t_{ik} according to Eq.(2);
    $u_{ic_1} = max(\{u_{ik}\}); t_{ic_2} = max(\{t_{ik}\})$;
    if $Sim(CP_i, L_{c_1}) < \tau$ && $Sim(CP_i, L_{c_2}) < \tau$ **then**
        if $c_1 = c_2$ **then** {Assign CP_i to the leader $\{L_{c_1}\}$; $N_u = N_l - 1$;}
        else {Assign CP_i to the leader $\{L_{c_1}\}$ and $\{L_{c_2}\}$; $N_u = N_l - 2$;}
        if $N_u > 1$ **then**
            Compute the λ according to Eq.(4);
            for $(k = 1; k \leq N_u; k++)$ **do**
                if $u_{ik} \geq \lambda$ **then** Assign CP_i to the leader $\{L_k\}$;
        end
    end
    end
    else
        $N_l = N_l + 1$;
        CP_i to be a new Leader, namely, $\{L_{N_l}\} = \{CP_i\}$;
    end
  end
Output all the $\{L_k\}$.
end

4   Experimental Results

In this section, we show several experiments to illustrate the ideas presented in the previous section.

The first experiment is on a simple manual data set, which involves four-separated clusters of 50 points (patterns) each. The FCM algorithm and PFL algorithm are implemented and used here. The results are shown in Fig.1, part (a) describes the data set.

Parts (b) and (c) of Fig.1 show the final clustering results obtained from the FCM algorithm and the PFL algorithm, respectively. Obviously, the FCM algorithm is difficult to label the vague boundaries correctly, and there are 10 points classed incorrectly, which are presented by the circles in the figure. But for the PFL algorithm, there are no points classed incorrectly. We can see, some boundary points are put into the different clusters, which is more reasonable in some cases such as web mining.

Part (d) of Fig.1 shows the manual data set when noise is added, and the noise points are denoted by asterisk. Parts (e) and (f) of Fig.1 show the final partition obtained from the FCM algorithm and the PFL algorithm when noise is added, respectively. In part (e), the elliptical areas are the resulting from the

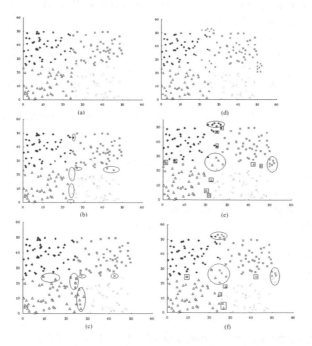

Fig. 1. Results on a manual data set: (a) the manual data set; (b) clustering result from the FCM algorithm; (c) clustering result from the PFL algorithm; (d) the manual data set when noise is added; (e) clustering result from the FCM algorithm when noise is added; (f) clustering result from the PFL algorithm when noise is added

noise points, and the rectangular areas are the inaccurate clustering results. In part (f), the elliptical areas are the resulting from the noise points, and the rectangular areas are the overlapping area for the boundaries. In fact, the noise points added are vague and uncertain, it is more reasonable to divide them into more than one cluster.

To compare the parts (e) with (f), the FCM increases the error rate of clustering, and there are some error labels in the noise areas. But there are no points labeled incorrectly in the PFL algorithm. The PFL algorithm divides the noise points to more than one cluster just as the reality.

In order to compare the algorithms, we have done another experiment, where some data sets from UCI repository [10] are used. The Iris data set contains 3 classes of 50 instances each, where each class refers to a type of iris plant. The Optdigits data set, optical recognition of handwritten digits data set, contains 10 classes and 64 dimensions. The K-Means algorithm, FCM algorithm, PFL algorithm are implemented and used here, and the CPU time and the accuracy rate are tested. The results of the experiments are shown in Table 1.

Table 1. Comparison of the CPU time and Results of the Algorithms

Database	$|\mathbf{X}|$	Algorithm K-Means		Algorithm FCM		Algorithm PFL	
		Accuracy rate	CPU(s)	Accuracy rate	CPU(s)	Accuracy rate	CPU(s)
MD 1	200	0.93	1.754	0.95	1.721	1.00	1.651
MD 2	200	0.92	1.723	0.94	1.712	1.00	1.661
Iris	150	0.87	2.725	0.89	2.734	0.99	2.239
Optdigits	500	0.69	5.532	0.74	5.334	0.82	3.506

In Table 1, $|\mathbf{X}|$ is the number of samples in the database, MD 1 is the manual data set, and MD 2 is the MD 1 when noise is added, which are used in the previous experiment and denoted in Fig.1.

From Table 1, we can find that Algorithm PFL developed in this paper are feasible to cluster. Algorithm PFL has the highest accuracy rate and the least CPU running time. Algorithm PFL not only improves the performance of clustering, but also decreases the time expense.

5 Conclusion

The clusters tend to have vague or imprecise boundaries in some fields such as web mining. There is a likelihood that an object may be a candidate for more than one cluster. Fuzzy clustering is sensitive to noises and possibilistic clustering is sensitive to the initialization of cluster centers and generates coincident clusters. Hence, based on combination of fuzzy clustering and possibilistic clustering, a novel possibilistic fuzzy leader (PFL) clustering algorithm is proposed in this paper, which can divide the data set into a set of overlapping clusters. To combat the time efficiency, the framework is based on leader clustering, there are much less iteration. In addition, a λ-cut set is defined to process the overlapping clusters adaptively. The comparison of experimental results shows that the new method is also more immune to noise and has better accuracy.

Acknowledgments

This work was supported in part by the China NNSF grant 60773113, the Science & Technology Research Program of the Municipal Education Committee of Chongqing of China grant KJ080510, and the Natural Science Foundation of Chongqing of China grant CSTC2009BB2082.

References

1. Asharaf, S., Murty, M.N.: An adaptive rough fuzzy single pass algorithm for clustering 1arge data sets. Pattern Recognition 36, 3015–3018 (2003)
2. Barni, M., Cappellini, V., Mecocci, A.: Comments on A possibilistic approach to clustering. IEEE Trans Fuzzy Systems 4(3), 393–396 (1996)
3. Bezdek, J.C.: Pattern recognition with fuzzy objective function algorithms. Plenum Press, New York (1981)
4. Chen, J., Lu, H., Song, Y., Song, S., Xu, J., Xie, C., Ni, W.: A Possibility Fuzzy Clustering Algorithm Based on the Uncertainty Membership. Journal of Computer Research and Development 45(9), 1486–1492 (2008)
5. Joshi, A., Krishnapuram, R.: Robust Fuzzy Clustering Methods to Support Web Mining. In: Proceedings of the Workshop on Data Mining and Knowledge Discovery, SIGMOD, pp. 15/1–15/8 (1998)
6. Krishnapuram, R., Keller, J.: A possibilistic approach to clustering. IEEE Trans Fuzzy systerms 1(2), 98–110 (1993)
7. Lingras, P.: Interval set c1ustering of web users with rough k-means. Journal of Intelligent Information System 23(1), 5–16 (2004)
8. Pal, R., Pal, K., Bezdek, J.C.: A possibilistic fuzzy c-means clustering algorithm. IEEE Tram Fuzzy Systems 13(4), 517–530 (2005)
9. Spath, H.: Cluster analysis algorithm for date reduction and classification of objects. Ellis Horwood Publ., Chichester (1980)
10. UCIrvine Machine Learning Repository: http://archive.ics.uci.edu/ml/
11. Wu, X., Zhou, J.: A novel possibilistic fuzzy c-means clustering. ACTA Electronica Sinica 10, 1996–2000 (2008)
12. Zadeh, L.A.: Fuzzy sets. Information and Control 8, 338–353 (1965)

Projected Gustafson Kessel Clustering

Naveen Kumar and Charu Puri

Department of Computer Science, University of Delhi, Delhi, India
nk.cs.du@gmail.com, cpuri.cs.du@gmail.com

Abstract. Fuzzy techniques have been used for handling vague boundaries of arbitrarily oriented cluster structures. However, traditional clustering algorithms tend to break down in high dimensional spaces due to inherent sparsity of data. In order to model the uncertainties of high dimensional data, we propose modification of objective functions of Gustafson Kessel algorithm for subspace clustering, through automatic selection of weight vectors and present the results of applying the proposed approach to UCI data sets.

Keywords: Gustafson Kessel algorithm, High dimensional data, Subspace clustering, Validity Measures.

1   Introduction

Clustering is a useful tool for analyzing data by quantitative determination of underlying structures [12]. Various similarity measures such as probability distributions, regression, correlation, hypothesis testing and distance methods have been used for discovering patterns. However, real world data sets often suffer from curse of dimensionality due to inherent sparsity of high dimensional data [17]. Traditional clustering algorithms fail in identifying hidden relationships of underlying structure due to the reason that the nearest neighbor of a pattern may be nearly as close as farthest neighbor, if distance is computed in full dimensional space[15]. To cope with the problem of high dimensional feature spaces, feature reduction and feature selection techniques have been used in the literature [12]. Feature reduction techniques project the whole feature space to a lower dimensional subspace so that cluster structures become apparent. However, feature reduction techniques cannot demonstrate patterns clustered differently in varying subspaces. Traditional feature selection techniques such as principal component analysis (PCA) have been used in multivariate statistics, methods based on singular value decomposition have been used in information retrieval to transform the attributes. Feature selection suffers from usability problem as it becomes hard to interpret the results intuitively [17]. Hence, there is a need for more generalized techniques that can be used to obtain meaningful clusters. Agrawal et al. introduced the concept of projective clustering [19] to generate clusters in distinctive subspaces of multidimensional space so that intra cluster similarity is maximized and inter cluster similarity is minimized. Elke et al. proposed an algorithm to discover cluster in arbitrarily oriented subspaces [5]. The

H. Sakai et al. (Eds.): RSFDGrC 2009, LNAI 5908, pp. 431–438, 2009.

pioneering approaches to projected clustering can be classified into three categories: grid based approaches such as CLIQUE [19] and MAFIA [9], approaches based on density connectivity such as SUBCLU [17], and partitioning and/or hierarchical approaches like PROCLUS [3]. CLIQUE is a grid based algorithm that partitions the whole data space into non-overlapping rectangular units. It uses an apriori-like method to recursively navigate through the set of possible subspaces in a bottom up way. It works on the principle that if a k-dimensional unit is dense, then all (k-1)-dimensional units will also be dense. It merges the dense units to form clusters over large subspaces. SUBCLU is a greedy algorithm that computes all density connected sets. It detects arbitrarily shaped and positioned clusters in subspaces. PROCLUS is based on the concept of k-medoid clustering. It explores the locality of space near medoids to determine relevant dimensions. It uses a greedy hill climbing technique to iteratively search for medoids. SCHISM[18] extends CLIQUE using a variable threshold in order to cope up with varying number of dimensions and makes use of grid based discretization for pruning. FIRES[16] is a generic framework based on approximate subspace cluster computation. EDSC[11] is a multi step filter and refinement algorithm meant for density based subspace clustering. DUSC[10]introduces the concept of dimensionality bias and performs density based clustering using statistical techniques for pruning. Classically, the clustering has been based on the disjointness condition that no two patterns belong to same cluster[9]. However, in real data sets a pattern may belong to various clusters and a dimension may be relevant to various clusters with varying degree of membership [1]. Hence, such situations require weakening of disjointness condition. Ruspini[6] developed the first fuzzy clustering algorithm based on least-squares optimization. Dunn[14] developed a fuzzy extension of this approach to clustering and proposed the fuzzy k-means algorithm. Bezdek [13] extended Dunn's formulation and proposed a generalization of conventional hard c-means clustering by fuzzy partitioning of data. However, fuzzy c-means cannot detect arbitrarily oriented clusters[12]. Gustafson and Kessel[4] proposed GK algorithm based on adaptive distance measure. In [2] Babuka et. al proposed Improved covariance estimation for Gustafson-Kessel clustering. We extend the Gustafson Kessel objective function for projective clustering which automatically detects the relevant cluster dimensions of high dimensional data set. Experimental results indicate that it enhances the efficiency of clustering solution by simultaneously pruning away the irrelevant subspaces. The structure of this paper is as follows: in section 2, we extend the Gustafson Kessel algorithm for subspace clusterting. In section 3 results based on UCI and synthetic data are presented, and finally section 4 contains conclusions and suggestions for future work.

2   Projected Gustafson Kessel Clustering

In this section, we introduce the necessary notations, review the GK algorithm [4] and formulate a new objective function for adapting it to subspace clustering. Fuzzy c-means[13] is subject to the constraint that it can detect only spherical clusters. The Gustafson-Kessel algorithm [4] associates each cluster with the cluster centre and its covariance. The main feature of Gustafson Kessel clustering is the local adaption of distance matrix in order to identify ellipsoidal clusters. Given a data set $X = \{x_1, x_2, ..., x_N\}$ in the d-dimensional space and the number of clusters k, the objective is to determine the cluster prototype $Z = \{z_1, z_2, ..., z_k\}$ and to partition the data set X into clusters. A fuzzy partition of the data set X can be represented by a $k \times N$ matrix $U = [\mu_{ij}]$, where μ_{ij} denotes the degree of membership with which j^{th} pattern belongs to the ith cluster, for $1 < i \leq k$, $1 \leq j \leq N$. The matrix U is called the fuzzy partition matrix. It satisfies the following conditions:

$$\mu_{ij} \in [0, 1], \ 1 \leq j \leq N, \ 1 \leq i \leq k, \tag{1}$$

$$\sum_{i=1}^{k} \mu_{ij} = 1, \ 1 \leq j \leq N, \tag{2}$$

The above constraints express the fact that the sum of memberships of a pattern over the set of clusters must be equal to 1. The fact that the number of clusters is at least two, is expressed by the following constraint:

$$0 < \sum_{j=1}^{N} \mu_{ij} < N, \ 1 \leq i \leq k. \tag{3}$$

The constraints expressed in (1), (2), (3) lead to the following fuzzy partition space for (X, k):

$$M^{fk} = \{U = [\mu_{ij}]_{k \times N} \in \Re^{k \times N} \ | \mu_{ij} \in [0, 1] \ \forall \, i, j; \sum_{i=1}^{k} \mu_{ij} = 1, \ \forall \, j; \ 0 <$$
$\sum_{j=1}^{N} \mu_{ij} < N, \ \forall \, i\}$ The GK algorithm uses the following objective function:

$J_m = \sum_{j=1}^{N} \sum_{i=1}^{k} \mu_{ij}^m d_{ij}^2$
where,
$d_{ij}^2 = (x_j - z_i)A_i(x_j - z_i)^T$

A_i being a symmetric, positive definite matrix. The exponent $m \in (1, \infty)$ is a fuzzification parameter. The above objective function is constrained to find the clusters in the the entire feature space and therefore cannot determine the respective natural subspaces of each cluster in high dimensional data set.

We associate with each cluster i a weight vector ω_i. Thus $W = [\omega_{ir}]_{k \times d}$ matrix where, ω_{ir} denotes the contribution of rth dimension to i^{th} cluster. The sum of contributions from all dimensions adds to 1 for any cluster. This is expressed by the constraint, where

$$\sum_{r=1}^{d} \omega_{ir} = 1, \ 1 \leq i \leq k, \tag{4}$$

$$\omega_{ir} \in [0, 1], \ 1 \leq i \leq k, \ 1 \leq r \leq d, \tag{5}$$

Also, as there should be at least two dimensions, we get the constraint:

$$0 < \sum_{i=1}^{k} \omega_{ir} < k, \ \forall \, r \tag{6}$$

Incorporating (4), (5), (6) in the fuzzy partition space for (X,d) we obtain the fuzzy partitioning space for (X,d) as follows: $M^{fd} = \{W = [\omega_{ir}]_{k \times d} \in \Re^{k \times d} \ |\omega_{ir} \in [0,1] \ \forall \, i,r; \ \sum_{r=1}^{d} \omega_{ir} = 1, \ \forall \, i; 0 < \sum_{i=1}^{k} \omega_{ir} < k, \ \forall \, r\}$ Now, we formulate a new objective function as:

$J_{\alpha,\beta} = \sum_{j=1}^{N} \sum_{i=1}^{k} \sum_{r=1}^{d} \mu_{ij}^{\alpha} \omega_{ir}^{\beta} d_{ijr}^{2}$
where,
$d_{ijr}^{2} = \sum_{s=1}^{d} (x_{jr} - z_{ir}) a_{rs} (x_{js} - z_{is})$
$A_i = [a_{rs}]_{d \times d}$

Parameters $\alpha \in (1, \infty)$, $\beta \in (1, \infty)$ are weighting components. These parameters control the fuzzification of μ_{ij} and ω_{ir}. The necessary condition for minimization of the Projected Gustafson Kessel objective function yields the following update equations:

$A_i = ((det(F_i)\rho_i))^{1/d} F_i^{-1}$

$\mu_{ij} = 1 / \sum_{l=1}^{k} \left[\frac{\sum_{r=1}^{d} (\omega_{ir})^{\beta} d_{ijr}^{2}}{\sum_{r=1}^{d} (\omega_{lr})^{\beta} d_{ilr}^{2}} \right]^{1/\alpha - 1}$

$\omega_{ir} = 1 / \sum_{l'=1}^{d} \left[\frac{\sum_{j=1}^{N} (\mu_{ij})^{\alpha} d_{ijr}^{2}}{\sum_{j=1}^{N} (\mu_{ij})^{\alpha} d_{ijl'}^{2}} \right]^{1/\beta - 1}$

3   Experimental Results

For evaluating the accuracy and efficiency of clustering, we compared PGK algorithm with PROCLUS and GK algorithms. We implemented these algorithms in MATLAB. Both real and synthetic datasets have been used. In order to minimize the effect of initial points in clustering, we repeated the experiments several times. The parameters for each method are optimized to achieve a fair comparison. In order to find the usefulness of clusters found by the proposed algorithm, an experimental study on UCI real datasets has been carried out. Four datasets from UCI machine repository have been chosen: Forest Fire, Alzehmir, Breast Cancer, Parkinson(www.ics.uci.edu/ mlearn/MLRepository.html). Chosen data sets have class labels assigned to instances. The class labels were removed during clustering and used later to measure the accuracy of clustering. Each of the chosen data sets comprises of real instances with no missing value.

To measure the cluster purity, we use clustering accuracy measure defined as follows[8]: $r = \sum_{i=1}^{c} x_i/n$ where x_i is the number of instances in cluster 1 and n is the number of instances in data set as measures based on distance are not relevant in the case of high dimensional data. In order to compare the cluster

Table 1. Data Sets

Data Sets	Instances	Attributes	Classes
Forest Fire	517	13	3
Alzehmir	45	8	3
Parkinson	197	23	2
Breast Cancer	569	32	2

Table 2. Comparison of PGK and GK on basis of Accuracy Measure

Data Sets	PGK	PROCLUS	GK
Forest Fire	0.7943	0.6941	0.5342
Alzehmir	0.8173	0.6324	0.5914
Parkinson	0.7492	0.7132	0.6498
Breast Cancer	0.8019	0.7302	0.6913

Table 3. Dimensions found by PGK and PROCLUS

#	Forest Fire		Alzehmir	
	PGK	PROCLUS	PGK	PROCLUS
1	3	2,3,4,12	5,6,7,8	1,5,6,7,8
2	12	2,4,12	1,5,6	1,5,7,6,8
3	2,3,12	1,2,3,4, 5,8,11,12	5,6,7	2,4,5, 6,7,8

Table 4. Dimensions found by PGK and PROCLUS

#	Breast Cancer		Parkinson	
	PGK	PROCLUS	PGK	PROCLUS
1	7,12,17	4,7,9,10 11,12,21	4,5,6,7	4,5,6,7, 8, 9, 10, 11
2	4,7,17	4,7,9, 12,14,19	5,6,7	5,6,7,8, 12,13,14

purity results of hard PROCLUS and soft partitioning PGK algorithms, we converted fuzzy assignments into hard ones by choosing the cluster with highest degree of membership for each data point. A data point x_i is said to belong to cluster C_l if: $l = arg\ max_{1 \leq j \leq k}\ \mu_{ij}$. GK is a full dimensional clustering algorithm, i.e it computes clusters giving each dimension equal weight. As all real data sets are labeled, we could determine the number of clusters for each data set. We also examined how well each of the algorithms determined the correct subspaces of each cluster by measuring their accuracy. We present the accuracy results for above UCI data. Table 2 shows that the PGK algorithm achieves highest accuracy when different algorithms are applied on UCI datasets. Tables 3and 4 gives the subspace dimensions associated with each cluster for Forest Fire, Alzehmir, Breast Cancer and Parkinson datasets respectively. PGK gives better clustering results as it forms the clusters in lower dimensional space with higher accuracy.

Column 1 of tables 3 and 4 represents number of clusters.

Cluster validity measures have been used to find whether a given fuzzy partition is the best fit of data with respect to various parameters[7][12]. There are various validity measures which helps in correctly determining the appropriate number of clusters. As no validity measure is perfect by itself, it is preferable to use these measures simultaneously or in combinations[12]. We have used Partition Coefficient, Classification Entropy and SSE in our experiments. Sum of the squared errors(SSE) is used as measure of compactness. Lower values of the measure indicate compact clusters. As we are looking for clusters in subspaces of the entire set of dimensions we modify the notion of sum of squared errors by considering the distance using only the dimensions relevant to a cluster as follows: SSE is defined as: $SSE = \sum_{i=1}^{k} \sum_{x_j \in c_i} d(x_j, z_i)$ Table 6 shows the

Table 5. PC and PE comparison

	PGK		GK	
	PC	PE	PC	PE
1	0.9310	0.1239	0.5275	0.6651
2	0.9964	0.0107	0.5918	0.7022
3	0.7236	0.2258	0.6031	0.5833
4	0.8327	0.4691	0.5107	0.6824

Table 6. SSE

	PGK	PROCLUS	GK
1	12.3579	25.4373	32.76534
2	2.9321	6.4598	8.4721
3	2.3579×10^3	5.0973×10^3	7.9432×10^3
4	3.5791×10^3	6.4972×10^3	9.5432×10^3

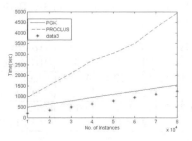

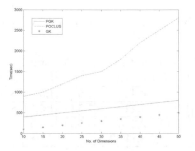

Fig. 1. Scalability as a function of number of instances

Fig. 2. Scalability as a function of number of dimensions

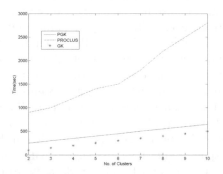

Fig. 3. Scalability as a function of number of clusters

modified SSE values on applying the different algorithms to various data. We observed that PGK algorithm achives lower SSE value and hence, more compact clusters in comparison to other algorithms. We also used cluster validity measures Partition Coefficient(PC) and Coefficient of entropy (PE)[20] in our experiments. Table 5 shows the values of these measures obtained for different UCI data sets on applying PGK and GK algorithm. It clearly shows that for each data set PGK achieves higher values for PC and lower values for PE as compared to GK. Thus, PGK scores over GK in the experiments we carried out.

Column 1 of Table 5 and 6 represents data sets: 1. Forest Fire, 2: Alzehmir, 3: Breast Cancer and 4:Parkinson.

We studied the scalability of PGK, GK and PROCLUS clustering algorithm on increasing number of data instances, clusters and dimensions. In the experiment on scalability we used synthetic data sets so that various parameters can be controlled. As in [19] relevant dimensions follow normal distribution and irrelevant dimensions follow uniform distribution. We tried various parameters and we discovered that parameters of PGK set as $\alpha=2$, $\beta=2$, $\epsilon = 1e - 5$ gives the best results. We observe that in all the experiments carried out the PGK scales significantly better than the PROCLUS however it is somewhat slower by a linear factor as compared to GK algorithm. For measuring the scalability of algorithm with increasing number of instances various datasets having 10,000 to 80,000 instances are generated with each data set having 5 clusters. Each data set has 15 dimensions out of which 10 follows the normal distribution. The results are presented in Figure 1. Scalability with the increasing number of attributes is measured using datasets containing 50,000 instances with 5 clusters and having 10 to 50 dimensions respectively. In each case 5 dimensions follows normal distribution and others follow uniform distribution. The results of experiment are presented in Figure 2. We consider datasets containing 50,000 instances with 15 dimensions from which 10 follow normal distribution and 5 follow uniform distribution. The number of clusters is varied from 2 to 10. The results are shown in Figure 3.

4   Conclusion and Future Work

We considered the problem of clustering for high dimensional dataset. Existing projected clustering algorithms do hard partitioning of the feature set which leads to loss of information. In this paper, we have addressed the issue of soft partitioning of feature space. Our main contribution is the adaptation of GK algorithm for subspace clustering in such a way that each dimension has a weight associated with each cluster. This algorithm searches for those subsets of feature vectors which are having high membership weights along one or more attributes. Different clusters may have different subspace preference weights. PGK determines the best feature weights for each cluster. It discovers soft partitions of data set in soft subspaces. It minimizes the objective function to determine prototype parameters, weight matrix and membership matrix, for each cluster. We tested the algorithm using various real world datasets from UCI Machine Learning Repository. For each data set chosen, the algorithm is able to discover clusters in appropriate subspace with high degree of accuracy. In the future work, we propose to investigate the extension of PGK for categorical data. By taking outliers into account, present model can be further improved.

Acknowledgment

This work was supported by University of Delhi, research grant No. Dean(R)/ R&D/2007/Ph-III/382 and UGC Research grant no. Ref. No. Sch/JRF/AA/16/ 2005-2006/7372. We express our thanks to Prof. Ajay Kumar Department of Mathematics, University of Delhi for valuable mathematical discussion.

References

1. Wiswedel, B., Berthold, M.R.: Fuzzy clustering in parallel universies. NAFIPS, 567–572 (2005)
2. Babuka, R., van der Veen, P.J.: Kaymak, Improved covariance estimation for Gustafson-Kessel clustering. In: FUZZ-IEEE 2002, vol. 2, pp. 1081–1085 (2002)
3. Aggarwal, C., Wolf, J., Yu, P., Procopiuc, C., Park, J.: Fast algorithms for projected clustering. ACM SIGMOD, 61–72 (1999)
4. Gustafson, D.E., Kessel, W.: Fuzzy clustering with a Fuzzy Covariance Matrix. In: Proc. IEEE-CDC, vol. 2, pp. 761–766 (1979)
5. Achtert, E., Böhm, C., David, J., Kröger, P., Zimek, A.: Robust Clustering in Arbitrarily Oriented Subspaces. SDM, 763–774 (2008)
6. Ruspini, E.H.: A New Approach to Clustering Information and Control, pp. 22–32 (1969)
7. Hoppner, F., Klawonn, F., Kruse, R., Runkler, T.: Fuzzy Cluster Analysis: Methods for Classification, Data Analysis, and Image Recognition. John Wiley & Sons, Chichester
8. Gan, G., Wu, J., Yang, Z.: PARTCAT: A Subspace Clustering Algorithm for High Dimensional Categorical Data. IJCNN, 4406–4412 (2006)
9. Nagesh, H., Goil, S., Choudhary, A.: MAFIA: Efficient and Scalable Subspace Clustering for Very Large Data Sets, Technical Report, Northwestern Univ. (1999)
10. Assent, I., Krieger, R., Müller, E., Seidl, T.: DUSC: Dimensionality Unbiased Subspace Clustering. In: ICDM 2007 (2007)
11. Assent, I., Krieger, R., Müller, E., Seidl, T.: EDSC: efficient density-based subspace clustering. In: Proceeding of the 17th ACM conference on Information and knowledge management (2008)
12. Abonyi, J.: Balazas Feil, Cluster Analysis for Data Mining and System Identification, Birkhauser
13. Bezdek, J.C.: Pattern recognition with Fuzzy Objective Function Algorithm. Plenum Press, New York (1981)
14. Dunn, J.: A Fuzzy Relative of the ISODATA Process and Its Use in Detecting Compact Well-Separated Clusters. J. Cybernetics 3, 32–57 (1974)
15. Beyer, K., Goldstein, J., Ramakrishnan, R., Shaft, U.: When is "nearest neighbor" meaningful? In: Beeri, C., Bruneman, P. (eds.) ICDT 1999. LNCS, vol. 1540, pp. 217–235. Springer, Heidelberg (1998)
16. Kelling, K., Peter, H., Kröger, P.: A Generic Framework for Efficient Subspace Clustering of High-Dimensional Data. In: ICDM, pp. 205–257 (2005)
17. Kailing, K., Kriegel, H.-P., Kroger, P.: Density-Connected Subspace Clustering for High Dimensional Data, pp. 246–257. SIAM, Philadelphia (2004)
18. Sequeira, K., Zaki, M.: SCHISM: A new approach for interesting subspace mining. In: Proc. IEEE ICDM, Hong Kong (2004)
19. Agrawal, R., Gehrke, J., Gunopolos, D., Raghavan, P.: Automatic Subspace Clustering of High Dimensional Data for Data Mining Applications. In: ACM SIGMOD (1998)
20. Xie, X.L., Beni, G.: A validity measure for fuzzy clustering. Pattern Analysis and Machine Intelligence 13, 841–847 (1991)

Improved Visual Clustering through Unsupervised Dimensionality Reduction

K. Thangavel[1], P. Alagambigai[2], and D. Devakumari[3]

[1] Department of Computer Science, Periyar University,
Salem - 11, Tamilnadu, India
drktvelu@yahoo.com
[2] Department of Computer Applications, Easwari Engineering College,
Chennai - 89, Tamilnadu, India
alagambigai@yahoo.co.in
[3] Department of Computer Science, Government Arts College,
Dharmapuri,Tamilnadu, India
ramdevshri@gmail.com

Abstract. Interactive visual clustering allows the user to be involved into the clustering through visualizing process via interactive visualization. In order to perform effective interaction in the visual clustering process, the efficient feature selection methods are required to identify the most dominating features. Hence, in this paper an improved visual clustering system is proposed using an efficient feature selection method. The relevant features for visual clustering are identified based on their contribution to the entropy. Experimental results show that the proposed method works well in finding the best cluster.

Keywords: Data Mining, Contribution to Entropy, Feature Selection, Visual Clustering.

1 Introduction

Clustering is a common technique used for unsupervised learning, for understanding and manipulating datasets [6]. It is a process of grouping the data into classes or clusters so that the objects within clusters have high similarity in comparison to one another, but are very dissimilar to objects in other clusters. Cluster analysis is based on a mathematical formulation of a measure of similarity. Generally the clustering algorithms deal with the following issues:

(i)   The definition of similarity of data items
(ii) The characteristics of clusters including size, shape and statistical properties
(iii) The computation cost and error rate of the result.

Many clustering algorithms are proposed regarding these issues [3]. Most of earlier clustering research have been focused on automatic clustering process and statistical validity indices. All the clustering algorithms such as K-Means almost exclude human interaction during the clustering process. Human experts do not monitor the whole clustering process and the incorporation of domain knowledge is highly complicated.

H. Sakai et al. (Eds.): RSFDGrC 2009, LNAI 5908, pp. 439–446, 2009.

In reality, no clustering algorithm is completed, until it is evaluated, validated and accepted by the user.

Visualization is known to be most intuitive method for validating clusters, especially clusters in irregular shape and it can improve the understanding of the clustering structure. Visual representations can be very powerful in revealing trends, highlighting outliers, showing clusters and exposing gaps [6, 7]. To incorporate visualization techniques, the existing clustering algorithms use the result of clustering algorithm as the input for visualization system. The better solution is to combine two processes together, which means to use the same model in clustering and visualization. This leads to the necessity of interactive visual clustering.

Interactive visual clustering has demonstrated great advantage in data mining, since it allows the user to participate in the clustering process by providing the domain knowledge and making better decisions based on his visual perception. This makes "clustering – analysis/evaluation" process to be efficient. VISTA, an interactive visual cluster rendering system, is known to be an effective model, which involves human in the clustering process. It allows the user to observe potential clusters interactively in a series of continuous visual tuning by the parameter called α. The change of α in dominating dimensions resulting in good cluster distribution. When the dimensionality of the dataset increases, identification of the dominating dimensions for visual tuning and visual distance computation process becomes tedious. One common approach to solve this problem is dimensionality reduction.

Dimension reduction or attribute selection aims at choosing a small subset of attributes that is sufficient to describe the data set. It is the process of identifying and removing as much as possible the irrelevant and redundant attributes. Sophisticated attribute selection methods have been developed to tackle the following three problems: (i) reduce classifier cost and complexity, (ii) To improve model accuracy (attribute selection), (iii) improve the visualization and comprehensibility of induced concepts [10].

At the pre-processing and post-processing phase, feature selection/extraction (as well as standardization and normalization) and cluster validation are as important as the clustering algorithms. The benefits of feature selection are twofold: it considerably decreases the computation time of the induction algorithm and increases the accuracy of the resulting mode [2]. All feature selection algorithms fall into two categories: (i) the filter approach and (ii) the wrapper approach [2, 11]. The filter approach basically pre-selects the dimensions and then applies the selected feature subset to the clustering algorithm. The wrapper approach incorporates the clustering algorithm in the feature search and selection. The wrapper approach divides the task into three components (i) feature search (ii) clustering algorithm and (iii) feature subset evaluation.

In this paper, we propose a framework to improve the visual clustering by applying "filter" based dimensionality reduction. In this approach, the most relevant features are identified for visual tuning according to its contribution to the entropy (CE), which is calculated on a leave-one-out basis. Then VISTA has been applied for visual clustering with reduced data.

The rest of the paper is organized as follows. In section 2, the background study is described. The related work is explored in section 3. The overview of visual cluster rendering system is discussed in section 4. The proposed work is discussed in section 5. The experimental analysis is presented in section 6. Section 7 concludes the paper with directions for future research work.

2 Related Work

Clustering of large data bases is an important research area with a large variety of applications in the data base context. Missing in most of the research efforts are means for guiding the clustering process and understand the results. Visualization technology may help to solve this problem since it allows an effective support of different clustering paradigms and provides means for a visual inspection of the results [4]. Since a wide range of users for different environment utilize the visualization models for clustering, it is essential to ease the human computer interaction. One way to ease the human computer interaction is to provide minimum number of features for clustering and analysis. There is large variety of visualization models are proposed during the past decade, but very few are deals with exploring the dataset with minimum features.

The goal of feature selection for clustering is to find the smallest feature subset that best uncovers "interesting natural" grouping (clusters) from data set. Feature selection has been extensively studied in the past two decades. Even though feature selection methods are applied for traditional automatic clustering, visualization models are not utilizing them much. This motivates to the proposed framework.

The issues related to feature selection for unsupervised learning can be found in [2, 11, 13]. Jennifer G. Dy and Broaley [2] proposed a wrapper based feature selection for unsupervised learning, which wraps the search around Expectation-Maximization clustering algorithm. Roy Varshavsky, et. al., [12] proposed a novel unsupervised feature filtering of Biological data based on maximization of Singular Value Decomposition (SVD) entropy. The features are selected based on, (i) simple ranking according to Contribution to the Entropy (CE) values (SR), (ii) forward selection by accumulating features according to which set produces highest entropy (FS1), (iii) forward selection by accumulating features through the choice of the best CCE out of the remaining ones (FS2), (iv) backward elimination (BE) of features with the lowest CE. This proposed work involves the feature selection based on simple ranking.

Interactive clustering differs from traditional automatic clustering in such a way that it incorporates user's domain knowledge into the clustering process. There are wide variety of interactive clustering methods are proposed in recent years [4, 6, 8]. while a very few of them concentrates on feature selection. Keke Chen and Liu, L. [6, 7] proposed VISTA model, an intuitive way to visualize clusters. This model provides a similar mapping such as star coordinates [5], where a dense point cloud is considered a real cluster or several overlapped clusters.

3 VISTA – Visual Cluster Rendering System

Chen and L. Liu [6], [7] proposed a dynamic visualization model; VISTA provides an intuitive way to visualize clusters with interactive feedbacks to encourage domain experts to participate in the clustering revision and cluster validation process.

The VISTA model adopts star coordinates [5]. A k-axis 2D star coordinates is defined by an origin $\vec{o}\,(x_o,\, y_o)$ and k coordinate $S_1,\, S_2,\, S_3,\, \ldots ,\, S_k$ which represents the k dimensions in 2D spaces. The k coordinates are equidistantly distributed on the circumference of the circle C, where the unit vectors are

$$\vec{S}_i = (\cos(\,2\pi i/k\,),\, \sin(\,2\pi i/k\,)),\quad i = 1, 2, 3,\, \ldots ,\, k \qquad (1)$$

The radius c of the circle C is the scaling factor to the entire visualization. Changing c will change the effective size and the detailed level of visualization. Let a 2D point $Q(x,\, y)$ represent the mapping of a k-dimensional max-min normalized (with normalization bounds [-1, 1]) data point $P\,(x_1,\, x_2,\, x_3,\, \ldots ,\, x_k)$ on the 2D star coordinates. $Q(x,\, y)$ is determined by the average of the vector sum of the k vectors $\alpha_i\, x_i\, \vec{S}_i$, $(i = 1, 2,\, \ldots ,\, k)$, where α_i are the k adjustable parameters. This sum can be scaled by the radius c. The VISTA mapping is adjustable by α_i. By tuning α_i continuously, we can see the influence of ith dimension on the cluster distribution through a series of smoothly changing visualizations, which usually provides important clustering clues. The dimensions that are important for clustering will cause significant changes to the visualization as the corresponding α values are continuously changed [7]. Even though the visual rendering is completed within few minutes, the sequential rendering becomes tedious when the number of dimensions is large. In most of the cases, the continuous change of α leads to different patterns, may resulting in incorrect clusters.

4   Improved Visual Clustering Framework

The block diagram of the proposed frame work is demonstrated in Fig .1. The basic idea of the proposed method is to identify important dimensions according to its contribution to the entropy (CE) by a leave-out basis. Features with high CE lead to entropy increase; hence they are assumed to be very relevant to our proposed method. The features of the second group are neutral. Their presence or absence does not change the entropy of the dataset and hence they can be filtered out without much information loss. The third group includes features that reduce the total Singular Value Decomposition (SVD) - entropy (usually $c < 0$). Such features may be expected to contribute uniformly to the different instances, and may just as well be filtered out from the analysis. The relevant features are then applied to VISTA model for clustering process. The step-by-step process of the proposed algorithm is mentioned here under,

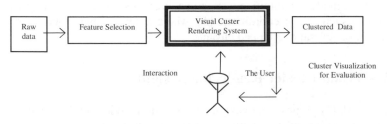

Fig. 1. Improved Visual Clustering Framework

Input : *n* **Data set with underlying Distribution**
Output : *K* **Partitions of Data sets**

Step 1: Find the Contribution to the entropy [12, 14] of the ith feature as

$$CE_i = E(A_{[nXm\,]}) - E(A_{[nXm\,-1]}) \qquad (2)$$

Where the Entropy

$$E = -\frac{1}{\log(N)}\sum_{j=1}^{N} V_j \log(V_j) \qquad (3)$$

And the normalized relative values $\qquad\qquad\qquad\qquad$ (4)

$$V_j = \frac{S_j^2}{\sum_k S_k^2}$$

Where S_j^2 is the eigen values of the nXn matrix $A A^i$

Step 2: Sort the features based on their contribution to the entropy..

Step 3: Group the features as

    i). $CE_i > $ C, features with high contribution

    ii). C > $CE_i > $ C features with average contribution

    iii). $CE_i < $ C features with low (usually negative) contribution

    Where C = Average $(CE\ )$

Step 4: Eliminate the dimensions with average and negative contribution, since they are irrelevant.

Step 5: Explore only the selected dimensions in VISTA.

Step 6: Perform interactive visual clustering with α- tuning until satisfactory results.

Fig. 2. Improved Visual Clustering method

5   Experimental Results and Discussion

Five well known dataset UCI machine learning datasets are used to show the effectiveness of the proposed work (http://www.ics.uci.eedu/~mlearn/). The quality of clusters is assessed by Jaccard coefficient proposed in [1]. The Jaccard coefficient validations are based on the agreement between clustering results and the "ground truth". The experiments are performed based on the domain knowledge obtained from automatic clustering results. The domain knowledge plays a critical role in the clustering process, which is the semantic explanation to the data groups. It often indicates a high level cluster distribution, which may be different from the structural clustering results. Initially the dataset is explored in VISTA. The initial alpha values are set as -0.5. For experimental purpose alpha variation is set as 0.01.

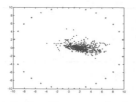   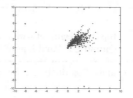

(a) Before Feature selection   (b) After Feature Selection

Fig. 3. Visualization of Breast Cancer Data Set

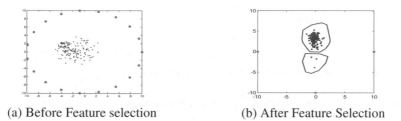

(a) Before Feature selection   (b) After Feature Selection

Fig. 4. Visualization of Hepatitis Data Set

The visualization of Breast Cancer data with the entire set of dimensions is shown in Fig. 3 a) and the visualization of dataset with selected features based on CE is shown in Fig. 3 b) with $\alpha_i = 0.5$. From the visualization results, it is observed the distribution of points obtained by the proposed method is quite different to that of the original data distribution obtained with original sample. The distribution of points with feature selection shows the cluster distribution effectively than original visualization. Since the number of features selected is very less, this eases the visual distance computation process and makes the human – computer interaction process to be more effective. Fig. 4 a) and b) shows the visualization of Hepatitis data set before and after feature selection. From the visualization results, it is observed that feature selection makes the data visualization be more effective than the entire dimension and the cluster distribution is also clearly identified. Similarly Fig. 5 a) and b) and Fig. 6 a) and b) show the visualization of Australian data set and Ionosphere dataset before and after feature selection respectively.

The visual clustering is performed by the user with domain knowledge. Even though the visual rendering performed sequentially the user may vary the α, which leads to different point distribution. And another important issue is every rendering may result in different point distribution. Hence, the visual clustering on each data set performed several times and the best and average results are considered for analysis. The results of visual clustering with entire set of dimensions and with selected features are represented in Table 1. From the results, it is observed that proposed method resulting in better cluster quality compared with clustering using the entire dimension for Breast cancer data set. For the other two data sets it shows similar results before and after feature selection.

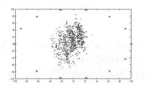
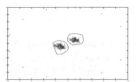

(a) Before Feature selection   (b) After Feature Selection

Fig. 5. Visualization of Australian Data Set

Fig. 6. Visualization of Ionosphere Data Set

Table 1. Comparison of number of features and Cluster quality

Datasets	No. of Records	No. of Clusters	No. of Features		VISTA with Entire dimension		VISTA with Selected Features	
			Before Selection	*After Selection*	*Best*	*Avg*	*Best*	*Avg*
Hepatitis	155	2	19	4	57.24	54.30	65.97	63.24
Australian	690	2	14	13	60.50	60.50	60.50	60.50
Breast Cancer	569	2	32	5	53.21	53.21	53.23	53.18
Dermotology	366	6	34	33	25.65	23.65	31.28	23.65
Ionosphere	351	2	34	16	45.10	43.39	46.15	44.06

6   Conclusion

Most of the feature selection process for clustering is focused on wrapper method. In this proposed feature selection method based on the filter model individual attributes are selected based on their CE value. The features that contain low contribution are considered as irrelevant features, and they are eliminated. The interactive clustering is performed only with the relevant features, thus reduces the number of iterations in the process of computing visual distance and eases the interactive process. The experimental result shows that, the proposed feature selection method for interactive visual cluster rendering system improves the performance of the clustering results. The identification of relevant features with different criteria needs further research.

Acknowledgments. The first author acknowledges for financial assistance received from UGC, New Delhi under grant No: (F-No. 34-105/2008, SR).

References

1. Jiang, D., Tang, C., Zhang, A.: Cluster analysis for gene expression data: a survey. IEEE Transactions on Knowledge and Data Engineering 16(11) (2004)
2. Dy, J.G., Broadly, E.C.: Feature Selection for unsupervised learning. J. Machine Learning Research 5, 845–889 (2004)
3. Guha., G., Rastogi., R., Shim, K.: CURE: An efficient clustering algorithm for large databases. In: Proc. of the ACM SIGMOD, pp. 73–84 (1998)
4. Hinnerburg, A., Keim, D., Wawryniuk, M.: HD-Eye:Visual Mining of High – Dimensional Data. IEEE Computer Graphics and Applications 19(5), 22–31 (1999)
5. Kandogan, E.: Visualizing Multi-dimensional Clusters. Trends and outliers using star coordinates. In: Proc. of ACM KDD, pp. 107–116 (2001)
6. Chen, K., Liu, L.: VISTA: Validating and Refining Clusters via Visualization. Information Visualization 4, 257–270 (2004)
7. Chen, K., Liu, L.: iVIBRATE: Interactive Visualization-Based Framework for Clustering Large Datasets. ACM Transactions on Information Systems 24, 245–294 (2006)
8. desJardins, M., MacGlashan, J., Ferraioli, J.: Interactive visual clustering. Intelligent User Interfaces, 361–364 (2007)
9. Tory, M., Moller, T.: Human Factors in Visualization Research. IEEE Transactions on Visualization and Computer Graphics 10(1) (2004)
10. Mithra, P., Murthy, C.A., Pal, S.K.: Unsupervised Feature Selection using Feature Similarity. IEEE Transactions on Pattern Analysis and Machine Intelligence 24(3), 301–312 (2002)
11. Jouve, P.-E., Nicoloyannis, N.: A filter feature selection method for clustering, pp. 583–593. Springer, ISMIS, Heidelberg (2005)
12. Vayshavsky, R., Gottlieb, A., Linial, M., Horn, D.: Noval Unsupervised Feature Filtering of Biological Data. In: Text Mining and Information extraction, pp. 1–7. Oxford University Press, Oxford (2004)
13. Thangavel, A.P.: Feature selection for Medical database using Rough System. International Journal on Artificial Intelligence and Machine Learning 6(1), 11–17 (2006)
14. Wall, M.E., Rechtsteiner, A., Rocha, L.M.: Singular value Decomposition and Principal component Analysis. In: Berrar, D., Dubitzky (eds.) A Practical approach to Microarray data analysisi, pp. 91–109. Kluwer, Dordrecht

An Improved Cluster Oriented Fuzzy Decision Trees

Shan Su, Xizhao Wang, and Junhai Zhai

Key Lab. of Machine Learning and Computational Intelligence,
College of Mathematics and Computer Science,
Hebei University, Baoding, 071002, China
susan119521@gmail.com

Abstract. In this paper, an improved cluster oriented decision trees algorithm shortly named ICFDT is presented. In this algorithm, fuzzy C-means clustering algorithm (FCM) without instance labels is used to split the nodes and two novel node expanding criteria are proposed. One criterion uses the ratio of homogenous samples in the node to split; the other splits the node by membership degree without labels. The experimental results in artificial and machine learning datasets show that our method can achieve better performance comparing to standard decision tree named C4.5.

Keywords: Decision tree, FCM, Cluster oriented decision trees, Node splitting criteria.

1 Introduction

Decision tree is a greedy algorithm which originating from conceptional learning theory. It is organized from up-to-down, and utilizes divide-and-conquer strategy. The main propose of decision tree is dividing the features space or attributes space of dataset into many small parts in which data are similar with each others. That is the similar data are put together, and then give them labels to illustrate their features. Therefore, decision trees are widely applied in machine learning, expert systems and other fields. Aiming at classifying the dataset and forming decision rules, decision trees algorithm may be one of the most suitable technologies [1].

Traditional decision tree algorithms operate on discrete attributes that assume a finite number of values. In the construction of tree, an attribute is chosen at a time. What's more, a most discriminative attribute is selected and then the tree grows by adding the node whose attribute's values are located at the branches starting from this node. The discriminative power is quantified by some criterion such as entropy, Gini index, etc[2,3]. The fundamental decision trees are predominantly applied in discrete class problems. And the continuous class problems are handled by regression trees[4].

In practical, the attribute's values are continuous, so the discretization of attributes is a must. However, the discretization mechanism should affect the accuracy of tree. According to the defects of traditional decision tree, some methods have been presented to solve this problem. Since L.A.Zadeh first presented the rough set theory in 1965[5], many researchers have combined the rough set theory and decision tree algorithm

H. Sakai et al. (Eds.): RSFDGrC 2009, LNAI 5908, pp. 447–454, 2009.

together, such as fuzzy ID3[6]and so on [7]. In order to solve the discretization problem[8]. Witold Pedrycz first presented clustered-oriented decision tree in 2005 [9]. Because fuzzy clustering is the central concept behind the generalized tree, they will be referred to C-fuzzy decision tree [10,11].

In the development of cluster oriented decision trees, at first we choose the leaf node with the highest heterogeneous value and treat it as a candidate for further refinement. Then the candidate is split into some clusters by clustering algorithm. The process repeats until the decision tree is accomplishment.

This paper is organized in the following manner. Section 2 is an introduction of fuzzy clustering algorithm named fuzzy C-means clustering algorithm. Section 3 provides a new type of cluster oriented decision tree algorithm. In this algorithm, two novel nodes expanding criteria and splitting stopping condition are introduced. A series comparative experiment is presented in Section 4. Section 5 summarizes the primary work of this paper.

2   Fuzzy C-Means Clustering Algorithm

Fuzzy C-means clustering algorithm (FCM) is a typical objective oriented clustering algorithm presented by Bezdek in 1981[12]. It expands from K-means algorithm and divides data set into C clusters by fuzzy logic. In K-means algorithm, each instance is only completely belongs to one cluster. However in FCM, the relationship between each instance and cluster is denoted by a membership function. The membership function is quantified by the similarity degree between sample and the prototype of cluster. This similarity degree is defined with the distance between sample and prototype. Accordingly, the clustering accuracy can be farther improved by the concept of fuzzy logic.

Let $X = \{x_1, x_2, \cdots, x_N\}$ be a training set which consists of N instances $\{x_n\}_{i=1}^{N}$ with n dimensions. As to the format of the dataset, it comes as a family of input-output pair $X = \{x(k), y(k) | k = 1, 2, \ldots N\}$, in which $x(k) = [x_1(k), x_2(k), \ldots x_n(k)]^T \in R^n$. The data points are clustered into C clusters by FCM, each cluster is characterized by prototype $f(i)$. Thus, the similarity degree between each data point and cluster are denoted by a partition matrix, $U = [u_{ik}], i = 1, 2, \cdots, C, k = 1, 2, \cdots, N$ where u_{ik} denotes the similarity between kth data point and ith prototype (here U denotes a family of C-by-N matrices). Likewise, u_{ik} denotes the membership degree between kth data point and ith prototype. The partition matrix U satisfies the following conditions:

1) $u_{ik} \in [0,1]$

2) $\sum_{i=1}^{C} u_{ik} = 1$

3) $\sum_{i=1}^{N} u_{ik} \in [0, N]$

FCM can achieve best clustering result by minimizing the objective function. In each iteration, the minimization of objective function is obtained through updating the partition matrix U and prototype vector of clusters f until satisfies the stopping conditions. The standard objective function J assumes the format

$$j = \sum_{i=1}^{C} \sum_{k=1}^{N} u_{ij}^{m} d_{ij}^{2} \tag{1}$$

with m is a fuzzification factor and is contained in the range of $[1,\infty]$, typically m = 2. The purpose of FCM is allocating the similar data points into a uniform cluster, and giving the cluster a score to illustrate its similarity degree higher than others. The value of objective function J is in inverse proportion to the clustering result. That is the lower value of objective function J, the better clustering result. In order to satisfying the condition that the minimization of objective function J, updating the partition matrix U assumes the format

$$u_{ik} = \frac{1}{\sum_{j=1}^{C} \left(\dfrac{d_{ij}}{d_{kj}} \right)^{2/(m-1)}} \tag{2}$$

The update of prototype is as follows:

$$f_{i} = \frac{\sum_{j=1}^{N} u_{ij}^{m} x_{i}}{\sum_{j=1}^{N} u_{ij}^{m}} \tag{3}$$

The series of iterations is started from a randomly initiated partition matrix and involves the calculations of the prototypes and partition matrices.

3 An Improved Cluster Oriented Decision Trees

3.1 Overall Architecture of Cluster Oriented Decision Trees

Cluster oriented decision trees utilize fuzzy clustering to classify the data points instead of entropy which is commonly used in traditional decision trees. The architecture of decision trees is clustering dataset X into c clusters. So that the dataset X are divided into c groups. Locating the dataset as the root node and putting c prototypes in c leaf nodes of tree. The process of clustering is assigning the every data in the c clusters. That is putting the data in the c leaf nodes. So each cluster is the subset of dataset X. The development of tree is guided by a settled heterogeneity criteria, which quantifies the variance of data points in node. And it is denoted by V. In each of iteration, choosing the leaf node with the highest value of V, clustering this node into to c clusters. This process is repeated until the stopping criteria are meet. The development of cluster oriented decision tree is shown in Fig.1.

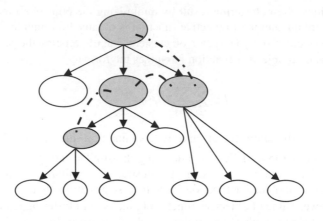

Fig. 1. Architecture of growing a cluster oriented fuzzy decision tree

3.2 Node Expanding Criteria

Clustering algorithm is one of unsupervised learning algorithm. The development of cluster oriented fuzzy decision trees is guided by fuzzy clustering. C-fuzzy decision tree is built with whole attributes and label which is treated as an attribute[9]. So the development of traditional C-fuzzy decision trees contains the target attribute. However, the improved cluster oriented decision trees don't contain the target attribute into the process of FCM. So, in the iteration, the label is removed. Thus, the computation time is lower than traditional C-fuzzy decision trees.

The improved cluster oriented decision tree has two novel node splitting criteria. One criterion considers the label of instance when splitting the nodes. The other criterion does not take the label in count, just splits the node based on the attributes. The two node expanding criteria are described as follows.

A. Assume a threshold ρ as the judgment of whether splitting the node. The heterogeneity criterion is denoted by V. The value of V is big, that is the data points in the node are not homogenous. So the node is inclined to expand. For ith node, there are n_i data points in this node. If the most of data points in this node are homogenous and the amount of these data is m_i, then the heterogeneity of ith node assumes the format

$$V_i = \frac{m_i}{n_i} \tag{4}$$

If $V_i \geq \rho$, it is shown that the data points in ith node are homogenous. This node is not to expand. Otherwise, the node is to split with FCM. Thus it can be seen that V_i denotes the ratio of homogenous instances in this node. This criterion should be understood easily.

B. FCM is one of unsupervised algorithm which is used during the process of splitting nodes. So the node expanding criterion which without target attribute is taken into account. In this criterion, the membership degree of data points in i*th* node is considered. The heterogeneity of i*th* node by the expression

$$V_i = \sum_{i=1}^{c} \sum_{j=1}^{n_i} u_{ij}(1 - u_{ij})$$ (5)

Where n_i denotes the number of data points in i*th* node, c denotes the number of clusters, u_{ij} denotes the numerical value of partition matrix. When the heterogeneity criteria are generalized to classical set, that is the instance is completely belongs to clusters or not. The value of u_{ij} is 1or 0, and the V_i is inclined to 0. The instances are most homogenous in node. Therefore, in fuzzy set, the value of V_i is bigger; the samples in this node are more heterogeneous. This node is subjected to expand.

3.3 Splitting Stopping Condition

There are two splitting stopping conditions of cluster oriented decision tree which are same to C-fuzzy decision trees[9]. One is stopping before splitting; the other is stopping after splitting. The condition of stopping before splitting denotes stopping expanding the node when amount of data points in this node is lower than c. This idea is easy to comprehend. Obviously, we expect the number of data is the multiple of c, such as $2c$, $3c$.

Stopping after splitting criterion estimates the structure of leaf node (the clusters by splitting the node) to decide whether expanding the node. There should be an index to represent the structure of leaf node. If the data obviously belong to the leaf node, the index should be high. It shows the structure of leaf node is better. Otherwise, if the membership degrees of a data point to c clusters are approximately equal, such as 1/c, the index would be low. It shows the structure of leaf node is worse. As mentioned above, we choose φ as the index of structure. For k*th* data point, the index of this data by the expression

$$\varphi_k = 1 - C^C \prod_{i=1}^{C} u_{ik}$$ (6)

When the data are completely belong to a cluster, the value of φ_k is 1. But when the membership degrees of data to each cluster are equal to 1/c, the value of φ_k is 0. Assuming the amount of data point in node is n1, and then the index of structure is as follows

$$\overline{\varphi} = \frac{1}{n_i} \sum_{k=1}^{n_i} \varphi_k = \frac{1}{n_i} \sum_{k=1}^{n_i} (1 - C^C \prod_{i=1}^{C} u_{ik})$$ (7)

Threshold of structure index assumes the format

$$\varphi = 1 - C^C (\frac{1}{C}(1+\alpha))^{C/2} (\frac{1}{C}(1-\alpha))^{C/2}$$ (8)

Growing the trees is constrained by these two conditions. When expanding a node with FCM, it must meet the first condition. The efficiency of splitting is judged by the second condition. If it doesn't satisfy the second stopping condition, we should return to the previous step.

When the nodes expanding are accomplished, the leaf nodes of trees are formed. Now we should give classified labels to the leaf nodes to represent the data in the node. We put the most general classified label as the label of leaf node. The leaf node is labeled by the most of instances with the same label.

4 Experiments

In this section, a series of experiments on an artificial dataset and two machine learning datasets have been conducted. In the experiment, we choose the second node splitting criterion as heterogeneity criterion. So in the whole development of decision trees, the instances labels don't take into account. This process is same to clustering algorithm. The experimental results illustrate that the effectiveness of the improved cluster oriented decision trees.

A. Experiment 1

This experiment concerns two-dimensional (2-D) artificial dataset. The dataset is generated random generator and visualized in Fig.2.

In this artificial dataset, 59 data are labeled by 3 classes. There are 19 1-class, 17 2-class and 13 3-class data in artificial dataset. And all data are linearly separable. The comparative analysis is detailed in Table 1.

From Table 1, we can learn that the improved cluster oriented fuzzy decision trees algorithm has better accuracy and smaller tree size comparing to C4.5. C4.5 could not

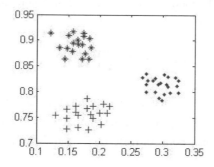

Fig. 2. Two-dimensional (2-D) artificial dataset

Table 1. The comparative analysis on artificial dataset

Decision trees	Accuracy	Depth	Leaf nodes
C4.5	98.3051%	3	3
ICFDT	100%	2	3

divide the dataset completely however our method can be effectiveness. Because of the FCM which we used in splitting the node is effective to cluster distribution.

B. Experiment 2

The following experiments are conducted on two machine learning dataset, Iris and Pima. The comparative experimental results are shown in Table 2 and Table 3.

Table 2. The comparative analysis on Iris

Decision trees	Accuracy	Depth	Leaf nodes
C4.5	96%	5	5
ICFDT	96%	4	5

Table 3. The comparative analysis on Pima

Decision trees	Accuracy	Depth	Leaf nodes
C4.5	73.83%	10	20
ICFDT	78.78%	5	15

As Table 2 and Table 3 reveals, we can learn that the improved cluster oriented fuzzy decision trees without target attribute has the better performance comparing to C4.5. And the tree size is smaller than C4.5.

5 Conclusion

This paper presents an improved cluster oriented fuzzy decision tree shortly named ICFDT. The basic ingredient of proposed decision tree model is fuzzy clustering without instances labels. The construction of cluster oriented tree is guided by successive refinements of the clusters forming the nodes of tree. To illustrate the effectiveness of improved cluster oriented fuzzy decision tree, it is implement on artificial and machine learning dataset. The experimental studies show the ICFDT has better performance and smaller tree size comparing to C4.5. And a lot of experiments should be conducted in future to show the improvement of ICFDT comparing to traditional cluster oriented fuzzy decision trees.

Acknowledgements. This research is supported by the Natural Science Foundation of Hebei Province (F2008000635), by the key project foundation of applied fundamental research of Hebei Province (08963522D), by the plan of 100 excellent innovative scientists of the first group in Education Department of Hebei Province.

References

1. Tom, M.: Mitchell: Machine Learning, pp. 38–56. China Machine Press, Beijing (2003)
2. Quinlan, J.R.: Induction of Decision Trees. Machine Learning 1, 81–106 (1986)
3. Quinlan, J.R.: C4.5: Programs for Machine Learning. Morgan Kaufmann, San Francisco (1993)
4. Breiman, L., Friedman, J.H., Olshen, R.A., Stone, C.J.: Classification and Regression Trees. Belmont, Wadsworth (1984)
5. Zadeh, L.A.: Fuzzy Sets. Information and Control 8, 338–353 (1965)
6. Weber, R.: Fuzzy ID3: A class of methods for automatic knowledge acquisition. In: Proc. 2nd Internet. Conf. on Fuzzy Logic Neural Networks, Iizuka, Japan, July 17-22, pp. 265–268 (1992)
7. Yuan, Y., Shaw, M.J.: Introduction of Fuzzy Decision Trees. Fuzzy Set and Systems 69, 125–139 (1995)
8. Pedrycz, W., Sonsnowski, Z.A.: Designing Decision Trees with the Use of Fuzzy Granulation. IEEE Trans. Syst., Man, Cybern. A, Syst., Humans 30(2), 151–159 (2000)
9. Pedrycz, W., Sonsnowski, Z.A.: C-Fuzzy Decision Trees. IEEE Trans. Syst., Man, Cybern. C, Applications and Reviews 35(4), 498–511 (2005)
10. Yang, S.-B., Chang, C.-C.: New C-Fuzzy Decision Tree with Classified Points. Journal of Electronic Imaging 17(3), 017–033 (2008)
11. Krishnamoorthi Makkithaya, N.V., Reddy, S., Dinesh Acharya, U.: Improved C-Fuzzy Decision Tree for Intrusion Detection. In: Proceedings and World Academy of Science, Engineering and Technology, vol. 32 (2008) ISSN 2070-3740
12. Bezdek, J.C.: Pattern Recognition with Fuzzy Objective Function Algorithms, pp. 13–93. Plenum, New York (1981)

Combining Naive-Bayesian Classifier and Genetic Clustering for Effective Anomaly Based Intrusion Detection

S. Thamaraiselvi, R. Srivathsan, J. Imayavendhan,
Raghavan Muthuregunathan, and S. Siddharth

Department of Information Technology, Anna University, Chennai
stselvi@annauniv.edu,
{srivathsmit,imayavendhanj,raghavan.mit,mitsiddharth}@gmail.com

Abstract. Network Intrusion detection systems have become unavoidable with the phenomenal rise in internet based security threats. Data mining technique based Intrusion Detection System, have the added advantage of processing large amount of data speedily. However, success rate is dependent on selecting the optimal set of features here. Given an optimal set of features and a good training data set, Bayesian classifier is known for its simplicity and high accuracy. On the other hand, clustering techniques have the flexibility to detect novel attacks even when training set is not present. Therefore, combining the results of both classification and clustering techniques can improve the performance of Intrusion Detection systems greatly. Our project aims at building flexible Intrusion Detection system by combining the advantages of Bayesian classifier and the genetic clustering algorithm. It was tested with KDD Cup 1999 dataset by supplying it with a good training set and a minimal one. In the first case, it produced excellent results, while in the second case it gave consistent performance.

Keywords: NIDS, intrusion detection, Anomaly, Genetic Algorithm, Feature selection, Naïve Bayesian classifier, Genetic clustering.

1 Introduction

There are two categories of intrusion detection systems – Misuse and anomaly. Misuse intrusion detection systems are signature-based systems and can detect only known attacks. Anomaly detection systems report deviation from normal profile as an attack. They have the ability to find novel attacks at the expense of high false positive rate. Data mining approaches, which can handle the large amount of data required by both these IDSs, have been used to increase the success rate of both these IDSs. Both supervised and unsupervised learning techniques have been used for designing IDSs.

Supervised techniques have high accuracy, but require a good training set. However, for detecting intrusions obtaining a training set is very difficult. As a compromise, in most cases a training set containing well-known attacks is obtained and used. Clearly, novel attacks cannot be detected in such cases. Applying unsupervised learning technique to IDSs has a major advantage as they do need a training set.

H. Sakai et al. (Eds.): RSFDGrC 2009, LNAI 5908, pp. 455–462, 2009.

Nevertheless, these techniques have low accuracy when compared to supervised learning techniques. Thus it can be inferred that supervised learning techniques provide better prediction rate when relevant data present in the training set. On the other hand, in cases where relevant information is absent in the training set and in cases where novel attacks have to be detected unsupervised learning techniques perform better.

In this paper, we define architecture for an IDS that combines the advantages of Naive Bayesian Classifier and Genetic Clustering algorithm. Furthermore, to increase efficiency, Feature selection using genetic algorithm is done to prune the feature set. Simulations were carried out to test the efficiency of our model by comparing it with the efficiency of each of the individual supervised and unsupervised technique. Results show that, our IDS performs better even in the absence of a good training set.

The rest of this paper is organized as follows. Section 2 explains various researches related to our IDS. Section 3 defines the architecture of our IDS and its significance. Section 4 discusses results obtained when tested with KDD Cup 1999 dataset. Section 5 concludes our work with suggestions on future work.

2 Related Work

Data mining techniques are widely used in IDS as it can handle large data set and allows automation of IDS. Minnesota intrusion detection system [6] efficiently used data mining techniques like outlier detection for anomaly detection. But, the major disadvantage of the MINDS system is that it requires a human analyst to verify network connections.

Pedro Domingos and Michael Pazzani in their comparative research [8] proved that the performance of Bayesian classification does not improve much by removing attribute dependencies. Following this research, our system uses the simple Bayesian classifier, which does not consider attribute dependencies. In Bayesian anomaly detection[2], if the value of maximum posterior probability for a data tuple is too low, it is considered as an anomaly. Such tuples are grouped based on the similarities in their attribute values and each group is given a label. The *anomalous* class used in our anomaly detection module is a simplification of this approach.

Among the many clustering techniques in vogue, Hae-sang-Park et al.'s k-means-like clustering for medoids [3] is very efficient. This clustering technique is similar to k-means clustering except that it uses medoids instead of centroids. Outliers will affect the results of k-means clustering, which is prevented in k-medoid clustering. Our IDS system uses this clustering technique and the clusters formed are further optimized by genetic algorithm.

Leonid Portnoy et al. [7] were the first to use an unsupervised (clustering) technique for detecting intrusions based on two assumptions. These assumptions form the basis for all clustering based intrusion detection. Liu et al. proposed a further improvement by using Genetic algorithm[5] for determining the number of clusters (k). But both these techniques have less accuracy when applied to intrusion detection, as these assumptions can go wrong in many cases.

Feature subset selection is an important part of IDS because it prunes irrelevant features and improves efficiency. Fuzzy-genetic approach for IDS by Fries in [1], selects an optimal subset of features using Genetic algorithm. In their feature selection, the number of features selected can be ignored while calculating fitness because it does not contributes much to the efficiency. Research on Feature subset Selection Bias [4], advocates that the usage of same dataset for feature selection and learning does not have adverse effects on classification. Hence, we intend to use the same dataset for both training and feature selection.

3   Architecture of Our IDS

The Fig. 1 shows the complete architecture of our IDS. There are three major phases in our Intrusion Detection System. In the first phase, the relevant features are selected using Genetic algorithm. These set of features are used by both Bayesian classification and Genetic clustering modules. In the second phase, clusters are formed by the Genetic clustering module and the Bayesian learns from the training set. In the third phase, the Bayesian classification module is given with packets from test data. Bayesian classifier classifies the input packet either under normal or known-attack classes or under *anomalous* class (if $M_p < 0.1$). If the input packet is classified under anomalous class, clustering results are considered for intrusion detection.

Our Intrusion Detection System has three main modules:

- Feature selection that selects relevant features.
- Bayesian classification module
- Genetic clustering technique- forms clusters of both training and test set.

3.1   Feature Selection Using Genetic Algorithm

The KDD cup 1999 dataset [9] we used has 41 features for each connection in the network. A possible solution can be obtained by choosing a subset of these 41 fea

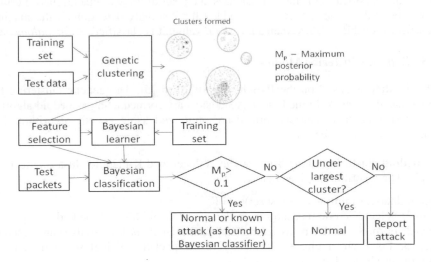

Fig. 1. Architecture of our IDS

Table 1. Thirteen Features selected by genetic algorithm

Feature name	Description
protocol_type	Type of protocol
service	Network service on destination
flag	Status of connection: normal or error
src_bytes	Number of data bytes from source to destination
land	1 if connection is from/to same port
num_compromised	number of "compromised" conditions
su_attempted	1 if "su root" command attempted
num_file_creations	number of file creation operations
num_shells	number of shell prompts
num_outbound_cmds	number of outbound commands in an ftp session
is_host_login	1 if login belongs to hot list
rerror_rate	% of connections with REJ errors
diff_srv_rate	% of connections to different services

tures. Thus, each chromosome to be used for Genetic Algorithm should have *41-bits* each bit representing the selection of a feature. Fitness function for the feature selection for a chromosome 's' is, accuracy of naïve Bayesian classifier. Genetic algorithm selected the *thirteen* features in Table 1.

3.2 Bayesian Classification Module

Although Bayesian classification is simple, its performance is comparable with state-of-the-art classification techniques [8]. The predefined set of classes used in our Bayesian classification is the normal class and a set of known-attacks class. Further, a new class called *anomalous* class is added for Anomaly detection. If the maximum posterior probability is less than a *threshold* value, it is classified as *anomalous* class.

3.3 Genetic Clustering Module

This module is based on the IDBGC algorithm in [5]. The genetic clustering technique has two stages. In the first stage, clusters are formed using k-medoid algorithm. In the second stage, clusters formed are optimized using genetic algorithm. These stages are explained below.

K-medoid clustering algorithm. Park et al developed K-medoid clustering technique in [3]. The algorithm is given below:

Input: dataset (training set + test set) and K
Step 1: Select K random packets from *dataset* as K initial set of medoids
Step 2: Form K clusters by assigning each packet in the *dataset* to its nearest medoid
Step 3: For each of clusters formed, find the packet which if selected as medoid minimizes cluster distance.
Step 4: Repeat steps 2 and 3 until the new medoids obtained are the same.

Genetic Optimization Step. As the number of possible attacks in the network is unknown, determining the value of K is not possible. So the clusters formed are optimized by combining some of them to form clusters with lesser intra-cluster distance and with greater inter-cluster distance [5] using genetic algorithm.

A chromosome contains K bits each representing one of the K clusters. If i[th] bit is '1' then it means that i[th] cluster is selected and if the bit is '0' then the cluster is not selected. The packets in the unselected clusters are assigned to the nearest medoid of the selected clusters. Thus, some clusters are combined to form a new set of clusters. Fitness is based on inter-cluster distance and intra-cluster distance. The fitness function we used for *l* clusters $(c_1, c_2, ..c_l)$ formed for a chromosome is given by,

$$fitness = \sum_{i=1}^{l} w_1 * c\_dis(c_i) + \sum_{i,j=1 \ to \ l, i \,!=j} w_2 * min\_dis(c_i, c_j)$$

In the above fitness function, c_dis (cluster distance) stands for the sum of distance of every point in cluster to the medoid and is considered as intra cluster distance. Also, min_dis (minimum distance) stands for the minimum distance between given two clusters and is considered as the inter cluster distance.

3.4 Detecting Attacks from Clusters Formed

After cluster formation using the above genetic clustering method, each cluster is categorized as normal or as an attack based on assumptions introduced by Leonid Portnoy et. al. in [7] it can be inferred that:

- Largest cluster formed contains only normal packets
- Other clusters contains only attack packets

3.5 Combining Results of Clustering and Classification

First, each data tuple is given as input to both Bayesian anomaly detection module and k-medoid clustering module. When Bayesian anomaly detector classifies the current tuple to any of the predefined set of classes (known attack set and normal) the tuple is assigned to that class. Otherwise, if Bayesian anomaly detector classifies a data tuple as *anomalous* class, unsupervised learning results are used. Thus if a packet is classified under *anomalous* class and is present in the largest cluster, then it is assumed to be normal. Similarly, if a packet is classified under anomalous class and is not present in the largest cluster it is considered as an attack.

3.6 Significance of IDS Architecture

The IDS architecture is more flexible. The basis for this IDS is the Bayesian classification module. A simple Bayesian classifier can classify the input data tuple only to any of the predefined set of classes (known attacks and normal). However, it fails to in the following two cases:

- The data tuple under consideration belongs to predefined set of classes but the Bayesian learner is given with insufficient training data to classify it.
- The current data tuple does not belong to any of the predefined set of classes.

In the above cases, our Bayesian classifier classifies the current tuple under *anomalous* class and the results of genetic clustering are considered. Thus our IDS is able to detect cases where relevant information is absent in training data and intelligently use results of genetic clustering technique in such cases.

Also, the results of genetic clustering are not used when relevant information is present in the training set (Maximum posterior probability > 0.1). This is because:

- Simple Bayesian classification is more accurate [8].
- Clustering has low accuracy than supervised learning techniques.

As a scope for further improvement, both the training set and the test data are given as input to clustering module. This increases the accuracy of clustering. The threshold value 0.1for M_p was assigned as threshold for M_p empirically.

4   Experimental Results and Discussion

To prove the efficiency of our IDS we compare the accuracy of IDS with individual Bayesian classification and the genetic clustering techniques. Several small datasets were obtained from KDD Cup dataset each containing about 10000 packets. Each dataset contains tuples of all attack types present in the dataset.

4.1   Results of Clustering and Bayesian Classification

Bayesian classification was tested using ten datasets by giving a good training set. Similarly, genetic clustering was also tested with ten datasets. Operating characteristic of the both these tests are shown in Fig. 2. The average accuracy for Bayesian classification with good training set is 99.4 % and the average accuracy of Genetic clustering is 60%.

4.2   Overall Performance of Our IDS

The overall performance of our IDS is very much dependent on the training set given as input. If the training set contains enough data, the performance of Bayesian

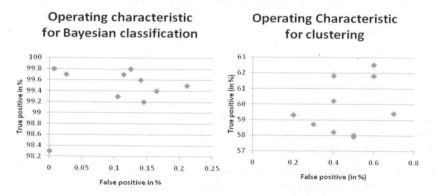

Fig. 2. Operating characteristic for Bayesian classification and genetic clustering for ten datasets

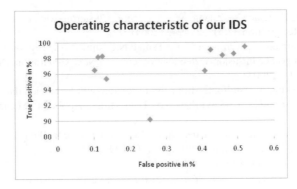

Fig. 3. Operating characteristic for our IDS for 10 datasets

classification is enough to detect attacks. Otherwise, the influence of clustering module is greater.

Ten datasets containing both known and novel attacks were obtained to test our IDS(Fig. 4) Operating *Average accuracy* of our IDS *was about 97%* and average *false positive rate was about 0.3%*.

5   Conclusion and Future Work

Obtaining a sound training set for IDS is fast becoming an impossible challenge and with the advent of newer and more virulent attacks in the network a flexible IDS is required. We have developed such an IDS by combining two popular techniques – Bayesian classification and Genetic clustering. Satisfactory results which substantiate its flexibility were obtained. Also, the use of both training set and test set for the clustering technique has led to a significant improvement in its accuracy.

Clustering techniques still require improvements to be used more effectively with IDS. Attacks like smurf involve large number of similar packets which are often categorized along with normal packets by clustering techniques. A more problem-specific distance metric with slackened assumptions about network packets must be used to increase the accuracy of these techniques.

References

1. Fries, T.P.: A fuzzy genetic approach for intrusion detection. In: Proceedings of the GECCO conference companion on Genetic and evolutionary computation, pp. 2141–2146 (2008)
2. Menzies, T., Allen, D., Orrego, A.: Bayesian Anomaly Detection. In: Workshop on Machine Learning Algorithms for Surveillance and Event detection at 23rd ICML, Pittsburgh (2006)
3. Park, H.-s., Lee, J.-s., Jun, C.-h.: A K-means-like Algorithm for K-medoids Clustering and Its Performance. In: Proceedings of the 36th CIE Conference on Computers and Industrial Engineering, pp. 1222–1223 (2006)
4. Singhi, S.K., Liu, H.: Feature Subset Selection Bias for Classification Learning. In: Proceedings of the 23rd International Conference on Machine Learning, Pittsburgh (2006)

5. Liu, Y., Chen, K., Liao, X., Zhang, W.: A Genetic Clustering Method for Intrusion Detection. Pattern Recognition 37(5), 927–942 (2004)
6. Ertoz, L., Eilertson, E., Lazarevic, A., Tan, P.N., Kumar, V., Srivastava, J., Dokas, P.: MINDS - Minnesota Intrusion Detection System. In: Next Generation Data Mining. MIT Press, Cambridge (2004)
7. Portnoy, L., Esking, E., Stolfo, S.: Intrusion Detection with Unlabeled data using clustering. In: Proceedings of ACM CSS Workshop on Data Mining Applied to Security, DMSA 2001 (2001)
8. Domingos, P., Pazzani, M.: On the Optimality of the Simple Bayesian Classifier under Zero-One Loss. Machine Learning 29, 103–130 (1997)
9. KDDCup 1999 Dataset (1999),
   http://kdd.ics.uci.edu/databases/kddcup99/kdd.html

Ant Colony Optimisation Classification for Gene Expression Data Analysis

Gerald Schaefer

Department of Computer Science
Loughborough University
Loughborough, U.K.
Gerald.Schaefer@ieee.org

Abstract. Microarray studies and gene expression analysis have received much attention over the last few years and provide promising avenues towards the understanding of fundamental questions in biology and medicine. In this paper we investigate the application of ant colony optimisation (ACO) based classification for the analysis of gene expression data. We employ cAnt-Miner, a variation of the classical Ant-Miner classifier, to interpret numerical gene expression data. Experimental results on well-known gene expression datasets show that the ant-based approach is capable of extracting a compact rule base and provides good classification performance.

1   Introduction

Microarray expression studies measure, through a hybridisation process, the levels of genes expressed in biological samples. Knowledge gained from these studies is deemed increasingly important due to its potential of contributing to the understanding of fundamental questions in biology and clinical medicine. Microarray experiments can either monitor each gene several times under varying conditions, or analyse the genes in a single environment, but in different types of tissue. In this paper, we focus on the latter where one important aspect is the classification of the recorded samples. This can be used to categorise different types of cancerous tissues as in [1] where different types of leukemia are identified, or to distinguish cancerous tissue from normal tissue as done in [2] where tumor and normal colon tissues are analysed.

One of the main challenges in classifying gene expression data is that the number of genes is typically much higher than the number of analysed samples. Also, is it not clear which genes are important and which can be omitted without reducing the classification performance. Many pattern classification techniques have been employed to analyse microarray data. For example, Golub *et al.* [1] used a weighted voting scheme, Fort and Lambert-Lacroix [3] employed partial least squares and logistic regression techniques, whereas Furey *et al.* [4] applied support vector machines. Dudoit *et al.* [5] investigated nearest neighbour classifiers, discriminant analysis, classification trees and boosting, while Statnikov *et al.* [6] explored several support vector machine techniques, nearest neighbour classifiers, neural networks and probabilistic neural networks. In several of these studies it has been found that no one classification algorithm is performing best on all datasets (although for several datasets SVMs seem to perform best) and that

H. Sakai et al. (Eds.): RSFDGrC 2009, LNAI 5908, pp. 463–469, 2009.

hence the exploration of several classifiers is useful. Similarly, no universally ideal gene selection method has yet been found as several studies [7,6] have shown.

In our previous work [8] we presented a fuzzy rule-based classification system to analyse microarray expression data. Gene expression data was described by fuzzy sets, and rules of combinations of these sets were employed to arrive at a classification. In our experiments, we showed our approach to afford good classification performance for this type of problem and outperforming CART [9], a classical rule-based classifier. However, similar to many rule-based approaches this method suffers from the course of dimensionality leading to a rule base consisting of a large number of rules. This was addressed through a rule splitting stage leading to a smaller number of simpler rules, but only partially so. In further work [10], we derived a more compact rule base using a genetic algorithm that assesses the fitness of individual rules and selects a rule ensemble that maximises classification performance.

In this paper we show how a classification system based on ant colony optimisation can be applied to the problem of analysing gene expression data. We employ cAnt-Miner [11], a variation of the classical Ant-Miner classifier [12], which is inherently capable of interpreting the numerical gene expression data. Experimental results on various well-known gene expression datasets show that the ant-based approach is capable of extracting a compact rule base for classifying gene expression levels while providing good classification performance.

2   Ant Colony Based Classification

2.1   Ant Colony Optimisation

Ant colony optimisation (ACO) [13] is a relatively recent computational intelligence paradigm that is inspired by the collective behaviour of natural ants. In ant colonies, each individual ant performs its own task independently, yet the various individual tasks are related, and through collaboration it is possible to solve complex problems. In particular, ants are capable of finding the shortest path between their nest and a food source based only on local information. They are furthermore capable to adapt to changes in the environment. To achieve this, ants communicate with each other by means of pheromone trails. Ants leave pheromone as they move around in the environment, while other ants can follow pheromone paths. Therefore, the more ants follow a certain trail, the more attractive this trail becomes to other ants, hence leading to the equivalent of a positive feedback loop where the probability of following a certain path is proportional its 'quality'.

Essentially ACO can be seen as an agent-based system that simulates ants in order to solve real world optimisation problems. Each agent represents an ant and acts in an environment which defines the search space. Each path followed by an ant describes one candidate solution in that search space. Once such a candidate solution is evaluated using an objective funtion, the amount of pheromone that is associated with the corresponding trail is modified according to the quality of the candidate solution. When ants decide which path to follow, a path with a higher amount of associated pheromone is more likely to be chosen. Using this strategy, ants will eventually converge towards the optimal solution (i.e. the shortest path) of the given problem.

Ant colony optimisation has been successfully applied to a variety of real world problems and has been shown to provide a robust and versatile method for optimisation [13].

2.2 ACO-Based Classification

Ant colony optimisation can also be employed for pattern classification as has been shown in [12] with the introduction of the Ant-Miner algorithm. The basic idea is to perform classification using a rule base and to optimise this rule base through ant colony optimisation. In Ant-Miner, each path constructed by an ant represents one rule of the rule base. Each such rule has the following form:

```
IF <term1 AND term2 AND ...>
THEN <class>
```

with each term being defined by a triple

```
<attribute, operator, value>
```

such as <Day = Monday> and `class` representing the consequent class, i.e. one of a set of predefined categories.

Algorithm 1 presents a high-level overview of the Ant-Miner algorithm. Ant-Miner starts with an empty rule base and successively adds rules one by one. To construct a new rule, an ant is initialised with an empty rule (i.e. no terms in the antecedent part) and adds one term at a time to the antecedent. Terms are added until a term would cause the rule to cover fewer than a preset number of training samples, or all possible attributes have been added. Once the rule has been constructed, a pruning step is applied to remove any irrelevant terms. Then, the consequent class of the rule is determined as the most frequent class of the covered training samples. Rule construction is continued until a predefined number of rules has been built or an already existing rule is recreated. Of the created rules, the best one is chosen based on a quality measure which is often defined as the product of sensitivity and specificity. This rule is then added to the rule base and the process is repeated until all but a predefined number of training samples are covered by the rule base.

2.3 cAnt-Miner

While Ant-Miner has been shown to provide good classification performance coupled with a compact rule base [12], one downside of the algorithm is that it is only capable of processing nominal data, i.e. data that can be described by a finite number of nominal or discrete values. Therefore as such it cannot be applied to handle continuous numerical data directly. The only way to cope with continuous data is hence to discretise them in a pre-processing step, e.g. using the C4.5-Disc method [14].

The cAnt-Miner algorithm [11] takes a different, integrated approach to ant-based classification. Discrete intervals are created on-the-fly, and hence no pre-processing step is required. This dynamic discretisation is directly incorporated into the rule construction stage of the Ant-Miner algorithm and consequently supports also terms that include < and ≥ operators. The discretisation itself is based on an entropy measure that describes the impurity of a collection of samples.

Algorithm 1. Pseudo-code of Ant-Miner algorithm

Initialise training set
Reset rule base
repeat
    Initialise all paths
    repeat
        Use ant to construct a rule
        Prune rule
        Determine consequent class of rule
        Update pheromones
    until (stopping criterion)
    Select best rule from constructed rules
    Add selected rule to rule base
    Eliminate training samples covered by selected rule
until (stopping criterion)

3   ACO Classification of Gene Expression Data

In our work, we apply the ant-based cAnt-Miner classifier to the problem of analysing gene expression data. In particular, we perform experiments on two popular gene expression datasets and compare the performance of our approach with the results obtained using the fuzzy rule-based classfier from [8] and the hybrid fuzzy/GA classifier from [10]. We found that cAnt-Miner is rather robust with respect to chosen parameters; we typically employ 3000 ants for rule discovery, and select the minimum number of samples covered per rule from the range [3;5], the maximum number of uncovered training samples from [1;3], while the number of rules used for testing convergence was set to 5.

The first dataset we inspect is the Colon dataset from [2]. This dataset is derived from colon biopsy samples. Expression levels for 40 tumor and 22 normal colon tissues were measured for 6500 genes using Affymetrix oligonucleotide arrays. The 2000 genes with the highest minimal intensity across the tissues were selected. We pre-process the data following [5], i.e. perform a thresholding [floor of 100 and ceil of 16000] followed by filtering [exclusion of genes with max/min < 5 and (max-min) < 500] and $\log_{10}$ transformation. We then select the top 50 respectively the top 100 genes as input to the classifiers.

The results on this datasets are given in Table 1, in terms of correctly classified samples (CR), falsely classified samples (FR), and classification accuracy. It can be seen that our ACO-based classifier matches the classification accuracy of the fuzzy classifier for the case of 100 selected genes, and outperforms the hybrid fuzzy/GA classifier for this configuration. Based on the top 50 genes, both the fuzzy and the hybrid fuzzy classifier perform slightly better than our cAnt-Miner approach. However, it should also be noted, that as reported in [8], CART [9], a conventional rule-based classifier, performs rather poorly on this dataset with a classification accuracy of only 77.42% respectively 72.58% based on 50 respectively 100 genes, and cAnt-Miner clearly outperforms this.

Also given in Table 1 is information on the complexity of the derived rule base in terms of the number of rules and the number of attributes per rule. For the fuzzy

Table 1. Classification performance on Colon dataset given in terms of number of correctly classified samples (CR), falsely classified or unclassified samples (FR), and classification accuracy. Also given are the number of rules in the rule base and the number of terms of each rule. Results are presented based on 50 and 100 gene expressions respectively.

n	method	CR	FR	Accuracy	#rules	#attributes/rule
50	fuzzy classifier [8]	53	9	85.48	$3^2 \cdot \binom{50}{2} = 11025$	2
	fuzzy/GA classifier [10]	52	10	83.87	20	up to 50
	ACO classifier	51	11	82.26	5	2
100	fuzzy classifier [8]	51	11	82.26	$3^2 \cdot \binom{100}{2} = 44550$	2
	fuzzy/GA classifier [10]	50	12	80.11	20	up to 100
	ACO classifier	51	11	82.65	5	2

classifier from [8] these are fixed; each rule has 2 attributes while the number of rules in the rule base depends on the number of fuzzy sets employed and the number of total attributes (i.e. the number of genes). With more than 10,000 respectively more than 40,000 rules, the generated rule bases are vast and hence the advantage of rule-based systems, namely interpretability, is clearly lost. The hybrid fuzzy/GA classification system has a fixed rule based size of 20 rules, however the number of attributes employed in each rule is variable and can reach the total number of attributes. Again, such rules are not easily interpreted and it is hence questionable whether useful insights can be inferred from such a rule base.

In contrast, our ACO-based classifier produces very compact rule bases with most instances consisting of only 5 rules of 2 attributes each. This is clearly much simpler than those generated by the other approaches, and hence, despite the slightly lower classification rate, represents the only system where true insights about the analysed data can be inferred from the generated rule base. An example of such a rule base in given in Figure 1.

```
IF geneexp002 < 2.8888973 AND geneexp011 >= 2.5680682
    AND geneexp052 >= 2.2540251 THEN 1
IF geneexp003 >= 2.5163602 AND geneexp059 < 2.5732967 THEN 0
IF geneexp094 >= 2.4165843 AND geneexp022 < 3.7536059 THEN 1
IF geneexp031 < 2.9691642 THEN 0
IF <empty> THEN 1
```

Fig. 1. Example of rule base generated by the ACO classifier

The second gene expression dataset we analysed is the Leukemia dataset reported in [1]. Here, bone marrow or peripheral blood samples were taken from 47 patients with acute lymphoblastic leukemia (ALL) and 25 patients with acute myeloid leukemia (AML). The ALL cases can be further divided into 38 B-cell ALL and 9 T-cell ALL samples and it is this 3-class division that we are basing our experiments on rather than the simpler 2-class version which is more commonly referred to in the literature.

Table 2. Classification performance on Leukemia dataset. The table is laid out in the same fashion as Table 1.

n	method	CR	FR	Accuracy	#rules	#attributes/rule
50	fuzzy classifier [8]	68	4	94.44	$3^2 \cdot \binom{50}{2} = 11025$	2
	fuzzy/GA classifier [10]	69	3	96.29	20	up to 50
	ACO classifier	67	5	93.06	4	2
100	fuzzy classifier [8]	71	1	98.61	$3^2 \cdot \binom{100}{2} = 44550$	2
	fuzzy/GA classifier [10]	68	4	94.44	20	up to 100
	ACO classifier	66	6	91.67	4	2

Each sample is characterised by 7129 genes whose expression levels where measured using Affymetrix oligonucleotide arrays. The same preprocessing steps as for the Colon dataset are applied, and again the top 50 respectively top 100 genes are extracted.

Results on the Leukemia dataset are given in Table 2, in the same form as for the Colon dataset. Again, the ACO classifier affords slightly lower classification performance compared to the fuzzy rule-based classifier and the fuzzy/GA classification system. However, compared to the CART classifier, which is reported to achieve a classification accuracy of only 65.28% based on 50 respectively 62.50% based on 100 genes [8], ant-based classification is clearly superior. With a rule base of typically only 4 rules with 2 attributes, our approach is clearly producing the most interpretable rule bases. Compared to this, the fuzzy/GA classifier has 20 rules with a varying number of attributes, while the pure fuzzy classifier relies on tenths of thousands of individual rules.

4   Conclusions

In this paper we have proposed the use of an ant colony optimisation-based classifier for the analysis of gene expression data. In particular, we employ the cAnt-Miner classification algorithm to interpret the continuous real number data of gene expression levels and show that our approach produces very compact rule bases of only a few short rules while providing classification performance similar to other, more complex classification systems. These small rule bases could then form a starting point to gain further insight into the analysis of the inspected datasets.

References

1. Golub, T.R., Slonim, D.K., Tamayo, P., Huard, C., Gaasenbeek, M., Mesirov, J.P., Coller, H., Loh, M.L., Downing, J.R., Caligiuri, M.A., Bloomfield, C.D., Lander, E.S.: Molecular classification of cancer: class discovery and class prediction by gene expression monitoring. Science 286, 531–537 (1999)
2. Alon, U., Barkai, N., Notterman, D., Gish, K., Ybarra, S., Mack, D., Levine, A.: Broad patterns of gene expression revealed by clustering analysis of tumor and normal colon tissues probed by oligonucleotide arrays. Proc. Natnl. Acad. Sci. USA. 96, 6745–6750 (1999)

3. Fort, G., Lambert-Lacroix, S.: Classification using partial least squares with penalized logistic regression. Bioinformatics 21(7), 1104–1111 (2005)
4. Furey, T., Cristianini, N., Duffy, N., Bednarski, D., Schummer, M., Haussler, D.: Support vector machine classification and validation of cancer tissue samples using microarray expression data. Bioinformatics 16(10), 906–914 (2000)
5. Dudoit, S., Fridlyand, J., Speed, T.: Comparison of discrimination methods for the classification of tumors using gene expression data. Journal of the American Statistical Association 97(457), 77–87 (2002)
6. Statnikov, A., Aliferis, C., Tsamardinos, I., Hardin, D., Levy, S.: A comprehensive evaluation of multicategory classification methods for microarray expression cancer diagnosis. Bioinformatics 21(5), 631–643 (2005)
7. Liu, H., Li, J., Wong, L.: A comparative study on feature selection and classification methods using gene expression profiles and proteomic patterns. Gene Informatics 13, 51–60 (2002)
8. Schaefer, G., Nakashima, T., Yokota, Y., Ishibuchi, H.: Fuzzy classification of gene expression data. In: IEEE Int. Conference on Fuzzy Systems, pp. 1090–1095 (2007)
9. Breiman, L., Friedman, J., Olshen, R., Stone, R.: Classification and Regression Trees. Wadsworth (1984)
10. Schaefer, G., Nakashima, T.: Data mining of gene expression data by fuzzy and hybrid fuzzy methods. IEEE Trans. on Information Technology in Biomedicine (to appear)
11. Otero, F., Freitas, A., Johnson, C.: cAnt-Miner: An ant colony classification algorithm to cope with continuous attributes. In: Dorigo, M., Birattari, M., Blum, C., Clerc, M., Stützle, T., Winfield, A.F.T. (eds.) ANTS 2008. LNCS, vol. 5217, pp. 48–59. Springer, Heidelberg (2008)
12. Parpinelli, R., Lopes, H., Freitas, A.: Data mining with an ant colony optimization algorithm. IEEE Trans. Evolutionary Computation 6(4), 321–332 (2002)
13. Dorigo, M., Stuetzle, T.: Ant Colony Optimization. MIT Press, Cambridge (2004)
14. Kovahi, R., Sahami, M.: Error-based and entropy-based discretization of contiuous features. In: 2nd Int. Conference on Knowledge Discovery and Data Mining, pp. 114–119 (1996)

A Comparative Performance Analysis of Multiple Trial Vectors Differential Evolution and Classical Differential Evolution Variants

G. Jeyakumar and C. Shunmuga Velayutham

Amrita School of Engineering
Amrita Vishwa Vidyapeetham, Coimbatore
Tamil Nadu, India
g_jeyakumar@ettimadai.amrita.edu,
cs_velayutham@ettimadai.amrita.edu

Abstract. In this paper we present an empirical , comparative performance, analysis of fourteen variants of Differential Evolution (DE) and Multiple Trial Vectors Differential Evolution algorithms to solve unconstrained global optimization problems. The aim is (1) to compare Multiple Trial Vectors DE, which allows each parent vector in the population to generate more than one trial vector, against the classical DE and (2) to identify the competitive variants which perform reasonably well on problems with different features. The DE and Multiple Trial Vectors DE variants are benchmarked on 6 test functions grouped by features – unimodal separable, unimodal nonseparable, multimodal separable and multimodal non-separable. The analysis identifies the competitive variants and shows that Multiple Trial Vectors DE compares well with the classical DE.

Keywords: Differential Evolution, Multiple Trial Vectors Differential Evolution, differential mutation strategies, probability of convergence, performance analysis, unconstrained global optimization.

1 Introduction

Differential Evolution (DE), proposed by Storn and Price [1,2], is a very simple but very powerful stochastic global optimizer for continuous search domain. It has been proven a robust global optimizer and has been successfully applied to many global optimization problems [3,4] and real-world applications. Like all Evolutionary Algorithms (EA's), DE is a stochastic population-based search method that employs repeated cycles of recombination and selection to guide the population towards the vicinity of global optimum. However, unlike other members of EA family, DE uses a *differential mutation* operation based on the distribution of parent solutions in the current population, coupled with recombination with a predetermined parent to generate a single trial vector followed by a one-to-one greedy selection scheme between the trial vector and the parent. Depending on the way the parent solutions are perturbed to generate a trial vector, there exists many trial vector generation strategies and consequently many DE variants.

H. Sakai et al. (Eds.): RSFDGrC 2009, LNAI 5908, pp. 470–477, 2009.

The conceptual and algorithmic simplicity of DE has attracted many researchers who are actively working on its various aspects. Adaptive mixing of perturbation techniques, multi-objective optimization, high dimensional optimization, diversity enhancement, to cite but a few examples, are some of the recent advances and ideas in DE literature [5]. Of particular interest is the generation of multiple trial vectors in place of single trial vector using differential mutation and recombination [6] followed by a tournament selection between the trial vectors and the predetermined parent, with an aim to increase the probability of the parent vector to generate a fitter trial vector.

Little research effort has been devoted to understand multiple trial vectors DE. In this paper we extend the idea of multiple trial vectors generation to fourteen variants of classical DE and have carried out an empirical analysis of the performance of these 14 variants of multiple trial vectors DE on six benchmark problems grouped by their modality and decomposability. A comparative performance analysis between the multiple trial vectors DE variants and their classical counterparts on the benchmark functions has also been carried out. Henceforth, in this paper we adopt an acronym *mtvDE* to refer the multiple trial vectors DE. Despite the fact that a very limited set of 6 benchmark problems will not guarantee reliable conclusion, the analysis indeed give insights about the efficacy of *mtvDE* variants and identifies competitive variants which perform reasonably well on problems with different features.

The remainder of the paper is organized as follows. Section 2 describes the *mtvDE*. Section 3 briefly reviews related works and Section 4 details the design of experiments. Section 5 discusses the simulation results and finally Section 6 concludes the paper.

2 Multiple Trial Vectors Differential Evolution

In the classical Differential Evolution algorithm, repeated cycles of differential mutation and crossover generate a single trial vector Z_i. A one-to-one tournament selection between the predetermined parent and trial vector is carried out after each differential mutation and crossover operation and the winner is placed in the new population.

To increase the probability of generating fitter trial vector by each parent, multiple trial vectors generation scheme has been proposed in [6,7]. The *mtvDE* works as follows. At each generation, each parent vector will create n_t trial vectors by repeated n_t cycles of differential mutation and crossover. After that, a tournament selection between n_t trial vectors and their corresponding parent vector is carried out and the winner is placed in the new population. The algorithmic description of *mtvDE* is depicted in "Fig. 1".

By extending the idea of multiple trial vectors generation to the seven com-

```
Population Initialization X(0) ← {x₁(0),...,x_NP(0)}
g ← 0
Compute { f(x₁(g)),...,f(x_NP(g)) }
while the stopping condition is false do
    for i = 1 to NP do
        for j = 1 to nₜ do
            MutantVector:y_{i,j} ← generatemutant(X(g))
            TrialVector:z_{i,j} ← crossover(x_i(g),y_{i,j})
        end for
        Survivor_i ← Tournament(z_{i,1}...z_{i,nₜ},x_i(g))
        x_i(g+1) ← Survivor
    end for
    g ← g+1
    Compute{ f(x₁(g)),...,f(x_NP(g))}
end while
```

Fig. 1. Description of *mtvDE* algorithm

monly used differential mutation strategies viz. *rand/1, best/1, rand/2 , best/2, current-to-rand/1, current-to-best/1* and *rand-to-best/1* and combining them with two commonly used crossover schemes (binomial and exponential), we get fourteen possible variants of *mtvDE*. Following the standard DE nomenclature used in the literature, the fourteen *mtvDE* variants can be written as follows. *mtvDE/rand/1/bin, mtvDE/rand/1/exp, mtvDE/best/1/bin, mtvDE/best/1/exp, mtvDE/rand/2/bin, mtvDE/rand/2/exp, mtvDE/best/2/bin, mtvDE/best/2/exp, mtvDE/current-to-rand/1/bin, mtvDE/current-to-rand/1/exp, mtvDE/current-to-best/1/bin, mtvDE/current-to-best/1/exp, mtvDE/rand-to-best/1/bin* and *mtvDE/rand-to-best/1/exp*. In this paper, an empirical comparative performance analysis between DE and *mtvDE* variants has been carried out.

3   Related Works

In [7], Storn explored the idea of multiple trial vectors generation for each parent, but the trial vectors were generated and compared with parent vector one after another till a better trial vector than its parent vector was found. Once a fitter trial vector was found, the differential mutation and recombination operations end.

Efren Menzura-Montes et. al. used multiple trial vectors generation in DE to solve constrained optimization problems in engineering design [6]. Five trial vectors were produced by each parent using */rand/1/bin* variant. Through a pre-selection mechanism, the best of the trial vectors was identified (based on feasibility or lowest sum of constraint violation) and made to compete against its corresponding parent vector.

Efren Menzura-Montes et. al. [8] empirically compared the performance of eight DE variants, involving arithmetic recombination along with binomial and exponential, on unconstrained optimization problems. They concluded *rand/1/bin, best/1/bin, current-to-rand/1/bin* and *rand/2/dir* as the most competitive variants. However, the potential variants like *best/2/*, rand-to-best/1/** and *rand/2/** were not considered in their study.

Babu and Munawar [9] compared the performance of ten variants of DE (excluding the *current-to-rand/1/** and *current-to-best/1/** variants of our variants suite) to solve the optimal design problem of shell-and-tube heat exchangers. They concluded *best/*/** strategies to be better than *rand/*/** strategies.

4   Design of Experiments

In this paper, we investigate the performance of *mtvDE* variants and compare them against classical DE variants, by implementing fourteen variants on a set of benchmark problems with high dimensionality and different features. We have chosen six test functions [8,10], of dimensionality 30, grouped by features - unimodal separable, unimodal nonseparable, multimodal separable and multimodal nonseparable. All the test functions have an optimum value at zero except for *f03*. The details of the benchmark functions are described in Table 1. In order to show the similar results, the description of *f03* was adjusted to have its optimum value at zero by just adding the optimal value for the function with 30 decision variables (12569.486618164879) [8].

Table 1. Description of the benchmark functions

f01 - Schwefel's Problem 2.21 $f_{sch}(x) = max_i\{|x_i|, 1 \le i \le 30\}; -100 \le x_i \le 100$	*f04* – Generalized Restrigin's Function $f_{Ras}(x) = \sum_{i=1}^{30}[x_i^2 - 10\cos(2\pi x_i) + 10]; -5.12 \le x_i \le 5.12$
f02 – Schwefel's Problem 1.2 $f_{schDS}(x) = \sum_{i=1}^{30}(\sum_{j=1}^{i} x_j)^2; -100 \le x_i \le 100$	f05 - Generalized Rosenbrock's Function $f_{Ros}(x) = \sum_{i=1}^{29} \|100(x_{i+1} - x_i^2)^2 + (x_i - 1)^2\|; -30 \le x_i \le 30$
f03 – Generalized Schwefel's Problem 2.26 $f_{Sch}(x) = \sum_{i=1}^{30}(x_i \sin(\sqrt{|x_i|})); -500 \le x_i \le 500$	f05 - Generalized Rosenbrock's Function $f_{Gri}(x) = \frac{1}{4000}\sum_{i=1}^{30} x_i^2 - \prod_{i=1}^{30} cos\left(\frac{x_i}{\sqrt{i}}\right) + 1; -600 < x_i < 600$

The parameters for all the DE and *mtvDE* variants are: population size NP = 60 and maximum number of generations = 3000 (consequently, the maximum number of function evaluations calculate to 180,000 in case of DE and 360,000 in case of *mtvDE*). The moderate population size and number of generations were chosen to demonstrate the efficacy of both DE and *mtvDE* variants in solving the chosen problems. The variants will stop before the number of generations is reached only if the tolerance error (which has been fixed as an error value of 1 x 10^{-12}) with respect to the global optimum is obtained. Following [8,11], we defined a range for the scaling factor, F ϵ [0.3,0.9] and this value is generated anew at each generation for all variants. We use the same value for K as F. In case of *mtvDE* , for all the variants , we set n_t to be 2 (i.e. 2 trial vectors are produced by each parent vector).

The crossover rate, CR, was tuned for each variant-test function combination. Eleven different values for the CR viz. {0.0, 0.1, 0.2, 0.3, 0.4, 0.5, 0.6, 0.7, 0.8, 0.9, 1.0} were tested for each variant-test function combination for DE and *mtvDE* separately. For each combination of variant-test function-CR value, 50 independent runs were performed. Based on the obtained results, a bootstrap test was conducted in order to determine the confidence interval for the mean objective function value. The CR value corresponding to the best confidence interval, of 95%, was chosen to be used in our experiment. The CR values obtained for each variant-test function combination for DE and *mtvDE* are listed as follows. In case of DE for *f01*{0.5, 0.9, 0.2, 0.9, 0.2, 0.9, 0.2, 0.9, 0.2, 0.9, 0.2, 0.9, 0.4, 0.9}, for *f02*{0.9, 0.9, 0.5, 0.9, 0.9, 0.9, 0.7, 0.9, 0.9, 0.9, 0.9, 0.9, 0.9, 0.9} for *f03*{0.5, 0.0, 0.1, 0.7, 0.2, 0.3, 0.7, 0.3, 0.4, 0.3, 0.8, 0.2, 0.8, 0.4} for *f04*{0.1, 0.9, 0.1, 0.9, 0.1, 0.9, 0.1, 0.9, 0.1, 0.9, 0.1, 0.9, 0.1, 0.9} for *f05*{0.9, 0.9, 0.8, 0.8, 0.9, 0.9, 0.6, 0.9, 0.1, 0.9, 0.1, 0.9, 0.8, 0.9} and for *f06*{0.1, 0.9, 0.1, 0.8, 0.1, 0.9, 0.1, 0.9, 0.1, 0.9, 0.2, 0.9, 0.1, 0.9 } and in case of *mtvDE for f01*{0.3, 0.9, 0.3, 0.9, 0.2, 0.9, 0.2, 0.9, 0.1, 0.9, 0.2, 0.9, 0.3, 0.9}, for *f02*{0.9, 0.9, 0.4, 0.8, 0.9, 0.9, 0.7, 0.9, 0.9, 0.9, 0.9, 0.9, 0.9, 0.9}, for *f03* {0.1, 0.3, 0.1, 0.7, 0.4, 0.8, 0.1, 0.3, 0.5, 0.3, 0.3, 0.1, 0.3, 0.1}, for *f04* {0.1, 0.9, 0.1, 0.9, 0.1, 0.8, 0.2, 0.9, 0.1, 0.9, 0.1, 0.9, 0.1, 0.9 }, for *f05* { 0.8, 0.9, 0.7, 0.8, 0.9, 0.9, 0.6, 0.9, 0.1, 0.9, 0.1, 0.9, 0.8, 0.9 }, and for *f06* {0.1, 0.9, 0.1, 0.9, 0.1, 0.9, 0.1, 0.9, 0.1, 0.9, 0.2, 0.9, 0.1, 0.9}.

As EA's are stochastic in nature, 100 independent runs were performed per variant per test function (by initializing the population for every run with uniform random initialization within the search space). For the sake of performance analysis among the variants, we present the mean objective function value (MOV) and the probability of convergence [12] for each variant-test function combination.

5   Results and Discussion

The simulation results for all the test functions are presented in Table 2. For the function *f01*, the result shows that the best results were provided by */rand/1/bin*, */best/2/bin*, */rand-to-best/1/bin*, *mtvDE/rand/2/bin* and *mtvDE/best/2/exp* variants. The worst performance was provided by */best/1/exp*, */current-to-best/1/exp*, */current-to-rand/1/exp* and *DE/rand/2/exp* variants. It is interesting to note that the best and worst performance for *f01* were provided by similar set of DE and *mtvDE* variants.

For the unimodal non separable function *f02*. The best performance was shown by */best/2/bin*, */best/2/exp*, *mtvDE/rand/*/bin*, *mtvDE/rand/1/exp* and m*tvDE/rand-to-best/1/** variants. The worst performing variants were */best/1/**, *DE/rand/2/exp*, */current-to-best/** and */current-to-rand/**. The top 4 variants for *f01* displayed similar high performance in the case of *f02* too as can be seen in Table 2.

In case of *f03* the best performance is provided by */best/1/bin* and *mtvDE/best/1/exp* variants. In case of *f04* */rand/1/bin* and */rand-to-best/1/bin* have once again emerged as best performing variants along with */rand/2/bin*. Similarly

Table 2. MOV values obtained for DE and mtvDE variants

Sno	Variant	Unimodal functions		Multimodal functions			
		f01- DE/ mtvDE	*f02* -DE /mtvDE	*f03* -DE /mtvDE	*f04*-DE / mtvDE	*f05*-DE / mtvDE	*f06*-DE /mtvDE
1	rand/1/bin	0.00 / 0.00	0.07 / 0.00	0.13 / 0.07	0.00 / 0.00	21.99 / 5.77	0.00 / 0.00
2	rand/1/exp	3.76 / 0.92	0.31 / 0.00	0.10 / 0.21	47.93 / 0.00	25.48 / 6.23	0.05 / 0.02
3	best/1/bin	1.96 / 11.99	13.27 / 44.32	0.00 / 0.00	4.33 / 0.00	585899.88/ 727417.29	3.72 / 4.74
4	best/1/exp	37.36 / 50.54	57.39 / 178.88	0.01 / 0.00	50.74 / 1.16	64543.84 / 37552.84	5.91 / 13.09
5	rand/2/bin	0.06 / 0.00	1.64 / 0.00	0.22 / 0.10	0.00 / 3.73	19.01 / 0.66	0.00 / 0.00
6	rand/2/exp	32.90 / 13.28	269.86 / 1.47	0.27 / 0.10	101.38 / 10.72	2741.32 / 204.31	0.21 / 0.08
7	best/2/bin	0.00 / 0.00	0.00 / 0.00	0.17 / 0.04	0.69 / 10.80	2.32 / 0.84	0.00 / 0.00
8	best/2/exp	0.05 / 0.00	0.00 / 0.00	0.08 / 0.02	80.63 / 13.56	1.12 / 0.84	0.03 / 0.02
9	current-to-rand/1/bin	3.68 / 0.59	3210.36 / 139. 42	0.14 / 0.06	37.75 / 14.28	52.81 / 28.84	0.00 / 0.00
10	current-to-rand/1/exp	57.52 / 49.20	3110.90 / 227.42	0.12 / 0.06	235.14 / 38.98	199243.32/ 47153.21	1.21 / 0.36
11	current-to-best/1/bin	3.71 / 0.23	3444.00 / 132.64	0.19 / 0.05	37.04 / 51.49	56.91 / 30.88	0.00 / 0.00
12	current-to-best/1/exp	56.67 / 49.68	2972.62 / 268.08	0.10 / 0.10	232.80 / 121/26	119685.68/ 41046.61	1.21 / 0.38
13	rand-to-best/1/bin	0.00 / 0.00	0.07 / 0.00	0.22 / 0.06	0.00 / 207.22	17.37 / 5.68	0.00 / 0.00
14	rand-to-best/1/exp	3.38 / 0.97	0.20 / 0.00	0.12 / 0.07	48.09 / 207.61	24.54 / 8.00	0.05 / 0.03

Table 3. Number of successful runs and probability of convergence for DE and *mtvDE* variants

DE / mtvDE							
Variant	f01	f02	f03	f04	f05	f06	Pc (%)
best/2/bin	100/100	100/100	1/19	47/32	38/79	100/96	64.33/71.00
rand-to-best/1/bin	100/100	79/100	0/8	100/100	0/4	100/100	63.17/68.67
rand/1/bin	100/100	73/100	4/4	100/100	0/7	100/100	62.83/68.50
best/1/bin	79/0	86/88	88/93	3/3	0/0	1/0	42.83/30.67
rand/2/bin	0/100	0/100	1/4	100/100	0/5	100/100	33.50/68.17
best/2/exp	1/100	100/100	17/40	0/0	29/79	44/32	31.83/58.50
best/1/exp	0/0	58/0	85/93	0/0	0/0	0/0	23.83/15.50
current-to-best/1/bin	0/0	0/0	3/11	0/0	0/0	96/100	16.50/18.50
current-to-rand/1/bin	0/0	0/0	2/5	0/0	0/0	96/100	16.33/17.50
rand-to-best/1/exp	0/0	10/100	6/15	0/0	0/6	69/71	14.16/32.00
rand/1/exp	0/0	4/100	7/12	0/0	0/7	68/76	13.17/32.50
rand/2/exp	0/0	0/4	2/4	0/0	0/0	3/0	0.83/1.33
current-to-best/1/exp	0/0	0/0	5/7	0/0	0/0	0/0	0.83/1.17
current-to-rand/1/exp	0/0	0/0	3/7	0/0	0/0	0/0	0.50/0.83

*/rand/2/exp, */current-to-rand/1/exp , */current-to-best/1/exp and */best/1/exp have once again displayed poor performance, as in the case of *f01* and *f02*, along with */best/2/exp*. The variants m*tvDE/best/2/bin*, m*tvDE/best/1/exp* and m*tvDE/rand/2/exp* have performed poorly than their counterparts.

Test function *f05* was not solved by any variant. However, */best/2/** and *mtvDE/rand/2/bin* variants have displayed relatively better performance. */rand/2/exp, */current-to-rand/1/exp* and */current-to-best/1/exp* were the consistent poor performing variants. Interestingly */best/1/** variants have also shown the worst performance in both the cases (DE and *mtvDE*). However in case of *f06*, 6 variants displayed best performance. */rand/*/bin, */rand-to-best/1/bin* and */best/2/bin* were the consistent best performing variants. This best performance is also shared by */current-to-best/1/bin* and */current-to-rand/1/bin* variants. As with *f05*, */best/1/** variants have shown a relatively poor performance.

Based on the overall results in Table 2 the most competitive variants were */rand-to-best/1/bin,*/best/2/bin* and */rand/1/bin*. The variants */rand/2/bin* and */best/2/exp* variants also showed good performance consistently. On the other hand, the worst overall performance were consistently displayed by variants */current-to-best/1/exp* and */current-to-rand/1/exp*. The variants */best/1/bin* and */rand/2/exp* were also displaying poor performance. */best/1/** variants show good performance for multi-modal separable function. It is worth noting that binomial recombination showed a relatively better performance over the exponential recombination. It is also worth noting that the relatively better performance of *mtvDE* over DE in all the test functions may largely be attributed to the increased number of function evaluations available to *mtvDE*. As a matter of fact, when in some representative runs, both DE and *mtvDE* were allowed precisely the same number of fitness evaluations, both displayed similar high performance capability.

Next in our experiment, the probability of convergence (Pc), i.e. the percentage of successful runs to total runs, is calculated for each variant-function combination. This measure identifies variants having higher convergence capability to global optimum. It is calculated as the mean percentage of number of successful runs out of total number of runs i.e. Pc=(nc/nt)% where nc is total number of successful runs made by each variant for all the functions and nt is total number of runs, in our experiment $nt = 6 * 100 = 600$.

The convergence probability for both DE and *mtvDE* variants were calculated separately and the results are shown in Table 3. The table presents the number of successful runs made by each variant for each function, the total number of successful runs and the probability of convergence. As can be seen from the table, the competitive variants identified earlier viz. */best/2/bin*, */rand-to-best/1/bin* and */rand/1/bin* have higher probability of convergence. Similar trend could be observed for */rand/2/bin* and */best/2/exp* variants. The worst performing variants */current-to-best/1/exp*, */current-to-rand/1/exp* and */rand/2/exp* were found to have the least probability of convergence.

6   Conclusion

In this paper, we have extended the idea of multiple trial vector generation to fourteen variants of classical DE and presented an empirical comparative performance analysis of these fourteen variants of *mtvDE* against those of classical DE. The variants were tested on 6 test functions of dimension 30, grouped by their modality and decomposability. The experiments identified */best/2/bin, */rand-to-best/1/bin* and */rand/1/bin* variants as the most competitive variants in terms of the mean objective function values. The worst performing variants were */current-to-best/1/exp*, */current-to-rand/1/exp* and */rand/2/exp*. In fact the calculation of probability of convergence reiterated the observation about the performance of above said variants. Our future work would involve validating above observations by testing DE and *mtvDE* variants on a larger suite of test functions and also testing *mtvDE* for performance variation with respect to the number trial vectors, n_t.

References

1. Storn, R., Price, K.: Differential Evolution – A Simple and Efficient Adaptive Scheme for Global Optimization Over Continuous Spaces, Technical Report TR-95-012, ICSI (March 1995)
2. Storn, R., Price, K.: Differential Evolution – A Simple and Efficient Heuristic Strategy for Global Optimization and Ccontinuous Spaces. Journal of Global Optimization 11(4), 341–359 (1997)
3. Price, K.V.: An Introduction to Differential Evolution. In: Corne, D., Dorigo, M., Glover, F. (eds.) New Ideas in Optimization, pp. 79–108. Mc Graw-Hill, UK (1999)
4. Price, K., Storn, R.M., Lampinen, J.A.: Differential Evolution: A Practical Approach to Global Optimization. Springer, Heidelberg (2005)
5. Chakraborty, U.K. (ed.): Advances in Differential Evolution. Studies in Computational Intelligence Series, vol. 143. Springer, Heidelberg (2008)
6. Montes, E.M., et al.: Multiple Trial Vectors in Differential Evolution for Engineering Design. Engineering Optimization 39(5), 567–589 (2007)
7. Storn, R.: System Design by Constraint Adaptation and Differential Evolution. IEEE Transactions on Evolutionary Computation 3(1), 22–34 (1999)

8. Mezura-Montes, E., Velazquez-Reyes, J., Coello Coello, C.A.: A Comparative Study on Differential Evolution Variants for Global Optimization. In: Genetic and Evolutionary Computation Conference, GECCO 2006, July 8-12 (2006)
9. Babu, B.V., Munawar, S.A.: Optimal Design of Shell-and-Tube Heat Exchanges by Different Strategies of Differential Evolution, Technical Report, PILANI -333 031, Department of Chemical Engineering, Birla Institute of Technology and Science, Rajasthan, India (2001)
10. Yao, X., Liu, Y., Liang, K.H., Lin, G.: Fast Evolutionary Algorithms. In: Rozenberg, G., Back, T., Eiben, A. (eds.) Advances in Evolutionary Computing: Theory and Applications, pp. 45–94. Springer, New York (2003)
11. Mezura-Montes, E.: Personal Communication (unpublished)
12. Feoktistov, V.: Differential Evolution In Search of Solutions. Springer, Heidelberg (2006)

Granular Computing for Text Mining: New Research Challenges and Opportunities

Liping Jing[1] and Raymond Y.K. Lau[2]

[1] School of Computer and Information Technology, Beijing Jiaotong University,
Beijing, China, 100044
lpjinghk@gmail.com
[2] Department of Information Systems, City University of Hong Kong,
Tat Chee Avenue, Kowloon Toog, Hong Kong
raylau@cityu.edu.hk

Abstract. As an emerging computational methodology, granular computing provides an effective strategy for solving many real world problems such as mining latent relationships from text. This paper examines the relationship between granular computing and text mining from a theoretical perspective. Firstly, we analyzes the granular structure of text data which is the key step for textual data representation. Secondly, some granule-based computational methods are described, especially on term-document and document-document similarity calculations. Finally, we highlight several potential research areas where the performance of text mining could be enhanced by applying the concepts of granular computing.

1   Introduction

With the rapid development of the Internet, the volume of semi-structured and unstructured textual data such as XML documents, e-mail messages, blog posts, academic papers has been under an exponential growth. Discovering useful knowledge from such huge volume of data has become a very challenging problem. Text mining, as a hot research topic, tries to extract knowledge from unstructured data by using techniques from data mining, machine learning, natural language processing, information retrieval, and knowledge management [1]. Text mining is a knowledge-intensive process in which a user interacts with a document collection by a suit of analysis tools, and finally identifies and explores some interesting patterns.

Because of the complexity of text data, say large volume, content diversity and complicated semantic, text mining is a tough research area. Recently, more and more techniques have been developed to address the research challenges, such as, text clustering, classification algorithms, and information extraction methods. Although, these techniques perform well at their corresponding step, a good text mining strategy should consider not only the mining analysis performance but also the information related to other steps, i.e., text preprocessing and results

H. Sakai et al. (Eds.): RSFDGrC 2009, LNAI 5908, pp. 478–485, 2009.

visualization steps. In this case, granular computing is helpful to open our eyes on systemically identifying patterns from text data.

Granular computing is a newly developed computational methodology which has been drawn much attention by researchers [2,3]. It concerns the processing of complex information entities called granules, which arises in the process of data abstraction and knowledge derivation from information. Generally speaking, information granules are collections of entities that originate at the numeric level and are arranged together due to their similarity, functional or physical adjacency, indistinguishability, coherency and etc. [3]. Granular computing has two stages: granule representation and computation with granules. Granule representation is a primitive part of the whole granular computing process; at this stage, granular computational method exploits structures in terms of granules, levels, and hierarchies based on multilevel and multiview representations. Each representation level can be viewed as a representation of a problem at a specific level of granularity. The relationship between levels can be interpreted in terms of abstraction, control, complexity, detail, resolution [2]. The ideas of granular computing, therefore, can be used to reexamine many classical problems in order to obtain new understandings and more insights.

In this paper, we apply the concept of granular computing to text mining for effective addressing text mining problems. For example, text data can be represented on different levels, say, character-level, word-level, phrase-level and even concept-level. Each level can be regarded as a granule in granular computing. Next, the structured information processing of granular computing can be used to efficiently mine the textual data. We focus on the computations of granules in the form of similarity calculations between term and document, or between documents. This kind of computations play an important role in text mining. For example, the similarity between query term and document collections is the basis of information retrieval, and the similarity between documents is the criterion of text classification. Several research challenges and opportunities by applying the ideas of granular computing to enhance text mining are also discussed.

The rest of this paper is organized as follows. Section 2 gives a granular structure of text data. The relationship between text mining and granular computing is analyzed in detail in Section 3. The future research challenges and opportunities of text mining with granular computing are given in Section 4. Finally, we provide a conclusion of our work in Section 5.

2   Granular Structures of Text

Heeman [4] pointed out that textual data has granular structures, for example, a document had low-level and high-level structures. A low-level structure reflects the hierarchical structure of the language content including characters, words, phrases, concepts, clauses, sentences, paragraphs, as shown in Fig.1(a). Characters are the individual component-level letters, numerals, special characters and even space. Words are the basic level of semantic richness. Phrases consist of words and can be extracted by using phrase-extraction techniques. Concepts are extracted by complex preprocessing routines and may be single words,

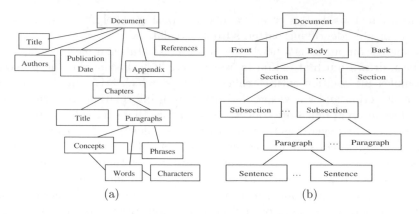

Fig. 1. The Granules in Text data in the view of (a) content and (b) local organization

multiword expressions, whole clauses or even larger units which are related to specific concepts identifiers (say, dictionaries, ontologies, and some knowledge resources). Each representation has its advantages and disadvantages. For example, Character-level includes useful and common positional information of document, however, it can not often be handled by text mining techniques because the feature space is fairly unoptimized. Meanwhile, concept-level has a good performance when handling synonymy, polysemy, and even hyponym and hypernym, so that it can support very sophisticated concept hierarchies and leverage the domain knowledge, but it is to large extent dependent on the domain knowledge and a complicated processing to apply such heuristics.

A high-level structure reflects the organization of a document. Subsections are made up by paragraphs, sections by subsections, and a document by sections, as shown in Fig.1(b). A book is made up by chapters and each chapter is made up by sections. A structured document or markup languages focus on a high-level representation of a document based on its logical organization. That is, a document is no longer treated as a stream of characters in a linear order. Typically, a markup language is used to label and to tag different parts of a document and, furthermore, to link different parts by hyperlinks.

3   Granular Computing for Text Mining

The goal of text mining is to extract key elements from large unstructured data sets, discover relationships and summarize the information. Text mining can be taken as an inference process via a pipeline of steps including preparing, parsing, representing, analyzing text data and visualizing analysis result, where each step may result in the failure of the mining process. These steps can be projected to the granular computing according to the text granular structure. For instance, a document can be represented in terms of characters, words, phrase, concepts or even topics. Different levels may indicate different meanings.

3.1   Granular Text Representation

A document collection X with n documents, in text mining, is usually represented as a vector set $X = \{X_1, X_2, \cdots, \cdots, X_n\}$, and each vector $X_i = \{x_{i1}, x_{i2}, \cdots, \cdots, x_{im}\}$ represents a document in m-dimensional space, i.e., there are total m terms in the whole collection [17]. x_{ij} gives the weight of the jth term in the ith document. Here, terms can be words, phrases or concepts. Also, a document can be represented as a sequence of small units (e.g., characters, words, phrases and sentences) in duplicate detection application [12].

The simplest approach obtaining terms from documents is to segment the whole document into smaller units according to space, comma or other tags. However, this approach can not exploit the real semantic information of the documents. Recently, researchers considered the document granular structure and extract more rich and flexible representation models. In this case, Information extraction (IE) [24] plays an important role. For example, IE can derive taxonomy by extracting phrases and their relationships [23,18], also such taxonomy patterns were used to represent document collection [8]. With the IE techniques, the document can be represented in different granular levels, say, phrase-level, concept-level and even taxonomy-level. Moreover, the document local organization in Fig.1(b), was used in structural IE. Structured IE uses a two-stage to finish structure IE tasks, so that, a plain text document can be structured into different target fields which makes the subsequent text analysis become easier.

3.2   The Computation of Granules

Similarity calculating is a main step in text mining, such as similarity between term and document in information retrieval, similarity between documents in text clustering/classification. Meanwhile, such similarity values are effected to a large extent by the document granular structures.

1) Term-Document Similarity
Scoring the similarity between a query and a document is a key component of information retrieval. Usually, information retrieval (e.g., search engine) only considers whether the query terms appear in the corresponding documents and then ranks the searched documents according to their static information, like, Google with PageRank [10]. Robertson et al. [9] firstly added query term importance information among the corresponding relevant documents in the search engine model BM25. [9] used $Sim_{term}(Q, D) = \alpha G(D, Q) + \beta \sum_{t \in Q} \omega_t \cdot T(D, t)$ to calculate the similarity, where $G(D, Q)$ is the static Pagerank of the document D, and $\sum_{t \in Q} \omega_t \cdot T(D, t)$ is used to evaluate how the importance of the query terms in the corresponding document.

Recently, Zhu et al. [6] introduced term position information into the similarity calculating for information retrieval, $Sim_{pos}(D, Q) = \alpha G(D, Q) + \beta_1 \sum_{t \in Q} \omega_t \cdot T(D, t) + \beta_2 \cdot X(D, Q)$. Among them, term-proximity term $X(D, Q)$ indicates how close query terms appear in a document. Similarly, Lv and Zhai [5] considered the term positional information in language model building.

Meanwhile, a document, intuitively, should be more relevant to a query if they share more concepts, and these concepts are recorded in the ontologies or other knowledge resource. Such concept-level information was recently used in information retrieval by Lau et al. [7] with $Sim_{concept}(D,Q) = S(D,Q) - \varphi|DC(D) - QC(Q)|$, where $S(D,Q)$ is the popular similarity value. $|DC(D) - QC(Q)|$ is used to estimate the concept-level distance between D and Q according to ontology.

All these similarity measures demonstrate that different document representations (i.e., granularity, or levels) may provide the users with different perspectives of the document contents. Therefore, integrating these different-level representations can handle the document information as much as possible.

2) Document-Document Similarity
Evaluating document similarity plays an important role in text mining. For example, text clustering groups together similar documents and separate dissimilar ones based on some similarity functions which take a pair of documents and produce a real value that is a measure of the documents' proximity.

In document duplication detection, each document is first chunked into smaller units, then each textual chunk is hashed down to a fingerprint with the attributes: fprint (the hashed fingerprint) and docID (the document ID). Finally, the document similarity can be approximated by comparing the fingerprint sets of documents for overlap using min-wise independent permutations as $Sim_{doc}(D_1, D_2) = (F_1 \bigcap F_2)/(F_1 \bigcup F_2)$, where F_1 and F_2 are the fingerprint sets for document D_1 and D_2 respectively. The fingerprint technique was early used in web page duplicate detection by Broder [11] based on character-level chunk, subsequently, word-level [12], phrase-level [13] and document-level [14] were developed.

For text clustering, document-document similarity is the basic criterion to group the document collection. Term-level vector, $X_i = \{x_{i1}, x_{i2}, \cdots, x_{ij}, \cdots, x_{im}\}$ is a typical document representation, where x_{ij} usually means the $TF \cdot IDF$ weight of the jth term in the ith document. Here, cosine similarity $Sim_{cosine}(D_1, D_2) = \frac{X_1 \cdot X_2}{||X_1|| \cdot ||X_2||}$ is usually used, in which, the semantic relationship between terms is ignored because it assumes that all terms are independent. Before [16], unique words were treated as the vector terms. In this case, different words are assumed independent to each other, which is not true in real world. Later, some researchers introduced phrase-level document-document similarity measure [16,15], where a multi-word phrase is treated as one term in data representation. Meanwhile, ontology (say, GO (www.geneontology.org/) and etc.) or other knowledge resources (say, Wikipedia (en.wikipedia.org/wiki/) and etc.) are used to extract concepts for document concept-level representation [19,20,21,22].

4   New Research Challenges and Opportunities

From the computational perspective, granular computing can solve a problem by systematically exploring the granular structures. This involves moving the perspective upwards and downwards along a hierarchy of granules. Such hierarchical analysis also can be applied in text mining, like Fig.2. Based on the content granular structures of text data (Fig.1(a)), the document collection can be represented in different content levels. So far, researches only focus on individual level, i.e, they do not consider mapping connections between granules or levels. Actually, there should be some connection between different granules, say, a phrase is consistent of several words.

One granular computing approach solving this problem is to integrate these different-level document representations, so that the final document representation can capture more rich and meaningful information, like the framework circled in dash rectangle (the lower section) in Fig.2, and then the general text analysis methodologies can be applied. The another method is to analyze the text data on different levels with the existing methodologies and then combine the analysis results, like the framework circled in red sold rectangle (the upper section) in Fig.2.

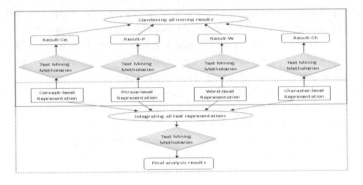

Fig. 2. Granular Computing Framework of Text Mining

Recently, mining precious knowledge from multi-resources becomes a hot research topic in many fields, so does in text mining [19,21,22]. In this case, how to effectively make use of multi resources is the key problem. Granular computing provides a reasonable strategy to integrate these multi resources by taking them as different granules. Here, we gave an example for identifying protein protein interaction (PPI) from bioinformatics literature, GO, UMLS (www.nlm.nih.gov/research/umls/), and other thesaurus. Proteins can be represented with terms in the bioinformatics literature, meanwhile, some terms may be related to one concept recorded in UMLS, therefore, we can use this concept replace the individual terms. One reason to do this is that concept can capture more information, the other reason is dimensional reduction because more than one terms can be replaced by one concept. Furthermore, the functional information of parts of proteins is annotated in GO, such strong protein relationships can be used as prior information to supervise PPI identification process.

Another main problem for text mining is to reasonably explain and show the mining results. Granular computing is helpful here. For example, in digital library, a large list of papers or books may be returned with a query, which is very terrible for a user if he or she is only interested in part of the results. A simple way is to provide a granule option so that the user can select which part he or she is interested in. For instance, the searched documents can be sorted by different granules (Title, Abstract, Full Text, etc.) in PubMed(www.ncbi.nlm.nih.gov/sites/entrez), or grouped by different topics in Vivisimo(vivisimo.com). Furthermore, combining concepts thesaurus and structures to exploit the mining results is an emergent and interesting research area. Finally, new methods should be explored to compute the semantic granularity (e.g., levels of specificity) of text.

5   Conclusions

In this paper, we provide a theoretical analysis of applying the concepts of granular computing to address the research challenges presented in text mining. Future research challenges and opportunities of conducting text mining with the help of granular computing are discussed. As a matter of fact, the ideas of granular computing have been applied to text mining research for some time. For example, researchers began to realize that it was helpful to represent document collection with different levels of abstraction such as words, phrases, and even concepts. Each level of abstraction is essentially a granule. However, the relationships among different granules have not been investigated fully. Granular computing will help researchers develop a more effective method for text mining by exploring different levels of information, and even the relationships among granules. Moreover, granular computing can offer a sound methodology of integrating multiple sources of data to enhance distributed text mining in the future.

Acknowledgement

We would like to appreciate Dr. Yao (Dept. of CS, University of Regina) for his wonderful suggestion and revision on this paper. The research work in this paper was supported by the National Natural Science Foundation of China (60905028,90820013,60875031), 973 project (2007CB311002) and Program for New Century Excellent Talents in University (NCET-06-0078).

References

1. Feldman, R., Sanger, J.: The text mining handbook. Cambridge University Press, Cambridge (2006)
2. Bargiela, A., Pedrycz, W.: Granular computing: an introduction. Kluwer Academic Publishers, Boston (2002)
3. Yao, Y.: Granular computing for data mining. In: Proceedings of SPIE Conf. on dta mining, instrusion detection, information assurance and data networks security, pp. 1–12 (2006)
4. Heeman, F.: Granularity in structured documents. Electronic publishing 5, 143–155 (1992)
5. Lv, Y., Zhai, C.: Positonal language models for information retrieval. In: Proceedings of the 32nd ACM SIGIR, Boston, USA (2009)

6. Zhu, M., Shi, S., Yu, N., Wen, J.: Can phrase indexing help to process non-phrase queries. In: Proceedings of CIKM, Napa Valley, CA, USA, pp. 679–688 (2008)
7. Lau, R., Lai, C., Li, Y.: Mining fuzzy ontology for a web-based granular information retrieval system. In: Proceedings of RSKT, Gold Coas, Australia (2009)
8. Li, Y., Zhou, X., Bruza, P., Xu, Y., Lau, R.: A two-stage text mining model for information filtering. In: Proceedings of CIKM, Napa Valley, CA, USA, pp. 1023–1032 (2008)
9. Robertson, S., Walker, S., Beaulieu, M.: Experimentation as a way of life: Okapi at TREC. Information Processing and Management 36, 95–108 (2000)
10. Brin, S., Page, L.: The anatomy of a large-scale hypertextual web search engine. Computer Networks and ISDN Systems 30(1-7), 107–117 (1998)
11. Broder, A.: Some applications of Rabin's fingerprinting method. In: Capocelli, R., De Santis, A., Vaccaro, U. (eds.) Sequences II: Methods in Communications, Security, and Computer Science, pp. 143–152. Springer, Heidelberg (1993)
12. Broder, A., Glassman, S., Zweig, G.: Syntactic clustering of the web. In: Proceedings of WWW, pp. 391–404 (1997)
13. Fetterly, D., Manasse, M., Najork, M.: Detecting Phrase-Level Duplication on the World Wide Web. In: Proceedings of ACM SIGIR (2005)
14. Yang, H., Callan, J.: Near-Duplicate Detection by Instance-level Constrained Clustering. In: Proceedings of ACM SIGIR (2006)
15. Hammouda, K., Kamel, M.: Efficient phrase-based document indexing for Web document clustering. IEEE Transactions on knolwedge and data engineering 16(10), 1279–1296 (2004)
16. Zamir, O.: Clustering Web Documents: A Phrase-Based Method for Grouping Search Engine Results. PhD Dissertation, Unviersity of Washington, USA (1999)
17. Salton, G., McGill, M.: Introduction to modern information retrieval. McGraw-Hill Computer Science Series. McGraw-Hill, New York (1983)
18. Liu, Y., Loh, H., Lu, W.: Deriving taxonomy from documents at sentence level. In: Emerging technologies of text mining, Information Science Reference, Hershey, New York, pp. 99–119 (2008)
19. Hotho, A., Staab, S., Stumme, G.: Wordnet improves text document clustering. In: Proceedings of the Semantic Web Workshop at the 26th ACM SIGIR (2003)
20. Jing, L., Zhou, L., Ng, M., Huang, J.: Ontology-based distance measure for text clustering. In: Proceedings of the 6th SIAM International conference on data mining (2006)
21. Zhou, X., Hu, X., Zhang, X.: IEEE Topic signature language mdels for Ad Hoc retrieval. IEEE transactions on knowledge and engineering 19(9), 1276–1287 (2007)
22. Hu, J., Fang, L., Cao, Y., Zeng, H., Li, H., Yang, Q., Chen, Z.: Enhancing text clustering by leveraging Wikipedia semantics. In: Proceedings of 31st ACM SIGIR, pp. 179–186 (2008)
23. Wu, S., Li, Y., Xu, Y.: Deploying approaches for pattern refinement in text mining. In: Proceedings of ICDM, pp. 1157–1161 (2006)
24. Sarawagi, S.: Information extraction. FnT Databases 1(3) (2008)

Polarity Classification of Subjective Words Using Common-Sense Knowledge-Base

Ashish Sureka, Vikram Goyal, Denzil Correa, and Anirban Mondal

Indraprastha Institute of Information Technology (IIIT), India
{ashish,vikram,denzilc,anirban}@iiitd.ac.in
http://www.iiitd.edu.in/

Abstract. Semantic orientation of a word indicates whether the word denotes a positive or a negative evaluation. We present an approach to compute semantic orientation of words using machine-interpretable common-sense knowledge. We employ ConceptNet (a large semantic network of commonsense knowledge) for determining the polarity or semantic orientation of a sentiment expressing word. We apply heuristics on certain pre-defined predicates expressing semantic relationship between two concepts for classifying words that have a positive or negative polarity and finding words that have similar polarity. The advantages of the proposed approach are that it does not require any pre-annotated training dataset or manually created seed list. The proposed solution relies on a lexical resource which is created by volunteers on the Internet and not by trained or specialized knowledge engineers. We test our approach on publicly available pre-classified sentiment lexicon and present the results of our experiments and also examine the tradeoffs and limitations of the proposed solution. We conclude that it is possible to determine polarity of words with high accuracy by exploiting a machine-understandable layman's knowledge and basic facts that ordinary people know about the world.

Keywords: Word-Level Polarity Classification, Common-Sense Knowledge Base, Sentiment Analysis, Opinion Mining.

1 Introduction

Semantic orientation of a word indicates whether the word denotes a positive evaluation (such as praise or positive opinion) or a negative evaluation (such as criticism or negative opinion) [8][7]. Semantic orientation of a word is also referred as the valence or polarity of a word and systems to automatically determine semantic orientation of a word has applications in the area of sentiment analysis, opinion mining, multi-perspective question and answering and filtering abusive messages. Opinion mining and sentiment analysis of a product review or any subjective statement is an area which has received significant interest in recent times and polarity determination of a word is fundamental to the problem of sentiment analysis (refer to a detailed survey on opinion mining and sentiment analysis by Bo Pang and Lillian Lee [5]). Polarity determination at world level

H. Sakai et al. (Eds.): RSFDGrC 2009, LNAI 5908, pp. 486–493, 2009.

(fine-grained analysis) forms a component of a larger system wherein polarity determination at sentence, paragraph or document level needs to be performed. There are two sub-problems within the problem of determining semantic orientation of a word. One sub-problem consists of computing the direction (positive or negative) and the other sub-problem consists of computing the intensity or strength (weak or strong) within the computed direction. For example, the word good is a weak positive word whereas excellent or fabulous or astonishing is a strong positive word. Similarly, the word bad is a weak negative word whereas horrible or terrible is a strong negative word. Automatically determining the semantic orientation of word is required for developing sentiment lexicon as it is tedious and time consuming to manually label all the words in a language with its polarity and intensity.

The earliest work to solve the problem of automatically determining the semantic orientation of a word was done by Hatzivassiloglou et al [8]. The basis of the approach by Hatzivassiloglou et al is that adjectives conjoined by words such as and or or share the same polarity whereas adjectives conjoined by words such as but will have opposite polarity or orientation. The methods consists of extracting pairs of adjectives using conjunctions like and, or, but, either-or, or neither-nor from 1987 Wall Street Journal Corpus (a document set consisting of 21 million words) and assigning similar or different polarities to adjectives based on the type of conjuctions. Turney et al. proposed a general strategy for inferring semantic orientation of a word based on their hypothesis that the semantic orientation of a word tends to correspond to the semantic orientation of its neighbors [7]. Neighborhood between words is determined using statistical association or statistical dependence between words (word co-occurrence). Kamps et al. use WordNet to measure semantic orientation of adjectives by exploiting the graph-theoretic model of WordNet's synonym relations [3]. Esuli et al. present a technique for determining the semantic orientation of terms through gloss classification (performs quantitative analysis of the glosses or definitions of terms given in on-line dictionaries) [1]. Wilson et al. presents an approach to recognizing contextual polarity of phrases (a two-step process that employs machine learning that begins with a large stable of clues marked with prior polarity and then identifies the contextual polarity of the phrases that contain instances of those clues in a corpus) [9].Takamura et al. present a technique that consists of constructing a lexical network by connecting similar or related words and adopting the Potts model for the probability model of the lexical network [6].

1.1 Paper Contributions

We propose a novel technique for determining the polarity of a word by making use of a semantic network of common-sense knowledge. Previous approaches compute semantic orientation of words in a corpus-driven manner by performing statistical analysis on a corpus or rely on lexical resources created by experts and trained knowledge engineers. Previous approaches also rely on a pre-annotaed training dataset or a seed list of pre-classified sentiment words for performing its task. In this paper, we present a new approach that differs from the previous approaches and has the following advantages. The main advantages of our solution is that it relies on a lexical resource (called as ConceptNet) that represents common-sense knowledge created by volunteers on the Internet (14,000 contributors from around the world as mentioned in the paper by Liu et al. [4]) and not by trained or specialized knowledge engineers. Also, the proposed approach does not require any pre-annotated training dataset or manually created seed list to perform its tasks. Creating training dataset of pre-classified words and manually building specialized lexical resources for sentiment analysis application requires trained and specialized people and can be a time-consuming as well as tedious process. The proposed solution overcomes the dependency on experts by automatically creating sentiment lexicon and computing semantic orientation of words based on common-sense knowledge created by ordinary people as volunteers and not specialized knowledge engineers. The proposed approach performs polarity classification of sentiment word belonging to any lexical category (adjective, adverb noun and verb) unlike some approaches that are able to perform polarity classification of words belonging to just adjectives. We present empirical results (based on experiments performed on publicly available test dataset and a standard benchmark for this task) which prove that it is possible to predict with good accuracy the polarity of a word by using laymans common-sense knowledge. The limitation of our approach is that the accuracy and coverage of the words is a function of the number of concepts, assertions, relations and quality of data in the common-sense knowledge-base. The work presented in this paper is a step in the direction of our research on investigating the usefulness of machine understandable commonsense knowledge in the application domain of sentiment analysis and opinion mining.

2   Solution Approach

We leverage ConceptNet (which is machine-interpretable semantic network representing common-sense knowledge) for polarity classification of words. The common-sense knowledge present in ConceptNet is collected from volunteers on the Internet since the year 2000 and represents facts that ordinary people knows about the world [2]. The data present in ConceptNet is contributed by ordinary people unlike lexical resources such as WordNet and FrameNet which are mainly created by trained and specialized knowledge engineers. As ConceptNet is a semantic network, it consists of nodes connected by edges. The nodes represent the concepts and the edges represent predicates. Predicates express semantic relationships between two concepts. Some relationships between concepts in the ConceptNet semantic network are: IsA, MadeOf, UsedFor, CapableOf, DesireOf, CreatedBy, InstanceOf, PartOf, PropertyOf and EffectOf [2]. In ConceptNet, an assertion is uniquely defined by five properties: language, relation, concept1, concept2 and frequency. The Language property defines the language an assertion is expressed in (such as English). The Relation property defines the relation or the name of the predicate that connects the two concepts in the assertion (such as IsA, PartOf). Concept 1 and Concept 2 define the first and the second argument

Table 1. Pre-defined pattern over assertions belonging to the Desires relation

Assertion Property	Value of the Assertion Property
Language	English
Relation	Desires
Concept 1	a person or human or everyone
Concept 2	Word whose polarity needs to be determined
Assertion Type	+1 or -1

of the relation (words and phrases). The Frequency property expresses how often the given concepts would be related by the given relation, ranging from never to always. Also for each assertion, there is a field which defines the assertion type. The value of the assertion type is +1 if the assertion makes a positive statement (such as Diamonds are pretty) and -1 if it makes a negative statement (such as a person doesn't want anxiety).

(Step 1). The first step of the proposed solution consist of checking if the word matches the pattern or structure defined in Table 1. The pattern is based on our hypothesis that if a *person* or *human* or *everyone* (as Concept 1) desires (Relation type as Desires) something (represented as Concept 2), then Concept 2 (in our case a sentiment expressing word whose polarity needs to be determined) will have positive connotation if the assertion type is positive (i.e. has a value of +1) and will have negative connotation if the assertion type is negative (i.e. has a value of -1). This step does not require any seed list or pre-classified sentiment word and has an advantage over approaches that depend on having a training dataset or manually created seed list. We validated our hypothesis by entering few terms on the web-based interface provided at the ConceptNet website. For example, some of the words which are expressed as Concept 2 and where the Concept 1 is person, Relation is Desires, Assertion Type is +1 are: accomplish (verb), admiration (noun), affection (adjective), beautiful (adjective), bliss (noun), clever (adjective), comfort (verb) etc. Similarly, some of the words which are expressed as Concept 2 and where the Concept 1 is person, Relation is Desires, Assertion Type is -1 are: agonize (verb), annoyance (noun), anxiety (noun), bad (adjective), boredom (noun), cancer (noun), confuse (verb), criminal (noun), criticism (adjective), damaging (adjective) etc. We noticed that some words fall into a category where Concept 1 is person (or human or everyone), Relation is Desires and Assertion Type is both +1 and -1. Since, there is a conflict in assertion type, we do not predict the polarity of such words and leave it blank to be computed in the next steps of the overall process.

(Step 2). The second step of the solution consists of checking a pattern based on *DefinedAs* relationship. The pattern is based on the hypothesis that two concepts connected to each other using a *DefinedAs* relation in the same assertion will have the same polarity (synonym or semantically similar relationship). Hence, if the polarity of one of the concept is known in such a relation then the polarity of the connected work can also be computed. This step uses the classifications from the previous step to perform classifications of unclassified words. The seed for this step comes from previous step and hence this step as well as the subsequent steps does not require any pre-created seed list or training dataset. Unlike Step 1 (which is applied once), Step 2 is executed repeatedly until there is no additional coverage between two consecutive steps. This is done because the first run of Step 2 may result in polarity determination of certain words that can help in predicting polarities of words which could not be determined during the first run of Step 2. For example, let us say that there are two assertions "A DefinedAs B" & "B DefinedAs C" where A,B & C are three concepts in the ConceptNet semantic network. If the polarity of A is known and B is unknown after Step 1, then at the end of the first run of Step 2, polarity of B can be determined. The polarity of concept C can be determined after the second run of Step 2. Thus, Step 2 is repeated as long as the coverage is increasing. We validated our hypothesis by entering few terms on the web-based interface provided at the ConceptNet website. Some illustrative examples of two concepts connected to each other using DefinedAs relation: (blossom, flower), (devil, Satan), (eliminate, exclude), (grotesque, bizarre), (indelicate, indecent), (savage, vicious), (advance, progress), and (whip, beat). The concepts in ConceptNet are natural language fragments and we noticed that often the relationship is of the type A DefinedAs Same B" and "A DefinedAs Opposite B" where A and B are concepts. For example, one of the assertions in ConceptNet is: "Advance DefinedAs same Progress" (can be interpreted as synonyms). Some illustrative examples on concepts having the same polarity that we have provided belong to the assertion type A DefinedAs Same B". This can be handled by locating the word *same* in the concept and removing it from the concept string for extracting the word whose polarity needs to be determined. We noticed several assertions of type A DefinedAs Opposite B (can be interpreted as antonyms). Such assertions can be handled by extracting the term *opposite* from the concept string and flipping the polarity of B i.e. applying the inference that concept Bs polarity is opposite to the polarity of concept A. Some illustrative examples of two concepts connected to each other using *DefinedAs* relation and where the assertion is of type "A DefinedAs Opposite B" are: (dawn, dusk) (selfishness, selflessness), (slow, fast), (abnormal, normal), (bad, good), (clean, dirty), (cruel, kind), (evil, good), (evil, nice), (happiness, sadness), (hard, soft), and (yes, no).

(Step 3 and Step 4). Similar to Step 2, the third and fourth step consists of classifying a word using the polarity of words computed from previous steps (viewed as pre-annotated dataset or seed list for this step) and exploiting the *IsA* and *HasProperty* predicate of ConceptNet. This is based on the hypothesis that Concepts (in our case sentiment expressing nouns, verbs, adverbs or adjectives) connected to each other using *IsA* relationship are semantically *related* (may not be *similar* as in the case of *DefinedAs* predicate) and share the same polarity. Similar to the previous Step, we check the value of assertion type (+1 or -1) and the presence of terms like *same* and *opposite* in the concept for computing the semantic orientation of an unclassified word connected to a word (whose polarity is known) through the *IsA* and *HasProperty* predicate. Step 2,3 and 4 are executed repeatedly (*DefinedAs* analysis followed by *IsA* analysis followed by *HasProperty* analysis) to traverse the semantic network and assign polarities of connected words by exploiting properties of pre-defined predicates. Some illustrative examples of words connected to each other in the Concept-Net semantic netowrk through *IsA* predicate and having posistive polarity are: (cleanliness, good), (faith, trust), (happiness, bliss), (heart, love), (urge, desire), (virtue, good) and (honor, virtue). Some illustrative examples of words having negative polarity and connected through *IsA* predicate are: (assault, crime), (die, tragedy), (fraud, deception), (fraud, cheat), (injury, damage), (kill, crime), (slay, kill) and (war, conflict).

3 Empirical Evaluation

The test data for validating our approach consists of the publicly available subjectivity lexicon which can be freely downloaded from the "MPQA Releases - Corpus and Opinion Recognition System" website of the University of Pittsburgh [1]. The subjectivity lexicon has been used in [9] as well several other work. The subjectivity lexicon consists of 2007 words that have an entry in the ConceptNet. The 2007 words belonging to our test dataset have been pre-classified as positive or negative in the subjectivity lexicon (a benchmark for the task of polarity classification of subjective words). Thus, the actual polarity of all the 2007 words is known in advance which can be compared to the predicted polarity from our approach to determine the accuracy of the proposed solution. Amongst a total of 2007 distinct words in the test dataset, 830 words have positive polarity and 1177 words have negative polarity.

After executing Step 1, we obtained the results presented in Table 2. As mentioned in Step 1, we assign polarities to words where there is no conflict of polarities i.e. words that have been assigned a single polarity only. For example, after executing Step 1, we noticed that there were 22 words which had both positive and negative assertion types. The words are: busy, dance, death, drunk, dying, enlightenment, fairness, faith, free, happiness, hunger, laugh, less, little, live, rich, ridicule, scared, screw, shelter, truth and war. Table presents total and category-wise coverage after executing Step 1. We noticed that our system was able to classify 550 words out of 2007 (coverage of 27.4%) after removing 22 words that had a conflict of polarity. Table 3 presents the confusion matrix. As shown in Table 3, the classification accuracy that we obtained was 95.45%. Step 1 resulted in good coverage (able to classify 27.40% of the words) and high classification accuracy (correctly predicted the polarity of 95.45% of the words that it was able to classify). The results obtained after Step 1 validates our hypothesis that the assertion in ConceptNet in which the relation type is *Desires* and the first concept is *person, human* and *everyone* can be used to infer the polarity of the second concept (the second argument of the *Desires* predicate).

After executing Step 2 once (i.e. applying *DefinedAs* predicate), we were able to correctly classify (with 100% accuracy) nine more words. The pair of words

---

[1] URL: http://www.cs.pitt.edu/mpqa/

Table 2. Total coverage and category-wise coverage after executing Step 1

	Positive & Negative	Positive	Negative
Test Data	2007	830	1177
Coverage Absolute	550	245	305
Coverage Percentage	27.40%	29.51%	25.91%

Table 3. Confusion matrix and classification accuracy after executing Step 1

	Predicted	
	Positive	Negative
Actual Positive	227	7
Actual Negative	18	298
Correct Classification	(227+298)/550 = 95.45%	
Incorrect Classification	(7+18)/550 = 4.54%	

Table 4. Confusion matrix and accuracy after executing Step 2,3 and 4

	Predicted	
	Positive	Negative
Actual Positive	288	23
Actual Negative	51	398
Correct Classification	(288+398)/760 = 90.26%	
Incorrect Classification	(23+51)/760 = 9.74%	

connected to each other using *DefinedAs* predicate and having *same* polarity were: (fancy, like), (gratitude, thank), (liberal, generous), (murky, dark), (paranoia, fear). The polarity of *like, thank, generous, dark* and *fear* were computed from previous step which resulted in correctly classifing the polarity of words fancy, gratitude, liberal, murky and paranoia in Step 2. In this step, the system was also able to correctly classify (with 100% accuracy) words connected using *DefinedAs* relationship but having opposite polarity (as implied by the presence of the word *opposite* in the concept): (cold, warm), (cruel, kind), (hard, easy) and (rich, poor). We noticed that in this step, the accuracy was 100% but the coverage was low. Table 4 presents the final results obtained after executing Steps 2,3 and 4 repeatedly (Step 2 followed by Step 3 followed by Step 4) until no further classifications were observed. As shown in Table 4, the approach correctly predicted 686 words from a total of 760 words that it could classify (an accuracy of 90.26%). The system was able to create a sentiment lexicon of 760 words from a common-sense knowledge base without using any training dataset or a seed list with an accuracy of around 90%.

4   Conclusions

This paper investigates the usefulness of commonsense knowledge for classifying polarity of sentiment expressing words as positive or negative. Evaluation on test data consisting of publicly available pre-annotated subjectivity lexicon shows that leveraging common-sense knowledge that is shared by the vast majority of people for determining semantic orientation determination of words is feasible. The main advantage of the system is that it does not require any training data, hand-crafted seed list or any external resource that is created by trained and specialized knowledge engineers. The accuracy and coverage of the words is a function of the number of concepts, assertions, relations and quality of data in the common-sense knowledge-base.

References

1. Esuli, A., Sebastiani, F.: Determining the semantic orientation of terms through gloss classification. In: Proceedings of the 2005 ACM CIKM International Conference on Information and Knowledge Management, pp. 617–624 (2005)
2. Havasi, C., Speer, R., Alonso, J.: ConceptNet 3: A Flexible Multilingual Semantic Network for Common Sense Knowledge. In: Proceedings of Recent Advances in Natural Languges Processing, Borovets (2007)
3. Kamps, J., Marx, M., Mokken, R.J., Rijke, M.D.: Using wordnet to measure semantic orientation of adjectives. In: Proceedings of the 4th International Conference on Language Resources and Evaluation, vol. IV, pp. 1115–1118 (2004)
4. Liu, H., Singh, P.: Commonsense Reasoning in and over Natural Language. In: Negoita, M.G., Howlett, R.J., Jain, L.C. (eds.) KES 2004. LNCS (LNAI), vol. 3215, pp. 293–306. Springer, Heidelberg (2004)
5. Pang, B., Lee, L.: Opinion mining and sentiment analysis. Foundations and Trends in Information Retrieval 2, 1–135 (2008)
6. Takamura, H., Inui, T., Okumura, M.: Extracting Semantic Orientations of Phrases from Dictionary. In: The Conference of the North American Chapter of the Association for Computational Linguistics, Rochester, pp. 292–299 (2007)
7. Turney, P.D., Littman, M.L.: Measuring praise and criticism: Inference of semantic orientation from association. ACM Transactions on Information Systems 21(4), 315–346 (2003)
8. Vasileios, H., Kathleen, M.R.: Predicting the semantic orientation of adjectives. In: Proceedings of the 35th Annual Meeting of the Association for Computational Linguistics and the 8th Conference of the European Chapter of the ACL, New Brunswick, pp. 174–181 (1997)
9. Wilson, T., Wiebe, J., Hoffmann, P.: Recognizing contextual polarity in phrase-level sentiment analysis. In: Proceedings of Conference on Empirical Methods in Natural Language Processing, Vancouver (2005)

A Weighted Hybrid Fuzzy Result Merging Model for Metasearch

Arijit De[1] and Elizabeth D. Diaz[2]

[1] TCS Innovation Labs-Mumbai, Tata Consultancy Services,
Thane (W), Mumbai 400601
[2] University of Texas of Permian Basin, Odessa, TX 79762
arijit.axd9142@gmail.com, diaz_e@utpb.edu

Abstract. Result merging of search engine results for metasearch is a well explored area. However most result merging models try to collate document rankings from the search engines whose results are being merged into a single ranking using some mathematical function. However, only a few models compare documents in pair wise comparisons during the process of result merging. In this paper, we propose a Weighted Hybrid Fuzzy Result Merging model that comprehensively compares search engines and documents in pairs before applying the result aggregation function. We compare and contrast the performance of our model with existing models for result merging.

1 Introduction

Before discussing our model, let us delve briefly into the metasearch environment. Metasearch engines are tools to carry out parallel and integrated searches through multiple databases/data repositories. Functionally, a metasearch engine dispatches a user query to search engines selected using a search engine selection strategy, select documents from result sets returned by the latter, using a document selection strategy and then returns the documents obtained in the form of a merged list for the user. Key functions include query dispatching, result retrieving, search engine and document selection and result merging. Result merging is a well explored area. Algorithms, models and strategies include the application of linear combination of document ranks [5,7,8], collaborative filtering, multi-criteria decision making techniques such as Borda Fuse [1], and of course fuzzy aggregation [3, 4, 11, 12] of search engine results based on Yager's operator [9,10,11] Most search engine result merging models employ a mathematical aggregation function to search engine ranks. Aslam and Montague [1] apply Borda-Fuse in a linear combination function and Weighted Borda Fuse in a weighted linear combination. Diaz [3,4] and De [11,12] apply Yager's [9,10,11] OWA and IGOWA operator as a fuzzy aggregation function. De [13] use information from pair-wise comparison of documents returned in determining the rank of the document(s) in the merged result list. The motive of our work has been to further explore the effects of pairwise comparisons in result merging and to observe how search engine importance weights affect result merging in the context of pair wise comparisons. We propose the Weighted Hybrid Fuzzy Result Merging model that first compares search engines pair wise and then documents through the

H. Sakai et al. (Eds.): RSFDGrC 2009, LNAI 5908, pp. 494–501, 2009.
© Springer-Verlag Berlin Heidelberg 2009

Analytical Hierarchy Process before using search engine importance weights to generate ordered weights that are used in result aggregation. This paper is organized as follows. In the next section we describe the OWA [3,4] and IGOWA [12] models for metasearch. Subsequently, we describe our proposed model for metasearch. Subsequently we describe our experiments, results and conclusions.

2 Related Work

Diaz [3, 4] develops a fuzzy result merging model OWA which is based on the fuzzy aggregation OWA operator by Yager [9, 10]. The OWA model uses a measure, positional value (PV) to quantify the rank of a document in a search engine result list to be merged. The positional value of a document d_l in the result list l_k returned by a search engine s_k is defined as $(n - r_{ik} + 1)$ where, r_{ik} is the rank of d_i in search engine s_k and n is the total number of documents in the result. Thus, higher the rank of a document in a result list, the larger the positional value of the document in that list. One key feature of the OWA model is that it provides two heuristics (H1 and H2) for handling missing documents. This is done by computing the positional value of a missing document in the result list and thereby effectively inserting the document in the result list in which it is missing. Diaz in [3] shows that the heuristic H1 provides the most effective way to handle missing documents. Let PV_i be the positional values for a document d in the ith search engine. Let m be the total number of search engines. Let r be the number of search engines in which d appears. Let j denote a search engine not among the r search engines where d appears. In heuristic H1 PV_j for all j is denoted by the average of the positional values of the documents in r search engines. In heuristic H2 PV_j for all j is denoted by the average of the positional values of the documents in all m search engines.

In computing the final score of a document in the merged list, the positional values are sorted in descending order and then are aggregated along with a set of ordered weights using Yager's OWA operator.

The IGOWA model [11] was a direct extension to the OWA model, but used search engine performance weights to compare the ordered weights used in OWA aggregation. Thus search engine performance weights could be used to affect the score/rank of a document in the merged list.

The Hybrid Fuzzy Result Merging model [14] is similar to the OWA and IGOWA models described in section 3.3 and 3.4 respectively. However it allows for pair wise comparisons by applying the Analytical Hierarchy Process, to compute the positional values. These are then aggregated using the OWA operator.

3 Proposed Weighted Hybrid Fuzzy Result Merging Model

Our proposed Weighted Hybrid Fuzzy Result Merging model is similar to the OWA model as it uses the concept of positional values and uses the heuristic H1 for handling missing documents. It is similar to the Hybrid Fuzzy Result Merging model as it applies the Analytical Hierarchy Process to search engine rankings and document rankings, to compute positional values based on document and search engine rankings.

These pair-wise positional values are then used in aggregation using the OWA operator. The ordered weights computed using search engine performance weights.

Saaty [14] proposed the Analytic Hierarchy Process, which outlines the mechanism for pairwise comparison for objects. Let us say we have two object O_i and O_j. We can compare the two objects and quantify the comparison as outlined by Saaty. If O_i is equally important as O_j then the pair-wise value is 1, if O_i is weakly more important than O_j then the pair-wise value is 3, O_i is strongly important than O_j then the pair-wise value is 5, O_i is very strongly important than O_j then the pair-wise value is 7. If O_i is absolutely more important than O_j then the pair-wise value is 9. Similarly if O_i is less important than O_j to a varying degrees then the value is 1/3, 1/5, 1/7, 1/9 respectively.

The Analytical Hierarchy process is a multi-criteria decision making process. In this process, Saaty [14] first proceeds to compare the criteria themselves pair-wise and gives each comparison a value as mentioned earlier. These values of pair-wise comparisons are then put into a matrix A. This matrix is then normalized by dividing each element with the sum of the member column and then averaging the normalized values by each row. This way, criteria scores are obtained. In the same way alternatives are compared with each other with respect to each criteria, a matrix is formed normalized and row averages are computed to obtain alternative scores with respect to each criteria. These are multiplied by criteria scores computed earlier. Thus for each criteria, alternative pairing we obtain a score.

To put this in context of metasearch engines, let us suppose we have m search engines and each return a set of n documents ranked in any specified order. Let us say we have a result of documents $D = \{d_1, d_2, \ldots d_n\}$ returned by the Search Engine SE_k that need to be ranked. Here the Search Engines are the criteria and the documents the alternatives. We can compare the documents pair wise and form a square matrix $A = [a_{ij}]$ where a_{ij} is a measure of how the two documents d_i and d_j compare pair wise using a scale of 1/9 to 9 as per the AHP scale mentioned earlier.

Let r_i and r_j be the ranks of two documents i and j respectively. a_{ij} is calculated as $((r_i \sim r_j) / n) \cdot 9$ and normalized to the nearest value in the number set [1,3,5,7,9] if $r_i > r_j$. and the reciprocal it $r_i < r_j$. Thus we can form the matrix $A = [a_{ij}]$ in conformance with table 1. The next step is to normalize A such that each element is divided by the sum of the column in which it belongs. The next step is to compute the document scores for search engine k by averaging of each row of the matrix. Thus obtaining a search engine-document score matrix $SE\text{-}DOC\text{-}SCORE_k = [s_{1k}, s_{2k}, \ldots s_{nk}]$ which are the document scores. We can proceed to compute the scores for all search engines. We can also similarly rank search engines in order of preference or performance and compute search engine scores $SE\text{-}SCORE = [ss_1, ss_2, \ldots ss_m]$ where m is the number of search engines. We can compute the final scores of the pth document DOC-$SCORE_p$ as $[ds_{p1}, ds_{p2}, \ldots ds_{pm}]$ where $ds_{p1} = s_{p1} * ss_1$ etc. Thus we can obtain a two dimensional matrix SCORE of m columns (one for each search engine) and n rows (one for each column).

With pair-wise comparisons taken care of using the AHP MCDM technique we now apply the OWA operator for aggregation. The OWA operator was an aggregation function employed in Fuzzy Multi-Criteria decision making. Let there be a set of n criteria. Let $a_1, a_2, \ldots, a_n$ be degree to which an alternative satisfies each of the n criteria. To combine these degrees to which an alternative satisfies multiple criteria

Table 2. Values in a Pair-Wise Matrix

	Pair-Wise Comparison			Normalized			Score
	se_1	se_2	Se_3	se_1	se_2	se_3	
se_1	1.00	3.00	7.00	0.68	0.69	0.64	0.67
se_2	0.33	1.00	3.00	0.23	0.23	0.27	0.24
se_3	0.14	0.33	1.00	0.10	0.08	0.09	0.09

the OWA operator F is applied as $F(a_1, a_2,..., a_n) = \sum_{j=1}^{n} w_j b_j$, where b_j is the jth largest a_j. The Ordered Weights can be calculated in different ways. Popularly they can be calculated using a Regular Increasing Monotone (RIM) quantifier guided approach specified by Yager [10]. Once again in our problem of metasearch the documents are the alternatives and the search engines are the criteria. However, here we consider the scores obtained through the AHP pair-wise comparison process as the degree to which a document satisfies the search engine. The OWA operator can be transformed in this context as follows:

$$SCORE(D_i) = \sum_{j=1}^{n} ds'_{ij} w_j(D_i) \text{ where } ds'_{ij} \text{ is the jth greatest } ds_{ij} \qquad (4)$$

We can compute the jth ordered weight for document D_i, $w_j(D_i)$ as per Yager [10] as follows:

$$w_j(D_i) = Q\left(\frac{\sum_{k=1}^{j} u_k}{T}\right) - Q\left(\frac{\sum_{k=1}^{j-1} u_k}{T}\right) \qquad (5)$$

Where $T = \sum_{k=1}^{n} u_k$ and Q is Regular Increasing Monotone quantifier of the form Q = r^α. Here α is a quantifier of parameter. Once we have obtained the ordered weights we can now calculate the score for the document, using the OWA operator. We can rank documents by their final scores.

Let us illustrate the working of the example. Let us consider search engines se_1, se_2, se_3 and 5 documents numbered d_1 through d_5. The search engines se_1, se_2 and se_3 return the documents in the order { d_2, d_3, d_1, d_4, d_5 }, { d_5, d_4, d_1, d_3, d_2 } and { d_1, d_5, d_2, d_3, d_4 }. Let us say the performance weights of the search engines se_1, se_2, se_3 be 0.6, 0.3 and 0.2. Based on the performance of search engines we can rank them in the following order $se_1 > se_2 > se_3$. We proceed first to apply pair-wise comparison to the search engines.

Performing pair-wise comparisons of documents based on their ranks in list returned by se_1 we obtain the following document scores:

Table 3. Pair-Wise Matrix for Comparing Documents in Search Engine List

	d_1	d_2	d_3	d_4	d_5
d_1	1.00	0.20	0.33	5.00	7.00
d_2	5.00	1.00	3.00	7.00	9.00
d_3	3.00	0.33	1.00	5.00	7.00
d_4	0.20	0.14	0.20	1.00	3.00
d_5	0.14	0.11	0.14	0.33	1.00

Table 4. Normalized Pair-Wise Matrix for Comparing Documents in Search Engine List

	d_1	d_2	d_3	d_4	d_5
d_1	0.11	0.11	0.07	0.27	0.26
d_2	0.54	0.56	0.64	0.38	0.33
d_3	0.32	0.19	0.21	0.27	0.26
d_4	0.02	0.08	0.04	0.05	0.11
d_5	0.02	0.06	0.03	0.02	0.04

Similarly we can perform pair-wise comparison for search engines se_2 and se_3. The final scores of each document with respect to each search engine are computed similarly and tabulated below.

Table 5. Final Document scores for Search Engines

	se_1	se_2	se_3
d_1	0.11	0.04	0.01
d_2	0.33	0.06	0.00
d_3	0.17	0.02	0.01
d_4	0.04	0.01	0.02
d_5	0.02	0.12	0.04

Proceeding to compute the ordered weights for document d_4 we first sort the positional values in descending order [0.04 (se_1), 0.02(se_3), 0.01 (se_2)]. Using the performance weights of the search engines [0.6, 0.1, 0.3] and a RIM quantifier of the form $Q = r^\alpha$ with $\alpha = 2$ we obtain the Ordered Weights as [0.07, 0.13, 0.79]. Then we can apply the OWA operator to obtain the score of the document as 0.07*0.04 + 0.13*0.02 + 0.79*0.01 = 0.02. The remaining document scores can be computed similarly. Thus scores for documents d_1 through d_5 would be 0.05, 0.12, 0.06, 0.02, 0.05 respectively. The merged ranking would be { d_2, d_3, d_5, d_1, d_4 }.

4   Experimental Results and Discussion

We use the Text REtrieval Conference (TREC) datasets, TREC 3 (ad hoc track), TREC 5 (ad hoc track) and TREC 9 (web track). Each dataset contains a set of systems (search engines) and 50 topics (queries). For each query and search engine there is a ranked result list of 1000 documents returned. The relevance information for the documents is provided along with the datasets. Since our Weighted Hybrid Result Merging model requires us to not only rank search engines for pair-wise comparisons using AHP but also require search engine performance weights. To obtain search engine performance weights for our data sets, we split the queries in each data set into two parts. The odd numbered queries in each data set are used to evaluate the performance of search engines. Based on the performance of search engines we rank them for pair-wise comparisons. The performance weights are then applied to compute ordered weights for OWA aggregation, based on equation 5. In our experiments we compare the OWA model [3], the IGOWA model [12] and our proposed Weighted Hybrid Fuzzy Model. All these models use the OWA aggregation operator. The ordered weights are computed in different ways but all of them involve the use of a RIM quantifier of the form r^α, where r is the input to the function and α is the parameter. In our experiments we vary the value of α as (0.25, 0.5, 1, 2, 2.5). For each value of α we perform a total 1000 trials of experiment in batches of 200. In each batch a specified number of search engines (picked randomly from the dataset) are merged. This number varies as 2,4,6,8 and 10. The average precision of the merged list are computed and at the end the mean of the average precision from all 1000 trials is computed. The average precision is the average of the precisions computed at points in the result list where a relevant document is found. The results for TREC 3, TREC 5 and TREC 9 are shown in table 6. For each of the datasets the average precision is measured while the OWA quantifier parameter α varies from 0.25 to 2.5. From the table it can be observed that for TREC 3, for all values of α the IGOWA model outperforms the OWA model. This fact was established by De [12]. However when the value of α is small (i.e., α = 0.25), both the OWA and the IGOWA model outperforms our proposed Weighted Hybrid Fuzzy Model. However, as the value of α is increased, our proposed model begins to outperform both the OWA and IGOWA model. In case of both the OWA and the IGOWA model there is a dip in performance as α goes from 0.25 to α = 1. As α moves beyond 1 the performance goes up. However, in case of our proposed model the performance goes up as we move from α = 0.25 to α = 2.5. Similar observations can be made with respect to the TREC 5 dataset. In case of the TREC 9 data set our proposed model outperforms the OWA model. At α = 0.25, the performance of the IGOWA model and our proposed Weighted Hybrid Fuzzy model remains the same. But the latter outperforms IGOWA at higher values of α. Yager shows that the degree 'orness' of a aggregation is $1/(1+\alpha)$. When using a quantifier guided approach and using a RIM quantifier to compute the ordered weights. Thus orness decreases as α increase from 0.25 to 2.5. At α = 1 the orness becomes half, in other words the conditions represent simple averaging. At this point the Average precision in case of the OWA model is the least. As α tends away from the, α = 1 point of inflection, average precision increases. Applying pair wise analysis using AHP tends to remove this dip in performance as exhibited by our Weighted Hybrid Fuzzy model. Table 6, illustrates these results.

Table 6. Final Document scores for Search Engines

Data Sets	Models	α				
		0.25	0.5	1	2	2.5
TREC 3	OWA	0.3004	0.2648	0.2323	0.2889	0.3293
	IGOWA	0.3278	0.2811	0.2493	0.3097	0.3494
	Weighted Fuzzy Hybrid	0.2786	0.3217	0.3438	0.3579	0.4152
TREC 5	OWA	0.2710	0.2791	0.2408	0.2805	0.3269
	IGOWA	0.3038	0.2744	0.2617	0.3291	0.3521
	Weighted Fuzzy Hybrid	0.2697	0.2996	0.3147	0.3310	0.3596
TREC 9	OWA	0.1577	0.1577	0.1578	0.1578	0.1578
	IGOWA	0.1782	0.1649	0.1649	0.1649	0.1799
	Weighted Fuzzy Hybrid	0.1781	0.1781	0.1781	0.1907	0.1961

5 Conclusions

In this paper we have proposed a Weighted Hybrid Fuzzy model for result merging for metasearch that compares search engines and documents pair-wise to determine the final position of the document in the merged list. The model is an extension of the IGOWA model [12] and the OWA [3] model for metasearch, as it applies the OWA operator for aggregation. The ordered weights for aggregation using the OWA operator are computed based on the performance weights of search engines whose result sets are being merged. The model first uses the Analytical Hierarchy Process (AHP) to do pair wise comparison of search engines, and documents in result lists returned by them, prior to aggregation using the OWA operator. It then proceeds to compute ordered weights based on search engine performances and uses these weights in the OWA operator for determining document positions in the merged list.

We compare our proposed model with the IGOWA and OWA models that it extends. In our experiments involving standard Text Retrieval Conference (TREC) datasets TREC 3, TREC 5 and TREC 9, we show that the performance of merging in terms of average precision of the merged result list improves significantly when using the Weighted Hybrid Fuzzy model over IGOWA and OWA. However our model requires some learning to determine search engine performance weights and also search engine rankings that are needed in search engine pair-wise comparison.

References

[1] Aslam, J., Montague, M.: Models for metasearch. In: Proceedings of the 24th annual international ACM SIGIR Conference on Research and Development in Information Retrieval, New Orleans, LA, USA, September 2001, pp. 276–284 (2001)

[2] Borda, J.C.: Memoire sur les elections au scrutiny. Histoire de l'Academie Royale des Sciences, Paris (1781)

[3] Diaz, E.D., De, A., Raghavan, V.V.: A comprehensive OWA-based framework for result merging in metasearch. In: Rough Sets, Fuzzy Sets, Data Mining, and Granular-Soft Computing, Regina, SK, Canada, September 2005, pp. 193–201 (2005)

[4] Diaz, E.D.: Selective Merging of Retrieval Results for Metasearch Environments, Ph.D. Dissertation, University of Louisiana, Lafayette, LA (May 2004)

[5] Hull, D.A., Pedersen, J.O., Schütze, H.: Method combination for document filtering. In: Proceedings of the 19th annual international ACM SIGIR Conference on Research and Development in Information Retrieval, Zurich, Switzerland, August 1996, pp. 279–287 (1996)

[6] Meng, W., Yu, C., Liu, K.: Building efficient and effective metasearch engines. ACM Computing Surveys 34, 48–84 (2002)

[7] Thompson, P.: A combination of expert opinion approach to probabilistic information retrieval, part 1: The conceptual model. Information Processing and Management 26(1), 371–382 (1990)

[8] Thompson, P.: A combination of expert opinion approach to probabilistic information retrieval, part 2: mathematical treatment of CEO model. Information Processing and Management 26(3), 383–394 (1990)

[9] Yager, R.R.: On ordered weighted averaging aggregation operators in multi-criteria decision making. Fuzzy Sets and Systems 10(2), 243–260 (1983)

[10] Yager, R.R.: Quantifier guided Aggregating using OWA operators. International Journal of Intelligent Systems 11(1), 49–73 (1996)

[11] De, A., Diaz, E.E., Raghavan, V.V.: On Fuzzy Result Merging for Metasearch. In: IEEE International Fuzzy Systems Conference, 2007. FUZZ-IEEE 2007, July 23-26, pp. 1–6 (2007)

[12] De, A., Diaz, E.D., Raghavan, V.: A Fuzzy Search Engine Weighted Approach to Result Merging for Metasearch. In: An, A., Stefanowski, J., Ramanna, S., Butz, C.J., Pedrycz, W., Wang, G. (eds.) RSFDGrC 2007. LNCS (LNAI), vol. 4482, pp. 95–102. Springer, Heidelberg (2007)

[13] De, A., Diaz, E.: Hybrid Fuzzy Result Merging for Metasearch Using Analytic Hierarchy Process. In: 28th North American Fuzzy Information Processing Society Annual Conference, NAFIPS (2009)

[14] Saaty, T.L.: Relative Measurement and its Generalization in Decision Making: Why Pairwise Comparisons are Central in Mathematics for the Measurement of Intangible Factors - The Analytic Hierarchy/Network Process. Review of the Royal Spanish Academy of Sciences, Series A, Mathematics 102(2), 251–318 (2007)

Comparing Temporal Behavior of Phrases on Multiple Indexes with a Burst Word Detection Method

Hidenao Abe and Shusaku Tsumoto

Department of Medical Informatics, Shimane University, School of Medicine
89-1 Enya-cho, Izumo, Shimane 693-8501, Japan
abe@med.shimane-u.ac.jp, tsumoto@computer.org

Abstract. In temporal text mining, some importance indices such as simple appearance frequency, tf-idf, and differences of some indices play the key role to identify remarkable trends of terms in sets of documents. However, most of conventional methods have treated their remarkable trends as discrete statuses for each time-point or fixed period. In order to find their trends as continuous statuses, we have considered the values of importance indices of the terms in each time-point as temporal behaviors of the terms. In this paper, we describe the method to identify the temporal behaviors of terms on several importance indices by using the linear trends. Then, we show a comparison between visualizations on each time-point by using composed indices with PCA and the trends of the emergent terms, which are detected the burst word detection method.

Keywords: Text Mining, Trend Detection, TF-IDF, Jaccard Coefficient, Linear Regression.

1 Introduction

In recent years, the development of information systems in every field such as business, academics, and medicine, and the amount of stored data have increased year by year. Accumulation is advanced to document data by not the exception but various fields. Document data provides valuable findings to not only domain experts in headquarter sections but also novice users on particular domains such as day trading, news readings and so on. Hence, the detection of new phrases and words has become very important. In order to realize such detection, emergent term detection (ETD) methods have been developed [1,2].

However, because the frequency of words was used in earlier methods, detection was difficult as long as the word that became an object did not appear. Usually, emergent or new concepts are appeared as new combination of multiple words, coinages created by an author, and words with different spellings of current words. Most conventional methods did not consider above-mentioned natures of terms and importance indices separately. This causes difficulties in text mining applications, such as limitations on the extensionality of time direction, time consuming post-processing, and generality expansions.

H. Sakai et al. (Eds.): RSFDGrC 2009, LNAI 5908, pp. 502–509, 2009.

After considering these problems, we focus on temporal behaviors of importance indices of terms and their temporal patterns. Temporal behaviors of the importance indices of extracted phrases are paid attention so that a specialist may recognize emergent terms and/or such fields. In order to detect various temporal patterns of behaviors of terms in the sets of documents, we have proposed a framework to identify the remarkable terms as continuous changes of multiple metrics of the terms [3].

In this paper, we propose an integrated for detecting trends of technical terms by combining automatic term extraction methods, importance indices of the terms, and trend analysis methods in Section 2. After implementing this framework as described in Section 3, we performed a comparison between the emergent words detected by a past emergent term detection study and the trends of terms based on the indices. In Section 4, the comparison is shown by using the titles of the two well-known AI related conferences: IJCAI and AAAI. Finally, in Section 5, we summarize this paper.

2   An Integrated Framework for Detecting Trends of Technical Terms Based on Importance Indices

In this section, we describe about the difference between conventional ETD methods and our proposal; detecting continuous temporal patterns of terms in temporal sets of documents.

As illustrated in Fig.1, conventional ETD methods focus on to find out discrete status of emergent terms as points defined by one or more importance indices in

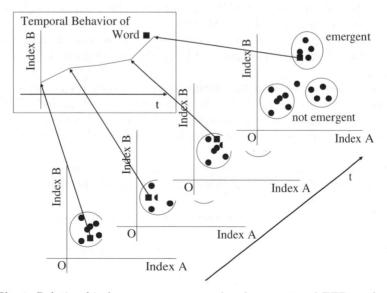

Fig. 1. Relationship between our proposal and conventional ETD methods

text mining. Since each point corresponds to each term, it is difficult to detect similar terms as emergent. Several conventional methods expand each point to regions by using frequent patterns with regular expressions for words [4] and so on. However, it is more important to capture temporal changes of each term as temporal trends. We considered about to introduce general temporal analysis framework for finding out remarkable trends of terms.

In order to find remarkable temporal trends of terms, we developed a framework for detecting various temporal trends of technical terms by using multiple importance indices consisting of the following three components:

1. Technical term extraction in a corpus
2. Importance indices calculation
3. Trend detection

There are some conventional methods of extracting technical terms in a corpus on the basis of each particular importance index [2]. Although these methods calculate each index in order to extract technical terms, information about the importance of each term is lost by cutting off the information with a threshold value. We suggest separating term determination and temporal trend detection based on importance indices. By separating these phases, we can calculate different types of importance indices in order to obtain a dataset consisting of the values of these indices for each term. Subsequently, we can apply many types of temporal analysis methods to the dataset based on statistical analysis, clustering, and machine learning algorithms.

First, the system determines terms in a given corpus. There are two reasons why we introduce term extraction methods before calculating importance indices. One is that the cost of building a dictionary for each particular domain is very expensive task. The other is that new concepts need to be detected in a given temporal corpus. Especially, a new concept is often described in the document for which the character is needed at the right time in using the combination of existing words.

After determining terms in the given corpus, the system calculates multiple importance indices of the terms for the documents of each period. Further, in the proposed method, we can assume the degrees of co-occurrence such as the χ^2 statistics for terms consisting of multiple words to be the importance indices in our method.

In the proposed method, we suggest treating these indices explicitly as a temporal dataset. The features of this dataset consist of the values of prepared indices for each period.

Fig.2 shows an example of the dataset consisting of an importance index for each year.

Then, the framework provides the choice of some adequate trend extraction method to the dataset. In order to extract useful time-series patterns, there are so many conventional methods as surveyed in the literatures [5,6]. By applying an adequate time-series analysis method, users can find out valuable patterns by processing the values in rows in Fig.2.

Term	Jacc_1996	Jacc_1997	Jacc_1998	Jacc_1999	Jacc_2000	Jacc_2001	Jacc_2002	Jacc_2003	Jacc_2004	Jacc_2005
output feedback	0	0	0	0	0	0	0	0	0	0
H/sub infinity	0	0	0.012876	0	0.00885	0	0	0	0.005405	0.003623
resource allocation	0.006060606	0	0	0	0	0	0	0	0	0
image sequences	0	0	0	0	0	0	0	0.004785	0	0
multiagent systems	0	0	0	0	0	0	0.004975	0	0	0
feature extraction	0	0.005649718	0	0.004484	0	0	0	0	0	0
images using	0	0	0	0	0	0.004673	0	0	0	0
human-robot interaction	0	0	0	0	0.004425	0	0	0	0	0
evolutionary algorithm	0	0.005649718	0	0.004484	0	0	0	0	0.002703	0.003623
deadlock avoidance	0	0	0	0	0.004425	0	0	0	0	0
ambient intelligence	0	0	0	0	0	0	0	0	0	0.003623
feature selection	0	0	0	0	0	0	0	0	0.002703	0
data mining	0	0	0	0	0.004425	0	0	0	0.002703	0

Fig. 2. Example of a dataset consisting of an importance index

3 Implementing the Integrated Framework for Detecting Temporal Trends of Technical Terms

As described in Section 2, the integrated framework for detecting temporal trends of technical terms consists of the three sub-processes. In order to implement the framework, we assigned each process described as follows.

Considering the difficulties of the term extraction without any dictionary, we apply a term extraction method that is based on the adjacent frequency of compound nouns. This method involves the detection of technical terms by using the following values for each composed nouns CN:

$$FLR(CN) = f(CN) \times (\prod_{i=1}^{L}(FL(N_i)+1)(FR(N_i)+1))^{\frac{1}{2L}}$$

where each CN consists of L words. Then, $f(CN)$ means the frequency of appearances of CN solely, and $FL(N_i)$ and $FR(N_i)$ indicate the frequencies of differences on the right and the left of each noun N_i. In the following experiment, we selected technical terms with this FLR score as $FLR(term) > 1.0$. This threshold is important to select adequate terms at the first iteration. However, since our framework assumes the whole process as iterative search process for finding required trends of terms by a user, the user can input manually selected terms in the other iterations. In order to determine terms in this part of the process, we can also use other term extraction methods and terms/keywords from users.

As for importance indices of words and phrases in a corpus, there are some well-known indices. Term frequency divided by inversed document frequency (tf-idf) is one of the popular indices used for measuring the importance of the terms. tf-idf for each term $term$ can be defined as follows:

$$TFIDF(term, D_{period}) = TF(term) \times log \frac{|D_{period}|}{DF(term)}$$

where $TF(term)$ is the frequency of each term $term$ in the corpus with $|D_{period}|$ documents. $|D_{period}|$ means the number of documents included in each period. $DF(term)$ is the frequency of documents containing $term$.

As another importance index, we use Jaccard's matching coefficient [7][1]. Jaccard coefficient can be defined as follows:

---

[1] Hereafter, we refer to this coefficient as "Jaccard coefficient".

$$Jaccard(term, D_{period}) = \frac{DF(w_1 \cap w_2 \cap ... \cap w_L)}{DF(w_1 \cup w_2 \cup ... \cup w_L)}$$

where $DF(w_1 \cap w_2 \cap ... \cap w_L)$ is equal to $DF(term)$, because each $term$ consists of w_i ($1 \leq i \leq L$). $DF(w_1 \cup w_2 \cup ... \cup w_L)$ means the frequency of documents that contains w_i. Jaccard coefficient is originally defined as the ratio of the probability of an intersection divided by the probability of a union in a set of documents. In this framework, we applied this index by defining as the ratio of the frequency of an intersection divided by the frequency of a union in each set of documents D_{period}. Each value of Jaccard coefficient shows strength of co-occurrence of multiple words as an importance of the terms in the set of documents.

In addition to the above two importance indices, we used simple appearance ratio of terms in a set of documents.

$$Odds(term, D_{period}) = \frac{DF(term)}{|D_{period}| - DF(term)}$$

where, $DF(term)$ means the frequency of the appearance of each term $term$ in each set of documents D_{period}.

As for the first step to determine temporal trends in the dataset, we apply the linear regression analysis technique in order to detect the degree of existing trends for each importance index. The degree of each term $term$ is calculated as the following:

$$Deg(term) = \frac{\sum_{i=1}^{M}(y_i - \bar{y})(x_i - \bar{x})}{\sum_{i=1}^{M}(x_i - \bar{x})^2}$$

where $\bar{x}$ is the average of the M time points, and $\bar{y}$ is the average of each importance index for the period. Simultaneously, we calculate the intercept $Int(term)$ of each term $term$ as follows:

$$Int(term) = \bar{y} - Deg(term)\bar{x}$$

4   Comparison of Visualization on the Time Points of Technical Terms and Their Temporal Trends

In this section, we compared the emergent words detected by a past emergent term detection study and the trends of terms based on the three importance indices.

In [8], it proposed a method to detect bursty words that occur with high intensity over a limited period. The analysis uses a probabilistic automaton whose states correspond to the frequencies of individual words. To the titles of several famous international conferences related to computer science, the bursty words are detected[2].

From AAAI and IJCAI titles, the method detected the following words as currently bursting: auctions, combinational, and reinforcement. In order to compare the bursty words in the past, we also compared the degrees and the intercepts of terms including the following words: language and objects. These two words are detected as bursty word from 1980 to 1983, and from 1998 to 2002 or later in the earlier work.

---

[2] They can be found in http://www.cs.cornell.edu/home/kleinber/kdd02.html

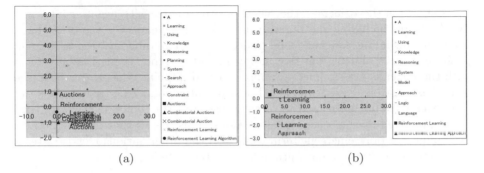

(a) (b)

Fig. 3. Visualization of the top ten frequent terms and the terms including the burst words for the titles; (a)AAAI from 1998 to 2000, (b)IJCAI from 1997 to 2001

4.1 Visualizing the Burst Terms by Using Multiple Importance Indices

By using top ten frequent terms and the terms including burst words for the period, we visualized these terms on two dimensions consisting of composed indices by using the first two components of the result of PCA (Principal Component Analysis).

As for AAAI titles, we visualize the top ten frequent terms and the terms including 'auction', 'combinational', and 'reinforcement' by using the values of the three indices and $DF(term, D_{period})$ from 1998 to 2000. As shown in Fig.3(a), the terms including the bust words are placed near the origin than the frequent terms.

For the titles of IJCAI, the top ten frequent terms and the terms including 'reinforcement' are visualized as shown in Fig.3(b) from 1997 to 2001.

As shown these figures, emergent terms are placed different regions of the popular terms. However, in order to identify emergent terms with the values of indices for the limited period, it is difficult to categorize as discrete statuses.

4.2 Temporal Trends of the Burst Terms by Detecting Linear Regression Technique

Table1 shows the terms including the five words with the degrees and intercepts of tf-idf, Jaccard coefficient and Odds. In order to eliminate specific paper, we selected the terms, which appeared more than two times.

Table 1. Degrees and intercepts of tf-idf, Jaccard coefficient, and Odds of the terms including language, objects, auctions, combinational, and reinforcement

	Term	TFIDF Deg	TFIDF Int	Jacc Deg	Jacc Int	Odds Deg	Odds Int
AAAI	Combinatorial Auctions	0.402	−2.039	0.052	−0.272	0.00027	−0.00119
	Combinatorial Auction	0.205	−0.950	0.011	−0.058	0.00015	−0.00063
	Auctions	0.123	−0.467	0.005	−0.021	0.00010	−0.00035
	Reinforcement Learning	0.748	−1.640	0.006	−0.001	0.00058	−0.00011
	Reinforcement Learning Algorithm	0.064	−0.199	0.000	−0.001	0.00003	−0.00007
IJICAI	Reinforcement Learning	0.576	−1.687	0.005	−0.010	0.00039	−0.00097
	Reinforcement Learning Approach	0.079	−0.113	0.001	−0.001	0.00005	−0.00010

As shown in these degrees, they show positive degree with negative intercepts for the terms including the bursty words in recent years. On the other hand, the terms including 'language' and 'objects' show two different trends. One has positive degrees and big positive intercepts. This trend means that the terms with these trends are assumed as important and popular issue in this field. The other has negative degrees and small positive intercepts. This trend means that the terms have not been used in recent years, and the topics shown with these terms appeared as the other representations. This means that our method can determine not only emergent terms, but also the other various trends of terms based on the multiple importance indices in Text Mining.

5   Conclusion

In this paper, we proposed a framework to detect remarkable trends of technical terms by focusing on the temporal changes of the importance indices. We implemented the framework by combining the technical term extraction method, the three important indices, and linear regression analysis.

The case studies show that the temporal changes of the importance indices can detect the trend of each phrase, according to the degree of the values for each annual set of the titles of the four academic conferences. Regarding to the result, our method can support to find out remarkable technical terms in documents based on the temporal changes of the importance indices.

In the future, we will apply other term extraction methods, importance indices, and trend detection method. As for importance indices, we are planning to apply evaluation metrics of information retrieval studies, probability of occurrence of the terms, and statistics values of the terms. To extract the trends, we will introduce temporal pattern recognition methods, such as temporal clustering [6,9]. Then, we will apply this framework to other documents from various domains.

References

1. Lent, B., Agrawal, R., Srikant, R.: Discovering trends in text databases. In: Proc. of the Third International Conference on Knowledge Discovery and Data Mining (KDD 1997), pp. 227–230 (1997)
2. Kontostathis, A., Galitsky, L., Pottenger, W.M., Roy, S., Phelps, D.J.: A survey of emerging trend detection in textual data mining. A Comprehensive Survey of Text Mining (2003)
3. Abe, H., Tsumoto, S.: Detecting temporal trends of technical phrases by using importance indices and linear regression. In: Proc. of the 18th International Symposium on Methodologies for Intelligent Systems (ISMIS 2009), pp. 251–259. Springer, Heidelberg (2009)
4. Mei, Q., Zhai, C.: Discovering evolutionary theme patterns from text: an exploration of temporal text mining. In: KDD 2005: Proceedings of the eleventh ACM SIGKDD international conference on Knowledge discovery in data mining, pp. 198–207. ACM, New York (2005)

5. Keogh, E., Chu, S., Hart, D., Pazzani, M.: Segmenting time series: A survey and novel approach. In: An Edited Volume, Data mining in Time Series Databases, pp. 1–22. World Scientific, Singapore (2003)
6. Liao, T.W.: Clustering of time series data: a survey. Pattern Recognition 38, 1857–1874 (2005)
7. Anderberg, M.R.: Cluster Analysis for Applications. In: Monographs and Textbooks on Probability and Mathematical Statistics. Academic Press, Inc., New York (1973)
8. Kleinberg, J.M.: Bursty and hierarchical structure in streams. Data Min. Knowl. Discov. 7(4), 373–397 (2003)
9. Ohsaki, M., Abe, H., Yamaguchi, T.: Numerical time-series pattern extraction based on irregular piecewise aggregate approximation and gradient specification. New Generation Comput. 25(3), 213–222 (2007)

A Comparative Study of Pattern Matching Algorithms on Sequences

Fan Min[1,2] and Xindong Wu[2,3]

[1]School of Computer Science and Engineering,
University of Electronic Science and Technology of China,
Chengdu 610054, China
[2]Department of Computer Science,
University of Vermont, Burlington, Vermont 05405, USA
[3]School of Computer Science and Information Engineering,
Hefei University of Technology, Hefei 230009, China
minfan@uestc.edu.cn, xwu@cems.uvm.edu

Abstract. In biological sequence pattern mining, pattern matching is a core component to count the matches of each candidate pattern. We consider patterns with wildcard gaps. A wildcard gap matches any subsequence with a length between predefined lower and upper bounds. Since the number of candidate patterns might be huge, the efficiency of pattern matching is critical. We study two existing pattern matching algorithms named Pattern mAtching with Independent wildcard Gaps (PAIG) and Gap Constraint Search (GCS). GCS was designed to deal with patterns with identical gaps, and we propose to revise it for the case of independent gaps. PAIG can deal with global length constraints while GCS cannot. Both algorithms have the same space complexity. In the worst case, the time complexity of GCS is lower. However, in the best case, PAIG is more efficient. We discuss appropriate selection between PAIG and GCS through theoretical analysis and experimental results on a biological sequence.

Keywords: Pattern matching, sequence, wildcard gap, constraint.

1 Introduction

Recent advances in biology and the Internet have attracted extensive research interests regarding pattern mining from sequences [1][2][3]. A DNA sequence is represented by a sequence S with a small alphabet $\sum =\{$A, C, G, T$\}$ [4][2]. RNAs have a slightly different alphabet {A, U, C, G}, and proteins have a larger alphabet with 20 characters. Lengths of these sequences typically range from a few thousand to a few million. For example, the H1N1 No. FJ984346 sequence [5] begins with ATGGAAGA and its length is 2,150.

A repetitive fragment of a sequence is represented by a pattern P. There are a number of different definitions of a pattern, with respective applications in biology sequences [2]. For the simplest case, a pattern is a subsequence. $P =$ GA has 2 matches in sequence ATGGAAGA, beginning at indices 4 and 7,

H. Sakai et al. (Eds.): RSFDGrC 2009, LNAI 5908, pp. 510–517, 2009.

respectively. We can introduce a special symbol ϕ, called **wildcard** [6] or **don't care** [7][8], to P that matches any character in the alphabet. $P = G\phi A$ also has 2 matches in ATGGAAGA, while the beginning indices are 3 and 4, respectively. An even more general definition allows wildcard gaps. $P = G[0,2]A$ is a pattern with 0 to 2 wildcards between G and A. It matches subsequences GA, GGA, GGAA, GAA of ATGGAAGA, and the total number of matches is 5. In this paper we adopt the last definition.

Frequently appearing patterns tend to be interesting. During the process of pattern mining, we need to compute the number of matches of each constructed pattern. Therefore, pattern matching is a core component of pattern mining. It is also an independent problem in many applications, where the pattern is specified by the user (see, e.g., [6][9][10]).

PAIG [10] and GCS [3] are two pattern matching algorithms. Both can compute the number of matches in polynomial time. However, since the number of patterns to be checked in the process of pattern mining might be huge, the efficiency difference is important in applications.

In this paper, we closely study these two algorithms. We compare the applicability of them, and the focus is their time complexities. Both theoretical analysis and experiments on biological sequences show that PAIG is more efficient when gaps in the pattern are not flexible, while GCS is more efficient for the other case. We can decide which algorithm to be employed for a new pattern according to the computational time of some existing patterns.

2   PAIG and GCS

In this section, we revisit PAIG and GCS using the following example: $S =$ AGAAGAGGAAGAA and $P = A[0,2]G[1,2]A[0,3]A$. An enhanced version of GCS is proposed. The length of S is denoted by L, the length of P without considering wildcard gaps is denoted by l, and the maximal gap length in P is denoted by W. For this example, $L = 13$, $l = 4$, and $W = 3 - 0 + 1 = 4$.

2.1   PAIG

The main idea of PAIG is to fill an $L \times (l - 1)$ matching table, as given by Table 1. The meaning of each cell in the table is explained as follows. The row index indicates the starting index of the match in the sequence; the column index indicates the length of the pattern; numbers in the cell indicate the ending indices of matches; and numbers in brackets indicate counters of respective matches. For example, $8(1),9(2)$ in row index 5 indicates that starting from index 5 of S, there are $1 + 2 = 3$ matches of $P_2 = A[0,2]G[1,2]A$, 1 ending at index 8 of S, and the other 2 ending at index 9. The final result is obtained through summing up all counters of the last column. It is 13 for the example.

Computational time is saved through skipping some empty cells. Once an empty cell is obtained, all remaining cells in the same row should also be empty. Therefore we should simply skip them and go to the beginning of the next row.

Table 1. Matching table of PAIG

index	S	P_1 =A[0,2]G	P_2 =A[0,2]G[1,2]A	P_3 =A[0,2]G[1,2]A[0,3]A
0	A	1(1)	3(1)	5(1)
1	G	-	-	-
2	A	4(1)	-	-
3	A	4(1),6(1)	8(1),9(1)	9(1),11(2),12(2)
4	G	-	-	-
5	A	6(1),7(1)	8(1),9(2)	9(1),11(3),12(3)
6	G	-	-	-
7	G	-	-	-
8	A	10(1)	12(1)	-
9	A	10(1)	12(1)	-
10	G	-	-	-
11	A	-	-	-
12	A	-	-	-

For example, starting from index 2, there is no match of P_2 =A[0,2]G[1,2]A, so there should be no match of P_3 =A[0,2]G[1,2]A[0,3]A, either. For the same reason, in row indices 1, 4, 6, 7, 10, 11 and 12 we only fill the first column. In applications, only a small fraction of the table should be filled.

2.2   GCS

GCS also employs a matching table to obtain the matching information, as given by Table 2. The matching table m is an $L \times l$ matrix of integers, where $m[i][j]$ is the number of matches of P_j ending at s_i. For example, $m[9][3] = 7$ is the number of matches of P ending at index 9.

Table 2. Matching table of GCS

index	S	A	[0,2]G	[1,2]A	[0,3]A
0	A	1	0	0	0
1	G	0	1	0	0
2	A	1	0	0	0
3	A	1	0	1	0
4	G	0	2	0	0
5	A	1	0	0	1
6	G	0	2	0	0
7	G	0	1	0	0
8	A	1	0	2	0
9	A	1	0	3	2
10	G	0	2	0	0
11	A	1	0	0	5
12	A	1	0	2	5

Algorithm 1. Gap constraint search (enhanced version)

Input: $S = s_0 s_1 \ldots s_{L-1}$, $P = p_0 g(N_0, M_0) p_1 \ldots g(N_{l-2}, M_{l-2}) p_{l-1}$
Output: the number of matches
Method: gcs

```
 1: mTbl = new int[2][L];
 2: for (j = 0; j ≤ L − 1; j ++) do
 3:    if sⱼ ≠ p₀ then
 4:       mTbl[0][j] = 1;
 5:    else
 6:       mTbl[0][j] = 0;
 7:    end if
 8: end for
 9: for (i = 1; i ≤ l − 1; i ++) do
10:    for (j = 0; j ≤ L − 1; j ++) do
11:       if sⱼ ≠ pᵢ then
12:          mTbl[i mod 2][j] = 0;
13:          continue;
14:       end if
15:       mTbl[i mod 2][j] = 0;
16:       for (k = max{0, j - M_{i−1} - 1}; k ≤ j - N_{i−1} - 1; k ++) do
17:          mTbl[i mod 2][j] += mTbl[(i - 1) mod 2][k];
18:       end for
19:    end for
20: end for
21: return ∑_{i=0}^{L−1} mTbl[i][(l - 1) mod 2];
```

Algorithm 1 is an enhanced version of GCS [3]. Lines 2 through 8 initialize the first column. Lines 9 through 20 compute elements of other columns. Finally, Line 21 sums up numbers of the last column and return it.

To obtain each element, one should first check whether or not the current character in S matches the ending character of the current pattern. This is done by Lines 11 through 14. For example, at position [6][3] of Table 2, since G does not match A, the element is simply set to 0. If these two characters are identical, the summing operation is undertaken, as indicated by Lines 16 through 18. The number of elements to be summed up is equal to the flexibility of the recent gap. For example, $m[11][3] = m[7][2] + m[8][2] + m[9][2] + m[10][2]$.

Two enhancements are made here. First, the algorithm is suitable for patterns where gaps are independent. Second, m is represented by an array with 2 columns instead of l, and the space complexity is reduced to $O(L)$ from $O(Ll)$. This is because the computation of $m[i][j]$ only relies on some elements in column $j-1$. A mod operation is employed to fulfill the task.

3   Comparisons

This section compares PAIG and GCS from a theoretical point of view. The focus is time complexities.

3.1   General Comparison

Generally, both PAIG and GCS employ incremental approaches through filling matching tables. In other words, they are dynamic programming oriented approaches [3]. This is why they are efficient.

The data structure of GCS is simpler, and it is a matrix of integers. The algorithm description is shorter.

Parallelization of both algorithms is easy. The sequence can be segmented, then run on different computers/CPUs, and the final result is obtained through summing up results of each segment. For PAIG, a number of characters *after* each segment should be kept, such that matches starting close to the end of the segment could be counted. For GCS, however, a number of characters *before* each segment should be kept for the same reason.

In PAIG, more information is recorded. Therefore the starting-ending pairs of matches are available. In GCS, only the ending points of matches are available. For this reason, PAIG can be easily revised to suit global length constraints, while GCS cannot. In the given example of Tables 1 and 2, if we require that each match have a length between 7 and 9, then according to the last column in Table 1, the following starting-ending pairs satisfy this requirement: (3, 9), (3, 11), (5, 11) and (5, 12). There are $1 + 2 + 3 + 3$ matches.

3.2   Complexity Comparison

PAIG and GCS have the same space complexity. While employing certain memory sharing techniques, the space complexity of PAIG is $O(lW)$. However, it is required that the sequence be segmented and read a number of times. If we read the whole sequence into the memory, the space complexity is

$$O(L + lW) = O(L), \tag{1}$$

since very long gaps or sequences are unreasonable. The space complexity of GCS is also $O(L)$, as discussed in Subsection 2.2. To reduce it further, we can fill the table row by row instead column by column. The mod operation is also needed to keep only W rows. In that case, the space complexity is $O(lW)$. However, this approach is more complex, and also requires the segmentation of the sequence. In most applications, reading the whole sequence into the memory is a better choice.

The time complexity of PAIG is [10]

$$O(L \times l \times lW \times W \times \log(lW)) = O(Ll^2W^2 \log(lW)). \tag{2}$$

For GCS, the complexity of Lines 2 through 8 is L, the complexity of Line 21 is also L, and the overall time complexity is only

$$L + \sum_{i=1}^{l-1} \sum_{j=0}^{L-1} (1 + f(p_i)W_{i-1})) + L = O(LlW), \tag{3}$$

where $f(p_i)$ is the frequency of p_i in S, and $W = \max_{0 \le i \le (l-2)} W_i$ is the maximal gap length.

Intuitively, GCS should be significantly more efficient than PAIG. However, this is not always true in applications. The reason is that complexities given above are for the worst case. In the following we will discuss the complexities for the best case. According to the analysis in Subsection 2.1, when we fill the matching table of PAIG, if one element is empty, all remaining cells in the same row are simply skipped. Therefore the time complexity of PAIG is

$$\Omega(L). \tag{4}$$

However, each element in the matching table of GCS should be filled, and the time complexity is

$$\Omega(Ll), \tag{5}$$

which could be observed from Lines 9 and 10 in the algorithm. We will discuss this issue in more detail through experiments.

4 Experiments

Our experiments were conducted on a number of DNA sequences downloaded from the National Center for Biotechnology Information website [5]. Since different sequences result in similar conclusions, we only discuss results of sequence New York/11/2009 (H1N1) No. FJ984346.

Fig. 1 compares the performance of PAIG and GCS. P1 = A([0,M]A)^9 stands for A[0,M]A[0,M]A[0,M]A[0,M]A[0,M]A[0,M]A[0,M]A[0,M]A[0,M]A. "Computing time" is the number of basic operations (comparison or addition). We chose it instead of running time since it is implementation independent. We observe that with the increase of M, the time of PAIG increases polynomially, and the time of GCS increases linearly. The time of PAIG exceeds that of GCS when $M \geq 4$ for P1, and $M \geq 7$ for P2. This is because that the number of matches increases faster on P1, and PAIG fills more cells of the matching table.

Fig. 2 shows the computing time for P3 which contains both A and C. PAIG performs worse when $M \geq 5$, where 5 is between 4 and 7. We also observe that the computing time of P3 is always between that of P1 and P2, as indicated in Fig. 1.

Fig. 3 shows the numbers of matches for different patterns. $N(P, S)$ increases exponentially with the increase of M. $N(P3, S)$ is always between $N(P1, S)$ and $N(P2, S)$.

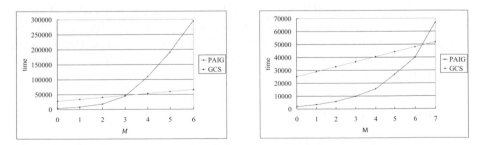

Fig. 1. Computing time for P1 = A([0,M]A)^9 and P2 = C([0,M]C)^9

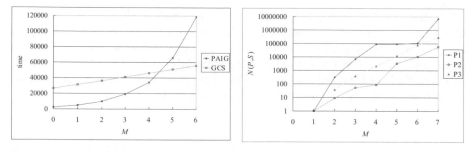

Fig. 2. Computing time for P3 = A([0,*M*] **Fig. 3.** Number of matches for different
C[0,*M*]A)^4[0,*M*]C　　　　　　　　　　patterns

The above results show a strong trend that PAIG is better when the pattern is
inflexible, and the number of matches is not big. GCS is better in the other case.
According to Figures 1 and 2, we can decide which algorithm is faster for many
patterns. For example, C[0,6]C[2,4]C[1,5]C[1,6]C[2,5]C[3,5]C[1,6]C[2,5]C[1,4]C is
less flexible than C([0,6]C)^9, therefore PAIG is faster. While the number of patterns to be checked is huge, we can choose a small partition of patterns to run
both algorithms, and then decide which algorithm to use for each pattern. How
to choose these representative patterns is application dependent. It is an issue
of pattern mining rather than pattern matching.

Figures 1 through 3 do not indicate that GCS is more scalable than PAIG.
With the increase of M, the pattern is more flexible, and the number of matches
is typically greater. However, the frequency of the pattern tends to be small,
and henceforth such patterns are not interesting. A typically employed gap is
[10, 12], which indicates that $M = 12 - 10 + 1 = 3$. We do not discuss this issue
further since it is out of the scope of this paper.

Patterns employed above have equal wildcard gaps $[0, M]$. The purpose is to
make patterns easier to describe. Both PAIG and GCS described in Section 2
can deal with independent wildcard gaps.

5　Conclusions

In this paper, we compared PAIG and GCS. PAIG can deal with global length
constraints while GCS cannot. For flexible patterns where the number of matches
is big, GCS is faster than PAIG. While in the other case PAIG is faster. However,
this does not indicate that GCS is more *scalable* than PAIG, since very flexible
patterns are usually not interesting. We indicated how to choose the more efficient algorithm for a given pattern. This is important in pattern mining where
many patterns are checked.

Acknowledgements

This research is supported by the National Natural Science Foundation of China
(NSFC) under grant No. 60828005.

References

1. Cole, R., Gottlieb, L.A., Lewenstein, M.: Dictionary matching and indexing with errors and don't cares. In: Proceedings of the 36th ACM Symposium on the Theory of Computing, pp. 91–100 (2004)
2. Zhang, M., Kao, B., Cheung, D.W., Yip, K.Y.: Mining periodic patterns with gap requirement from sequences. In: Proceedings of ACM SIGMOD, Baltimore Maryland, pp. 623–633 (2005)
3. Zhu, X., Wu, X.: Discovering relational patterns across multiple databases. In: Proceedings of IEEE 23rd International Conference on Data Engineering (ICDE 2007), pp. 726–735 (2007)
4. Coward, E., Drabløs, F.: Detecting periodic patterns in biological sequences. Bioinformatics 14(6), 498–507 (1998)
5. National Center for Biotechnology Information, http://www.ncbi.nlm.nih.gov/
6. Fischer, M.J., Paterson, M.S.: String matching and other products. In: Karp, R.M. (ed.) Complexity of Computation. SIAM-AMS Proceedings, vol. 7, pp. 113–125 (1974)
7. Manber, U., Baeza-Yates, R.: An algorithm for string matching with a sequence of don't cares. Information Processing Letters 37(3), 133–136 (1991)
8. Akutsu, T.: Approximate string matching with variable length don't care characters. IEICE Transactions on Information Systems E79-D(9), 1353–1354 (1996)
9. Cole, R., Hariharan, R.: Verifying candidate matches in sparse and wildcard matching. In: Proceedings of the 34th Annual ACM Symposium on Theory of Computing (STOC 2002), pp. 592–601. ACM, New York (2002)
10. Min, F., Wu, X., Lu, Z.: Pattern matching with independent wildcard gaps. In: PICom (accepted, 2009)

BPBM: An Algorithm for String Matching with Wildcards and Length Constraints[*]

Xiao-Li Hong[1], Xindong Wu[1,2], Xue-Gang Hu[1], Ying-Ling Liu[1], Jun Gao[1], and Gong-Qing Wu[1]

[1] School of Computer Science and Information Engineering,
Hefei University of Technology,
Hefei 230009, China
[2] Department of Computer Science, University of Vermont,
Burlington, VT 05405, USA
hongxl1984@126.com, xwu@cs.uvm.edu

Abstract. Pattern matching with wildcards and length constraints under the one-off condition is a challenging topic. We propose an algorithm BPBM, based on bit parallelism and the Boyer-Moore algorithm, that outputs an occurrence of a given pattern P as soon as the pattern appears in the given sequence. The experimental results show that our BPBM algorithm has an improved time performance of over 50% with the same matching results when compared with SAIL, a state-of-the-art algorithm of this matching problem. The superiority is even more remarkable when the scale of the pattern increases.

Keywords: Pattern matching, wildcards, length constraints, bit-parallelism.

1 Introduction

Pattern matching with wildcards owns a significant impact on many search applications such as text indexing, biological sequence analysis and data mining. For example, in the biology field, the DNA sequence TATA is a common promoter that often occurs after the sequence CAATCT within 30–50 wildcards [1, 9]. In addition, mining frequent patterns with wildcards from sequences has been an active research topic in data mining. However pattern matching is the key issue to such an efficient mining algorithm. We also can find many examples about patterns with wildcards in biological sequences and their corresponding motivations in [6]. There are many research efforts on the problem of pattern matching with wildcards already [4, 5, 7, 8, 9, 10]. In many existing research efforts, they specified the same number of wildcards between every two consecutive characters in a given pattern P, or fixed the total number of wildcards in P. Navarro and Raffinot [10] addressed a more flexible pattern matching problem. That is the user is allowed to specify a different number of wildcards between each two consecutive characters in P (e.g., $P = Ag(0,2)Cg(0,3)G$).

---

[*] This research is supported by the National Natural Science Foundation of China (NSFC) under grant 60828005 and the 973 Program of China under award 2009CB326203.

H. Sakai et al. (Eds.): RSFDGrC 2009, LNAI 5908, pp. 518–525, 2009.

An efficient algorithm Gaps-BNDM, based on bit-parallel technique, was also proposed. Unfortunately, this algorithm only presents the initial or final character positions for each occurrence of P. But a number of wildcards in P may cause a significant variance of the matching subsequence. For example, given a sequence T=ACCGG and a pattern P=Ag(0,2)Cg(0,3)G, there are 4 matching subsequences (i.e., {1, 2, 4}, {1, 2, 5}, {1, 3, 4}, {1, 3, 5}) of P in T. But with the above algorithm only 2 matches can be found. Note that for the same position 1 in T, there are 4 different matches. So, it is necessary to present all matching positions for each occurrence.

Chen et al. [3] considered a similar problem as Navarro and Raffinot [10]. The main difference is that they considered global length constraint for a pattern. Yet proved by Chen et al. is that the number of matches may be exponential with respect to the maximal gap range and pattern length (a gap means a sequence of wildcards with length constraint). They think it might not be necessary to find 'all' such matches, and proposed a new challenging problem: finding all matching subsequences with a one-off condition. The algorithm SAIL was put forward to present all matching positions for each occurrence as well as finding all occurrences under the one-off condition. At the same time SAIL consumes much time especially when the pattern is long. In this paper, we propose a BPBM algorithm which use both advantages of the Boyes-Moore algorithm [2] and bit-parallel technique [10]. With the same accuracy compared to SAIL, our algorithm improves the time significantly.

The remainder of the paper is organized as follows. Section 2 provides our problem statement. Section 3 presents the design of our BPBM algorithm. Section 4 analyzes the complexity of BPBM, and shows our comparative experimental results with SAIL. Section 5 draws our conclusions.

2 Problem Statement and Related Work

In this section, we adopt the problem definition proposed by Chen et al [3]. We use Σ to denote the alphabet. A wildcard (denoted by ϕ) is a special symbol that matches any character in Σ. A gap is a sequence of wildcards with length constraint. We use $g(N, M)$ to represent a gap whose size is within the range $[N, M]$.

A pattern is denoted by $P = p_1 g_1(N_1, M_1)...g_{m-1}(N_{m-1}, M_{m-1})p_m$, where $p_i \in \Sigma$, and $p_i \neq \phi$, $1 \leq i \leq m$. $g_i(N_i, M_i)$ is a gap between p_i and p_{i+1} and it indicates the local length constraint. If there's no wildcard between two adjacent characters in P, $g_i(0,0)$ will be used. We define the length of a pattern P, denoted by $|p|$=m. Note that $|P|=|g|+1$ is mandatory. L is used to denote the size of a pattern:

$$L = m + \sum_{i=1}^{m-1} M_i \ . \tag{1}$$

When we indicate the i-th position in P, it include the wildcards (i.e., $1 \leq i \leq L$). We use G_N, G_M to denote the minimum and maximum length constraints for P (i.e., a global length constraint), the following inequality should be satisfied:

$$m + \sum_{i=1}^{m-1} N_i \le G_N \le G_M \le m + \sum_i^{m-1} M_i . \tag{2}$$

Given a sequence $T = t_1...t_j...t_n$, if there exists a subsequence of position indices $\{ i_1,...,i_k,...,i_m \}$ where $t_j \in \Sigma$, $t_j \ne \phi$, $1 \le j \le n$, $1 \le i_k \le n$, $1 \le k \le m$, $i_{k-1} < i_k$ such that $t_{i_k} = p_k$, $N_{k-1} \le i_k - i_{k-1} - 1 \le M_{k-1}$, $G_N \le i_m - i_1 + 1 \le G_M$, the subsequence $\{ i_1,...,i_k,...,i_m \}$ is an occurrence of P in T, and the substring $t_{i_1} t_{i_1+1}...t_{i_k}...t_{i_m}$ is a matching subsequence. The second condition above is a local constraint and the third condition is about global constraint.

We apply a one-off condition and an on-line optimization in this paper. That is, every character in T can only be used at most once for matching p_i ($1 \le i \le m$) and an optimal occurrence should be output under the one-off condition. Through the example which is from Chen et al [3], we explain the one-off condition and optimization. Given $P = A$ $g_1(0,1)$ G $g_2(0,1)$ C, $T = AAGGCC$, $G_N = 3$ and $G_M = 5$. Sequence $\{1, 3, 5\}$ is an occurence. After it has been used to match one occurrence of P, they can not be used again under the one-off condition. Therefore P occurs only twice in T. When the first C at position 5 comes, four possible occurrences of P exist, which are $\{1, 3, 5\}$, $\{2, 4, 5\}$, $\{1, 4, 5\}$ and $\{2, 3, 5\}$. Which one of them is the optimal occurrence? As we can see, another C in position 6 will come. The only possible occurrence for this position is $\{2, 4, 6\}$. As a result, $\{1, 3, 5\}$ is the only optimal occurrence for C in position 5 under the one-off condition. We have taken the two issues into consideration when designing BPBM.

3 Algorithm Design

The BPBM algorithm proposed in this paper draws on the Boyer-Moore algorithm for the security thinking of a moving window, and also uses bit parallelism. We set the size of the window as G_M (i.e., the maximal length of a possible matching subsequence), but at the beginning, we set it as G_N . In each window, we read characters backwards. We use two nondeterministic automatons (NFA) to simulate the transfer of the states in the search process. We use u to represent the string that has been read in the current search window. The first NFA is used to identify all the suffix of P. When the state moves to the terminal state, it means u can match P. We then check the length of u. If it satisfies the global constraint of P, we output the occurrence. The second NFA is used to identify all the substrings of P (except the suffix). When the state moves to the terminal state, it means that the string u currently read is the prefix of P. Then we save the location of the prefix. BPBM now in accordance with the first NFA determines whether there is a successful match. When the first NFA doesn't have an active state, we calculate the secure moving distance of the searching window as follows:

1) If the second NFA has an active state, that is, u is a substring of P, then the window moves forward by 1 distance unit. (Because of P's complexity, it's hard to calculate the position of u in P, so we simply move 1 unit.)

2)   If the second NFA doesn't have an active state, but there exists a string v that is in the suffix of u and also in the prefix of P, find the v that is the longest in all such strings, and move the window forward by a minLen-length (v) distance. Here length (v) is the length of v. If length (v) is longer than G_N, the window is moved forward by 1 unit.

3)   If the second NFA doesn't have an active state and u dosen't have any suffix that is in the prefix of P, then the window is moved forward by a G_N distance.

The NFA we explained above allow the existence of wildcards between consecutive characters in P, and are constructed with reference to Navarro and Raffinot [10]. The NFA we construct here only consider the local constraints of P, and the global constraints will be considered in the realization. Figures 1 and 2 show the NFAs for the identification of the suffix of pattern P and a substring of pattern P (except the suffix) respectivly.

As with other algorithms which use bit-parallel technique, we keep state of the search in a bit mask. But we use two bit masks to record the state of the search, which are D1 and D2. D1 and D2 correspond to the two NFAs respectively. Also a table B, which for each character c in Σ stores a bit mask, is created. This mask sets the bits corresponding to the positions where the reversed pattern has the character c or a wildcard. Each time we position the window in the sequence T, we initialize D1, D2 as $0^{L-1}1$, $1^{L-1}0$ respectively. Note that the number of states (except the initial state 0) in the NFA also equals to the size in the given pattern. If the i-th bit is 1, then it denotes that the i-th state is active, and 0 means the state is inactive currently. For each new character read in the window, we update D1 and D2 as follows. First $D1=D1\&B[t_j]$ (t_j is the current character). If $D1 \neq 0^L$, then an operation will be used to simulate an ε-transition: $D1 \leftarrow D1 | ((F - (D1 \& I)) \& {\sim}F)$. The operation is proposed by Navarro and Raffinot for solving the ε-transition in the NFA. They created a bit mask I which has 1 in the "gap-initial" states, and another mask F that has 1 in the "gap-final" states. A "gap-initial" state is a state from where an ε-transition leaves. For each "gap-initial" state S_j corresponding to a gap $g(N, M)$, then its related "gap-final" state should be $S_{j+M-N+1}$. We then use the same way to update D2.

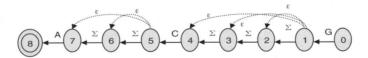

Fig. 1. NFA for recognizing suffix of pattern: A g_1 (0,2)C g_2 (0,3)G

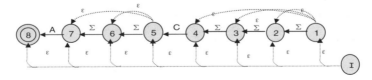

Fig. 2. NFA for recognizing substring of pattern: A g_1 (0,2)C g_2 (0,3)G (except suffix)

There are two phases in the BPBM algorithm, Preprocessing and Search. The Preprocessing initializes a table B and two arrays of I, F, which is similar to Gaps-BNDM. The difference is that we add an exceptional character ϕ into table B to mark the character has been used. B[ϕ] has 1 in the positions where the reversed pattern has wildcards. The procedure Search finds all occurrences of P. The main idea has been introduced at the beginning of this section. Now the problem is that under the one-off condition, BPBM should determine which occurence is an optimal one if there are multiple occurences of P_m's position. As discussed in Section 2, here we select the smallest position within the local constraints to guarantee optimum. In order to achieve this strategy, we don't report an occurrence of P when we find it, but mark it and continue to search until there's no state is active in D1. Then we come back to find the optimal occurrence of what we have found. Next, the specific positions for each occurence should be found and outputed. To this end, we should save all the search states. We add an array D whose length is G_M to save the value of D1 every

```
Search(T : t₁...tₙ, array B, I, F, G_M, G_N)
(1)   pos ← G_N, shift ← 0/* pos:the end of the window, shift:moving distance*/
(2)   While pos<n Do
(3)       len ← 0, D1 ← 0^{L-1}1, D2 ← 1^{L-1}0/*len counts the characters have been read*/
(4)       While D1≠0^L AND len<G_M Do
(5)           len ← len+1
(6)           D1 ← D1&B[t_{pos-len+1}], D1 ← D1|((F-(D1&I))&~F)
(7)           D[len] ← D1
(8)           If D2≠0^L Then
(9)               D2 ← D2&B[t_{pos-len+1}], D2 ← D2|((F[i]-(D2&I[i]))&~F[i])
(10)              If D2&10^{L-1}≠0^L Then ispre ← true, prenum ← len
(11)              D2 ← D2<<1/*End of If, ispre, prenum are used to mark prefix*/
(12)          If D1&10^{L-1} AND len≥G_N Then/*find a possible match*/
(13)              If D1<<1=0^L Then call PrintOcc(occ, D, len, pos) , jump to line 17;
(14)              Else find ← true, lentemp ← len/*save a possible match*/
(15)          D1 ← D1<<1/*End of While*/
(16)      If  find ← true Then call PrintOcc(occ, D, lentemp, pos)/*output the latest found match*/
(17)      If D2=0^L AND  ispre=true Then shift ← (G_N-prenum>0)?G_N-prenum:1
(18)      Else If D2=0^L  Then shift ← G_N
(19)      Else shift ← 1/*End of If*/
(20)      pos ← pos+shift/*End of While*/
PrintOcc(array occ[], array D[], num, pos)
(21)  i ← 2, posflag ← 0
(22)  Cout<<pos-num+1/*print p₁'s position in T*/
(23)  t[pos-num+1] ← φ, num ← num-1
(24)  While num>0
(25)      posflag ← posflag+1
(26)      If D[num]&occ[i]=occ[i] AND N_{i-1} ≤ posflag ≤ M_{i-1} Then
(27)          i ← i+1, t[pos-num+1] ← φ, Cout<<pos-num+1, posflag ← 0
(28)      num ← num-1/*End of While*/
```

Fig. 3. Pseudo code for procedure Search

time when reading a new character (shown on line 7 of Figure 3). We use an array occ with the length of m to represent P: occ[i], which is expressed by a mask $o_L o_{L-1} \cdots o_1$ and represents the position of P_i in P. The mask in occ[i] has the j-th bit set 1 if and only if the j-th in P is P_i. For example, when $P= \text{A}g_1(0,1)\text{C}g_2(0,1)\text{G}$, occ[1]=$10^4$, occ[2]=$0^2 10^2$ and occ[3]=$0^4 1$, the value of array occ can be initialized in the procedure Preprocessing. In the procedure Search, the sub-procedure of PrintOcc is called to output the specific positions for each optimal occurrence of P (shown on lines 13, 16 of Figure 3).

The procedure PrintOcc first outputs position $pos\text{-}num+1$ which is the optimal position of p_1, and marks it as ϕ (num is the size of the matching subseqence). Then sub-procedure PrintOcc searches forward to find the optimal position of another pattern character p_i ($2 \leq i \leq m$) by considering the number range g_{i-1}. So on line 21, variable i is initialized as 2. In order to satisfy the local constraints, we use variable $posflag$ to record the local wildcards' number. On line 26 of Figure 3, the equation D[num]&occ[i]=occ[i] accounts for t[pos-num+1]= p_i, and the condition $N_{i-1} \leq posflag \leq M_{i-1}$ accounts for the number range of wildcards between p_{i-1} and p_i. If both are satisfied, we have found the optimal position of p_i, and also we set t[pos-num+1] as ϕ. Figure 3 presents the Search procedure of BPBM.

4 Complexity Analysis and Experiments

In this section we will analyze the time complexity of our algorithm compared to SAIL. We analyze the Preprocessing procedure at first. The procedure creates bit masks for all characters in Σ, and each bit mask has L bits. So the time complexity of Preprocessing is $O(L|\Sigma|)$. In the procedure Search, the time complexity is $O(kG_M + n)$ where k is the number of p_m's occurrences in T. Each time to find an occurrence we need to call the procedure PrintOcc, whose time complexity is $O(G_M)$, because of the variable num, which controls the number of cycles with the maximum value as G_M. Accordingly, the time complexity of BPBM is $O(L|\Sigma|+n+2kG_M)$. However, in an actual implementation of BPBM, when the pattern P's size L is greater than a machine word w, the algorithm needs $\lfloor l/w \rfloor$ times of the machine word to store each bit mask which is used to keep record of the state of the search. Therefore, the actual time complexity of BPBM is $O((L|\Sigma|+n+2kG_M)\cdot\lfloor L/w \rfloor)$. In comparison, SAIL [3], which solves the same matching problem as this paper, has a complexity of $O(n+G_M \cdot k \cdot m \cdot g)$ where g is the max{ $M_j - N_j$ } ($1 \leq j \leq m-1$). Notice that k and n is the same order of magnitude. Therefore the time complexity of SAIL is about $mg/2\lfloor L/w \rfloor$ times compared to BPBM. Moreover, BPBM inherit the advantage of the Boyes-Moore algorithm and can skip some characters in the search process.

We have used a Pentium IV, 1024 MB, Windows XP and programming language C. Our experiments aim at comparing BPBM against SAIL. The tested data is over gene sequences from http://www.noncode.org/index.htm. We choose four RNA sequences as our testing text (AB114186, AF400501, AF222981, and AF252279). We first preprocess these sequences, and truncate them into 1000, 5000, 10000 and 30000 characters. And in order to simulate a large amount of data search, we have duplicated 10 times for each sequence. Fig. 4 shows four groups of experiments. In each group, a set of patterns with the same sizes are searched in four sequences, and we compute the average time of all patterns. In (a) and (b), we select the patterns whose size L is shorter than w, and in (c) and (d) the size is longer than w.

For all experiments shown in Figure 4, BPBM can get the same matching results compared to SAIL. We can find that BPBM has an improved time performance at least 50% faster than SAIL for each pattern. Specially, when the length of patterns increases, the time performance of BPBM can be 10% of SAIL. When the size of P increases, the occurrences of P in T become fewer, and so our algorithm consumes less time as shown in (c) and (d).

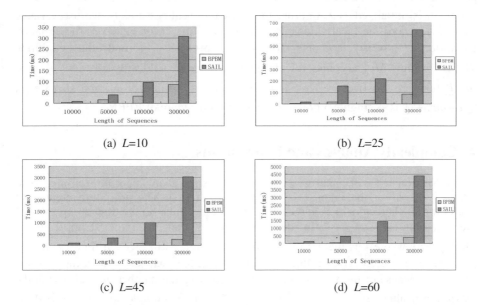

(a) $L=10$ (b) $L=25$

(c) $L=45$ (d) $L=60$

Fig. 4. Efficiency between SAIL and BPBM with fixed sequences

5   Conclusion

In this paper, we have addressed a challenging problem of pattern matching with wildcards, where the users can not only specify the local length constraint but also specify the global length constraint. With our proposed approach, a search engine can efficiently find the optimal occurrences of P in a given sequence under the one-off condition. Also the specific matching positions for each occurrence of P can be presented. BPBM is a solution for exact pattern matching, and in real world applications, it's more than often that a pattern does not repeat exactly. How to extend BPBM to solve approximate pattern matching, where some errors may be allowed for the matching of the patterns, will be our future work.

References

[1] Akutsu, T.: Approximate string matching with variable length don't care characters. IEICE Trans. Info. Syst. E79-D(9), 1353–1354 (1996)

[2] Boyer, R.S., Moore, J.S.: A fast string searching algorithm. CACM 20(10), 762–772 (1977)

[3] Chen, G., Wu, X., Zhu, X., Arslan, A.N., He, Y.: Efficient String Matching with Wildcards and Length Constraints. Knowledge and Information Systems 10(4), 399–419 (2006)

[4] Cole, R., Gottlieb, L., Lewenstein, M.: Dictionary matching and indexing with errors and don't cares. In: Proceedings of the 36th ACM Symposium on the Theory of Computing, pp. 91–100. ACM Press, New York (2004)

[5] Fischer, M.J., Paterson, M.S.: String matching and other products. In: Karp, R.M. (ed.) Complexity of computation, vol. 7, pp. 113–125. Massachusetts Institute of Technology, Cambridge (1974)

[6] Gusfield, D.: Algorithms on strings, trees, and sequences–Computer science and computational biology. Cambridge University Press, Cambridge (1997)

[7] Kalai, A.: Efficient pattern-matching with don't cares. In: Proceedings of the 13th ACM-SIAM Symposium on Discrete Algorithms, Society for Industrial and Applied Mathematics, pp. 655–656. Society for Industrial and Applied Mathematics, Philadelphia (2002)

[8] Kucherov, G., Rusinowitch, M.: Matching a set of strings with variable length don't cares. In: Proceedings of the 6th Symposium on Combinatorial Pattern Matching, pp. 230–247. Springer, Heidelberg (1995)

[9] Manber, U., Baeza-Yates, R.: An algorithm for string matching with a sequence of don't cares. Inf. Proc. Lett. 37(3), 133–136 (1991)

[10] Navarro, G., Raffinot, M.: Fast and Simple Character Classes and Bounded Gaps Pattern Matching, with Applications to Protein Searching. J. Computational Biology 10(6) (2003)

Author Index

Printing: Mercedes-Druck, Berlin
Binding: Stein + Lehmann, Berlin